Phase Transitions in Surface Films 2

NATO ASI Series

Advanced Science Institutes Series

*A series presenting the results of activities sponsored by the NATO Science Committee,
which aims at the dissemination of advanced scientific and technological knowledge,
with a view to strengthening links between scientific communities.*

The series is published by an international board of publishers in conjunction with the
NATO Scientific Affairs Division

A	Life Sciences	Plenum Publishing Corporation
B	Physics	New York and London
C	Mathematical and Physical Sciences	Kluwer Academic Publishers
D	Behavioral and Social Sciences	Dordrecht, Boston, and London
E	Applied Sciences	
F	Computer and Systems Sciences	Springer-Verlag
G	Ecological Sciences	Berlin, Heidelberg, New York, London,
H	Cell Biology	Paris, Tokyo, Hong Kong, and Barcelona
I	Global Environmental Change	

Recent Volumes in this Series

Series B: Physics

Phase Transitions in Surface Films 2

Edited by

H. Taub

University of Missouri–Columbia
Columbia, Missouri

G. Torzo

Università di Padova
Padova, Italy

H. J. Lauter

Institut Laue-Langevin
Grenoble, France

and

S. C. Fain, Jr.

University of Washington
Seattle, Washington

Plenum Press
New York and London
Published in cooperation with NATO Scientific Affairs Division

SEP/AE
PHYS

Proceedings of the NATO Advanced Study Institute and
International Course on
Phase Transitions in Surface Films,
held June 19–29, 1990,
in Erice, Sicily, Italy

Library of Congress Cataloging-in-Publication Data

NATO Advanced Study Institute and International Course on Phase
   Transitions in Surface Films (1990 : Erice, Italy)
     Phase transitions in surface films 2 / edited by H. Taub ... [et
   al.].
        p.   cm. -- (NATO ASI series. Series B, Physics ; v. 267)
     "Proceedings of the NATO Advanced Study Institute and
   International Course on Phase Transitions in Surface Films, held
   June 19-29, 1990, in Erice, Sicily, Italy"--T.p. verso.
     "Published in cooperation with NATO Scientific Affairs Division.
     Includes bibliographical references and index.
     ISBN 0-306-44005-9
     1. Surfaces (Physics)--Congresses. 2. Thin films--Congresses.
   3. Monomolecular films--Congresses. 4. Surface chemistry-
   -Congresses. 5. Phase transformations (Statistical physics)-
   -Congresses.   I. Taub, H. II. North Atlantic Treaty Organization.
   Scientific Affairs Division. III. Title. IV. Series.
   QC173.4.S94N378 1990
   530.4'27--dc20   91-23686
                                                               CIP

ISBN 0-306-44005-9

© 1991 Plenum Press, New York
A Division of Plenum Publishing Corporation
233 Spring Street, New York, N.Y. 10013

Printed in the United States of America

Preface

The Advanced Study Institute (ASI) entitled "Phase Transitions in Surface Films" was held at the Ettore Majorana Centre for Scientific Culture in Erice, Sicily from June 19 to June 29, 1990. It reviewed the present understanding (experimental and theoretical) of phase transitions of surfaces, interfaces, and thin films as well as the related structural and dynamical properties of these systems.

From its inception, this ASI was envisioned as a sequel to one of the same title organized eleven years earlier by J. G. Dash and J. Ruvalds which was also held at the Ettore Majorana Centre. The previous ASI reflected the progress which had been made in understanding quasi two-dimensional (2D) states of matter, particularly adsorbed monolayers, and the phase transitions which occur in them. At that time, the field was barely ten years old. The modern field to which we are referring here can be traced to the landmark experiments of A. Thomy and X. Duval. Beginning in 1967, they published a series of papers presenting evidence from vapor pressure measurements of 2D phases of krypton and other gases adsorbed on polycrystalline (exfoliated) graphite. Their work led to a large number of thermodynamic and scattering experiments on physisorbed films. This in turn motivated a great deal of theoretical interest in 2D systems and their phase transitions.

In the years following the first ASI, there has been continued rapid progress in this field as impressive as that which occurred earlier. To begin with just a few examples, phenomena such as the surface roughening transition and surface melting of bulk solids which have now been extensively investigated were barely touched upon at the previous ASI. The structure and phase transitions of multilayer films and monolayers adsorbed on fluid substrates are now being pursued experimentally with the same rigor as were the monolayer-on-solid systems of Thomy and Duval. In addition, considerable progress has been made by several diffraction techniques in structure determinations of monolayer phases modulated by a crystalline substrate. The study of surface dynamics has greatly benefitted from the results obtained from high-resolution neutron scattering which has revealed new features in the excitation spectra of helium films on graphite, has allowed the corrugation in the adatom-substrate potential of various monolayers on graphite to be probed quantitatively, and has elucidated surface melting by measurements of molecular diffusion in multilayer films.

The first Institute had dealt largely with experiments performed with films adsorbed on high-quality polycrystalline graphite substrates which had only recently become available. In the ensuing years, polycrystalline substrates of different symmetry than graphite, such as MgO, have stimulated a variety of new experiments. However, some of the most exciting results have been achieved not so much by using new samples but with familiar systems-- single-crystal surfaces, films adsorbed on single-crystals, and monolayers adsorbed on water--which are now accessible to newly developed or improved experimental techniques. These techniques include synchrotron x-ray diffraction, helium atom scattering, ellipsometry, and high-resolution electron diffraction.

As before, the experimental advances have spurred intense theoretical activity in several new areas: phase transitions involving modulated structures, surface roughening and reconstruction, wetting and film growth, and the statistical mechanics of membranes. Also, computer simulation of surface phenomena has developed greatly in the last ten years and played an increasingly important role in the interpretation of experiments. The reader is referred to Michael Wortis' summary at the end of the volume for an overview of theoretical developments.

The large growth in subject breadth since the earlier ASI posed some problems in selecting topics which would fit into a nine-day program even with as many as six hour-long talks per day. Eventually, the subject matter was divided somewhat arbitrarily into the following topics:

1) monolayers on crystalline substrates;
2) wetting phenomena, film growth, and multilayer structures;
3) surface reconstruction, roughening, and melting;
4) excitations and diffusion at surfaces;
5) monolayers adsorbed on liquid surfaces and the theory of membranes.

Some of the talks overlapped with more than one of these topics. A corresponding growth in the number of researchers in the field made for difficult decisions in selecting among many excellent candidates for speakers. Fortunately, the organizers were aided in the choice of topics and speakers by an Advisory Committee composed of M. Bienfait, J. G. Dash, E. Tosatti, J. A. Venables, J. Villain, and M. Wortis.

Our approach in organizing the ASI was to have lecturers present two or three talks to introduce and review recent developments in their respective areas. The lecturers were supplemented by single-topic speakers who discussed in greater detail particularly interesting results of their own research. Student participants had an opportunity to contribute posters in two sessions each lasting a few days. Poster authors and titles are listed below. The combination of lectures, single-topic talks, and poster sessions proved successful in stimulating interactions among the 97 participants.

Several of the talks presented are not included in this volume due to the publication of the material elsewhere or other commitments of the speakers. Nevertheless, they had an important impact on discussions and the interested reader is encouraged to consult the appropriate references. For this purpose, these talks are listed below.

For the organizers as well as for many other participants who attended the first Institute on this subject, the commitment to the Ettore Majorana Centre as the meeting location was as strong as to the excellence of the scientific program. The beauty of Erice and its surroundings and the warmth of its people had left an indelible impression which was only enhanced by our revisit. The efficiency of the Centre's staff in implementing the lecture program, the travel and living arrangements, and the excursions was greatly appreciated. We are especially grateful to S. A. Gabriele, P. Savalli, and J. Pilarski for their warm hospitality and untiring efforts.

We would like to thank several organizations for their financial assistance which made this ASI possible. Foremost among these is the Scientific Affairs Division of NATO which awarded the grant providing most of the living and travel expenses of many of the participants. In addition, the Commission of European Communities awarded a grant for a study entitled "Phase Transitions in Surface Films." A grant from the Consiglio Nazionale delle Ricerche of Italy provided both living and travel funds. Living expenses were also funded by the Sicilian Regional Government through the Ettore Majorana Centre, National Instruments, Inc. and Apple Computer, Inc. Travel expenses of six U. S. students were paid by the U. S. National Science Foundation. The Department of Physics and Astronomy, University of Missouri-Columbia assisted with organizational expenses.

Stephanie Coureton and Donna O'Neil at the University of Missouri-Columbia provided much of the secretarial assistance in organizing the Institute and in the preparation of this volume.

Finally, we are deeply indebted to R. D. Diehl, Y. Larher, and F. Toigo for assisting us in the organization of the meeting.

H. Taub, G. Torzo, H. J. Lauter, S. C. Fain, Jr.
Editors

UNPUBLISHED PAPERS

G. Reiter, "X-Ray Determination of the Corrugation Potential in 2D Liquids Intercalated in Graphite"

Y. Larher, "Wetting and Roughening: an Adsorption Isotherm Approach"

J. F. van der Veen, "Atomic Structure of Semiconductor Surfaces and Interfaces"

E. Tosatti, "Surface Melting Theory"

Michael E. Fisher, "The Universal Critical Adsorption Profile and its Extraction from Experiments"

H. van Beijeren, "Model Calculations of Equilibrium Crystal Shapes"

S. Leibler, "Statistical Mechanics of Membranes"

M. H. W. Chan, "Superfluid Transition of Thin ^{4}He Films in Porous Substrates"

POSTER SESSIONS

P. Dai, S.-K. Wang, H. Taub, J. E. Buckley, S. N. Ehrlich, and J. Z. Larese, "X-Ray Diffraction from a Liquid Crystal Film Adsorbed on Single-crystal Graphite"

V. L. Eden and S. C. Fain, Jr., "Ethylene Monolayers on Graphite: a Low-Energy Electron Diffraction Study"

S. C. Fain, Jr., J. Cui, and W. Liu, "H_2, HD and D_2 Monolayers on Graphite: Low-Energy Electron Diffraction Results"

M. Hamichi, A. Zerrouk, G. Raynerd, and J. A. Venables, "Diffraction Profiles of Xenon Adsorbed on Graphite"

G. Mistura, "Third Sound Study of a Thin, Layered Superfluid Film"

T. Moeller, V. L. P. Frank, H. J. Lauter, H. Taub, and P. Leiderer, "Temperature Dependence of the Phonon Gap of CD_4 and N_2 in the Commensurate Phase Adsorbed on Graphite"

D. B. Pengra and J. G. Dash, "Edge Melting of Monolayer Neon"

D. B. Pengra, D.-M. Zhu, and J. G. Dash, "Surface Melting of Strained and Unstrained Layers: Kr, Ar, and Ne"

A. Razafitianamaharavo, A. Abdelmoula, N. Dupont-Pavlovsky, B. Croset, and J. P. Coulomb, "Krypton Adsorption on (0001) Graphite Preplated with Cyclohexane or Carbon Tetrachloride: Miscibility and Displacement Transition"

R. M. Suter, R. Gangwar, and R. Hainsey, "Thermodynamics and Structure near the Second and Third Layering Transitions of Krypton on Graphite"

U. G. Volkmann, "Ellipsometric Measurements of Kr, Ar, Xe, Nitrogen, and Dichlordiflourmethane on Graphite"

S. Zhang and A. D. Migone, "Heat Capacity Studies of Ethane Adsorbed on Graphite at High Coverages"

Y. Carmi, E. Polturak, and S. G. Lipson, "Roughening Transition in Dilute ^{3}He-^{4}He Mixture Crystals"

G. Giugliarelli and A. L. Stella, "Lattice-Gas Model for Adsorption on Fractally Rough Surfaces"

G. Langie, "Density Functional Theory of Melting"

C. J. Walden, B. L. Györffy, and A. O. Parry, "Wetting Transitions in a Vector Model with Cubic Anisotropy"

H. Xu, "A Molecular Dynamics Simulation of Disorder on Stepped Solid Surfaces"

U. Albrecht, H. Dilger, P. Leiderer, and K. Kono "Growth and Thermal Annealing of H_2-Films"

T. Angot and J. Suzanne, "Structures and Phase Transitions in Rare Gases and Nitrogen Monolayers on MgO(100) Single-Crystal Surfaces"

H. S. Youn and G. B. Hess, "Multilayer Physisorption of Xe on HOPG"

N. D. Shrimpton, "Effects of Substrate Corrugation on the Creation of Dislocations in Monolayer Argon"

J. Z. Larese and Q. M. Zhang, "Neutron Diffraction Studies of Multilayer Films on Graphite: Layering and Melting"

J. R. Dennison, T. Will, and W. N. Hansen, "Vibrational Dynamics of Adsorbed Films Using the Fourier Transform Infrared Spectroscopy Metal Light Pipe Technique"

C. H. F. Glover, R. W. Godby, A. Qteish, and M. C. Payne, "How Do Steps Affect Surface Phase Transitions and Surface Diffusion"

B. Joós, N. C. Bartelt, and T. L. Einstein, "Distribution of Terrace Widths on a Vicinal Surface within the 1-D Free-Fermion Model"

A. Khater, "Phonons on Rough Surfaces"

J. M. Phillips, "A Comparison of the Layer-by-Layer Melting of Argon and Methane Multilayers"

J. M. Phillips, "Commensurability Transitions Driven by Coverage and Temperature in Multilayers"

L. Wilen and E. Polturak, "Film Growth and Wetting of Curved Surfaces"

H. M. van Pinxteren, B. Pluis, J. Meerman, A. J. Riemersma, J. W. M. Frenken, and J. F. van der Veen, "High-Temperature Behavior of Pb(001)"

S. Chandavarkar and R. D. Diehl, "Bond-Orientational Order in the Incommensurate Fluid Phase of K/Ni(111)"

R. Chiarello, J. Krim, and C. Thompson, "Wetting Behavior of Adsorbed Water Films on Au Surfaces"

E. Conrad and Y. Cao, "Thermal Disordering of Ni(110) and Ni(001) Surfaces: A Study by High-Resolution LEED"

D. Fisher and R. D. Diehl, "Alkali Metal Adsorption on the {100} and {111} Surfaces of Nickel"

V. H. Etgens, M. Sauvage-Simkin, R. Pinchaux, J. Massies, N. Grelser, N. Jedrecy, and S. Tatarenko, "Adsorbate Structures of the Te/GaAs(100) Interface: an *In Situ* Grazing Incidence X-Ray Diffraction Study"

A. Hoss, U. Romahn, H. Zimmermann, M. Noid, P. von Blanckenhagen, W. Schommers, and O. Meyer, "A Study of the Temperature Dependence of the Structure of the Au(110) Surface"

Z. Y. Li, S. Chandavarkar, and R. D. Diehl, "The Order-Disorder Transition of p(2x2) Overlayers on Ni(111)"

R. B. Phelps and P. L. Richards, "Heat Capacity of Adsorbed H_2 on Evaporated Ag Films"

B. Salanon, H. J. Ernst, F. Fabre, and J. Lapujoulade, "Thermally Activated Defects on Crystal Surfaces"

B. N. Dev, G. Materlik, F. Grey, and R. L. Johnson, "On the Possibility of Using X-Ray Standing Waves in the Study of Phase Transitions in Surfaces and Interfaces"

A. C. Mitus, A. Z. Patashinskii, and S. Sokolowski, "Computer Simulations of Local Structure of a Two-Dimensional Liquid"

A. Khater, and H. Grimech, "Defect Dynamics in Disordered Surfaces"

P. Eng, T. Tse, and P. Stephens, "Surface X-Ray Diffraction Ordering Kinetics of Pb on Ni(001)"

Contents

WETTING PHENOMENA, MULTILAYER STRUCTURES, AND FILM GROWTH

SUMMARY

PHASE TRANSITIONS OF MONOLAYER FILMS ADSORBED ON GRAPHITE

M. H. W. Chan

Department of Physics
Pennsylvania State University
University Park, PA   16802

Intensive structural and thermodynamic studies in the last decade revealed that the phase diagrams of monolayer film physisorbed on graphite, despite the vast number of adsorbates, are found to fall into three general categories.

In this lecture I shall try to summarize what has been learned in the subject of phase transition of monolayer films adsorbed on graphite in the last decade.   In view of the extensive activities, it would not be possible to give a comprehensive review of the subject.   I will restrict the discussion to experimental results on melting and liquid–vapor transition of films of relatively simple molecules adsorbed on graphite. The emphasis of this lecture is to bring out the similarities that exist in the various monolayer systems.   It is meant to be a supplement to other recent review articles [ref. 1] with slightly different emphasis.   In this manuscript I shall also try to provide a reasonably complete set of references so that interested readers can pursue the subject further.

Ne [ref. 2], Xe [ref. 3,4], CD_4 and CH_4 [ref. 5,6] are some adsorbates that form an incommensurate solid prior to melting that have phase diagrams resembling three dimensional bulk systems.   The schematic monolayer phase diagram of these systems is shown in Fig. 1.   In these systems, the melting transition is first order and occurs at a triple point T_3 when the surface coverage is below monolayer completion.   For $T>T_3$, there is a liquid–vapor coexistence region that terminates at a two-dimensional critical point T_c.   This coexistence boundary in the coverage–temperature plane of CH_4 on graphite was mapped out by locating the position of the peak of heat capacity traces taken at closely spaced surface coverages[6]. This coexistence boundary was found to be well described by a power law of the form,

$$(n_l - n_v) = B \left[\frac{(T_c - T)}{T_c}\right]^\beta$$

where T_c, n_l and n_v are the critical temperature and the liquid and vapor

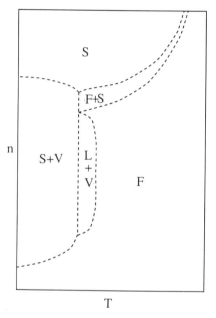

Figure 1: Schematic coverage (n) vs temperature (T) phase diagram of Physisorbed Monolayers that undergo first order melting at a two-demensional triple point. S, L, V, F. stands for solid, liquid, vapor and fluid phases, respectively. Dashed lines stand for first order phase boundaries.

densities, respectively. β was found to be 0.13 ± .02. This is in good agreement with the value of 0.125 expected for the two dimensional Ising model. The liquid–vapor boundary for Xe [ref. 7], C_2H_4 [ref. 8], and Ar [ref. 9] on graphite were also found to have a shape similar to that of CH_4 on graphite.

There has been a great deal of interest in the melting behavior of Xe on graphite.[4] As the surface coverage is increased into the compressed monolayer region, the melting temperature increases rapidly with increasing coverage from the triple point at 100K. Concomitant to the increase in T_m, recent synchrotron x-ray studies, including experiments that employ single-crystal graphite surfaces, claim to find evidence of a crossover from first-order to continuous melting behavior. The tricritical temperature T_{tc}, above which the melting is interpreted to be continuous, is placed by the x-ray studies to be near 125K.[4] This claim is based primarily on the observation of bond-orientational or hexatic-like order in the adsorbed film over a temperature range of 1K. The existence of a hexatic fluid is predicted by the dislocation pair unbinding two dimensional melting theory of Kosterlitz, Thouless[10], Halperin, Nelson and Young[11] (KTHNY). However, it has been noted that the hexatic-like orientational order can also be a consequence of the sixfold symmetric graphite potential. Two vapor-pressure isotherm studies also interpret their results to be consistent with possible crossover to a continuous melting transition. However, the value for T_{tc} is placed at 155 [ref. 12] and 147K [ref. 13] respectively. The placement of T_{tc} in the vapor-pressure isotherm studies is based on the apparent rounding of the vertical step of isotherms. A recent complementary heat capacity and vapor pressure isotherm study[7] indicated, on the other hand, that the melting of monolayer Xe on graphite is always first order up to the highest possible melting temperature at 186.7K. This conclusion is based on the persistent sharp heat capacity peak that appears at melting as the surface coverage is raised from submonolayer to highly-compressed monolayer coverage. With increasing total surface coverage, the liquid-like upper layers become more effective in stabilizing the bottom layer resulting in an increase in T_m of the bottom layer. This effect saturates as the upper layer approaches "infinite" thickness giving rise to a maximum temperature of 186.7K. The conclusion of a first order melting transition of monolayer X_e on graphite is consistent with the result of a number of large scale computer simulation studies.[14] There are indications that the melting behavior of monolayer Ne [ref. 2] CH_4 and CD_4 [ref.5] on graphite follows the evolution as shown in Fig. 1.

The phase diagrams of monolayer Ar and C_2D_4, as shown in Fig. 2, resemble that of Fig. 1 except that the solid-liquid coexistence region is either exceedingly narrow or simply not present. Heat capacity study of C_2D_4 and C_2H_4 on graphite find a broad and small peak at the melting temperature at submonolayer coverages, consistent with a continuous interpretation.[8] Such a conclusion is confirmed in a subsequent quasi-

inelastic neutron diffraction measurement.[15]   Computer simulation study
indicates that this continuous transition is related to the gradual
tilting up of the long molecular axis[16] making space available for other
molecule rather than the unbinding of dislocation pairs as predicted by
the KTHNY model.   The melting of ethane on graphite also appears to be of
similar mechanism.[17]

The melting transition of Ar on graphite is unique among the various
physisorbed systems.   Results from a host of experimental studies,

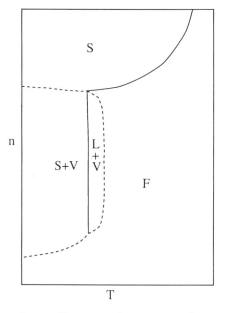

Figure 2:   Schematic phase diagram of systems that undergoes continuous
melting transition at submonolayer coverages.   Solid lines locate possible
continuous phase boundaries, dashed lines, first order boundaries.

including neutron diffraction[18], heat capacity[19], vapor pressure
isotherm[20], and x—ray scattering[21] found evidence that the melting in both
the submonolayer and monolayer regime is continuous.   The temperature
dependence of the correlation length of the Ar overlayer as temperature
approaches the melting temperature T_m was found in an x—ray study to be
consistent with the KTHNY prediction.[22]   The finding that T_m is below the

maximum of the broad anomaly[19] (half width of 6K) is also consistent with the KTHNY theory.   However, a subsequent high precision heat capacity experiment found an additional small but sharp peak on the low temperature side of the broad anomaly.[9] The temperature of the sharp peak appears to coincide with T_m as determined in the x-ray experiment.[21] The small peak is consistent with an "abrupt" density change of about 0.2% over a temperature range of 0.3K. This was interpreted as a signature of weakly first order melting transition. Two recent high precision synchrotron x-ray studies failed to resolve the "mystery" of this sharp peak.[22,23] The authors failed to find evidence of an abrupt "jump" in the lattice parameter.   Instead, the correlation length was found to change over a temperature range of 0.3K from 900Å to 200Å in one study and 1500Å to 400Å in the other.[22,23] The experiment on a single crystal surface confirmed the earlier LEED results[24] that the symmetry axis of solid Ar overlayer is rotated with respect to the graphite. This X-ray study reported that the rotation angle evolves smoothly through T_m.

Recently Zhang and Larese[25] have re-examined the x-ray results and found a small and narrow peak in the temperature derivative of the lattice constant. They have also carried out a new high precision vapor pressure isotherm study between 59 and 84K at the monolayer completion region. They found evidence of a two step melting process near the liquid-solid boundary at constant temperature. Based on the available experimental result, therefore, it is not clear whether the melting of Ar on graphite is continuous and describable by the KTHNY dislocation unbinding process. What is reasonably certain is that the unique melting behavior is related to the particularly strong effect of corrugation in the graphite adsorption potential experienced by the Ar overlayer.[26] It would be interesting to follow the melting behavior of a highly compressed Ar monolayer since the corrugation effect is expected to diminish. Indeed, there is an indication that the melting transition of a compressed Ar monolayer is first order like.[27]

The phase diagram of light molecules such as ^{3}He, ^{4}He, H_2 and D_2 on graphite have been determined in a series of heat capacity measurements,[28], complemented by neutron diffraction and LEED studies.[29] It is striking that the monolayer phase diagrams of these systems are found to be exceedingly similar to each other. It is shown schematically in Fig. 3. This phase diagram is dominated by a solid phase that is in registry with the graphite potential, forming a $(\sqrt{3} \times \sqrt{3})$ solid phase. The order-disorder transition from the $(\sqrt{3} \times \sqrt{3})$ phase to the fluid phase have been elegantly shown to exhibit critical behavior belonging to the

3-state-Potts universality class.[28,30] The commensurate-incommensurate transition upon the increase of surface coverage at low temperature have recently been shown in diffraction experiments to proceed via intermediate striped phase and domain-wall fluid regions.[29]

The monolayer phase diagrams of Kr [ref. 31,32], N_2 [ref. 33], and CO [ref. 34] on graphite are also dominated by the commensurate ($\sqrt{3}$ x $\sqrt{3}$)

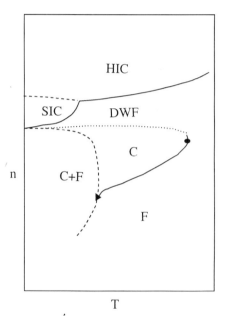

Figure 3: Phase diagram of ^{4}He, ^{3}He, H_2 and D_2 on graphite. SIC and HIC stands for an uniaxial incommensurate phase with stripes of domain walls and a hexagonally symmetric incommensurate phase, respectively. C, F and DWF stands for commensurate ($\sqrt{3}$ x $\sqrt{3}$), fluid and domain wall fluid phases. Solid lines stand for continuous transition, dashed lines, first order transitions; the position and nature of the phase transition along dotted lines are less certain. Solid triangle and circle locate positions of 3-state-Potts tricritical and 3-state-Potts critical points.

phase at low temperature. In the submonolayer coverage region, the solid patches sublimate into a homogenous fluid phase. According to the incipient triple point model,[31,33] the ($\sqrt{3}$ x $\sqrt{3}$) solid, being stabilized by the graphite adsorption potential, persist up to a temperature that is higher than the "expected" liquid–vapor critical temperature thus suppressing the existence of the liquid phase and also the two dimensional triple and critical points. Using this model, the sublimation boundary and the heat capacity vs temperature trace near the incipient critical

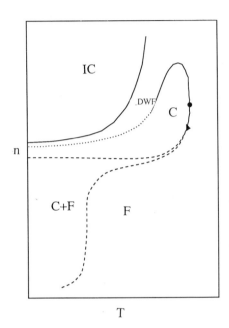

Figure 4: Phase diagram of monolayer Kr, N_2 and CO on graphite, C, F, IC and DWF stands for commensurate ($\sqrt{3}$ x $\sqrt{3}$) solid, fluid, incommensurate solid and domain wall fluid phase, respectivly. Solid triangle and circle locate positions of 3–state–Potts tricritical and 3–state–Potts critical points. Solid and dashed lines stand for continuous and first order phase boundaries. The nature of the transition along the dotted line is less certain.

point can be fitted quite well for both the K_r and N_2 on graphite systems.[31,33]

At high monolayer coverages, the melting (or the sublimation) temperature increases rapidly with coverage and the commensurate solid + fluid region narrows.[31–34] There have been a number of reports of the termination of the coexistence region at a tricritical point, T_{tc}. There are discrepancies, however, concerning the position of the tricritical point.[35]   In a recent vapor pressure isotherm study, T_{tc} for N_2 on graphite is found to be near 85.4K, very close to the maximum temperature of the commensurate phase.[33]   This finding is consistent with the most refined x-ray measurement of Kr on graphite where T_{tc} is found to be near 130K, also very close to the maximum temperature of the commensurate $(\sqrt{3} \times \sqrt{3})$ phase.[32]   The x-ray study also found evidence of a re-entrant fluid phase or the domain wall fluid phase that is sandwiched between the commensurate and compressed incommensurate solid phase.   These results are confirmed in a recent combined heat capacity and vapor pressure isotherm study of CO on graphite.[34]   The location of the commensurate solid, incommensurate solid, fluid and re-entrant fluid phase and the entire submonolayer and monolayer phase diagram of CO on graphite is exceedingly similar to that of Kr on graphite and also the N_2 on graphite systems.[32,34]   In addition, the heat capacity study yields, at appropriate coverages, critical exponents consistent with those expected for two-dimensional three-state-Potts tricritical and two-dimensional three-state-Potts critical phase transitions.

The phase diagram for these three systems is shown schematically in Fig. 4.   It should be pointed out that this phase diagram is topologically similar to that of Fig. 3 the phase diagram for the helium and hydrogen systems.   The "distortion" of the commensurate phase in Fig. 4, i.e. the three-state-Potts critical melting transition (in this region the "melting" is actually an order-disorder transition) appears at a coverage that is nearly 30% higher than that expected for the commensurate phase, is related to the fact that a second fluid layer of substantial density is needed to stabilize this "perfect" $(\sqrt{3} \times \sqrt{3})$ bottom layer.

In conclusion, we have learned a great deal about the phases and phase transitions of different monolayer films physisorbed in graphite in the past decade and it is difficult not to be impressed by the simplicity and elegance of these systems.

ACKNOWLEDGEMENTS

The author wishes to thank his colleagues both at Penn State and elsewhere for educating him and sharing of ideas with him and for fruitful and enjoyable collaborations.   In particular, he wishes to thank M. Bretz, who introduced him to the subject of monolayer phase transitions on graphite and M. Cole, J. G. Dash, S. Fain, R. B. Griffiths, L. Passell M. Schick, N. D. Shrimpton, W. A. Steele, O. E. Vilches, M. Wortis, and particularly M. E. Fisher who provided the author with many instructive tutorials. He wishes also to thank those who worked with him in the laboratory, in particular, A. D. Migone, H. K. Kim, Q. M. Zhang, Y. P. Feng, M. R. Bjurstrom, K. D. Miner, A. J. Jin and Z. R. Li. Research done in the author's laboratory is supported in part by the National Science Foundation primarily under grants DMR-8113262, DMR-8419262, DMR-8718771. He is grateful to N. D. Shrimpton for drawing the schematic phase diagrams for this paper.

References

1. R. J. Birgeneau and P. M. Horn, Science $\underline{232}$, 329 (1986); K. J. Strandburg, Rev. Mod.Phys. $\underline{60}$, 161 (1988); N. D. Shrimpton, M. W. Cole, W. A. Steele and M. H. W. Chan, in Surface Properties of Layered Materials, edited by G. Benedek, Khuwer Acad. Publishers, Dordrecht, The Netherlands, (to be published).
2. G. B. Huff and J. G. Dash, J. Low Temp. Phys. $\underline{24}$, 155 (1976); R. E. Rapp, E. P. deSouza and E. Lerner, Phys. Rev. B$\underline{24}$, 2196 (1981); F. Hanono, C.E.N. Gatts and E. Lerner, J. Low Temp. Phys. $\underline{60}$, 73 (1985); C. Tiby, H. Wiechert and H. J. Lauter, Surf. Sci. $\underline{119}$, 21 (1982).
3. J. A. Litzinger and G.A. Stewart, in Ordering in Two-Dimensions, edited by S. Sinha, p. 267 (North Holland, NY 1980).
4. P. A. Heiney et al., Phys. Rev. B$\underline{28}$, 6416 (1983); T. F. Rosenbaum et al., Phys. Rev. Lett. $\underline{50}$, 1791 (1983); P. Dimon et al., Phys. Rev. B$\underline{31}$, 437 (1984); E. D. Specht et al., J. Phys. (Paris), Lett. $\underline{45}$, L561 (1985); S. E. Nagler et al., Phys. Rev. B$\underline{32}$, 7373 (1985).
5. P. Vora, S. K. Sinha and R. K. Crawford, Phys. Rev. Lett. $\underline{43}$, 7041 (1979).
6. H. K. Kim and M. H. W. Chan, Phys. Rev. Lett. $\underline{53}$, 170 (1984); H. K. Kim, Q. M. Zhang and M. H. W. Chan, Phys. Rev. B$\underline{34}$, 4699 (1986).
7. A. J. Jin, M. R. Bjurstrom and M. H. W. Chan, Phys. Rev. Lett. $\underline{62}$, 1372 (1989).
8. H. K. Kim, Q. M. Zhang and M. H. W. Chan, Phys. Rev. Lett. $\underline{56}$, 1579 (1986).
9. A. D. Migone, Z. R. Li and M. H. W. Chan, Phys. Rev. Lett. $\underline{53}$, 810 (1984).
10. J. M. Kosterlitz and D. J. Thouless, J. Phys. C$\underline{6}$, 1181 (1973).
11. D. R. Nelson and B. I. Halperin, Phys. Rev. B$\underline{19}$, 2457 (1979); A. P. Young, Phys. Rev. B$\underline{19}$, 1855 (1979).
12. C. Tessier, Doctor of Science Thesis, L' Universite de Nancy I, 1983 (unpublished).
13. N. J. Collela and R. M. Suter, Phys. Rev. B$\underline{34}$, 2052 (1986); R. Gangwar, N. J. Collela and R. M. Suter, Phys. Rev. B$\underline{39}$, 2459 (1989).
14. S. W. Koch and F. F. Abraham, Phys. Rev. B$\underline{27}$, 2964 (1983); F. F. Abraham, Phys. Rev. Lett. $\underline{50}$, 978 (1983); Phys. Rev. B$\underline{28}$, 7338 (1983); Phys. Rev. B$\underline{29}$, 2606 (1984).
15. J. Z. Larese, L. Passell, A. Heidemann, D. Richter and J. P. Widested, Phys. Rev. Lett $\underline{61}$, 432 (1988).
16. M. A. Moller and M. L. Klein, Chem. Phys. $\underline{129}$, 235 (1989).
17. S. Zhang and A. D. Migone, Phys. Rev. B$\underline{38}$, 12039 (1988).
18. H. Taub, K. Caneiro, J. K. Kjems, L. Passell and J. P. McTague, Phys. Rev. B$\underline{16}$, 455 (1978); C. Tiby and H. Lauter, Surf. Sci. $\underline{117}$, 277 (1982).
19. T. T. Chung, Surf. Sci. $\underline{87}$, 348 (1979).
20. Y. Larher, Surf. Sci. $\underline{134}$, 469 (1983).

21. J. P. McTague, J. Als—Nielsen, J. Bohr and M. Nielsen, Phys. Rev. B$\underline{25}$, 7765 (1982).

22. K. L. D'Amico, J. Bohr, D. E. Moncton and G. Gibbs, Phys. Rev. B$\underline{41}$, 4368 (1990).

23. M. Nielsen, J. Als—Nielsen, J. Bohr, J. P. McTague, D. E. Moncton and P. W. Stephens, Phys. Rev. B$\underline{35}$, 1419 (1987).

24. C. G. Shaw, S. C.Fain, Jr., and M. D. Chinn, Phys. Rev. Lett. $\underline{41}$, 955 (1978).

25. Q. M. Zhang and J. Z. Larese, unpublished.

26. G. Vidali and M. W. Cole, Phys. Rev. B$\underline{29}$, 6736 (1984).

27. D. M. Zhu and J. G. Dash, Phys. Rev. B$\underline{38}$, 11673 (1988); D. M. Zhu, Ph.D. Thesis, University of Washington, Seattle, 1988.

28. M. Bretz et. al. Phys. Rev. A$\underline{8}$, 1589 (1973), S. V. Hering, S. W. Van Sciver and O. E. Vilches, J. Low Temp. Phys. $\underline{25}$, 793 (1976), [^{4}He and ^{3}He]; R. E. Ecke, Q. S. Shu, T. S. Sullivan and O. E. Vilches, Phys. Rev. B$\underline{31}$, 448 (1985); T. A. Rabedeau, Phys. Rev. B$\underline{39}$, 9643 (1989) [^{4}He]; S. W. Van Sciver and O. E. Vilches, Phys. Rev. B$\underline{18}$, 285 (1978) [^{3}He]; F. A. B. Chaves et al., Surf. Sci. $\underline{150}$, 80 (1985); H. Freimuth and H. Wiechert, Surf. Sci. $\underline{162}$, 432 (1985); F. C. Motteler and J. G. Dash, Phys. Rev. B$\underline{31}$, 346 (1985) [H_2]; F. Freimuth and H. Wiechert, Surf. Sci. $\underline{178}$, 716 (1986) [D_2]

29. J. Cui, S. C. Fain, J. Vac. Sci. Techn. A$\underline{5}$, 710 (1987) Phys. Rev. B$\underline{39}$, 8628 (1989); J. Cui, S. C. Fain, H. Freimuth, H. Wiechert, H. P. Schildberg and H. J. Lauter, Phys. Rev. Lett. $\underline{60}$, 1848 (1988); H. Friemuth, H. Wiechert and H. J. Lauter, Surf. Sci. $\underline{189-190}$, 548 (1987); H. J. Lauter, H. P. Schildberg, H. Godfrin, H. Wiechert and R. Haensel, Can. J. Phys. $\underline{65}$, 1435 (1987); H. P. Schildberg, H. J. Lauter, H. Freimuth, and H. Wiechert, Jpn. J. Appl. Phys. $\underline{26}$, Supp. 26—3, 343 (1987); 345 (1987); H. J. Lauter, H. Godfrin, V. L. P. Frank and H. P. Schildberg, Proceed. of 19th Int. Conf. on Low Temp. Phys. Brighton, U.K. (Aug. 1990).

30. Michael Bretz, Phys. Rev. Lett. $\underline{38}$, 501 (1977); S. Alexander, Phys. Lett. $\underline{54A}$, 353 (1975); R. B. Potts, Proc. Comb. Philos. Soc. $\underline{48}$, 196 (1952).

31. D. M. Butler, J. A. Litzinger, A. J. Stewart and R. B. Griffiths, Phys. Rev. Lett. $\underline{42}$, 1289 (1979).

32. E.D. Specht, A. Mak, C. Peters, M. Sutton, R. J. Birgeneau, K. L. D'Aminco, D.E. Moncton, S. E. Naylor and P. M. Horn, Z. Phys. B, Cond. Matt. $\underline{69}$, 347 (1987), and references therin.

33. A. D. Migone, M. H. W. Chan, K. J. Niskanen and R. R. Griffiths, J. Phys. C. $\underline{16}$, L 1115 (1983); M. H. W. Chan, A. D. Migone, K. D. Miner and Z. R. Li, Phys. Rev. B$\underline{30}$, 2681 (1984).

34. Y. P. Feng and M. H. W. Chan, Phys. Rev. Lett. $\underline{64}$ 2148 (1990).

35. A. Terlain and Y. Larher, Surf. Sci. $\underline{93}$, 64 (1980); Y. Larher and A. Terlain, J. Chem. Phys. $\underline{72}$, 1052 (1980); Y. Larher, J. Chem. Phys. $\underline{68}$, 2257 (1978); R. M. Suter, N. J. Colella and R. Gangwar, Phys. Rev. B$\underline{31}$, 627 (1985).

COMPUTER SIMULATIONS OF TWO DIMENSIONAL SYSTEMS

David P. Landau

Center for Simulational Physics
The University of Georgia
Athens, GA 30602, U.S.A.

INTRODUCTION

The study of phase transitions in two dimensional systems has long been a topic of interest in statistical physics in large part because of the exact solutions of several two dimensional Ising models which appeared almost half a century ago. Activity in this area was heightened even further by the realization that adsorbed monolayers represented some of the best pseudo—two dimensional systems in nature and that modern experimental techniques are providing an ever increasing amount of information about a wide variety of physical systems.

Much of the interest in adsorbed monolayers centers about order—disorder transitions which are observed to occur between states which are commensurate with the substrate as a function of temperature and chemical potential. Other transitions occur involving states which are not in registry with the underlying substrate and which pose a special challenge to our understanding. The behavior of adsorbed monolayers has often been described by relatively simple models which generally place adatoms in a very strong periodic potential due to the substrate and interacting via near neighbor two— and three—body interactions. In such situations it is reasonable to constrain adatoms to lie on sites on a lattice; attempts to solve these "lattice gas" models using analytic techniques have been marginally successful, however computer simulations have proven to be quite effective.[1] Other models allow continuous motion of the adatoms and employ potentials with true long range character. Simulations are also quite effective in dealing with this class of models, although they do face special problems.

The purpose of the current manuscript is to provide an introduction to the use of simulation techniques to study models for adsorbed monolayers, to survey some of the

Phase Transitions in Surface Films 2
Edited by H. Taub *et al.*, Plenum Press, New York, 1991

results, and to compare simulations data directly with experimental data to see what new insights the simulations have produced. The literature is already voluminous and no attempt is made to be complete; our goal is rather to present a relatively broad selection of existing work to provide the reader with a view of the possibilities which simulations offer. Simulational data are being produced at a rapid rate and the reader will continually find studies which are not mentioned here to be of great value.

SIMULATION METHODS

Monte Carlo Methods

Monte Carlo simulations are computer experiments in which states in phase space are randomly visited and their properties used to provide estimates for the actual expectation values of various thermodynamic parameters. Monte Carlo methods have been fully described in a number of other references[2] so we will not provide a lengthy description here. For purposes of simplicity we shall consider a system with discrete distributions of states. From statistical mechanics we know that in the canonical ensemble the value of any thermodynamic quantity A is then given by a Boltzmann weighted average over all states

$$<A> = \frac{\sum_{\mu} A_{\mu} \exp(-E_{\mu}/kT)}{\sum_{\mu} \exp(-E_{\mu}/kT)} \quad , \tag{1}$$

where E_{μ} is the energy of state μ, A_{μ} is the value of the parameter A in state μ, T is the temperature, and k is Boltzmann's constant. In practice the number of such states is so enormous that the actual evaluation of this average is impossible. The Monte Carlo method generates a markov chain of states, each with a corresponding value of A_{μ}, and provides an estimate for $<A>$ from a relatively modest number of configurations. In the simplest case one proceeds through the system examining one particle at a time and allowing it to change its state with probability

$$W(\sigma_i \rightarrow \sigma_i') = \exp(-\Delta E/kT), \ \Delta E>0, \tag{2}$$
$$= 1 \ , \ \Delta E<0$$

where ΔE is the change in energy between the old and new states. The desired expectation value is then a simple arithmetic average over the values of A generated through this process:

$$<A>_M = \frac{1}{M} \sum_\mu A_\mu .$$

(3)

The accuracy of the results depend upon the total number of configurations generated and improves with increasing sampling. It is important to remember, however, that this approach does not follow the true time development of the system; "Monte Carlo time" is merely a fictitious time resulting from the solution of a stochastic master equation. Since successive states are correlated, particularly near a phase transition, quite long Monte Carlo runs are necessary if one is to obtain high accuracy. The properties of the models also depend on the size of the system used and the nature of the boundary conditions, although in many cases these limitations are also present in the physical systems studied experimentally. Monte Carlo simulations can be carried out in the grand–canonical and microcanonical ensembles as well and with different kinds of particle "moves". Monte Carlo simulations have been used extensively for the investigation of a wide variety of problems and the details and pitfalls are well understood.

Molecular Dynamics Methods

A very different approach is taken by the deterministic molecular dynamics method. Newton's laws are written for each particle in a classical system and then integrated numerically for each particle to yield the true time development of the system. This is accomplished by determining the instantaneous velocity $\vec{v}_i$ and acceleration $\vec{a}_i$ (by explicit determination of the total force acting on the particle) at time t and then calculating the new position $\vec{r}_i$ and velocity $\vec{v}_i$ of the particle at a time $t+\Delta t$. In the simplest application of this method the values after time step Δt has elapsed are

$$\vec{r}_i(t+\Delta t) = \vec{r}_i(t) + \vec{v}_i(t)\Delta t + 1/2\, \vec{a}_i(t)(\Delta t)^2$$

(4a)

$$\vec{v}_i(t+\Delta t) = \vec{v}_i(t) + \vec{a}_i(t)\Delta t$$

(4b)

The force acting on the particle is then updated and the procedure repeated. Of course, in practice more sophisticated integration techniques are actually used and their relative merits have been discussed elsewhere.[3] The time integration interval Δt must be kept small or the method becomes unstable; since systems may evolve only slowly, molecular dynamics may sample a relatively small region of phase space. This may not be the most efficient method for studying equilibrium behavior but the technique does produce the correct time dependence. Molecular dynamics simulations suffer from the same concerns about finite lattice size and boundary conditions that apply to Monte Carlo studies.

Although molecular dynamics simulations are usually microcanonical, it is now possible to perform these simulations under a wide variety of constraints.

LATTICE MODELS

Lattice Gas–Ising Models

In the case of strong substrate corrugation potential, the probability distribution for adsorbed atoms is strongly peaked about the positions of the potential minima. In the asymptotic limit (of infinitely strong corrugation potential) we can consider the minima of this potential to form an effective periodic lattice of possible adsorption sites for

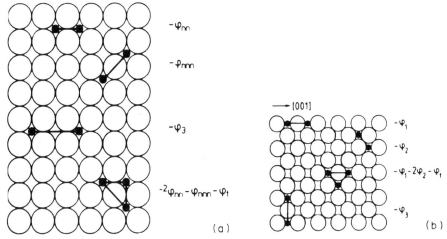

Fig. 1 Schematic illustration of interactions φ between adsorbate atoms on (a) the square lattice, and (b) the centered rectangular lattice. Open circles represent the top layer of the substrate and filled circles are adsorbed particles. The actual geometries of physical systems may be different. (From Binder et al.[4])

adatoms and all deviations from these positions will be ignored. This simple "lattice gas" model thus has site occupation variables c_i where $c_i=1$ if site i is occupied and $c_i=0$ if the site is empty. The binding energy of each adatom to the substrate is ϵ and interactions between atoms on neighboring sites are φ_{nn}, φ_{nnn} etc. for nearest–neighbors (nn–), next–nearest–neighbors (nnn–), etc. Figure 1 shows a schematic representation of a few near–neighbor couplings for simple substrates.[4] In addition, there may be multi–body interactions and Fig. 1 also shows the inclusion of three—body interaction terms φ_t. Note that the total interaction between a trio of adatoms on neighboring sites includes not only the three—body term explicitly but also all of the two—body couplings between the particles. The total Hamiltonian for the lattice—gas system is then

$$\mathscr{H} = -\epsilon \sum_i c_i - \sum_{i \neq j} \varphi_{ij} \, c_i c_j - \sum_{i \neq j \neq k} \varphi_t \, c_i c_j c_k \; . \tag{5}$$

The total occupation of the lattice is given by the coverage

$$\theta = \frac{1}{N} \sum_i c_i \; . \tag{6}$$

where N is the number of sites. If the adsorbed layer is in thermal equilibrium with the surrounding vapor, it is the pressure which is the independent variable and it is therefore preferable to use the grand canonical ensemble, subtracting a term μN_a (where N_a is the total number of adatoms) yielding the Hamiltonian

$$\mathscr{H}' = \mathscr{H} - \mu N_a = -(\epsilon + \mu) \sum_i c_i - \sum_{i \neq j} \varphi_{ij} \, c_i c_j - \sum_{i \neq j \neq k} \varphi_t \, c_i c_j c_k \tag{7}$$

where the chemical potential μ is now the independent variable. A simple transformation from the c_i to spin variables $\sigma_i = \pm 1$

$$\sigma_i = 1 - 2c_i \; , \tag{8}$$

allows us to re—express the Hamiltonian as a simple Ising model

$$\mathscr{H}_I = -H \sum_i \sigma_i - \sum_{i \neq j} J_{ij} \, \sigma_i \sigma_j - \sum_{i \neq j \neq k} J_t \, \sigma_i \sigma_j \sigma_k + H_o \tag{9}$$

where H_o contains a number of constant terms resulting from the transformation. The exchange constants J_{ij} in the Ising model are related to the lattice gas interaction parameters by

$$J_{ij} = \varphi_{ij}/4 + \sum_{k(\neq i,j)} \varphi_t \tag{10a}$$

$$J_t = -\varphi_t/8 \tag{10b}$$

and the "magnetic field" and chemical potential are related by

$$H = -[(\epsilon+\mu)/2 + \sum_{j(\neq i)} \varphi_{ij}/4 + \sum_{i\neq k(\neq j)} \varphi_t . \tag{11}$$

The magnetization per particle m of the Ising model is trivially related to the coverage

$$m = 1 - 2\theta . \tag{12}$$

One particularly important consequence of this transformation is that it reveals symmetries which may not be immediately obvious when the model is viewed in the lattice gas picture. For example, the Ising model without three–spin coupling is invariant under the transformation

$$H, \{\sigma_i\} \to -H, \{-\sigma_i\} \tag{13}$$

and this is equivalent to

$$\theta \to (1 - \theta) . \tag{14}$$

Thus, the phase diagrams for systems with two–body coupling will be symmetric in the (T,θ) plane about $\theta = 1/2$, and the adsorption isotherms are antisymmetric around the point $\theta = 1/2$, $\mu = \mu_c$, where μ_c is the chemical potential corresponding to $H = 0$. The inclusion of three body couplings will destroy this symmetry.

The nature of the critical behavior associated with phase transitions between states of different symmetry in lattice gas models has been explored in great detail by Domany et al.[5] As a result the expected critical exponents are confined to a relatively small set of different possibilities as long as the lattice gas restriction is present.

Another important advantage of the Ising model formulation is that it readily allows simulation at constant chemical potential μ so that one may work in either the canonical or grand canonical ensemble to determine phase diagrams. This flexibility is not always available in the experiments but can be quite useful in interpreting data that are somewhat ambiguous in one ensemble.

Phase Transitions and Critical Phenomena in Lattice Gas Models

Among the classic adsorption systems are noble gases adsorbed on exfoliated graphite. Data from specific heat measurements of He[4] adsorbed on two different substrates with unequal microcrystal size are shown in Fig. 2; both the positions and the magnitudes of the peaks differ for the two different substrates.[6] This is easily understood as a result of Monte Carlo simulations[7] of finite triangular lattices with periodic boundary conditions. The simplest suitable set of interactions which produces the correct $\sqrt{3} \times \sqrt{3}$ ordered state includes a nn–repulsion J_{nn} and an attractive coupling between adatoms in nnn–sites J_{nnn}. Figure 2 demonstrates that the rounding arises quite naturally as a consequence of finite size effects. Although the characteristics of the physical systems are

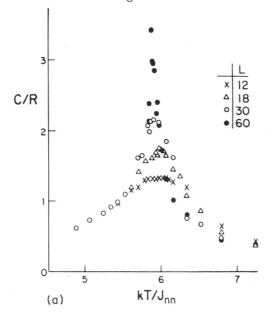

(a)

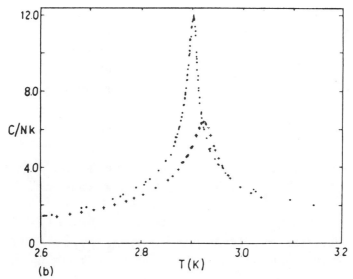

(b)

Fig. 2   (a) Specific heat of the triangular lattice gas vs temperature for $H/J_{nn} = 2.43$, $J_{nnn}/J_{nn} = -1$ (From Landau[7]); (b) Specific heat of He^4 adsorbed on graphoil (+) and on UCAR–ZYX (·) (From Bretz[6])

more complicated than those of the simulated model, the shift and rounding of the specific heat peak from the behavior of an infinite system is unmistakable. As a result of specific heat measurements carried out at constant coverage for Kr on graphite, a phase diagram was proposed which included a single phase transition at low densities and two transitions at high coverages between ordered and disordered phases. From the Monte Carlo data[7] taken with the chemical potential as the independent variable the behavior is transparent: there is a single phase boundary but it contains a tricritical point (see Fig. 3). When viewed in the coverage–temperature plane, however, the phase boundary merely opens up into a coexistence region. At 50% coverage the phase diagram shows unusual characteristics with a floating phase between the ordered and disordered regions and separated from them by Kosterlitz–Thouless–like transitions; but this behavior will probably be destroyed in physical systems by further neighbor coupling, a transition to an incommensurate phase, or by 2nd layer promotion.

Bartelt et al.[8] calculated the structure factor for a $\sqrt{3} \times \sqrt{3}$ monolayer using a simple lattice gas–Ising model with repulsive nn–coupling on a triangular lattice.

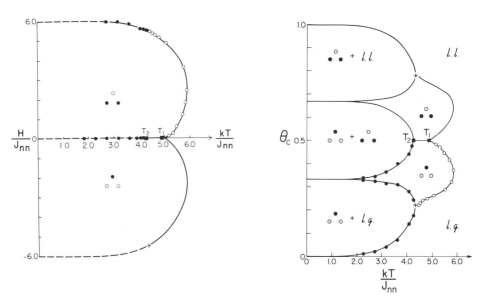

Fig. 3   Phase diagrams for the triangular lattice gas model with interaction $J_{nnn}/J_{nn} = -1$. (Left) field–temperature plane, open circles show 2nd order transitions, closed circles are 1st order transitions, and plus signs show the tricritical points. T_1 and T_2 mark the lower and upper extent of the zero field floating phase; (right) Coverage–temperature plane. (Data are shown only for $\theta \le 0.5$.) the lattice gas and lattice liquid phases are marked l.g. and l.l., respectively. (From Landau[7])

In terms of the Ising representation the structure factor is given by

$$S(q) = \frac{1}{N} \sum_{i,j} <\sigma_i \sigma_j> \exp[i\vec{q}\cdot(\vec{r}_i - \vec{r}_j)] \tag{14}$$

where N is the total number of sites in the lattice. In Fig. 4 we show contour plots of the LEED intensity I resulting from Monte Carlo calculations for lattices of 3888 sites. They discuss how finite–resolution LEED measurements allow the extraction of critical exponents by varying instrumental resolution. They applied these ideas to the Monte Carlo data, and from the temperature dependence of the structure factor data provided by the simulations they extracted "effective" critical exponent estimates for β, γ, and ν finding values which are within 10% of the expected values. This study also provides some justification for the identification of the inflection point of the I vs T curve as the critical temperature T_c.

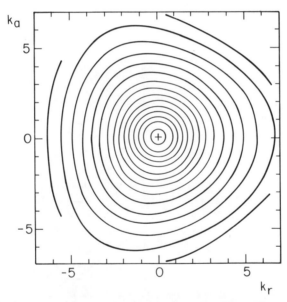

Fig. 4 Contour plot of the structure factor of a $\sqrt{3} \times \sqrt{3}$ monolayer in a triangular lattice gas with nn–repulsion φ_1 at a temperature $kT/\varphi_1 = 0.355$ (about 5% above T_c). Contour increments are separated by 0.1 on a logarithmic scale beginning with 3.2 at the outermost contour. The center of the surface Brillouin is to the left. The units of k_r and k_a, the radial and azimuthal components of $\bar{k}$ are in units of $\pi/27a$ where a is the lattice constant. (After Bartelt et al.[8]).

Doyen et al.[9] simulated a simple Ising model on a square lattice for comparison with data for the order–disorder transition in H/W(100). They considered nn–repulsive interactions only and simulated 30x30 lattices at 50% coverage. (This lattice size was roughly compatible with the experimentally encountered coherence width of 50–100 A.) They calculated the equivalent intensity of a LEED spot for the c(2x2) structure with 50% coverage as a function of temperature and found rather good agreement with experiment.[10] LEED intensities as well as specific heat for several other coverages were determined and several equilibrium configurations of particles were displayed.

Behm et al.[11] measured LEED diffraction patterns for H monolayers on Pd(100) surfaces and extracted a phase diagram from the temperature dependence of the LEED intensities. Choosing the temperature at which the intensity had dropped to one–half the low temperature value, they constructed the phase diagram shown in Fig. 5. Because of the lack of symmetry about $\theta = 0.5$, we see that two–body interactions alone cannot describe the phase diagram. A Monte Carlo study[12] of a lattice gas model with both nn– and nnn–two–body interactions and three–body interactions added as well did not produce a phase diagram which agreed well with experiment, as shown in Fig. 5.

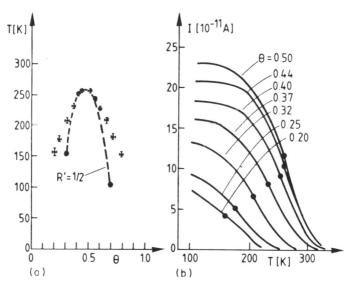

Fig 5 (a) Experimental phase diagram for H/Pd(100). Crosses are from $T_{1/2}$ where the LEED intensity has dropped to half the low temperature value. The dashed curve is from Monte Carlo data for R = $\varphi_{nnn}/\varphi_{nn}$ = −1 and R_t = φ_t/φ_{nn} = 1/2. (From Binder and Landau[12]). (b) LEED intensities as a function of temperature for different coverages θ. (From Behm et al.[11])

The interpretation of the LEED intensities is not unambiguous in that the intensities do not go to zero but rather show high temperature "tails" akin to those shown in finite lattice simulations. Indeed, as shown in Fig. 6, if the LEED intensities are extracted from the Monte Carlo data generated for this model we see that they too show high temperature "tails". Using the same criterion to locate the phase transitions as was applied in the experimental, work we find a phase diagram which is quantitatively different from that obtained from the data in other, more sophisticated ways. The simulations thus point out quite clearly the limitations in the interpretation of the experimental data.

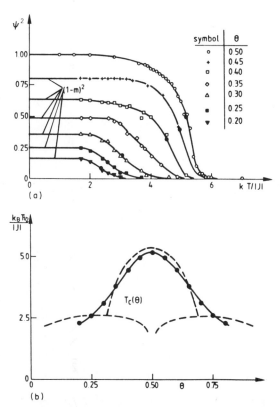

Fig. 6 (a) Squared order parameter for the c(2x2) structure as a function of temperature for constant coverage θ. Open circles show data for R=−1, R_t=0. Solid circles show where the value has dropped to 1/2 of the low temperature value. (b)Phase boundary in the coverage–temperature plane (solid circles) extracted from part (a) compared with the correct boundaries. (From Binder and Landau[12])

Another adsorption system for which a square lattice is appropriate is chlorine on Ag(100) which was studied by LEED and by Monte Carlo simulation by Taylor et al.[13] Since the experiment indicated that nn–sites were not occupied, the simulations were done with infinite nn–repulsion, i.e. a hard square model. The resulting relation between the structure factor and the coverage from the Monte Carlo simulation agree quite well with the experimental results (the c(2x2) LEED beam height) shown in Fig. 7 . From the coverage dependence of the intensity they extract a "Fisher renormalized" exponent $\beta/(1-\alpha) = 0.12$ which is consistent with the Ising value of $1/8$. The overall agreement between the Monte Carlo results and experimental data strongly supports the appropriateness of this "hard square" lattice gas model.

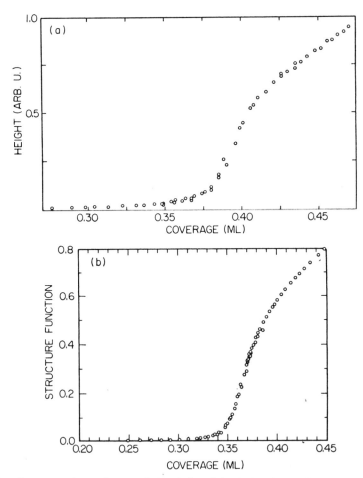

Fig. 7 Coverage dependence of the (1/2,1/2) LEED beam height at 300K (a) experimental value at 300 K, (b) results from Monte Carlo simulation on a 72x72 hard square model. (From Taylor et al.[13])

Monte Carlo simulations have also been used to study phase diagrams of square lattice Ising–lattice gas models with a wide range of finite nn– and nnn–interactions as a function of temperature and chemical potential.[14] For repulsive nn–interactions and sufficiently repulsive nnn–coupling a highly degenerate "row–shifted" state was discovered. Whether or not it is stable at finite temperature is unclear. Zero field studies[15] (i.e., θ=0.5), including 3nn–interactions revealed structures with larger unit cells including (4×2) and (4×4). Counterparts in physical systems have not yet been identified.

The ordering of H on Fe(110) has also been studied using lattice gas models and Monte Carlo methods. From LEED data Imbihl et al.[16] constructed the phase diagram shown in Fig. 8 . In addition to the disordered "lattice gas" and "lattice fluid"

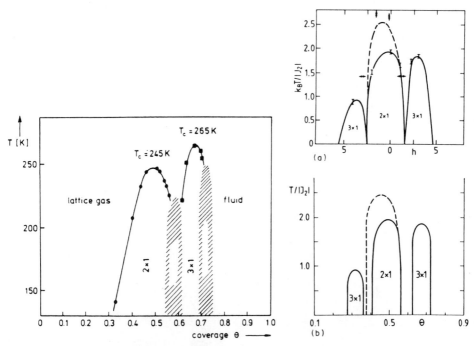

Fig. 8   (Left) Phase diagram for H/Fe(110) as determined from LEED measurements.   Dots are experimental values; shaded regions are "incommensurate". (From Imbihl et al.[16]) (Right) Phase diagram for the centered rectangular lattice gas model with φ_1=0, φ_3/φ_2 = 1/3, and φ_t/φ_2= –1/3 as determined by Monte Carlo simulations. The solid curves are the phase boundaries and the dashed curve shows disorder points. (From Kinzel et al.[17])

phases, they found (2x1) and (3x1) commensurate, ordered phases. The shaded regions in Fig. 8 were identified as incommensurate phases or phases composed of antiphase domains. Instead of occurring at the Bragg position the peak in the LEED beam profile is split and "satellites appear". A lattice gas model[17] was used which included nnn−, and 3nn−two−body coupling as well as a three−body term on an anisotropic triangular lattice (centered rectangular net), as shown in Fig. 1b. The resultant phase diagram, also shown in Fig. 8, has substantial similarity to the experimentally determined one. The low density (3x1) phase has not been found experimentally, but the relative positions and shape of the other phase boundaries are quite reasonable. The structure factor $S(\vec{q})$ was calculated from the Monte Carlo configurations with $\vec{q}$ being in the direction of the 3nn interaction, and here too there are similarities with the experimental result(see Fig. 9). For $kT/|J_2| = 1.4$ a single peak occurs at the commensurate position for $\theta < 0.53$ while for larger θ the peak splits. The similarities as well as the differences with respect to the experimental LEED beam profile are clearly displayed in Fig. 9. Since recent work[18] has shown that the actual adsorption sites are not those shown in Fig. 1b, the coverage axis in Fig. 8b is not actually correct. A Monte Carlo study of a revised model is still needed.

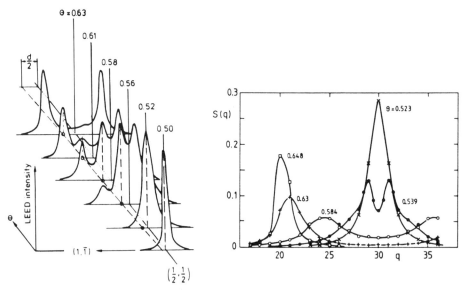

Fig. 9 (Left) Angular LEED beam profiles for H/Fe(110) at 200 K. (From Imbihl et al[13]). (Right) Monte Carlo results for the structure factor $S(q)$ as a function of wavevector q (in units of $2\pi/60$) for coverages near 1/2 at $kT/|J_2|=1.4$ for the centered rectangular model of Fig. 8. (From Kinzel et al.[17])

Lattice gas models have also been invoked in an attempt to describe the behavior of O/W(110). LEED data showed the formation of a 2x1 structure for less than 50% coverage and Ertl and Schillinger[19] simulated a simple model with three different types of two–body interactions. (The "nearest neighbors" in the $[\bar{1}11]$ and $[1\bar{1}1]$ were considered to have the opposite sign!) They calculated the p(2x1) LEED intensity vs temperature for 50% coverage, but given the differing experimental results available at that time the results were rather inconclusive. Ching et al.[20] used a more sophisticated hamiltonian, including four different two–body interactions and two kinds of three–body couplings, and also determined the behavior of different lattice sizes. By comparing with newer experimental data[21] they concluded that there was a 2nd order transition involving 10^2–10^3 sites, an estimate which agreed well with the mean size of regions of coherent scattering inferred from the width of the diffraction peaks. This comparison is shown in Fig. 10. The accuracy is such that the choice of all the interaction parameters is not really unique but the approximate magnitudes are expected to be correct.

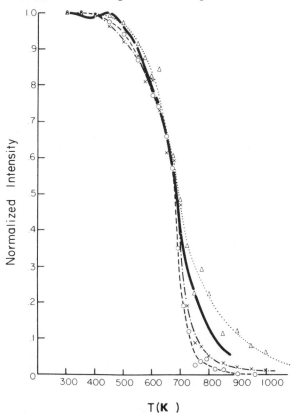

Fig. 10 Normalized p(2x1) LEED intensity for coverage θ=0.5. The solid curve is from experiment and Monte Carlo data for LxL lattices with periodic boundary condition are shown by (Δ) L=10; (x) L=20; (o) L=30. The broken curves are guides to the eye. (From Ching et al.[20])

In an effort to explain the behavior of Ag/W(110) Stoop[22] included two–body interactions out to 5th nearest neighbors and added two different types of attractive three–body interactions in his lattice gas model. (Groundstate calculations were performed in order to restrict the possible choices of interactions to those which could yield the correct ordered states.) Monte Carlo simulations were performed on 40x40 lattices and the resultant coverage–temperature phase diagram was compared with experimental results in order to determine which interactions gave the best fit. Although no unique set of interaction parameters could be extracted, it was clear that three–body interactions were essential.

A complicated system which has been modeled using pairwise interactions only, but up to 6th nearest neighbors, is Si/W(110) which was simulated by Amar et al.[23] They carried out Monte Carlo simulations on 60x60 lattices as well as mean field and transfer matrix calculations and obtained a complex phase diagram, shown in Fig. 11. The phase diagram contains p(2x1), 5x1, and 6x1 commensurate phases as well as an

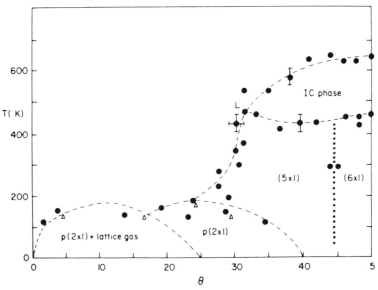

Fig. 11 Phase diagram of a lattice gas model for Si/W(110) in the temperature–coverage plane. Dots are from Monte Carlo calculations and triangles are transfer matrix values. The point labelled L is the Lifshitz point. The dotted line line shows the transition region between (5x1) and (6x1) and the dashed lines are guides to the eye for other phase boundaries. (From Amar et al.[23])

incommensurate phase, and there is a low temperature first order p(2x1) to disorder transition, which yields a coexistence region in coverage–temperature space. There is an apparent disagreement between these results and the experimentally determined maximum transition temperature of the p(2x1) ordered phase, but this may well simply be due to the inadequacy of some of the interaction parameters used or even the lack of three body couplings. Experimental measurements are needed to test the complete phase diagram.

When adsorbed on metallic surfaces alkali atoms, e.g., Na/W(110), they form permanent dipole moments. The adsorption occurs in the long bridge sites on W(110) giving rise to a centered rectangular lattice. Roelofs and Kriebel[24] performed a Monte Carlo simulation at fixed chemical potential assuming dipole–dipole interactions which were truncated at 13th nearest neighbors with a mean–field–like coupling representing the longer range interactions. They considered 30x30 and 60x60 systems with periodic boundary conditions with the specific sizes chosen so as to allow many different structures to "fit" without misfit seams. They found a complicated phase diagram which include both simple, commensurate phases as well as "intermixed" phases due to kinked domain wall formation (incommensurate floating phases?) as shown in Fig. 12. Experimentally determined phase diagrams show five different commensurate phases, but there are both quantitative as well as qualitative differences found with respect to the results of

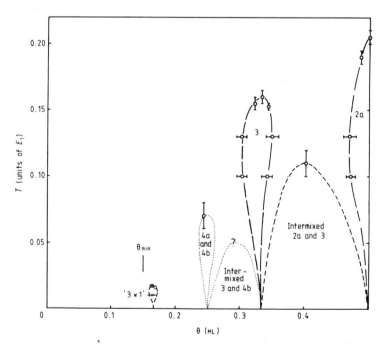

Fig. 12 Phase diagram for the repulsive dipole system in the temperature–coverage plane. Data are from Monte Carlo simulations using a truncated dipole–dipole coupling. (From Roelofs and Kriebel[24])

simulations. However, small adjustments to some of the interactions, i.e. the inclusion of non–dipolar interactions, would alter this result.

Multilayer Adsorption

Although strictly speaking this topic does not fit into the context of two dimensional phase transitions, it is nonetheless a natural extension of both the experimental and simulational studies already reviewed. Adsorption isotherms have been measured for many adsorbate/substrate systems,[25] and some of the most striking behavior is that which is found for the adsorption of noble gases on graphite. Fig. 13 shows the "risers" which are seen for Kr on graphite: at low temperatures a sequence of quite sharp, vertical steps was found, but at higher temperatures these steps became smeared out. Kim and Landau[26] studied a simple lattice gas model with a z^{-3} adatom–substrate attractive interaction (z being the height above the substrate in lattice constants) and nearest neighbor adsorbate–adsorbate attraction. They found a sequence of vertical steps (also shown in Fig. 13) at low temperatures resulting from a series of layering transitions, and they found that at higher temperatures these steps became smeared out as the critical points were approached or even eventually passed. Multilayer adsorption can, in principle, show a great diversity of phase transitions including the simultaneous formation of multiple layers or even wetting in which the thickness of the adsorbing film diverges. Wagner and Binder[27] even considered a model in which the interactions within the first layer were different than in all higher lying layers. They find a region of chemical potential for which an ordered phase occurs, but for higher values of chemical potentials a closely spaced sequence of layering transitions appears. Wetting and layering transitions have been studied[28] extensively in nn– simple cubic lattice gas–Ising models; but because of the appearance of capillary wave excitations, finite lattice size effects are quite pronounced and the complete phase diagram is still not fully determined. The addition of more distant neighbor interactions offers the potential for intriguing complexity.

Orientational Phase Transitions

Our previous presentation has treated models appropriate for the onset of positional order of adatoms; a simple and obvious extension of this discussion is the consideration of orientational order of adsorbed molecules on substrates with strongly corrugated potentials. In such cases lattice models are still appropriate, but one must consider the orientation of simple multipoles at each lattice site instead of the question of whether a site is occupied or not. Tang et al.[29] performed constant pressure molecular dynamics simulations of 400 diatomic molecules interacting through an atom–atom Lennard–Jones potential appropriate for O_2 molecules. They found a first–order transition from the ferroelastic (centered rectangular) phase to the paraelastic phase

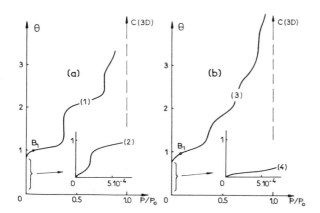

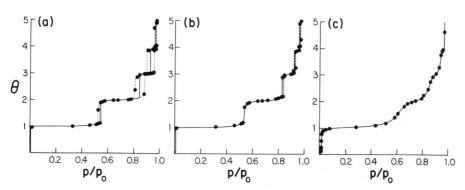

Fig. 13 Adsorption isotherms as a function of pressure. (top) Argon on graphite for (a)T=64 K, and (b) T=77 K. (From Thomy et al.[25]). (Bottom) Monte Carlo data for a lattice gas model with a substrate–adatom potential of z^{-3} for reduced temperature $\tau = T/3\varphi_{nn}$ (a) $\tau = 0.3$; (b) $\tau = 0.32$; (c) $\tau = 0.40$. The solid curves are equilibrium values and the dashed curves show hysteresis due to rapid absorption or desorption. (From Kim and Landau[26])

induced by the formation of a large density of localized herringbone–like defects. Jin et al.[30] performed constant pressure molecular dynamics simulations of the same model for O_2 on graphite, and from the time dependence of the density–density correlation functions they computed the scattering function $S(q,\omega)$. They found an order–disorder transition at $T^* = kT/\epsilon = 0.37$ and found a softening of the LA phonon frequency and of the elastic constant as the transition temperature is approached.

O'Shea and Klein[31] used Monte Carlo simulations to study quadrupoles on small (6x6 and 12x12) triangular nets and found that for free layers there was a pinwheel structure at low temperature, i.e., below where the herringbone structure forms. (When comparing with experimental results one should not forget that the electrostatic quadrupole–quadrupole coupling is much stronger for N_2 than for O_2.) They extended the simulations to molecular bilayers[32] including both nn– and nnn– quadrupolar interactions and 36 molecules per layer. They determined the behavior of both in–plane (herringbone) transitions as well as the behavior of free layers.

Klenin and Pate[33] used much larger quadrupolar systems with up to 2000 lattice sites. They found two transitions bracketing an intermediate state which was an admixture of nonrotationally invariant phases rather than a single long range ordered state. Isolated pinwheels were a commonly observed localized defect. Using a simplified model of planar quadrupoles on a triangular lattice, Mouritsen and Berlinsky[34] carried out extensive simulations with systems containing up to 10,000 sites. They concluded that there was a single 1st order transition which becomes smeared out in small lattices.

Harris et al.[35] examined a system of planar quadrupoles on a triangular lattice with vacancies. They not only studied the phase transition but also in the process helped inspire experimental work on CO/Ar and N_2/Ar mixtures on graphite.[36]

Molecular dynamics simulations were performed by Talbot et al.[37] for N_2 on graphite. Each molecule was modeled as two Lennard–Jones particles fixed rigidly .1.1Å apart. They found a rotational transition from the herringbone structure to a plastic crystal at 33K as compared with the experimental value of 27K.

The next degree of complexity which enters the study of orientational order is the inclusion of another rotational symmetry axis, i.e. the consideration of tetrahedral molecules which may have several possible orientations in the direction perpendicular to the plane as well as orientational order within the plane. O'Shea and Klein[38] used Monte Carlo simulations of small (6x6 and 12x12) systems of classical octopoles arrayed in a triangular net with periodic boundary conditions. Octopole–octopole interactions were restricted to nn–pairs. The maximum rotational "jump" was adjusted to maintain an acceptance rate of about 50%. They considered both free standing layers, for which the angular anisotropy of the graphite potential is assumed to be negligible and the molecules interact only with their neighbors, and "tripod" layers, for which the graphite potential is assumed to be so strong that the molecules must sit as stable tripods on the surface, being free only to rotate about an axis normal to the surface. In the case of the free standing layer they observed three phases, a low temperature ordered structure in which alternate rows are either stable or unstable tripods with respect to the surface, a high temperature rotationally disordered phase, and an intermediate partially ordered state. The tripod layer shows a single phase transition between a high temperature rotationally disordered phase and a low temperature strongly hindered, apparently ferro–rotational state. Their analysis of data for methane on graphite leads them to conclude that the tripod model is probably correct. A similar model for methane bilayers was simulated by Maki and O'Shea[39] with 36 molecules per layer. For the tripod bilayer model they find little correlation between the behavior of the two layers, but the lower layer seems to provide a weak field which favors unstable tripod orientations in the upper layer. One transition in the lower layer and two in the upper layer are found. The free standing bilayer apparently shows a low temperature phase with both layers ferro–rotationally ordered, an intermediate partially disordered state, and a high temperature disordered phase.

MODELS WITH FINITE SUBSTRATE POTENTIALS

Melting of Adsorbed Monolayers

When the corrugation depth of the model used becomes finite, the adatoms can move continuously on the surface and the lattice restriction is removed. As a consequence of this removal the number of degrees of freedom of the system is greatly increased and more realistic interaction potentials must be adopted. These interactions are necessarily long range (although in practice they are truncated at some finite particle separation), and because of the additional complications inherent in the model Monte Carlo simulations are possible only for smaller systems and for shorter times than in the case of lattice models. The possibility of continuous particle translation, however, opens up the opportunity to apply molecular dynamics methods to the models under consideration.

The most commonly used interparticle potential is the Lennard–Jones interaction $V_{LJ}(r)$ given by

$$V_{LJ} = \frac{\epsilon}{4} \sum_{ij} [(\frac{\sigma}{r_{ij}})^{12} - (\frac{\sigma}{r_{ij}})^6] \tag{15}$$

where r_{ij} is the distance between atoms and σ and ϵ are parameters describing the equilibrium inter–particle distance and the strength of the well depth, respectively.

Usually the potential is truncated at distances of about 2.5σ for computational simplicity.

Molecular dynamics simulations have been used to study the melting of the c(2x3) commensurate structure for monolayer Ar physisorbed on the (100) surface of MgO. Alavi and McDonald[40] used Lennard–Jones potentials for Ar–Ar, Ar–O, and Ar–Mg interactions for 192 atoms in a rectangular box with periodic boundary conditions. The temperature dependence of the order parameters shows that melting occurs via second layer promotion at 5–10% above the experimentally observed transition temperature. At lower than monolayer coverage the simulations show that order is lost in directions parallel to the (10) Mg–Mg channels. This behavior is the same as that deduced from the shifting and broadening of neutron diffraction peaks.

Abraham[41] carried out molecular dynamics simulations for submonolayer argon on graphite using a substrate of 400 A^2 with 1680 argon atoms. He found a continuous melting transition which is apparently in agreement with experimental findings. The density of the liquid phase just beyond the transition is quite similar to the commensurate density for argon.

Molecular dynamics simulations of Xenon on graphite have a more complex history. Early studies[42] suggested that the transition was first order, but later simulations concluded[43] that it is essentially continuous because of 2nd layer promotion. Temporal density fluctuations are reduced in simulations of larger, 2304 atom systems and the first layer remains as a solid–mixture at the transition.[44] The density–temperature behavior is continuous for simulations at constant temperature, coverage, and substrate area, and discontinuous for the constant temperature, coverage, and spreading pressure studies. The computed temperature dependence of the inverse correlation length, shown in Fig. 14, is in excellent agreement with experiment. The good agreement which is obtained with a flat substrate suggests that the horizontal characteristics of the substrate are unimportant.

Lennard–Jones models for adsorption of Kr on graphite have also been simulated. Bhethanabotla and Steele[45] compared the results obtained by molecular dynamics using the potential for bulk Kr and using a potential with reduced well depth resulting from substrate mediated effects. The latter potential was quite effective in producing good agreement with experiment.

Multilayer films in models with continuous particle position have been studied by Phillips[46] and Phillips and Hruska.[47] The interplay between the different layers is particularly interesting and CH_4 multilayers on graphite behaves differently than Ar multilayers.

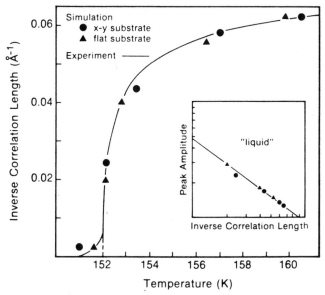

Fig. 14 Inverse correlation length for Xe/graphite and (inset) the peak amplitude as a function of inverse correlation length. Simulation data were from molecular dynamics calculations with Lennard–Jones potentials. (From Abraham[44])

Commensurate–Incommensurate Transitions

Because of the difference in the position of the minimum in the Kr–carbon potential and the Kr–Kr potential for Kr on graphite, at sufficiently low temperatures the ordered Kr monolayer coverage becomes incommensurate with the substrate. Abraham et al.[48] performed large scale molecular dynamics simulations to study this incommensurate phase. They used systems with up to 161,604 particles (corresponding to graphite substrate dimensions of up to 1700A) with appropriate Lennard–Jones 12:6 potentials for Kr–Kr and Kr–carbon interactions. The simulations were carried out at both constant temperature T and at constant coverage. As shown in Fig. 15, at low temperatures they found a honeycomb network of domain walls, each surrounding a region of commensurate structure; the hexagons differ in size and shape, thus verifying the picture proposed by Villain.[49] The walls also turn out to be "Heavy" as was predicted by Kardar and Berker.[50] The striped phase which had also been proposed was not seen nor was two phase coexistence. They also did not see a "fluid" phase resulting from an instability of the structure with respect to dislocations. With increasing T the domains become increasingly irregular and the domain walls thicken.

Schoebinger and Abraham[51] continued this work using a potential proposed by Vidali and Cole[52] and putting 20,736 particles inside a parallelogram shaped box with periodic boundary conditions. They found that the width of domain walls at low temperature is independent of coverage although domain size decreases rapidly with increasing coverage. They found that the wall interaction energy is negative and conclude that the C–IC (commensurate–incommensurate) transition is 1st order.

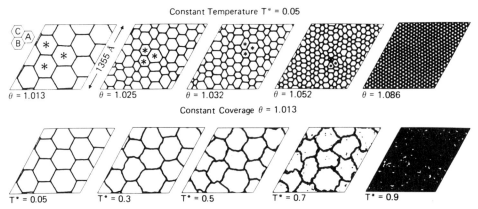

Fig. 15 "Snapshots" of equilibrium configurations of the incommensurate phase generated by molecular dynamics for 103,041 Kr atoms on graphite. The reduced temperature is in units of the Lennard–Jones well depth. (From Abraham et al.[48])

In a separate simulation Koch and Abraham[53] studied the C–IC transition in free clusters. Two different systems sizes were studied, 144 and 2304 atoms, and the molecular dynamics simulations were run for between 5×10^4 and 10^5 time steps. For the smaller lattice the C–IC transition was smooth and reversible, but for the larger system pronounced hysteresis was observed. In agreement with the prediction of Gordon and Villain[54] they find that thermal expansion stabilizes the commensurate phase at high temperatures.

Houlrik et al.[55] considered the same model using extensive Monte Carlo simulations and varying both the lattice size as well as the corrugation potential. They also find clear indication of a first order C–IC transition for sufficiently large lattice and very pronounced finite size effects. Both the shift in transition temperature as well as the smearing of the transition decrease rapidly as the cluster increases in size.

This discussion would not be complete without mention of the question of melting of a two dimensional solid in the absence of any substrate potential. In fact, one of the earliest molecular dynamics simulations was of the melting of a system of hard disks.[56] Halperin and Nelson[57] and Young[58] have developed a theory for the melting of a two dimensional "solid" via two second order transitions of the Kosterlitz–Thouless type, as the temperature is increased. The first transition occurs when the dislocation pairs present in the "solid" phase (with algebraic decay of positional order and long range orientational order) unbind leading to a "hexatic" state (with exponential decay of translational order and an algebraic decay of six–fold orientational order). This hexatic state then undergoes a second order transition, due to the appearance of free disclinations, to an isotropic liquid. Attempts to verify the KTHNY predictions have let to considerable discussion and controversy. McTague et al.[59] carried out constant volume Monte Carlo simulations of systems with up to 2500 particles and an r^{-6} repulsion and found evidence for dislocation mediated melting. Based on topological considerations, they identified a defect structure which they associated with the hexatic phase, see Fig. 16, and found that the orientational order parameter decays roughly algebraically in this region.

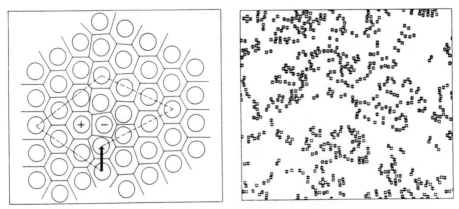

Fig. 16 Defect structure for 2d melting. (Left) Dirichlet domain construction illustrating a 7–5 pair. The arrow shows the Burgers vector. (Right) Defect structure in the "oriented fluid" at $T^{*}=0.1525$ produced by Monte Carlo simulations. Both free dislocations and grain boundaries are apparent. (From McTague et al.[59])

Abraham[60] performed isothermal–isobaric Monte Carlo simulations of a two dimensional Lennard–Jones system and concluded instead that there was a single first order transition. Abraham used systems with 256 and 529 particles and found hysteresis in the density as a function of temperature. His interpretation was that the "hexatic" phase observed in the constant volume simulations was merely the two–phase coexistence region associated with a first order transition. Snapshots of the lattice at different times supported this interpretation.

Other Monte Carlo[61] and Molecular dynamics[62] simulations have been performed for 2d models with a variety of potentials, but the limitations on both system size and maximum simulation observation time have limited the resolution of all these studies. Thermodynamic properties, elastic constants, defect behavior, and nnn–bond orientational order have all been examined in order to clearly identify the nature of the melting transition. For example, Strandburg et al.[63] studied the local bond–orientational order for both hard disk and Lennard–Jones systems. Their results supported the idea of coexisting phases with regions of about 50 atoms involved. In contrast, Udink and van der Ellsken[64] studied large systems, with 12,480 particles, using molecular dynamics and analyzing the data within the framework of using finite size scaling; they concluded that the behavior was in better agreement with KTHNY theory than was previously believed when sufficiently large systems are examined. Systems which have been proposed as physical realizations of two dimensional solids undergoing melting include Xe/Ag(111) and Xe/graphite. (Recent experimental results[65,66] at relatively high temperatures and densities suggest the presence of a hexatic phase.) Phillips and Bruch[67] carried out Monte Carlo simulations for Xe on Ag(111) using several different interaction potentials and voiced concern about the failure of the two dimensional approximation. They also find a curious degradation in agreement with experiment when an improved Hamiltonian containing many–body terms is used. This problem is certain to remain fertile ground for simulations in the future; a completely unambiguous description of 2d melting is still needed.

SUMMARY AND OUTLOOK

Extensive progress has been made in understanding phase transitions in adsorbed monolayers through the use of computer simulation techniques. Simulations can be used to calculate thermodynamics properties, and scattering profiles, and even offer the opportunity to view explicit atomic positions. In fact in many cases simulations have already been used to study systems of the same spatial extent as those which are found in physical systems. The choice of appropriate potentials remains a topic for further consideration. The most complete existing data are for lattice gas models and these results have already demonstrated the importance of using multiple interaction terms. In this regard, the inclusion of three–body interactions is particularly important.

There are clearly other properties of adsorbed monolayers which have been examined using computer simulations but which we have not discussed here. We have not considered coadsorption of multiple species[68] and the additional complication of the phase diagram which can result. Certainly the study of time dependent properties[69] including domain formation is an exciting field in which simulations in lattice gas models has already played a major role. Wetting phenomena[70] in two dimensions have been considered for several different systems. Surface reconstruction due to the addition of adsorbed monolayers is another area which merits substantial effort. The consideration of quantum effects in adsorbed monolayers has only just begun[71] and is certain become more important as is the examination of adsorption on imperfect substrates and/or the inclusion of substantial impurity effects.

As computer power increases still further, the utility of simulations will become enormous. They will soon be used not only to explain experimental data but also to predict which experiments might yield the most exciting results.

ACKNOWLEDGMENTS

The author wishes to express his appreciation to Prof. K. Binder for a long and fruitful collaboration including, in part, simulations of models for adsorbed monolayers. We are also indebted to Dr. A. M. Ferrenberg, Prof. S. C. Fain, Jr., and Prof. J. M. Phillips for helpful comments and suggestions. This work was supported in part by the National Science Foundation.

REFERENCES

1. Reviews which discuss, at least in part, the application of computer simulations to models for adsorbed monolayers include: K. Binder and D. P. Landau, in: "Advances in Chemical Physics," ed. K. P. Lawley, John Wiley & Sons Ltd (1989); E. Bauer, in: "Structure and Dynamics of Surfaces II," ed. W. Schommers and P. von Blanckenhagen, Springer, Berlin (1987); K. J. Strandburg, Rev. Mod. Phys. 60:161 (1988); F. F. Abraham, Rep. Prog. Phys. 45:1113 (1982).
2. "Monte Carlo Methods in Statistical Physics," ed. K. Binder, Springer, Berlin (1979); "Applications of the Monte Carlo Method in Statistical Physics," ed. K. Binder, Springer, Berlin (1984).

3. See e.g., H. J. C. Berendsen, "Enrico Fermi" summer course, 1985; M. P. Allen and D. J. Tildesley, in: "Computer Simulation of Liquids," Clarendon Press, Oxford (1987). Implementation of molecular dynamics methods for vector supercomputers has been discussed by D. Rapaport, in: "Computer Simulation Studies in Condensed Matter Physics II," ed. D. P. Landau, K.K. Mon, and H.–B. Schüttler, Springer, Berlin (1990).

4. K. Binder, W. Kinzel, and D. P. Landau, Surf. Sci. 117:232 (1982).

5. E. Domany, M. Schick, J. S. Walker, and R. B. Griffiths, Phys. Rev. B18:2209 (1978).

6. M. Bretz, Phys. Rev. Lett. 38:501 (1977).

7. D. P. Landau, Phys. Rev. B27:5604 (1983).

8. N. C. Bartelt, T. L. Einstein, and L. D. Roelofs, Phys. Rev. B35:1776 (1987); N. C. Bartelt, T. L. Einstein, and L. D. Roelofs, Phys. Rev. B32:2993 (1985).

9. G. Doyen, G. Ertl, and M. Plancher, J. Chem. Phys. 62:2957 (1975).

10. P. J. Estrup, in: "The Structure and Chemistry of Solid Surfaces," ed. G. A. Somorjai, Wiley, New York (1969).

11. R. J. Behm, K. Christmann, and G. Ertl, Surf. Sci. 99:320 (1980).

12. K. Binder and D. P. Landau, Surf. Sci. 108:503 (1981).

13. D. E. Taylor, E. D. Williams, R. L. Park, N. C. Bartelt, and T. L. Einstein, Phys. Rev. B32:4653 (1985).

14. K. Binder and D. P. Landau, Phys. Rev. B21:1941 (1980).

15. D. P. Landau and K. Binder, Phys. Rev. B31:5946 (1985).

16. R. Imbihl, R. J. Behm, R. J. Christmann, G. Ertl, and T. Matsushima, Surf. Sci. 117:257 (1982).

17. W. Kinzel, W. Selke, and K. Binder, Surf. Sci. 121:13 (1982).

18. W. Moritz, R. Imbhil, R. J. Behm, G. Ertl, and T. Matsushima, J. Chem. Phys. 83:1959 (1985).

19. G. Ertl and D. Schillinger, J. Chem. Phys. 66:2569 (1977).

20. W. Y. Ching, D. L. Huber, M. G. Lagally, and C.–C. Wang, Surf. Sci. 77:550 (1978).

21. T.–M. Lu, G.–C. Wang, and M. G. Lagally, Phys. Rev. Lett. 39:411 (1977).

22. L. C. A. Stoop, Thin Solid Films 103:375 (1983).

23. J. Amar, S. Katz, and J. D. Gunton, Surf. Sci. 155:667 (1985).

24. L. D. Roelofs and D. L. Kriebel, J. Phys. C20:2937 (1987).

25. See for example, A. Thomy, X. Duval, and J. Regnier, Surf. Sci. Rep. 1:1 (1981) and references therein.

26. I. M. Kim and D. P. Landau, Surf. Sci. 110:415 (1981).

27. P. Wagner and K. Binder, Surf. Sci. 175:421 (1986).

28. K. Binder, D. P. Landau, and S. Wansleben, Phys. Rev. B40:6971 (1989); K. Binder and D. P. Landau, Phys. Rev. B37:1745 (1988); A. Patrykiejew, K. Binder, and D. P. Landau, to be published.

29. S. Tang, S. D. Mahanti, and R. K. Kalia, Phys. Rev. Lett. 56:484 (1986).

30. W. Jin, S. D. Mahanti, and S. Tang, Sol. State Commun. 66:877 (1988).

31. S. F. O'Shea and M. L. Klein, Chem. Phys. Lett. 66:381 (1979).

32. S. F. O'Shea and M. L. Klein, Phys. Rev. B25:5882 (1982).

33. M. A. Klenin and S. F. Pate, Phys. Rev. B26:3969 (1982).

34. O. G. Mouritsen and A. J. Berlinsky, Phys. Rev. Lett. 48, 181 (1982).

35. A. B. Harris, O. G. Mouritsen, and A. J. Berlinsky, Can. J. Phys. 62, 915 (1984).

36. H. You and S. C. Fain, Jr., Phys. Rev. B34, 2840 (1986).

37. J. Talbot, D. J. Tildesley, and W. A. Steele, Mol. Phys. 51, 1331 (1984).

38. S. F. O'Shea and M. L. Klein, J. Chem. Phys. 71:2399 (1979).

39. K. Maki and S. F. O'Shea, J. Chem. Phys. 73:3358 (1980).

40. A. Alavi and I. R. McDonald, Mol. Phys. 69:703 (1990).

41. F. F. Abraham, Phys. Rev. 28:7338 (1983).

42. S. W. Koch and F. F. Abraham, Phys. Rev. B27:2964 (1983).

43. F. F. Abraham, Phys. Rev. Lett. 50:978 (1983).

44. F. F. Abraham, Phys. Rev. 29:2606 (1984).

45. V. Bhethanabotla and Steele, J. Chem. Phys. 92:3285 (1988).

46. J. M. Phillips, Langmuir 5, 571 (1989).

47. J. M. Phillips and C. D. Hruska, Phys. Rev. B39, 5425 (1989).

48. F. F. Abraham, W. E. Rudge, D. J. Auerbach, and S. W. Koch, Phys. Rev. Lett. 52:445 (1984).

49. J. Villain, in: "Ordering in Strongly Fluctuating Condensed Matter Systems," ed. T. Riste, Plenum, New York (1980).

50. M. Kardar and A. N. Berker, Phys. Rev. Lett. 48:1552 (1982).

51. M. Schoebinger and F. F. Abraham, Phys. Rev. B31:4590 (1985).

52. G, Vidali and M. W. Cole, Phys. Rev. B29:6736 (1984).

53. S. W. Koch and F. F. Abraham, Phys. Rev. B33:5884 (1986).

54. J. Villain and M. B. Gordon, Surf. Sci. 125:1 (1983).

55. J. M. Houlrik, D. P. Landau, and S. J. Knak Jensen, (to be published).

56. B. J. Alder and T. E. Wainwright, Phys. Rev. 127:359 (1962).

57. B. I. Halperin and D. R. Nelson, Phys. Rev. Lett. 41:121 (1978); D. R. Nelson and B. I. Halperin, Phys. Rev. B19:2457 (1979).

58. A. P. Young, Phys. Rev. B19:1855 (1979).

59. J. P. McTague, D. Frenkel, and M. P. Allen, in: "Ordering in Two Dimensions," ed. S. K. Sinha, North Holland, Amsterdam (1980).

60. F. F. Abraham, in: "Ordering in Two Dimensions," ed. S. K. Sinha, North Holland, Amsterdam (1980).

61. See e.g., J. Tobochnik and G. V. Chester, Phys. Rev. B25:6778 (1982).

62. See e.g., J. Q. Broughton, G. H. Gilmer, and J. D. Weeks, Phys. Rev. B25:4651 (1982); A. D. Novaco and P. A. Shea, Phys. Rev. 26:284 (1982); S. Toxvaerd, Phys. Rev. Lett. 51:1971 (1983); A. D. Novaco, Phys. Rev. B35:8621 (1987);

A. F. Bakker, F. C. Bruin, and H. J. Hilhorst, <u>Phys. Rev.Lett.</u> 52:449 (1981).

63. K. J. Strandburg, J. AA. Zollweg, and G. V. Chester,<u>Phys. Rev.</u> B30:2755 (1984).

64. C. Udink and J, van der Elsken, <u>Phys. Rev.</u> B35:279 (1987).

65. N. Greiser, G. A. Held, R. Frahm, R. L. Greene, P. M. Horn and R. M. Suter, <u>Phys. Rev. Lett.</u> 59, 1706 (1987).

66. R. Gangwar, N. J. Colella and R. M. Suter, <u>Phys. Rev.</u> B39, 2459 (1989).

67. J. M. Phillips and L. W. Bruch, <u>J. Chem. Phys.</u> 83, 3660 (1985); J. M. Phillips and L. W. Bruch, <u>Phys. Rev. Lett.</u> 60, 1681 (1988).

68. See e.g., H.–H. Lee and D. P. Landau, <u>Phys. Rev.</u> B20:2893 (1979).

69. See e.g., A. Sadiq and K. Binder, <u>Surf. Sci.</u> 128:350 (1983); J. D. Gunton and K. Kaski, <u>Surf. Sci.</u> 144:290 (1984); K. Kaski, S. Kumar, J. D. Gunton, and P. A. Rikvold, <u>Phys. Rev.</u> B29 (1984); K. Kaski, S. Kumar, J. D. Gunton, and P. A. Rikvold, <u>Surf. Sci.</u> 152/153:859 (1985); S. J. Knak Jensen, <u>J. Phys.</u> C17:4055 (1984); H. C. Fogedby and O. G. Mouritsen, <u>Phys. Rev.</u> B37:5962 (1988).

70. See e.g., I. Sega, W. Selke, and K. Binder, <u>Surf. Sci.</u> 154:331 (1985); W. Selke <u>in</u>: "Multicritical Phenomena," ed. R. Pynn and A. Skjeltorp, Plenum (1984); W. Selke, <u>Ber. Bunsenges. Phys. Chem.</u> 90:232 (1986).

71. See e.g., F. F. Abraham and J. Q. Broughton, <u>Phys. Rev. Lett.</u> 59:64 (1987); J. Q. Broughton and F. F. Abraham, <u>J. Phys. Chem.</u> 92:3274 (1988).

MODULATED STRUCTURES OF ADSORBED RARE GAS MONOLAYERS

Klaus Kern and George Comsa

Institut für Grenzflächenforschung und Vakuumphysik
Forschungszentrum Jülich, Postfach 1913, D-5170 Jülich, FRG

ABSTRACT

The structure of rare gas adlayers is determined by the interplay between the mutual interaction of the rare gas atoms and the interaction of
the rare gas atom with the substrate. Depending on the relative magnitude
of these interactions different types of structures of the first monolayer
appear: if the mutual interaction is dominant the rare gas adlayer is
incommensurate with the substrate. When the interactions are of the same
order, a large variety of phases (incommensurate -I-, commensurate -C-,
higher order commensurate -HOC) appear depending on temperature, coverage
and the relative structure of the substrate and the of the bulk rare gas
crystal. By varying the temperature and/or the coverage a whole series of
phases are visited from commensurate, over striped and hexagonal
incommensurate, to hexagonal incommensurate rotated, which ends in a high
order commensurate phase. In between there are both first and second order
phase transitions. On the other hand, when the lock-in forces of the
substrate dominate, all or a fraction of the adatoms will always lock into
preferential adsorption sites and commensurate or high order commensurate
phases, repsectively, are energetically favored with respect to true
floating incommensurate structures.

1. INTRODUCTION

The investigation of physisorbed rare gas adlayers has proven to be a
powerful tool in the understanding of elementary surface processes, such as
adsorption and desorption, surface melting or wetting [1]. In addition,

physisorption systems have acquired model character in the study of structural and dynamical properties of adsorbed layers and thin films [2], providing model systems of 2D phases and their mutual transitions.

In a simplified picture the structure of an adsorbed layer is governed
by the competition between the lateral adatom-adatom interaction and the
surface corrugation potential of the underlying substrate. On the one hand,
the lateral interaction between the adatoms will tend to establish an
adlayer structure determined by the natural adlayer interatomic distance,
i.e. incommensurate with the substrate. On the other hand, the lateral
variation of the substrate-adatom potential (corrugation) will try to force
the adatoms to occupy energetically favoured adsorption sites, hence
leading to a commensurate structure. In the case where the lateral adatom
interaction and the corrugation of the adatom-substrate potential have
about the same magnitude, the kind of adlayer structure (commensurate or
incommensurate) will largely depend on the structures symmetries, and the
ratio of the lattice constants as well as on the actual conditions, such as
coverage, spreading pressure, and surface temperature. Varying these
conditions, structural phase transitions between different commensurate and
incommensurate phases may occur and complex phase diagrams are obtained as
for instance in the case of physisorbed rare gases on graphite [3] and metal
surfaces [2].

Moreover, enhanced fluctuation effects due to the reduced
dimensionality add to the fascination of these delicate physical systems.
While phase fluctuations dominate low-temperature behavior of 2D-phases,
amplitude fluctuations dominate at higher temperatures when approaching the
critical temperature. Indeed, phase fluctuations, in form of long-
wavelength phonons, are responsible for the supression of a genuine long-
range order in two-dimensional solids, at all temperatures $T > 0K$.
Amplitude fluctuations, which are always present and dominate at high
enough temperatures, appear in 2D systems in the form of defects, in
particular as dislocations in 2D solids. Such dislocations (often termed
domain walls or solitons) result by adding or removing a half-infinite row
of atoms from an otherwise perfect lattice. They play a central role in
phase transitions of quasi two-dimensional systems, in the melting
transition as well as in the registry-disregistry transition.

The dominance of fluctuations in lower-dimensional systems can also be
understood by simple arguments. The order of a phase is thermodynamically
determined by the free energy, i.e. by the competition between energy and
entropy. In three-dimensional systems each atom has a large number of

nearest neighbors (12 in a fcc crystal), thus the energy term stabilizes an ordered state, local fluctuations being of minor importance. In one dimension, however, each atom has only two nearest neighbors. Here the entropy term dominates the energy term, and even very small local fluctuations destroy the order. In a close packed two-dimensional system each atom has six nearest neighbors and, depending on temperature, energy and entropy may be in balance. As amplitude fluctuations, i.e. topological defects, can be excited thermally, the two-dimensional systems seem to be ideally suited for studying defect-mediated phase transitions.

High resolution scattering of thermal He-atoms is a particularly appropriate tool for the study of rare gas layers on metal surfaces. This is not only because structural and dynamical information are both accessible in great detail, but in addition, due to the extreme sensitivity of the He scattering with respect to defects and impurities, it allows for a very accurate characterization of the substrate and of the adlayer morphology during the layer growth; last but not least thermal energy He has no influence whatsoever even on very unstable adlayer phases.

We will review first briefly the main experimental features of the application of He-scattering for the investigation of physisorbed rare-gas layers. Then we will illustrate with a few examples the structural richness of the rare-gas monolayers adsorbed on Pt(111). Finally, we will present the picture of the lattice dynamics of these systems.

2. THERMAL He-SCATTERING AS A PROBE OF ADSORBED LAYERS

The basic capabilities of He-scattering for the investigation of surface structure were already apparent in the early pioneering work in Otto Stern's laboratory in Hamburg in the thirties (ref. 4). However, it was not until the advent of the nozzle-beams that He-beams have become a highly efficient surface investigation tool. The nozzle-beam sources lead simultaneously to a dramatic increase of both the intensity and the monochromaticity of the beams, an achievement comparable only to that of the lasers for light beams. The three main approaches in He-surface scattering allow for an almost exhaustive characterization of rare gas layers: 1) diffraction – for structure (ref. 5), 2) inelastic scattering – for dynamics (ref. 6,7), 3) and diffuse elastic scattering in combination with interference – for thermodynamics, island formation, defect site occupation, degree of adlayer and substrate perfection (ref. 8-11).

A modern He surface-scattering spectrometer (see e.g. ref. 12) provides a highly monochromatic ($\Delta\lambda/\lambda \approx 0.7\%$), intense ($> 10^{19}$ He/sterrad. sec) and collimated ($\Delta\Omega \approx 10^{-6}$ sterrad) He-beam, with energies in the range 10-100 meV. The corresponding wavelength range (1.5-0.4 Å) is well suited for structure determinations which – due to the extreme sensitivity for the outermost layer – can be extended even to shallow, large period adlayer bucklings and matter waves. The resolution attained so far is of the order of 0.01 Å^{-1}. This is certainly inferior to X-ray performances, but – due again to the surface sensitivity – the signal/background ratio for the adlayer diffraction peaks is incomparably larger.

The narrow energy spread of the beam in conjunction with a high quality time-of-flight (TOF) system leads to an overall energy resolution for inelastic measurements of less than 0.4 meV at 18 meV beam energy. This and the relatively high cross-section for inelastic events in the range below 15-20 meV makes He-scattering an ideal instrument for the investigation of even detailed features of rare-gas layer phonons (at least of those perpendicularly polarized).

3. MONOLAYER STRUCTURES ON CORRUGATED SUBSTRATES

Atoms adsorbed on a periodic substrate can form ordered structures. These structures may be either in or out of registry with the structure of the substrate. It is convenient to describe this ordering by relating the Bravais lattice of the adlayer to that of the substrate surface. Park and Madden [13] have proposed a simple vectorial criterion to classify the structures. Let $\vec{a}_1$ and $\vec{a}_2$ be the basis vectors of the adsorbates and $\vec{b}_1$ and $\vec{b}_2$ those of the substrate surface; these can be related by

$$\begin{bmatrix} \vec{a}_1 \\ \vec{a}_2 \end{bmatrix} = G \begin{bmatrix} \vec{b}_1 \\ \vec{b}_2 \end{bmatrix} \tag{1}$$

with the matrix

$$G = \begin{bmatrix} G_{11} & G_{12} \\ G_{21} & G_{22} \end{bmatrix} \tag{2}$$

$\vec{a}_1 \times \vec{a}_2$ and $\vec{b}_1 \times \vec{b}_2$ are the unit cell areas of the adlayer and substrate surface, respectively; det G is the ratio of the two areas. The relation between the two ordered structures is classified by means of this quantity as follows:

i) det G = integer

the structure of the adlayer has the same symmetry class as that of the substrate and is in registry with the latter; the adlayer is termed commensurate.

ii) det G = irrational number

the adlayer is out of registry with the substrate; the adlayer is termed incommensurate.

iii) det G = rational number

the adlayer is again in registry with the substrate. However, whereas in i) all adlayer atoms are located in equivalent high symmetry adsorption sites, here only a fraction of adatoms is located in equivalent sites; the adlayer is termed high-order commensurate.

In Fig. 1, we show a simple one-dimensional model illustrating this classification. The periodicity of the substrate surface is represented by a sinusoidal potential of period b and the adlayer by a chain of atoms with nearest neighbor distance a.

Assuming that the structural mismatch between adlayer and substrate is not too large (< 10-15%), the nature of the adlayer ordering on the substrate is largely determined by the relative interaction strength e_1/V_c, where e_1 is the lateral adatom interaction in the layer and V_c, the modulation of the adsorbate-substrate potential parallel to the surface. When the diffusional barrier V_c is large compared to the lateral interaction e_1 commensurate structures will be formed. On the other hand, when the lateral adatom interactions dominate, incommensurate structures will be favored. Only when the competing interactions are of comparable magnitude, may both registry and out of registry structures be stabilized by the complex interplay of these interactions and other parameters.

We have recently measured the energetics of the adsorption of the rare gases Ar, Kr and Xe on the Pt(111) surface by means of thermal He-scattering. In table 1, we summarize the pertinent values. From inspection of the relevant quantities in the table we can deduce that rare gas-monolayers on Pt(111) appear to be well suited to study structural 2D solid-solid transitions.

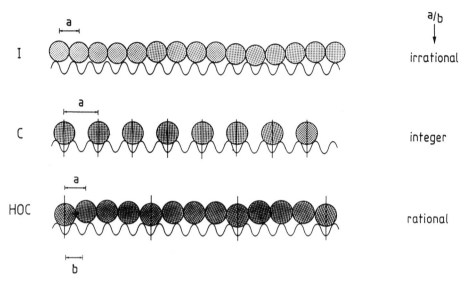

Fig. 1. One dimensional model of a physisorbed monolayer. The substrate is represented by a sinusoidal potential of period _b_ and the adlayer by a chain of atoms with lattice constant _a_.

Table 1. Characteristic energies (meV) of rare gas-adsorption on Pt(111) in the monolayer range.

	Xe	Kr	Ar
isosteric heat q_{st} at $\Theta \to 0$	277	128	78
lateral attraction e_ℓ	43	26	17
diffusional barrier V_c	~ 30	~ 10–20	~ 10–20

4. THE COMMENSURATE-INCOMMENSURATE (CI) TRANSITION IN 2D

Only when the lateral adatom interaction and the substrate corrugation are comparable, an ordered commensurate (C) adlayer can undergo a transition into an incommensurate (I) phase as a function of coverage or temperature. The CI-transition is driven by the formation of line defects, so called misfit dislocations as was demonstrated first by Frank and van der Merwe [14]. These authors studied a linear chain of atoms with a lattice constant a placed in a sinusoidal potential of amplitude V and periodicity b. The mutual interactions of atoms in the chain a represented by springs with a spring constant K. The calculations reveal that for slightly different lattice parameters of chain and substrate, i.e. for a weakly incommensurate adlayer, the lowest energy state is obtained for a system which consists of large commensurate domains separated by regions of bad fit. The regions of poor lattice fit are dislocations with Burgers vectors parallel to the length of the chain.

In two-dimensional systems domain walls are lines. In a triangular lattice there are three equivalent directions and therefore, domain walls

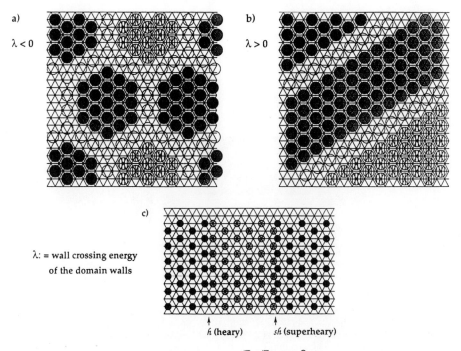

a) $\lambda < 0$

b) $\lambda > 0$

c)

λ: = wall crossing energy of the domain walls

h (heavy) sh (superheavy)

Fig. 2. Domain wall systems of an ($\sqrt{3} \times \sqrt{3}$) R30° phase on a triangular substrate lattice; the walls are of the "light" type.

can cross. Using Landau theory, Bak, Mukamel, Villain and Wentowska (BMVW) [15] have shown that it is the wall crossing energy, Λ, which determines the symmetry of the weakly incommensurate phase and the nature of the phase transition. For attractive walls, $\Lambda < 0$, a hexagonal network of domain walls (HI) will be formed at the CI transition because the number of wall crossings has to be as large as possible. This C-HI transition is predicted to be first order. For repulsive walls, $\Lambda > 0$, the number of wall crossing has to be as small as possible, i.e. a striped network of parallel walls (SI) will be formed in the incommensurate region. The C-SI transition should be continuous. The striped phase is expected to be stable only close to the CI phase boundary. At large incommensurabilities the hexagonal symmetry should be recovered in a first order SI-HI transition. In fig. 2 we summarize the possible domain wall structures. Superheavy and heavy walls are characteristic for those systems in which the incommensurate phase is more dense than the commensurate phase while for light and superlight walls the opposite holds.

The most completely studied examples of the CI transition in 2D adlayer systems occur in the Kr monolayer on the basal (0001) plane of graphite [16] and in the Xe monolayer on the Pt(111) surface [17]. Here we will discuss briefly the physics of the Xe/Pt system. Below coverages of $\Theta_{Xe} \simeq 0.33$ (Xe adatoms per one Pt-substrate atom) and in the temperature range 60-99 K the xenon condenses in a $(\sqrt{3}x\sqrt{3})R30^\circ$ commensurate solid phase. This phase has very

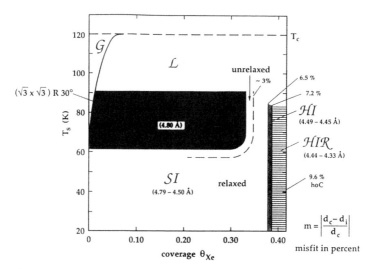

Fig. 3. Schematic phase diagram of monolayer Xe on Pt(111). C, SI, HI, HIR denote the commensurate $(\sqrt{3}x\sqrt{3})$ R30°, the striped incommensurate, the hexagonal incommensurate and the hexagonal incommensurate rotated 2D solid phases. G and L denote the 2D gas and liquid, respectively.

sharp diffraction peaks, characteristic for coherent Xe-domains which are about 800 Å in size. As the Xe-coverage is increased above 0.33 the relatively loosely packed Xe-structure undergoes a transition from the commensurate $\sqrt{3}$ structure to an incommensurate striped solid phase with superheavy walls (see fig. 2). This weakly incommensurate solid is able to accomodate more Xe atoms than the commensurate phase by dividing into regions of commensurate domains separated by a regularly spaced array of striped denser domain walls. Increasing coverage causes the commensurate domains to shrink and brings the walls closer together. The domain walls are thus a direct consequence of the system's efforts to balance the competition between the lateral Xe-Xe and the Xe-Pt interactions. The C-SI transition can also be induced by decreasing the temperature below ~ 60 K at constant coverage (Θ_{Xe} < 0.33); the driving force for this temperature induced CI-transition being anharmonic effects [18].

The usual measure for the incommensurability of an I-phase is the misfit m = $(a_c-a_I)/a_c$, where a_c is the lattice parameter of the commensurate phase and a_I that of the incommensurate structure. For striped I-phases, the misfit has of course uniaxial character, being defined only along the direction perpendicular to the domain walls. Qantitative measurements of the misfit during the C-SI transition of Xe on Pt(111) [17] have revealed a power law of the form:

$$m = \frac{1}{\ell} \propto (1-T/T_c)^{0.51\pm0.04} \qquad (3)$$

i.e., the distance between nearest neighbor walls ℓ scales with the inverse square root of the reduced temperature. This square root dependence is based on entropy mediated repulsing meandering walls and is in accord with theoretical predictions [19].

With increasing incommensurability the domain wall separation becomes progressively smaller until at a critical misfit of ~ 6.5% the Xe domain wall lattice rearranges from the striped to the hexagonal symmetry (fig. 2) in a first order transition [20]. A further increase of the incommensurability by adding more and more Xe eventually results in an adlayer rotation to misalign itself with the substrate in order to minimize the increasing strain energy due to the defect concentration. This continuous transition to a rotated phase (HIR) follows a power law $\varphi \propto (m-0.072)^{1/2}$ starting at a critical separation between nearest neighbor walls $\ell_c \simeq$ 10 Xe-row distances.

Novaco and McTague [21] have shown that these adlayer rotations for monolayers far from commensurability are driven by the interconversion of longitudinal stress into transverse stress. These authors also showed that the rotational epitaxy involves mass density waves (MDW) [also known as static distortion waves (SDW)], i.e. a periodic deviation of the position of monolayer atoms from their regular lattice sites. Indeed, it is the combination of rotation and small displacive distortions of the adatom net which allows the adlayer to minimize its total energy in the potential relief of the substrate. In a diffraction experiment, these mass density waves should give rise to satellite peaks.

Fuselier et al. [22] have introduced an alternative concept to explain the adlayer rotation: the "coincident site lattice." They pointed out that energetically more favorable orientations are obtained for rotated high-order commensurate structures. The larger the fraction of adatoms located in high-symmetry, energetically favorable sites, the larger the energy gain and the more effective the rotated layer is locked. It turns out that the predictions of the coincident site lattice concept for the rotation angle versus misfit agrees well with the Novaco-McTague predictions.

The experimental results do not allow so far to decide whether the Novaco-McTague mechanism involving MDW or the "coincident in lattice" concept involving HOC structures, or even both have the determining role in driving the adlayer rotation. In particular, no mass density wave satellites have been observed in electron and x-ray diffraction experiments from rotated monolayers [23] so far. In He diffraction scans of rotated Xe monolayers on Pt(111) we have, however, observed satellite peaks at small Q-vectors. Originally we assigned these peaks to a higher-order commensurate superstructure [8]. However, Gordon [24] pointed out that these satellites could be due to the MDW. Recently we have shown that both MDW as well as high-order commensurate buckling satellites are present in the rotated Xe monolayers on Pt(111)[2]; the arguments are recalled in the following three paragraphs. The distinction between the two types of satellites is straightforward. As pointed out by Gordon, the wave vector, Q, of the MDW satellites should be subject to the following relation:

$$Q \approx (8\pi/a_{Xe}^R) \ (m/\sqrt{3}) \ (1+m/8), \qquad (4)$$

with m the misfit, and a_{Xe}^R the lattice of the rotated Xe layer. For not too large misfits, this MDW satellite should appear in the same direction as the principal reciprocal lattice vector of the Xe layer, i.e., in the $\bar{\Gamma} \bar{M}_{Xe}$ direction. On the other hand, according to its particular structure (Fig. 3, Ref. 8), the commensurate buckling should have its maximum amplitude in the $\bar{\Gamma} \bar{K}_{Xe}$ direction. Moreover, these commensurate buckling satellites should only be present at the particular coverages where a certain high-order commensurability becomes favorable, in the present case at monolayer completion (m=9.6%) (see also ref. 25), whereas the MDW satellites should be present in the entire misfit range where the Xe layer is rotated (7.2%-9.6%).

In fig. 4 we show the dispersion of the MDW-satellites deduced from a series of diffraction scans, taken in the $\bar{\Gamma} \bar{M}_{Xe}$-direction, and compare them with Gordon's prediction for the MDW given above. The data follow qualitatively the predicted dependency; the agreement becomes quanti- tatively at misfits > 8%. The reason for the better agreement at large mis- fits is due to the fact that Gordons analysis of the MDW (similar to Novaco-MacTague's model calculations) have been performed in the linear response approximation of the adsorbate-substrate interaction; this approximation is only justified at large misfits, where the adlayer topography corresponds rather to a weakly modulated uniform layer than to a domain wall lattice [26].

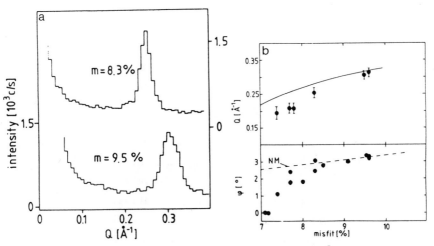

Fig. 4. a) Polar He diffraction scans of rotated Xe monolayers on Pt(111) taken along the $\bar{\Gamma}\bar{M}_{Xe}$ azimuth at misfits of 8.3% and 9.5%.
b) Dispersion of the mass density wave sattelites with misfit m. The solid line is Gordon's relation.

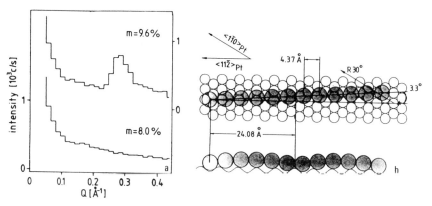

Fig. 5. a) Polar He-diffraction scans of rotated Xe monolayers on Pt(111) taken
along the $\bar{\Gamma}\bar{K}_{Xe}$ azimuth at misfits 8% and 9.6%.
b) Upper and side view of a 3.3° rotated domain at m = 9.6%.

In Fig. 5 we show scans like in Fig. 4 but now measured in the $\bar{\Gamma}\,\bar{K}_{Xe}$
direction at small Q for rotated Xe layers of misfit 8% and 9.6%. At variance
with the scans in the $\bar{\Gamma}\,\bar{M}_{Xe}$ direction (fig. 4), a satellite peak is observed
only for the complete Xe monolayer (m=9.6%). Being present only at a particular
misfit this peak does not originate from a MDW but from the buckling of a HOC-
structure. The location of this satellite peak at $Q = 0.28$ Å^{-1} corresponds to a
buckling period of 23 Å and can be ascribed to a high-order commensurate
structure shown in fig. 5b and described in detail in ref. 8.

5. THE SUBSTRATE CORRUGATION OF CLOSE PACKED METAL SURFACES

The discovery of the existence of various solid Xe phases on Pt(111) was
actually a surprise because the hexagonal close packed Pt(111) surface was
considered essentially non corrugated. As already stated in the introduction,
the various solid phases of adsorbed monolayers arise as a result of competing
interactions; lateral adatom interaction versus the corrugation of the holding
potential. In particular the commensurate $\sqrt{3}$ phase was unexpected; at 60 K the
"natural" two-dimensional Xe-lattice parameter in the absence of any substrate
would be about 8% smaller than the $\sqrt{3}$ distance of the Pt substrate (4.80 Å). In
order to stabilize the commensurate Xe-phase the corrugation of the substrate
has to compensate this substantial strain and to counterbalance the attractive
lateral Xe-interaction, which is $e_{\ell}^{Xe-Xe/Pt} = 43$ meV per atom.

It is indeed still a widespread belief that for rare gases adsorbed on
close packed metal surfaces the lateral corrugation is negligible. This belief

52

originates in the very low corrugation of the interaction potential as deduced from He-diffraction data. This argument, although found often in literature, is wrong and thus misleading. From small corrugations deduced from He-diffraction experiments it can not be inferred, that the lateral corrugation felt by **all** rare gases adsorbed on close packed metal surfaces is likewise small.

First, in diffraction experiments the He-atoms probe the corrugation of the repulsive interaction potential in the region of positive energies, while adsorbed atoms feel the corrugation of the bottom of the attractive potential well, i.e. corrugations at two different locations on the potential curve. Second, and this is even more important, the interaction of the heavier rare gases Xe, Kr, Ar or even Ne differs appreciably from the interaction of He with the same substrate (e.g. the binding energy for He and Xe on Pt(111) are about 5 meV and 311 meV respectively!).

It has been shown by Steele (27), that the physisorption potential of rare gas atoms on a crystal surface, is represented, in a good approximation, by a Fourier expansion in the reciprocal lattice vectors G of the substrate surface

$$V(r,z) = V_0(z) + V_{mod}(r) = V_0(z) + \sum_G V_G(z) \exp{(iGr)} \quad (5)$$

evidencing nicely the lateral modulation of the adsorption energy. Here z (> 0) is the distance of the adatom perpendicular to the surface and r the coordinate parallel to the surface. $V_0(z)$ is the mean potential energy of an adatom at a distance z, V_G the principal Fourier amplitude. Owing to the rapid convergence of the Fourier series, the second term is usually already one order of magnitude smaller than the first order term, and for the basal plane of graphite Gr(0001) and the fcc(111) surface we obtain:

$$V(\vec{r}) = V_0(z) \mp V_G \{\cos{(2\pi s_1)} + \cos(2\pi s_2) + \cos 2\pi(s_1 + s_2)\} \quad (6)$$

respectively. Here s_1 and s_2 are the dimensionless coordinates of the atoms in the substrate surface unit cell.

If we assume a 12-6 Lennard-Jones pair potential to represent the interaction between various atoms of the adsorbate/substrate system, the corrugation in the unit cell of a particular surface is entirely determined by the ratio σ/a, by the magnitude of the binding energy V_0 and by the sign of V_G, with σ being the Lennard-Jones diameter of the adatom and a being the nearest neighbor distance in the substrate surface.

The location of the energetically most favorable adsorption site is determined by the sign of V_G. For $V_G > 0$, the energy minima on a fcc(111) surface are at three fold hollow sites (H) and the barrier of diffusion across the bridge sites (B) is given by $V_C = V_{H-B} = V_G$. For $V_G < 0$, on the other hand, the energy minima are at on top sites (T) and the barrier of diffusion $V_C = V_{T-B} = -8\ V_G$. From intuition it was always believed that the sign of the Fourier amplitude has to be positive, i.e., the preferred adsorption site of the rare gas atom would be the threefold hollow sites. In this picture the energy difference between adsorption of a Xe-atom in a hollow and in a bridge position, V_{H-B}, amounts to about ~ 0.01 V_o ($\approx$ 3 meV) for the (111) surface of Pt.

A detailed analysis of the diffraction patterns in the striped incommensurate phase of Xe/Pt(111) by Gottlieb [28] came to the remarkable conclusion that the preferred adsorption sites of the Xe atoms are on top of platinum atoms rather than in threefold hollow sites, i.e. the sign of V_G is negative. From his analysis he deduced the values $V_C = V_{T-B} = -V_G \approx$ 6-8 meV for the Xe/Pt(111) system.

The Lennard-Jones potential is only a crude approximation of the interaction between a rare gas atom and a metal surface. Drakova et al. [29] demonstrated recently in a self consistent Hartree-Fock calculation, that the corrugation of the rare gas – transition metal surface potential is substantially enhanced by the hybridization between occupied rare gas orbitals and empty metal d-orbitals which particularly in the case of transition metal elements like Pd or Pt gives a strong increase of the corrugation.

The corrugation enhancement due to orbital hybridization as well as the negative sign of the Fourier amplitude V_G has been demonstrated by Müller [30] in a very recent _ab initio_ cluster calculation. The Pt(111) surface was modelled by a Pt_{22} cluster. The total energy of the Xe-Pt_{22} system was calculated in the Kohn-Sham scheme with the local-density approximation for exchange and correlation, and a localized muffin-tin orbital basis including s,p and d orbitals in all metal sites. In his ab-intio calculations Müller finds the on top position of Xe as the energetically most favorable location with a binding energy of 307 meV. The corrugation is determined to be $V_C = V_{T-B} \approx$ 22 meV.

This large corrugation is consistent with experimental data of Kern et al. [31] estimated the corrugation of the Xe/Pt(111) system to about 10% of the binding energy which is ~ 30 meV. A value which is compatible with the value of the diffusion barrier of Xe on W(110) (also a rather close packed surface) which has been measured to 47 meV [32]. It is thus no more a surprise but a logical consequence that Xe on Pt(111) can form a commensurate $\sqrt{3}$-phase.

6. HIGH ORDER COMMENSURATE PHASES

Another interesting concept describing structural phase transitions of adsorbed layers has been developed by Aubry [33]. The basic idea is that any ad-layer structure incommensurate with the substrate can be approximated within any given accuracy by a so called high-order commensurate (HOC) structure. These HOC structures are characterized by a (large) commensurate unit cell hosted by several adlayer atoms. Since a fraction of the adatoms will always lock into preferential adsorption sites these HOC (locked) phases should be energetically favoured with respect to a true incommensurate (floating) phase. If the corrugation potential is sufficiently large, it is expected that the phase diagram of the adlayer is composed of several HOC phases. The corresponding (first order) phase transitions will involve discrete jumps of the interatomic distance moving from one HOC phase to the other. Such a stepwise variation of the lattice parameter within a series of HOC phases is called a "devil's staircase". Of course, this simple picture is modified at elevated surface temperature when thermal fluctuations become important and can destabilize these HOC phases with large unit cells.

It is generally accepted that on the same substrate the absolute magnitude of the corrugation increases with the size of the rare gas adatom, while the corrugation energy decreases relative to the binding energy of the adatom as well as relative to the lateral adatom interaction [34]. Thus, within our simple model of competing interactions (corrugation versus lateral attraction) we expect on the Pt(111) surface a gradual transition from the floating Xe monolayer with its rich diversity of incommensurate domain wall phases to locked Kr or Ar layers which are dominated by the lock-in forces of the substrate, favoring HOC-phases instead of incommensurate domain wall phases.

Until very recently, there has been no convincing experimental evidence for the existence of high order commensurate physisorbed layers. This appeared to support the widespread belief that "experimentally it is impossible to distinguish between a high order-C structure and an incommensurate structure". This belief is certainly legitimate, if the only accessible experimental information is the ratio of the adlayer and substrate lattice basis vectors. Indeed, because one can find always one rational number within the confidence range of any experimental irational number, i.e. the basis vectors supplied by

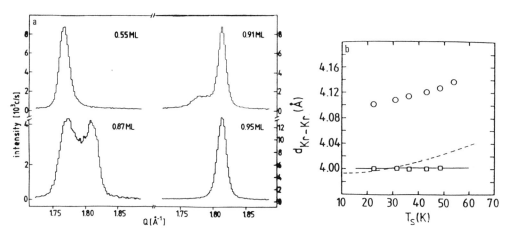

Fig. 6. a) Polar He-diffraction scans of the $(1,0)_{Kr}$-diffraction order from Kr adlayers on Pt(111) at various Kr submonolayer coverages at $T_s = 25$ K. b) Kr-layer lattice spacing vs T_s for the ( ) high (0.95 ML) and (o) low (0.5 ML) coverage phase; temperature dependence of the lattice spacing of (———) Pt-substrate, and of (– – –) bulk Kr.

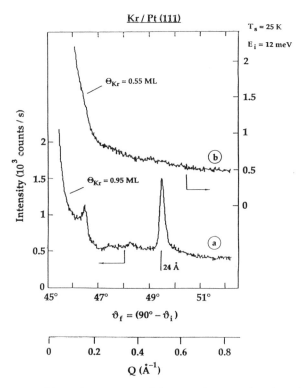

Fig. 7. Polar He diffraction scans of Kr monolayers in the vicinity of the specular peak ($Q = 0 Å^{-1}$); i) high (0.95 ML) and ii) low (0.5 ML) coverage phse, taken along the $\bar{\Gamma}\bar{K}_{Kr}$ -azimuth.

the most refined experiment are always compatible with a high (enough) order
commensurate phase. There are, however, two other experimentally accessible
parameters which allow a unequivocal distinction between a high order
commensurate "locked" and an incommensurate "floating" layer. First: the
superstructure formed by the atoms located in equivalent, high symmetry sites.
These stronger bound atoms being located "deeper" in the surface than the
others, the adlayer is periodically buckled. Because of the extreme sensitivity
of He-scattering to the surface topography, this superstructure, which
characterizes high order commensurate layers, is directly accessible to high
resolution He-diffraction experiments. Second: the thermal expansion. Indeed, a
"floating" layer is expected to thermally expand very much like the
corresponding rare gas bulk crystal, while a "locked" layer has to follow by
definition the substrate at which it is locked. The thermal expansion of rare
gas solids being at least ten times larger than that of substrates normally
employed, the destinction between high order commensurate "locked" and
incommensurate "floating" becomes straightforward. This very sharp criterion
requires that the "locking" is strong enough to withstand temperature
variations over a sufficiently large range ($\geq$ 10 K) to allow for reliable
thermal expansion measurements.

Figure 6a) shows a series of polar He-scans of the $(1,1)_{Kr}$ diffraction
peak taken at 25 K along the $\bar{\Gamma M}_{Kr}$ direction of the Kr monolayers adsorbed on a
Pt(111) surface at coverages between 0.5 and 0.95 ML [25]. The sequence is characteristic for a first order phase transition from a hexagonal solid phase —
with wavevector Q = 1.769 Å^{-1} (d_{Kr} = 4.10Å) to one with Q = 1.814 Å^{-1} (d_{Kr} =
4.00 Å), below and above 0.8 ML, respectively. During the phase transition the
intensity diffracted from one phase increases at the expense of the other.

The question concerning the incommensurate "floating" versus high order
commensurate "locked" nature of the two Kr-phases has been adressed by looking
at their thermal expansion behavior and by searching for superstructure
satellites. In fig. 6b) the measured Kr-Kr interatomic spacing versus temperature is shown for submonolayer films of coverage 0.5 ML and 0.95 ML. The
difference is striking. The low coverage phase shows a variation with temperature very much like bulk Kr (dashed) and is thus an incommensurate
"floating" phase. On the contrary, the lattice parameter of the high coverage
phase is – that of the Pt substrate (solid) – constant within experimental
error in the same temperature interval; accordingly, this Kr-phase is high
order commensurate "locked".

This assignment is supported by inspection of fig. 7, where polar scans
(He-beam energy 12 meV) in the $\bar{\Gamma K}_{Kr}$ -direction of the "floating" and of the

"locked" Kr-layer are shown. The scans differ substantially: the locked scan clearly evidences the presence of a superstructure, while the floating one does not. The superstructure peak at $Q = 0.532 \pm 0.022$ Å^{-1} corresponds to 1/5 of the Pt-substrate principal lattice vector, i.e. the Kr-layer forms a $(5 \times 5)R0^\circ$ HOC phase.

In contrast to Kr, Ar adsorbs on the clean Pt(111) surface in a hexagonal solid phase aligned with respect to the substrate[42]. In the submonolayer regime

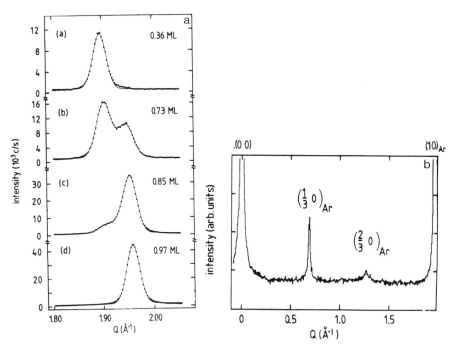

Fig. 8. a) Polar He diffraction scans of the $(1,0)_{Ar}$ -diffraction order for different Ar coverages b) Polar He diffraction scan of the high coverage ($\Theta = 0.91$ ML) Ar monolayer on Pt(111) in an extended Q-range.

a structural phase transition of the physisorbed Ar layer occurs at a coverage $\Theta \approx 0.75$ ($\Theta = 1$ refering here to full monolayer coverage): At lower coverage ($\Theta < 0.6$) the first order diffraction peak of the Ar-phase is centered at $Q = 1.90$ Å^{-1}, corresponding to a lattice parameter $a = 3.81$ Å. Upon increase of the Ar coverage a second first order diffraction peak at $Q = 1.96$ Å^{-1} emerges,

corresponding to an Ar-phase with lattice parameter a = 3.70 Å, i.e. slightly
compressed with respect to the first. While the intensity of the first peak
continuously decrease upon further Ar-adsorption, the second peak at Q = 1.96
Å^{-1} becomes more intense. At $\Theta \simeq 0.75$ the two diffraction peaks have about
equal intensities while above $\Theta \geq 0.9$ the first peak has vanished and only the
second one is observed. The discontinuous change of the diffraction peak
position and the observed phase coexistence indicate that this phase transition
is of first order.

The Ar phase at higher coverage shows no significant thermal expansion in
contrast to the large expansion of bulk Ar, indicating that the adlayer is a
locked commensurate phase. In addition, superstructure peaks are observed in
this high coverage phase at 1/3 and 2/3 of the wave-vector position of the
first order Ar diffraction peak (fig. 8b). Thus the period of the commensurate
unit cell equals three interatomic Ar distances (~ 11 Å). From the known
lattice parameter of the Pt(111) substrate a_{Pt} = 2.77 Å it follows that the Ar
phase at higher coverage ($\Theta \geq 0.75$) is a (4x4)R0° HOC-phase with a commensurate
unit cell of length $3xa_{Ar}$ = 4 x a_{Pt} = 11.08 Å. The analysis of the low coverage
phase is more difficult. Although further investigation is needed for a final

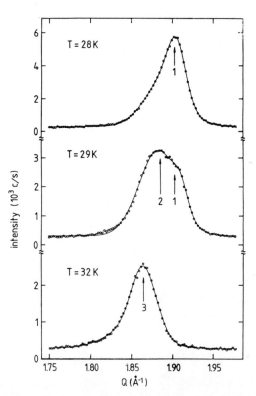

Fig. 9. Polar He-diffraction profiles of the $(1,0)_{Ar}$ peak in the low Ar
coverage regime (Θ = 0.36 ML) at different surface temperatures.

conclusion, our experimental data indicate that also the low coverage Ar phase ($\Theta \leq 0.75$) is partially locked. The low coverage Ar phase ($\Theta_{Ar} \approx 0.36$ ML) was prepared by dosing Ar at low surface temperature (~ 20 K) and successively briefly annealing the adlayer. Depending on the annealing temperature and the annealing time either an Ar lattice parameter of $a_{Ar} \simeq 3.81$ Å (1) or a slightly larger value of $a_{Ar} \simeq 3.83$ Å (2) was obtained.

The fact that there are indeed discrete jumps in the Ar lattice parameter as expected for a transition between phases in registry with the substrate is demonstrated in fig. 9. Here the polar diffraction profiles recorded from an Ar adlayer in the low coverage phase ($\Theta = 0.36$ ML) for different surface temperature are shown. The three spectra (a)-(c) indicate the coexistence and the discontinuous transition between (partially) locked structures; they are characterized by the lattice parameters $a_{Ar} \simeq 3.81$ Å ($a_{Ar}/a_{Pt} = 11/8$), $a_{Ar} \simeq 3.84$ Å ($a_{Ar}/a_{Pt} = 18/13$) and $a_{Ar} \simeq 3.88$ Å ($a_{Ar}/a_{Pt} = 7/5$).

6. MONOLAYER FILMS AND DYNAMICS OF COMPETING INTERACTIONS

When adsorbing a layer of atoms (for example rare gas atoms) on a substrate surface we are not dealing only with static interaction effects which determine the structure of the adlayer but also with dynamical interactions between collective excitations (for example phonons) of adlayer and substrate.

The phonon spectrum of a crystal surface consists of two parts (fig. 10): The bulk bands, which are due to the projection of bulk phonons onto the two-dimensional Brillouine zone of the particular surface, and the specific surface phonon branches [7]. A surface phonon is defined as a localized vibrational excitation with an amplitude which has wavelike characteristics parallel to the surface and decays exponentially into the bulk, perpendicular to the surface. Of particular interest is the lowest frequency mode below the transverse bulk band edge. In this mode, the atoms are preferentially vibrating in the plane defined by the surface normal and the propagation direction, i.e. in the sagittal plane. This wave is the famous Rayleigh wave.

Adsorbing a layer of densely packed atoms on the substrate surface adds three additional phonon modes to the system; two in-plane modes and one mode with polarization perpendicular to the surface. The frequency and dispersion of these modes is governed by the "spring constants" which couple the adatoms laterally and to the substrate. In the following we concentrate on the perpendicular mode. In physisorbed systems, the electronic ground state of the adsorbate is only weakly perturbed upon adsorption. The physisorption potential

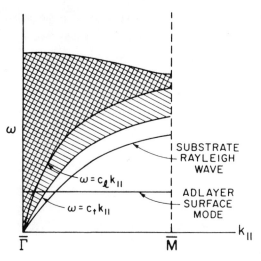

Fig. 10. Schematic normal mode spectrum of a rare gas monolayer physisorbed on
a single crystal surface [36].

is rather flat and shallow, i.e. the spring constant of the vertical adatom-
motion is weak and the corresponding phonon frequencies are low compared to the
substrate Rayleigh wave at the zone boundary.

The first systematic theoretical and experimental exploration of the
dynamics of rare gas monolayers on metal surfaces has been performed on Ag(111)
[35,36]. The lattice dynamical calculations were based on a simple model – Barker
pair potentials – to account for the lateral adatom interactions and a rigid
holding substrate. The calculations have supplied dispersion curves fully
adequate to account for the available experimental data.

As expected from any model involving only central forces between adatoms
and a rigid substrate, the three modes of the monolayer decouple, and the
perpendicular mode is dispersionless, i.e. the motion of the adatoms
perpendicular to the surface acts like an Einstein oscillator. Because the
perpendicular surface atom motions dominate the inelastic He cross sections,
this dispersionless mode has been, indeed, observed in the experiment [8,35].

It is noteworthy that the most general conclusion, emerging from this
first systematic exploration, has been that "coupling between adatom and
substrate atom motions is potentially more important than modest variations in
the nature of the adatom – adatom potential" [35]. Consistently, Hall, Mills and

Black continued their exploration by allowing now the substrate atoms to move and by focussing on the coupling between the substrate and adlayer modes [36]. The results of the calculations show that while near the zone boundary, where the substrate phonon frequencies are well above those of the adlayer, the influence of the substrate adlayer coupling is small. Near the zone center $\bar{\Gamma}$, significant anomalies introduced by the coupling are expected. These are twofold:

1) A dramatic hybridization splitting around the crossing between the dispersionless adlayer mode and the substrate Rayleigh wave (and a less dramatic one around the crossing with the $\omega = c_\ell Q_\parallel$ line – due to the Van Hove singularity in the projected bulk phonon density of states); 2) A substantial line-width broadening of the adlayer modes in the whole region near $\bar{\Gamma}$ where they overlap the bulk phonon bands of the substrate: the excited adlayer modes may decay by emitting phonons into the substrate, they become leaky modes. These anomalies were expected to extend up to trilayers even if more pronounced for bi- and in particular for monolayers. More recent experimental data of Gibson and Sibener [37] confirm qualitatively these predictions at least for monolayers. The phonon line-widths appear to be broadened around $\bar{\Gamma}$ up to half of the Brillouin zone. The hybridization splitting could not be resolved, but an increase of the inelastic transition probability centered around the crossing with the Rayleigh wave and extending up to 3/4 of the zone has been observed and attributed to a resonance between the adatom and substrate modes.

Recent measurements performed on Ar, Kr and Xe-layer on Pt(111) [38-40] with a substantially higher energy resolution ($\Delta E \leq 0.4$ meV) have now confirmed in every detail the theoretical predictions on the coupling effects.

A series of He energy loss spectra taken from a full Ar-monolayer on Pt(111) at different scattering angles are plotted over an energy-loss range from -2 to -6 meV in fig. 11. In these spectra the various features of the dynamical coupling between the adlayer and Pt substrate can be distinguished. As a reference, spectrum (d) taken at $\vartheta_i = 35^\circ$ corresponds to the creation of an Ar monolayer phonon with wave vector $Q=0.78$ Å^{-1}, i.e., close to the edge of the 2D Ar Brillouin zone [$Q(\bar{M}) = 0.98$ Å^{-1}]. Since at these large Q values the Pt Rayleigh wave and the Pt bulk phonons have much higher energies than the Ar adlayer mode, no coupling of the Ar phonons to the Pt substrate is expected. Indeed, the small linewidth $\Delta E=0.32$ meV of the Ar loss peak at -4.8 meV is determined only by the instrumental resolution ΔE_{instr} of the spectrometer without any measurable additional broadening. Spectrum (a) exhibits two-phonon loss peaks. The feature at -2.9 meV is a signature of the Pt Rayleigh wave

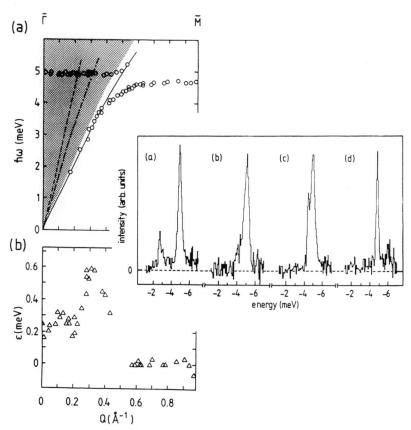

Fig. 11. Experimental dispersion curves from a monolayer Ar on Pt(111) (perpendicular polarized mode) and He TOF spectra (inset) taken at different incident angles ϑ; along the $\bar{\Gamma}\bar{M}$ direction of the Ar unit cell: (a) 40.4°, (b) 38.5°, (c) 37.7°, (d) 35°. The energy of the primary He-beam was 18.3 meV.

dynamically coupled to the Ar adlayer mode. Furthermore, the Ar loss at −5.0 meV is much broader ($\Delta E \approx$ 0.41 meV) than in spectrum (d). This difference is due to the fact that the Ar phonon in (a) is created at the center of the Brillouin zone $\bar{\Gamma}$(Q=0.0 $\overset{o}{A}{}^{-1}$) where an effective damping of the Ar phonons due to the coupling to the projected bulk phonon bands of the Pt substrate occurs. This so called "radiative damping" of the adlayer phonon results in a linewidth broadening $\epsilon \equiv [(\Delta E)^2 - (\Delta E_{instr})^2]^{1/2}$ of 0.2-0.3 meV. It is instructive to estimate the corresponding lifetime shortening due to this radiative damping. Using the Heisenberg uncertainty principle $\epsilon \Delta t = h$ the lifetime becomes $\Delta t \sim 3 \times 10^{-12}$ s. Relating this value to the time scale of the Ar-Pt vibration ($\hbar\omega$ = 5.0 meV, i.e., ν = 1.2 THz), a mean phonon lifetime of about three vibrational periods is obtained. Besides the radiative damping also the hybridization of the Ar adlayer mode and the Pt Rayleigh wave is observed experimentally. This is shown in spectrum (c) of Fig. 11. At these scattering conditions (ϑ = 37.7^{o}) corresponding to a wave vector Q=0.45 $\overset{o}{A}{}^{-1}$ of the Ar phonon the crossing of the adlayer mode and the Pt Rayleigh wave should occur. Instead, due to the hybridization of the two modes this crossing is avoided and a phonon doublet with an energy splitting of $\sim$ 0.8 meV is observed (note also the similar heights of the two peaks). The additional line width-broadening in the wavevector range Q=0.25 and 0.35 $\overset{o}{A}$ which is apparent in spectrum (b) of fig. 11 (compare with spectrum (a) and (d)) is not an artifact but can be attributed to the coupling with the projected bulk phonon band edge; the "van Hove anomaly".

More details of the dynamical coupling between an adsorbed layer and the substrate and of the phonon spectrum of thin physisorbed films (1-25 mono-layers) can be found in Ref. 38 41.

REFERENCES

1. Sinha, S.K. (Ed.), Ordering in Two Dimensions (North Holland, Amsterdam, 1980)
2. K. Kern and G. Comsa, in "Chemistry and Physics of Solid Surfaces VII", ed. by R. Vanselow and R.F. Howe, (Springer, Heidelberg, 1988), p. 65
3. R.J. Birgenau and P.M. Horn; Science 232, 329 (1986)
4. I. Estermann and O. Stern; Z. Phys. 61, 95 (1930)
5. T. Engel, K.H. Rieder: Structural Studies of Surfaces with Atomic and Molecular Beam Diffraction, Springer Tracts Mod. Phys., Vol. 91 (Springer, Berlin, Heidelberg 1982)
6. J.P. Toennies: J. Vac. Sci. Technol. A 2, 1055 (1984)
7. K. Kern and G. Comsa; Adv. Chem. Phys. 76, 211 (1989)
8. K. Kern, R. David, R.L. Palmer, G. Comsa: Phys. Rev. Lett. 56, 2823 (1986)
9. A.M. Lahee, J.R. Manson, J.P. Toennies, Ch. Wöll: Phys. Rev. Lett. 57, 471 (1986)
10. J.W. Frenken, J.P. Toennies, and Ch. Wöll; Phys. Rev. Lett. 60, 1727 (1988)
11. G. Comsa, B. Poelsema, Appl. Phys. A 38, 153 (1985)
12. R. David, K. Kern, P. Zeppenfeld, G. Comsa: Rev. Sci. Instr. 57, 2771 (1986)

13. R.L. Park and H.H. Madden; Surf. Sci. 11, 188 (1968)
14. F.C. Frank, J.H. van der Merwe; Proc. Roy. Soc. A 198, 216 (1949)
15. P. Bak, D. Mukamel, J. Villain, K. Wentowska; Phys. Rev. B 19, 1610 (1979)
16. E.D. Specht, A. Mak, C. Peters, M. Sutton, R.J. Birgeneau, K.L. D'Amico,
    D.E. Moncton, S.E. Nagler, P.M. Horn; Z. Phys. B 69, 347 (1987)
    S.C. Fain, M.D. Chinn, R.D. Diehl; Phys. Rev. B 21, 4170 (1980)
17. K. Kern, R. David, P. Zeppenfeld, R.L. Palmer, G. Comsa; Solid State Comm.
    62, 361 (1987)
18. M.B. Gordon, J. Villain; J. Phys. C18, 391 (1985)
19. V.L. Pokrovsky, A.L. Talapov; Sov. Phys. JETP 51, 134 (1980)
20. K. Kern; Phys. Rev. B35, 8265 (1987)
21. A.D. Novaco and J.P. McTague; Phys. Rev. B 19, 5299 (1979)
22. C.R. Fuselier, J.C. Raich and N.S. Gillis; Surf. Sci. 667 (1980)
23. C.G. Shaw, S.C. Fain and M.D. Chinn; Phys. Rev. Lett. 41, 955 (1978)
    K.L. D'Amico et al.; Phys. Rev. Lett. 53, 2250 (1984)
24. M.B. Gordon; Phys. Rev. Lett. 57, 2094 (1986)
25. K. Kern, P. Zeppenfeld, R. David and G. Comsa; Phys. Rev. Lett. 59, 79
    (1987)
26. H. Shiba; J. Phys. Soc. Jpn. 48, 211 (1980)
27. W. Steele; Surf. Sci. 36, 317 (1973)
28. J.M. Gottlieb, to be published
29. D. Drakova, G. Doyen, and F. v. Trentini; Phys. Rev. B 32, 6399 (1985)
30. J.E. Müller; Phys. Rev. Lett. 65, 3021 (1990)
31. K. Kern, R. David, P. Zeppenfeld, G. Comsa; Surf. Sci. 195, 353 (1988)
32. J.R. Chen and R. Gomer; Surf. Sci. 94, 456 (1980)
33. S. Aubry; in "Solitons and Condensed Matter Physics", (Springer,
    Heidelberg, 1978), p. 264
34. G. Vidali and M.W. Cole; Phys. Rev. B29, 6736 (1984)
35. K.D. Gibson, S.J. Sibener, B.M. Hall, D.L. Mills, and J.E. Black; J. Chem.
    Phys. 83, 4256 (1985)
36. B.M. Hall, D.L. Mills, and J.E. Black; Phys. Rev. B32, 4932 (1985)
37. K.D. Gibson and S.J. Sibener; Faraday Discuss. Chem. Soc. 80, 203 (1985)
38. K. Kern, P. Zeppenfeld, R. David, and G. Comsa; Phys. Rev. B 35, 886
    (1987)
39. B. Hall, D.L. Mills, P. Zeppenfeld, K. Kern, U. Becher, and G. Comsa;
    Phys. Rev. B 40, 6326 (1989)
40. P. Zeppenfeld, U. Becher, K. Kern, R. David, and G. Comsa; Phys. Rev. B
    41, 8549 (1990)
41. K. Kern, U. Becher, P. Zeppenfeld, B. Hall, and D.L. Mills; Chem. Phys.
    Lett. 167, 362 (1990)
42. P. Zeppenfeld, U. Becher, K. Kern, G. Comsa; to be published

SIGNATURES AND CONSEQUENCES OF THE SUBSTRATE CORRUGATION

L. W. Bruch

Physics Department
University of Wisconsin-Madison
Madison, Wisconsin 53706

ABSTRACT

Information on the amplitudes of the spatially periodic terms in the adatom substrate potential energy is reviewed for a series of adsorbates on graphite. The use of data for the structure and dynamics of the ordered adlayer is discussed. Generalizing interaction models to reflect the anisotropic polarizability of the substrate may account for the results for adsorbed inert gases, but does not appear to be sufficient to treat molecular adsorbates.

I. INTRODUCTION

Physical adsorption includes a range of phenomena in the monolayer regime which arise from the competition between the adsorbate-adsorbate and adsorbate-substrate interactions.[1,2] The amplitudes of the spatially periodic terms in the adatom-substrate potential energy V_h, here termed the substrate corrugation, were poorly known until recently; most information came from estimates based on modeling[3] of V_h and on the occurrence of commensurate adlayer lattices.[4] Measurements of the structure and dynamics of several adsorption systems now give data quite directly related to the substrate corrugation.[5-9] The purpose of this paper is to review the analysis of such data for information on the corrugation and the consequences for ordered adlayer structures of He, H_2, Ne, CH_4, N_2, and Kr adsorbed on the basal plane surface of graphite (Gr).

The Fourier decomposition of the holding potential

$$V_h(\underset{\sim}{r},z) = V_o(z) + \Sigma_g \, V_g(z) \, \exp(i \, g \cdot \underset{\sim}{r}) \qquad (1)$$

expresses the dependence on lateral displacement ($\underset{\sim}{r}$) through the reciprocal lattice vectors g of the substrate surface. For Gr, the origin of $\underset{\sim}{r}$ is taken at the center of the surface honeycomb cell. Here the sum is truncated at the first shell of reciprocal lattice vectors $|g| = g_o$ and the functions $V_o(z)$ and $V_{go}(z)$ are treated near the minimum ($z = z_o$) of V_h.

The leading amplitude V_{go} is negative for inert gases on graphite, and the minimum energy adsorption site is at $r = 0$. This is also the site for most models of N_2 on graphite, but the minimum energy site for methane is above a surface carbon atom ("tripod down" configuration). By contrast, V_{go} is positive for inert gas adsorption on the (111) face of an fcc lattice, if the minimum of V_h is at a 3-fold site, and the sign change has drastic consequences in the modulated adlayer structures.[10]

If the amplitudes V_g are large, sites of the adlayer minimum energy configuration are at minima of V_h. The more usual situation in physical adsorption is that the energy scale in V_{go} is similar to or smaller than the scale set by the lateral interactions. The ground state then may be a commensurate lattice, such as the $\sqrt{3}$ R 30° lattice on Gr, where the periodicity is set by the substrate periodicity, or a modulated lattice, where the spacings are mostly set by the adatom pair potentials. One test of holding potential models is whether the value of V_{go} suffices to stabilize known commensurate structures; simple models for Kr/Gr are insufficient to give the $\sqrt{3}$ R 30° lattice as the ground state.[11] Another test is to reproduce the domain wall structures observed[5,12] in uniaxial incommensurate (striped) lattices and the chemical potential increment required to create walls in a commensurate lattice.[13]

For the $\sqrt{3}$ R 30° commensurate lattice of a monatomic adsorbate of mass M on graphite, the angular frequency for parallel motions is nonzero at zero wavevector and is given by

$$\omega_o = g_o \{ - 3 \ V_{go} \ / \ M\}^{0.5} \tag{2}$$

in the classical harmonic lattice approximation, the Brillouin zone center gap.[14] Eq.(2) is the leading approximation for deriving V_{go} from a measurement of ω_o for a near-classical adlayer, such as Kr/Gr. The analysis for molecules is more complex, as discussed in Sec. III. Even for the monatomics, anharmonic effects[15] and a possible hybridization of the adlayer and substrate dynamics[16,17] give corrections to Eq.(2).

The organization of the presentation is: the inert gases and molecular hydrogen are reviewed in Sec. II; nitrogen and methane, with coupling of translational and librational motions, are reviewed in Sec. III. The anharmonic perturbation theory for the zone center frequency gap is summarized in Appendix A; the work of Mills et al.[16] and of DeWette et al.[17] on adlayer-substrate hybridization is summarized in Appendix B.

II.   SPHERICAL ADSORBATES ON GRAPHITE

The substrate corrugation affects the stability of the $\sqrt{3}$ R 30°
commensurate lattice, the domain wall structure in the uniaxial
incommensurate (striped) lattice of molecular hydrogen, and the value of
the zone center frequency gap. For the monolayer lattices of helium and
of molecular hydrogen, which are quantum solids, Eq.(2) must be
generalized.[13,18]

The zone center gap for the commensurate monolayer is closely
related to the breaking of center of mass translational symmetry by the
spatially periodic external potential from the substrate. Especially
for a monatomic (spherical) adsorbate in a lattice with one atom per
Bravais unit cell, the force constant for the center of mass motion can
be estimated from the variation of the ground state energy under lateral
displacements of the center of mass. This is a version of the
Born–Oppenheimer approximation, and takes the lattice to be in its
ground state for the displaced center of mass. As applied, the
functional form of the ground state wave function is not generalized for
the lower symmetry of the holding potential away from r = 0; a similar
generalization[19] for the 2D shear modulus gave only small shifts in
results. The quantum mechanical generalization of Eq.(2) is

$$\omega_o(QM) = g_o \{ -3 \, \nabla_{go} \, \langle \exp(i \, g \cdot r) \rangle \, / \, M \}^{0.5} \tag{3}$$

where ∇_{go} is the z-averaged corrugation amplitude and $\langle ... \rangle$ denotes an
average over lateral displacements with the ground state wave function
of the commensurate monolayer. Eq.(3) is implicit in Novaco's
self–consistent– phonon theory[20] of a quantum monolayer solid.
Calculations[18] for the expectation value in Eq.(3) also give the lateral
mean square displacement $\langle r^2 \rangle$ and provide a test of using[5] the
Debye–Waller approximation for the intensity of adlayer Bragg
reflections to estimate $\langle r^2 \rangle$ from neutron scattering data. Even for the
helium monolayer, the Debye–Waller approximation leads to values of $\langle r^2 \rangle$
accurate to 10%.

A.   Helium/Graphite

A considerable unification of scattering and thermodynamic data for
adsorbed helium was described already by Cole, Frankl, and Goodstein.[21]
An amplitude ∇_{go} = -0.28 meV (-3.25 K) fitted to selective adsorption
resonances accounted for departures of the low density helium specific
heat from the 2D equipartition value of 1 k_B/atom. Later[22] the
substrate corrugation was included in a calculation of the second virial
coefficient of the 2D gas. The accuracy of the Jastrow–McMillan
approximation for correlations in the ground state of the dense helium
layer also is now better understood.[23]

With this background, one might expect the zone center gap of the
commensurate He/Gr monolayer would be given accurately by calculations
using the atom scattering ∇_{go}. The result of a 2D calculation is quite

encouraging: for ^{3}He/Gr, the calculated gap[18] is 16 K while the
inelastic neutron scattering gives 13 K;[24] the net effect of anharmonic
corrections is likely to be a lowering of the calculated frequency. The
neutron data[24] also give a measure of the band width of the phonons (the
dispersion across the Brillouin zone), but that has not been treated in
the calculations.[18] A major remaining puzzle is the absence[25] of a clear
signature of this gap in specific heat data. There are no diffraction
data for striped domain structures of helium on graphite.

Both ^{3}He/Gr and ^{4}He/Gr show condensation transitions from dilute 2D
gas to commensurate monolayer solid,[26] but it was difficult[27] to
reproduce the observed self-binding of the solid with smaller values of
V_{go}. The 2D ground state energy per atom for the ^{4}He/Gr lattice
calculated[18] with the -3.25 K amplitude is -2.5 K, which is stable
relative to the ground state energy,[23] -0.8 K, of 2D liquid ^{4}He.
Systematic studies by Miller and Nosanow[28] of self-binding in 2D quantum
gases leave it marginal whether there are self-bound liquids of 2D ^{3}He.
The recent calculations[18] of the commensurate lattice energy of ^{3}He/Gr
give an energy +0.8 K/atom, but indicate self-binding for a 20% larger
value of the corrugation amplitude. Thus, the model calculations now
are close to reproducing the observed stability of the He/Gr
commensurate lattices. The larger self-binding for the ^{4}He/Gr
commensurate lattice does not disrupt the agreement between
thermodynamic and scattering determinations of the total binding energy
which was discussed by Cole et al.[21]

A related quantum monolayer is[29] He/MgO(001). The likely candidate
for a commensurate layer would be a square lattice of side 4.21 Å. The
total binding energy is much less than for He/Gr and estimates of $\bar{V}_{go}$
are very uncertain; it is an open question, in theory and experiment,
whether the commensurate lattice is stable.

B. Hydrogen/Graphite

Molecular hydrogen is assumed to be spherical for the modeling and
the corrugation amplitude $\bar{V}_{go}$ is estimated by isotropic
molecule-substrate atom interaction models fitted to the selective
adsorption resonances. The averaging over perpendicular (z) motions has
a significant effect on the result. Values used by Gottlieb and Bruch[13]
are -6.4 K for both H_2 and D_2 and by Novaco[20] are -7.7 K (H_2) and -8.1 K
(D_2).

The commensurate $\sqrt{3}$ R 30° lattices of H_2/Gr and of D_2/Gr are
stable even with the -6.4 K value; calculations[13] give a positive energy
for domain walls in the uniaxially incommensurate lattice. The measured
zone center frequencies are in good agreement with calculations based on
these corrugation estimates, Table II.1.

The hydrogen/graphite system provides the only example of a
striped, uniaxially incommensurate, structure observed for a spherical
adsorbate on graphite. Because of the large compressibility of the

quantum solid, the domain walls are rather narrow. The values for the wall width fitted to the relative intensities of neutron diffraction peaks agree fairly well[13] with values from calculations with a Jastrow-McMillan trial function.

C. Krypton/Graphite

At low temperatures, krypton condenses on graphite in the $\sqrt{3}$ R $30°$ commensurate lattice. Triangular incommensurate aligned and rotated phases are observed, as well as novel fluid phases.[2,30] However, in spite of an apparently continuous commensurate-to-incommensurate transition of the monolayer solid, no uniaxially incommensurate (striped) phase is observed. Specific heat data for the commensurate solid do not extend to low enough temperatures to isolate an Einstein term arising from the zone center gap, but there is a preliminary report of a signal for the zone center gap in inelastic neutron scattering.[31]

A long-standing problem in modeling Kr/Gr is that, with realistic models of the Kr-Kr interactions, the isotropic atom-atom models for the holding potential do not[4,11] yield large enough values of V_{go} for the commensurate lattice to be the ground state of the monolayer. Shrimpton et al.[1,4] summarized the situation as follows: for the lowest potential energy state of the monolayer solid to be the commensurate lattice, the corrugation amplitude must satisfy $V_{go} < -9.8$ K; including zero point energy eases the requirement to $V_{go} < -7.0$ K. The amplitude resulting from isotropic atom-atom models[3] generally is about $- 5$ K, although Vidali, Cole, and Klein[32] constructed a model with V_{go} near $- 8$ K. One proposed resolution is that the commensurate lattice may be stabilized at intermediate temperatures by thermal expansion from the ground state.[2,11]

If the atom-atom models are generalized to include anisotropy terms,[33] it is rather easy to obtain corrugation amplitudes in the required range. For instance, generalizing the model of Crowell and Steele,[34] isotropic Lennard-Jones (12,6) interactions with $\varepsilon = 67.5$ K and $\sigma = 3.49$Å, the corrugation amplitude is larger, as shown in Table II.2.

Table II.1. Zone center frequency gap for molecular hydrogen on graphite

V_{go}	H_2	D_2
-6.4 K[a]	39.1 K	31.2 K
-7.7 K, -8.1 K[b]	46.6	36.9
n-scattering[c]	47.3	40.2

[a]Calculations, References 13 and 18.
[b]Calculations, Reference 20.
[c]Experiments, Reference 5.

A preliminary estimate[31] of the zone center gap from inelastic
neutron scattering is ω_o = 0.2 THz = 10 K. In the harmonic
approximation this corresponds to V_{go} = -6 K, which is near the value
required for stability of the commensurate layer. The anharmonic
corrections to Eq. (2) are likely to be small for krypton/graphite at
low temperatures. With Lennard-Jones (12,6) parameters ε = 159 K and
σ = 3.60Å for the krypton-krypton potential, the frequency from the
anharmonic perturbation theory summarized in Appendix A is 1% less than
its value in the harmonic approximation. Because of internal
cancellations in the anharmonic terms which arise from the 3D character
of the monolayer, this reduction is smaller than in a mathematical 2D
self-consistent-phonon theory;[15] the 2D theory leads to a 10% reduction
in the effective frequency from the harmonic value. The effect of
dynamic coupling to the substrate (Appendix B) in the Brillouin zone
average sampled by inelastic neutron scattering is quite uncertain; the
calculations of de Wette et al.[17] indicate it may cause a lowering of up
to 10 % in the apparent zone center gap. Such problems will have to be
treated in more detail as the inelastic scattering experiments are
analysed for precise information about the substrate corrugation.

D. Neon/Graphite

Neon condenses on graphite at low temperatures in a commensurate
lattice with four atoms per Bravais cell.[37] The lattice constant is $\sqrt{7}$
times the graphite lattice constant (average neon nearest neighbor
spacing 3.25Å) and the axes are oriented at 10.9° to the 30° axes of the
graphite plane. There are three observations which reflect the
substrate corrugation: (1) the coincidence of the neon spacing and the
graphite length in neutron diffraction experiments,[37] (2) an Einstein
specific heat behavior at low temperatures for the registry solid,[38] and
(3) the Novaco-McTague rotation of the axes of the nonregistry solid
relative to the graphite axes.[39]

Table II.2. Corrugation amplitude for krypton on graphite[a,b]

γ_A	γ_R	$V_{go}(z_o)$	$V_{go}(z_A)$
0.0	0.0	-4.5 K	-5.4 K
0.4[c]	0.0	-5.3 K	-6.5 K
0.4	-0.54[d]	-7.3 K	-9.5 K
0.4	-1.05[e]	-9.2 K	-12.8 K

[a] γ_A and γ_R are anisotropy coefficients defined in Reference 33.
[b] The heights z_o and z_A correspond to the minimum of the laterally
    averaged potential V_o and of the total holding potential V_h
    respectively. The change in height z caused by zero point energy may
    lead to a 5 % reduction in the magnitude of V_{go}.
[c] The value γ_A = 0.4 is set, Reference 35, by data on the polarizability
    anisotropy of the graphite but γ_R is an adjustable parameter.
[d] This value was fitted to helium data, Reference 35.
[e] This value was fitted to nitrogen data, Reference 36.

Because the zero point energy dilates a 2D solid of neon to a near-neighbor spacing of 3.23 to 3.27 Å, depending on the model assumed,[40] little information on V_{go} is derived from the occurrence of the commensurate lattice. Also, the values of the Novaco-McTague rotation angle, calculated by perturbation theory, are insensitive to V_{go}.

The zone center frequency gap inferred from the specific heat data[38] is 3.5 K. A Born-Oppenheimer approximation[14] for the four-atom cell, similar in spirit to Eq.(3), leads to a gap of 1.9 K. The calculation has $V_{go} = -3.0$ K, derived from an isotropic atom-atom model fit to gas-surface virial data, and includes effects of the 3D modulations of atoms in the Bravais cell. The 1.9 K value incorporates four parameters from the net atom-substrate potential. Thus, the connection between the zone center gap and the amplitude V_{go} is more complex than for the simple Bravais cell cases.

Apparently the isotropic atom-atom model for interaction with the graphite leads to an underestimate of the corrugation for Ne/Gr. When anisotropy is included in the parameterized form of Carlos and Cole,[35] as described in Sec. II.C for krypton, the calculated zone center gap increases and reaches the specific heat value. At the overlayer height corresponding to the minimum of V_0, the gap values for the sequence of cases in Table II.2 are 1.9 K, 2.7 K, 3.8 K and 7.6 K. The value 3.8 K, for the set $\gamma_A = 0.4$ and $\gamma_R = -0.54$, is in the range of the 3.5 K gap from the specific heat data.

Thus, there is some indication that corresponding states scaling is applicable for the anisotropy terms and that the fit for helium/graphite may apply to the other inert gases. For spherical adsorbates on graphite, the present situation is that generalization of the atom-atom models of the holding potential by inclusion of anisotropy will probably suffice to account for observed and inferred substrate corrugations.

III.   MOLECULAR ADSORBATES ON GRAPHITE

The modeling of molecular adsorbates is more complex than for the inert gases because (1) the lower symmetry of the molecule admits permanent electrostatic multipolar interactions, (2) the molecule has several degrees of freedom besides the center of mass coordinates, (3) an extended molecule has several effective centers for the intermolecular forces, (4) the molecule may extend beyond one substrate Bravais lattice cell, and (5) the adlayer unit cells for registry structures frequently contain several molecules. The discussion here is narrowed to two small molecules, methane and nitrogen: they show monolayer registry phenomena similar to those for the inert gases, Sec. II, but there are serious difficulties in modeling the substrate corrugation.

At low temperatures, both methane/graphite and nitrogen/graphite
condense as commensurate monolayer solids with the molecular centers of
mass in the $\sqrt{3}$ R 30° lattice. Modeling yields this lattice as the
ground state, with orientational ordering, even with substrate
corrugations given by Steele's constructions.[41,42] For both systems, the
calculated zone center frequency gaps are much smaller than the values
from inelastic neutron scattering. The results of the calculations, for
the four cases of anisotropy coefficients introduced in the discussion
for krypton in Sec. II.C, are listed in Table III.1; the models are
described in this Section.

A.   Methane/Graphite

The methane molecule has high, tetrahedral, symmetry and is
frequently treated as a sphere. Conclusions on stability of the
commensurate lattice are similar for the spherical model[41] and the
tetrahedral model,[41,43] but the consequences for the zone center gap are
quite different. The tetrahedral model consists of the Severin and
Tildesley set[44] of Lennard-Jones (12,6) atom-atom interactions for

Table III.1.   Calculated zone center frequency gap (in K) for molecular
adsorbates on graphite

γ_A	γ_R	CD_4	N_2(a)	N_2(b)
0.	0.	9.0	8.0	8.9
0.4	0.	9.8	8.7	8.3
0.4	-0.54	9.7	9.5	7.7
0.4	-1.05	8.8	9.7	---

[a]Steele's holding potential model, Reference 50.
[b]ε_{NC} = 39.5 K, σ_{NC} = 3.0 Å.

methane to methane and methane to graphite. The spherical model has
direct molecule-molecule interactions with Lennard-Jones (12,6)
potential parameters ε = 137 K and σ = 3.6814 Å; the molecule to
graphite carbon parameters are then ε = 61.9 K and σ = 3.54 Å, from
combining rules. The monolayer lattice of minimum static potential
energy for the tetrahedral model is incommensurate, but the commensurate
lattice is stabilized by zero point energy and it remains stable when a
substrate mediated interaction is included.

The adsorption site in the commensurate lattice differs for the two
models: the tetrahedron is atop a surface carbon atom; the sphere is
above the center of a surface carbon hexagon. The difference is
important in calculating vibration force constants and the zone center
gap. For the sphere the corrugation amplitude V_{go} is -4.6 K at the
minimum of the holding potential and - 3.9 K at the height set by
including the zero point energy of perpendicular (z) motions for CD_4.
The corresponding values for the zone center gap, using Eq.(2), are 17
and 16 K and are in good agreement with the preliminary value[45] from

74

inelastic neutron scattering, 0.3 Thz (14 K). However, this appears to be fortuitous, because the zone center gap for the tetrahedral model, at the overlayer height determined with the inclusion of zero point energy, is 9.0 K.

The coupling of the center of mass translation to the libration coordinates has a large effect on the calculated zone center gap for the tetrahedron. If the coupling to the librational coordinates is suppressed, the 9 K value increases to 11.3 K. If, in an attempt to force a larger gap, the overlayer height is held fixed and the substrate amplitudes V_{go} are scaled up by a factor f, there is very little increase: because of the offsetting effect of coupling to the libration, the gap increases by only 1% (at f=1.15) and then decreases with further increase of f. Thus it is inadequate to characterize the failure to fit the scattering result as a failure in the scale of the corrugation.

Including anisotropy terms in the adatom-substrate atom pair potentials, in the same manner as done for the inert gases, does not increase the zone center gap appreciably for the tetrahedron. For the results shown in Table III.1, the anisotropy coefficients are taken to be the same for both the admolecule hydrogen and carbon interactions with the substrate, and the overlayer height is optimized in each case. The registry energy itself increases monotonically with the cases in Table III.1; the nonmonotonic trend for the zone center frequency gap reflects the coupling of center of mass and librational coordinates.

There have been no estimates of the effect of partial orientational disorder of the methane layer on an observed gap.

No modulated monolayer structures of the methane/graphite have been reported, but there is a prediction[43] of a Novaco-McTague rotation[46] of about 1° for an incommensurate triangular lattice with nearest neighbor spacing 4.12 Å. As for the inert gases, predictions with the Novaco-McTague perturbation theory are insensitive to the scale of the corrugation.

B. Nitrogen/Graphite

At low temperatures, molecular nitrogen condenses on graphite in a commensurate 2-in herringbone lattice, which is a rectangular lattice with two molecules per Bravais cell.[47] The centers of mass form a triangular lattice and the angle of the molecular axes relative to the rectangular axes is determined by neutron diffraction.[48] The length and energy scales of a succession[47] of monolayer lattices are fairly well reproduced with calculations[42] based on quasiharmonic lattice dynamics.

The zone center frequency gap determined by inelastic neutron scattering from the commensurate 2-in herringbone monolayer solid is 19 K (0.4 THz), but modeling scarcely gives half this value.[6] The intralayer interactions are based on the site-site X1 model of Murthy et al.[49] and various approximations are used for the interaction of the molecule with the graphite. The work of Hansen et al.[6] has powder averages of the calculated spectra, to mimic the experimental conditions, but Table III.1 simply lists the zone center ($q = 0$) gap for the adlayer with overlayer height and herringbone angle which minimize the zero temperature Helmholtz free energy.

In case (a) of Table III.1 Steele's parameters[50] for atom-atom interactions of nitrogen with substrate carbon, $\varepsilon_{NC} = 31.0$ K and $\sigma_{NC} = 3.36$ Å, are used. The isotropic model has $\omega_\parallel = 8.0$ K. The registry potential energy more than doubles with the increase of anisotropy in the cases of Table III.1, but the gap increases by less than 25%. The coupling of center of mass translation to the librational coordinates operates here, as in the methane/graphite cases. If, as in the methane work, a scaling factor f is applied to the corrugation amplitude and adjusted to maximize the gap frequency for the re-optimized lattice, the gap frequency increases by only 10%.

Another route to increase the substrate corrugation in the model, and perhaps the zone center gap, is to reduce[51] the length scale in the adatom-substrate potential. Case (b) in Table III.1 has $\varepsilon_{NC} = 39.5$ K and $\sigma_{NC} = 3.0$ Å. The amplitude V_{go} doubles, but the zone center gap does not increase significantly.

The minima of the one molecule holding potential have the molecular center of mass displaced from the center of the carbon hexagon for the cases with larger corrugation and anisotropy. This occurs already in the model of Joshi and Tildesley,[36] with a monolayer ground state still in the 2-in herringbone lattice inferred from general features of the diffraction patterns.[47]

Thus the size of the zone center gap for nitrogen/graphite remains a stubborn puzzle. Suggested experimental artifacts, such as a long wavelength cutoff by pinning of the edges of modest size monolayer solid islands, do not appear to cause large enough increases in the frequency to account for the observations. To resolve the problem by using much larger values of V_{go} would have the difficulty that the observed bilayer solid is incommensurate with the graphite.[52] Present modeling[42] does not give a fully incommensurate bilayer herringbone lattice and commensurate structures would be even more favorable with a larger V_{go}.

IV. CONCLUDING REMARKS

This review emphasizes those inert gases and small molecules which condense, at low temperatures, in commensurate lattices on the basal plane surface of graphite. Values for the leading amplitude of the spatially periodic component of the adatom substrate potential at the equilibrium overlayer height, V_{go}, may be inferred from the stability of the commensurate lattice relative to incommensurate and uniaxially commensurate structures and from data which reflect the dynamics of the commensurate lattice. Inelastic neutron scattering measurements of the

Brillouin zone center gap in the frequency spectrum of the commensurate lattices continue to be refined;[59] one purpose of this paper is to discuss the theoretical developments now needed in order to derive precise values for V_{go} from such data. The adsorption of argon and of xenon on graphite are not directly addressed here, although knowledge of V_{go} is needed for a full understanding of the monolayer phase diagrams.

It is difficult to cite firm values for V_{go} for most of the cases discussed here. However, for the spherical adsorbates a reasonable model based on data now available is: augment the adatom substrate potential parameters of Steele,[3] for atom-atom Lennard-Jones potentials fit to the low coverage heats of adsorption, with anisotropy terms having the coeffients ($\gamma_A = 0.4$ and $\gamma_R = -0.54$) of the fit by Carlos and Cole[35] for helium/graphite.

Similar data are available for adsorption on other substrates, but modeling the origin of V_{go} is difficult. The occurrence of commensurate lattices and the modulation of incommensurate lattices are used to infer values of V_{go} for adsorption of xenon on platinum,[10] but without a model for V_{go}. For adsorption on ionic crystals, effects of relaxations and ionic charge rearrangements at the surface[60] present additional complications.

APPENDIX A.   ANHARMONIC CORRECTIONS

There is a simple and direct relation between the zone center frequency gap and the corrugation amplitude V_{go} for the case of a classical monatomic adlayer with one atom per Bravais lattice cell in the harmonic lattice approximation.[14] Corrections to this leading approximation which arise from cubic and quartic anharmonic terms in the holding potential are discussed here for Kr/Gr. The three-dimensional character of the adsorbate motion is retained, to include an important contribution to the quartic anharmonicity which is omitted by approximations for mathematical two dimensions.[15]

The anharmonic zone center frequency is defined to be the peak frequency of the one-phonon inelastic scattering cross section.[53] The cubic anharmonicity is treated by second order perturbation theory.[54] The first order perturbation term from the quartic anharmonicity is equal to the first correction term in an expansion of the self consistent phonon approximation.[15,20] For the case of zero wavevector, compact expressions may be given for the peak frequency and peak width.

The Fourier decomposition of the adatom-substrate potential is presented in Eq.(1); as in the main text, the sum is truncated at the first nonvanishing shell, $g = g_o$. The origin of the vector r is (1) at the center of a honeycomb cell, for the basal plane graphite surface, or (2) atop a substrate atom, for the (111) surface of an fcc lattice. The anharmonic coefficients a_i and b_i are the terms in the expansion

$$V(r,z) = A + B \underset{\sim}{u}^2 + a_1 (u_y^3 - 3 u_x^2 u_y) + a_2 \xi \underset{\sim}{u}^2$$

$$+ a_3 \xi^3 + b_1 (\underset{\sim}{u}^2)^2 + b_2 \xi^4 + b_3 \underset{\sim}{u}^2 \xi^2$$

$$+ b_4 \xi (u_y^3 - 3 u_x^2 u_y), \tag{A.1}$$

where $\underset{\sim}{u}$ denotes lateral displacement from the holding potential minimum (honeycomb center for Gr and three-fold site for fcc(111)) and ξ denotes perpendicular displacement. The y axis is taken to be parallel to one of the primitive substrate reciprocal lattice vectors. The coefficients in this expansion can be expressed in terms of derivatives of V_o and V_{go}; a_1 and b_4 are zero for the graphite case.

For the monolayer, the harmonic angular frequencies for wavevector

Table A.1. Zone center parallel frequency gap (in K)
for Kr/Gr, perturbation shifts

Case	harmonic	cubic[d]	quartic[e]
Isotropic[a]	9.11	-0.044	+0.039
"He"[b]	11.92	-0.078	+0.039
"N_2"[c]	13.81	-0.111	+0.062

[a]Crowell and Steele isotropic Lennard–Jones model, Reference 34.
[b]Helium anisotropy parameters, Reference 35.
[c]Nitrogen anisotropy parameters, Reference 36.
[d]From Eq.(A.2).
[e]From Eq.(A.3).

k have polarization along the z-axis ("$\perp$") or parallel to the r-plane, with polarization index α and β. The zero wavevector frequencies are denoted $\omega_\perp$ and $\omega_\parallel$. Then for adatoms of mass m and averages $\langle \ldots \rangle$ over wavevectors in the adlayer first Brillouin zone, the cubic(3) and quartic(4) shifts of $\omega_\parallel$ are given by

$$\omega_\parallel(3) = (\hbar/4m^3\omega_\parallel)\{2a_2^2\Sigma_\alpha\langle(\omega_{k\perp}+\omega_{k\alpha})/(\omega_{k\perp}\omega_{k\alpha})/$$

$$[\omega_{o\parallel}^2 - (\omega_{k\perp}+\omega_{k\alpha})^2]\rangle$$

$$+ 9 a_1^2 \Sigma_{\alpha,\beta}\langle(\omega_{k\alpha}+\omega_{k\beta})/(\omega_{k\alpha}\omega_{k\beta})/[\omega_{o\parallel}^2 - (\omega_{k\alpha}+\omega_{k\beta})^2]\rangle; \tag{A.2}$$

$$\omega_\parallel(4) = (\hbar/2m^2\omega_\parallel)\{4b_1\Sigma_\alpha\langle1/\omega_{k\alpha}\rangle + b_3\langle1/\omega_{k\perp}\rangle\}. \tag{A.3}$$

The self-consistent phonon theory of Hakim et al.[15] has only the b_1 term in Eq.(A.3). For the Kr/Gr model used by deWette et al.[17] the b_3 term is of opposite sign and outweighs the b_1 term, so that the quartic frequency shift is positive. This result persists when anisotropy terms are included in the calculation of V_{go}, Table A.1.

The perturbation theory leads to a shift of overlayer from the minimum potential position. For both the Gr and fcc cases, the in-plane symmetry is high enough that there is no average displacement parallel to the surface. However, the average shift in the z-direction is

$$\langle \xi \rangle = (\hbar/2m^2\omega_{\perp}^2)\{3a_3\langle 1/\omega_{k\perp}\rangle + a_2 \Sigma_\alpha \langle 1/\omega_{k\alpha}\rangle\}. \tag{A.4}$$

For the Kr/Gr cases, $\langle \xi \rangle$ is -0.02 Å. There does not appear to be a generalization of self-consistent phonon theory including cubic anharmonic terms for this low-symmetry case.

APPENDIX B.   DYNAMIC ADLAYER-SUBSTRATE COUPLING

Dynamical coupling of the adlayer to the substrate frequently is neglected in physical adsorption theory, in part because the force constants for relative motions of the adlayer are much smaller than those for motions within the substrate. However, at long wavelengths the frequencies of the Rayleigh wave of the substrate surface are comparable to the frequencies of perpendicular adatom motion and of parallel motion of the commensurate adlayer. Hybridization of the perpendicular motion with the Rayleigh mode was observed[55,56,16] with inelastic atom scattering and modeled[16] with a continuum approximation for the substrate response. This hybridization and a mixing of the parallel adlayer motion with substrate motions were also treated in the lattice dynamics of commensurate monolayers of Kr/Gr and Xe/Gr.[17]

The dispersion of the Rayleigh mode is negligible for Pt(111) at the hybridization wavevectors,[57] but for Gr it is significant.[58]

Mixing of the adlayer perpendicular motion with the substrate Rayleigh and bulk modes leads to a damping of the adlayer mode at small wavevectors and to a shift of its resonant frequency.[16] Treating the substrate as an isotropic elastic continuum, the half-width-at-half-maximum (γ) of the $q = 0$ resonance and the decrease of its peak frequency are[16]

$$\gamma = (n M \omega_{\perp}^2)/(2 \rho c_1)$$

and $\Delta\omega = -(\gamma^2/(2\omega_{\perp}))$. \hfill (B.1)

Here n is the area density of the adlayer, ρ is the bulk substrate mass density, and c_1 is the speed of longitudinal sound in the substrate; M is the adatom mass and $\omega_{\perp}$ is the angular frequency of motion relative to the rigid substrate.

For the Kr/Pt(111) monolayer, with $\omega_\perp$ = 3.9 meV, the values from Eq.(B.1) are γ = 0.15 meV and $\Delta\omega$ = -0.003 meV. Doubling[16] the coupling to the substrate, so that γ = 0.3 meV and $\Delta\omega$ = -0.012 meV, the shift remains a small correction. The perpendicular frequency is higher at the zone center than at the zone boundary, where it is lowered more by the hybridization.

There are only numerical results for the shifts arising from the dynamic coupling of the parallel adlayer motion to the substrate. DeWette et al.[17] find for commensurate monolayers of Kr and of Xe on Gr that the hybridization occurs at very small wavevectors and that it is primarily a mixing with rather stiff shear horizontal motions of the anisotropic graphite. Extrapolations of the frequencies to zero wavevector are about 10 % lower than the gap frequency for the static substrate.

ACKNOWLEDGMENTS

This work was supported in part by the National Science Foundation through grant No. DMR-88-17761.

REFERENCES

1. N. D. Shrimpton, B. Joos, and B. Bergersen, Phys. Rev. B 38, 2124 (1988).
2. J. Villain and M. B. Gordon, Surface Sci. 125, 1 (1983); M. B. Gordon and J. Villain, J. Phys. C 18, 3919 (1985).
3. W. A. Steele, Surface Sci. 36, 317 (1973).
4. N. D. Shrimpton, B. Bergersen, and B. Joos, Phys. Rev. B 29, 6999 (1984); R. J. Gooding, B. Joos, and B. Bergersen, Phys. Rev. B 27, 7669 (1983); N. D. Shrimpton, Ph. D. thesis, University of British Columbia, 1987 (unpublished).
5. H. Freimuth, H. Wiechert, H. P. Schildberg, and H. J. Lauter, Phys. Rev. B 42, 587 (1990).
6. F. Y. Hansen, V. L. P. Frank, H. Taub, L. W. Bruch, H. J. Lauter, J. R. Dennison, Phys. Rev. Lett. 64, 764 (1990).
7. V. L. P. Frank, H. J. Lauter, and P. Leiderer, Phys. Rev. Lett. 61, 436 (1988); H. J. Lauter, V. L. P. Frank, P. Leiderer, and H. Wiechert, Physica B 156 and 157, 280 (1989).
8. K. Kern, R. David, P. Zeppenfeld, R. Palmer, and G. Comsa, Solid State Commun. 62, 391 (1987).
9. K. Kern, R. David, P. Zeppenfeld, and G. Comsa, Surface Sci. 195, 353 (1988).
10. J. M. Gottlieb (to be published).
11. F. F. Abraham, W. E. Rudge, D. J. Auerbach, and S. W. Koch, Phys. Rev. Lett. 52, 445 (1984); S. W. Koch and F. F. Abraham, Phys. Rev. B 33, 5884 (1986).

12. J. Cui, S. C. Fain, Jr., H. Freimuth, H. Wiechert, H. P. Schildberg, and H. J. Lauter, Phys. Rev. Lett. 60, 1848 (1988).
13. J. M. Gottlieb and L. W. Bruch, Phys. Rev. B 40, 148 (1989).
14. L. W Bruch, Phys. Rev. B 37, 6658 (1988).
15. T. M. Hakim, H. R. Glyde, and S. T. Chui, Phys. Rev. B 37, 974 (1988).
16. B. Hall, D. L. Mills, and J. E. Black, Phys. Rev. B 32, 4932 (1985); B. Hall, D. L. Mills, P. Zeppenfeld, K. Kern, U. Becher, and G. Comsa, Phys. Rev. B 40, 6326 (1989).
17. E. De Rouffignac, G. P. Alldredge, and F. W. De Wette, Phys. Rev. B 24, 6050 (1981); F. W. de Wette, B. Firey, E. de Rouffignac, Phys. Rev. B 28, 4744 (1983).
18. J. M. Gottlieb and L. W. Bruch, Phys. Rev. B 41, 7195 (1990).
19. L. W. Bruch and J. M. Gottlieb, Phys. Rev. B 37, 4920 (1988).
20. A. D. Novaco, Phys. Rev. Lett. 60, 2058 (1988).
21. M. W. Cole, D. R. Frankl, and D. L. Goodstein, Rev. Mod. Phys. 53, 199 (1981).
22. Z.-C. Guo and L. W. Bruch, J. Chem. Phys. 77, 1417 (1982).
23. P. A. Whitlock, G. V. Chester, and M. H. Kalos, Phys. Rev. B 38, 2418 (1988).
24. V. L. P. Frank, H. J. Lauter, H. Godfrin, and P. Leiderer, in "Phonons 89", World Scientific (1989).
25. F. C. Motteler, Ph. D. thesis, University of Washington, 1986 (unpublished).
26. M. Bretz, J. G. Dash, D. C. Hickernell, E. O. McLean, and O. E. Vilches, Phys. Rev. A 8, 1589 (1973); M. Schick, in "Phase Transitions in Surface Films", J. G. Dash and J. Ruvalds, eds., Plenum, New York (1980); see also D. S. Greywall and P. A. Busch, Phys. Rev. Lett. 65, 64 (1990).
27. A. D. Novaco and C. E. Campbell, Phys. Rev. B 11, 2525 (1975).
28. M. D. Miller and L. H. Nosanow, J. Low Temp. Phys. 32, 145 (1978).
29. T. S. Sullivan, A. D. Migone, and O. E. Vilches, Surface Sci. 162, 461 (1985); C. Schwartz, M. Karimi, and G. Vidali, Surface Sci. 216, L342 (1989).
30. E. D. Specht, A. Mak, C. Peters, M. Sutton, R. J. Birgeneau, K. L. D'Amico, D. E. Moncton, S. E. Nagler, and P. M. Horn, Z. Phys. B69, 347 (1987).
31. H. Taub, (private communication).
32. G. Vidali, M. W. Cole, and J. R. Klein, Phys. Rev. B 28, 3064 (1983).
33. G. Vidali and M. W. Cole, Phys. Rev. B 29, 6736 (1984).
34. A. D. Crowell and R. B. Steele, J. Chem. Phys. 34, 1347 (1961).
35. W. E. Carlos and M. W. Cole, Surface Sci. 91, 339 (1980).
36. Y. P. Joshi and D. J. Tildesley, Mol. Phys. 55, 999 (1985).
37. C. Tiby, H. Wiechert, and H. J. Lauter, Surface Sci. 119, 21 (1982).
38. G. B. Huff and J. G. Dash, J. Low Temp. Phys. 24, 155 (1976).
39. S. Calisti, J. Suzanne, and J. A. Venables, Surface Sci. 115, 455 (1982).
40. L. W. Bruch, J. M. Phillips, and X.-Z. Ni, Surface Sci. 136, 361 (1984).

41. J. M. Phillips and M. D. Hammerbacher, Phys. Rev. B 29, 5859 (1984); J. M. Phillips, Phys. Rev. B 29, 5865 (1984).
42. S. E. Roosevelt and L. W. Bruch, Phys. Rev. B 41, 12236 (1990).
43. L. W. Bruch, J. Chem. Phys. 87, 5518 (1987).
44. E. S. Severin and D. J. Tildesley, Mol. Phys. 41, 1401 (1980).
45. T. Moeller, H. J. Lauter, V. L. P. Frank, and P. Leiderer, in "Phonons 89", World Scientific (1989).
46. A. D. Novaco and J. P. McTague, Phys. Rev. Lett. 38, 1286 (1977).
47. R. D. Diehl and S. C. Fain, Jr., Surface Sci. 125, 116 (1983).
48. R. Wang, S.-K. Wang, H. Taub, J. C. Newton, and H. Schechter, Phys. Rev. B 35, 5841 (1987).
49. C. S. Murthy, K. Singer, M. L. Klein, and I. R. McDonald, Mol. Phys. 41, 1387 (1980).
50. W. A. Steele, J. Phys. (Paris) 38, C4-61 (1977).
51. G. Cardini and S. F. O'Shea, Surface Sc. 154, 231 (1985).
52. S.-K. Wang, J. C. Newton, R. Wang, H. Taub, J. R. Dennison, and H. Schechter, Phys. Rev. B 39, 10 331 (1989).
53. A. A. Maradudin and A. E. Fein, Phys. Rev. 128, 2589 (1962).
54. V. N. Kashcheev and M. A. Krivoglaz, Sov. Phys. Solid State 3, 1107 (1961).
55. K. D. Gibson and S. J. Sibener, Faraday Discuss. Chem. Soc. (London) 80, 203 (1985).
56. J. P. Toennies and R. Vollmer, Phys. Rev. B 40, 3495 (1989).
57. U. Harten, J. P. Toennies, Ch. Wöll, and G. Zhang, Phys. Rev. Lett. 55, 2308 (1985); V. Bortolani, A. Franchini, G. Santoro, J. P. Toennies, Ch. Wöll, and G. Zhang, Phys. Rev. B 40, 3524 (1989).
58. E. De Rouffignac, G. P. Alldredge, and F. W. De Wette, Phys. Rev. B 23, 4208 (1981).
59. H. J. Lauter, V. L. P. Frank, H. Taub, and P. Leiderer, in "19th International Conference on Low Temperature Physics", Brighton, UK (1990).
60. A. Frigo, F. Toigo, M. W. Cole, and F. W. Goodman, Phys. Rev. B 33, 4184 (1986).

EPITAXIAL ROTATION AND ROTATIONAL PHASE TRANSITIONS

François Grey[1],[2] and Jakob Bohr[2]

[1]Max Planck Institute for Solid State Research
D-7000 Stuttgart 80, FRG

[2]Physics Department, Risø National Laboratory
DK-4000 Roskilde, Denmark [1]

Abstract

When one crystalline material grows on another, the high-symmetry directions of the adsorbate usually align with high symmetry directions of the substrate. In some cases, though,the adsorbate undergoes an epitaxial rotation: the high symmetry directions of the adsorbate are rotated some apparently arbitrary angle with respect to those of the substrate. We discuss several experimental examples of this phenomenon, and review briefly the theoretical efforts to understand and predict epitaxial rotation. We argue that a simple symmetry principle seems to underly the theoretical results, and that it explains many otherwise curious experimental observations. The principle leads in a straightforward way to the consideration of several types of rotational phase transition, which we discuss qualitatively.

1. Introduction

How do crystals fit together? The theoretical literature on the subject of crystalline interface formation is abundant, and new additions are made regularly. Yet there seems to be no simple, unifying explanation of why various sorts of crystals fit together the way they do. In the absence of such a general theory, two approaches to understanding epitaxy are popular. The first is to develop models which are detailed enough to give quantitatively accurate predictions of optimum interface configurations. Such models can become quite intricate, and often involve so many unknown parameters that they are difficult to apply to all but a few real systems.

The alternative, which we shall pursue in this article, is to try to formulate a few simple principles which give some physical insight into a wide variety of experimental results. The predictions that can be made on the basis of such principles are of a more qualitative nature. Such principles are to be viewed more as 'rules-of-thumb' rather than as the offspring of rigorous theoretical analysis.

In section 2, we review briefly two principles which are frequently used for determining energetically favourable interfaces. In section 3, we discuss the phenomenon of epitaxial rotation and introduce a simple symmetry principle to explain it. We use the symmetry principle to interpret recent experimental results on epitaxial rotation in section 4. Finally, in section 5 we discuss qualitatively different sorts of rotational phase transition in terms of the the symmetry principle [2].

2. Moiré Patterns and Domain Walls

A concept which pervades much of the theoretical literature on interface formation is that of the "coincidence site lattice". This means finding a relative orientation of two crystals where the lattices match exactly at the interface, so that nodes on the two lattices coincide periodically. A criterion is needed to choose between the infinite number of possible coincidence site lattices. Intuitively, it seems reasonable that the energetically most favourable interfaces will be those with the smallest coincidence site lattice unit cells, since this represents the "best fit" of the two lattices. Let us call that the coincidence site lattice principle.

A problem with coincidence site lattices is that they represent a highly discontinuous function of crystal orientation, and are therefore difficult to handle analytically. To avoid this problem, it helps to focus on the moiré pattern generated by the overlapping of the two crytal lattices on either side of the interface. The moiré pattern evolves in a continuous manner as a function of relative orientation of the two crystals. Further, if the interface structure is allowed to relax, the dark fringes of the moiré pattern can be interpreted as a network of misfit dislocations or "domain walls". Domain walls tend to repel, so a condition for stability is to maximize their separation as a function of the relative orientation of the crystals. Let us call this the domain wall repulsion principle.

It is not always obvious which moiré fringes to consider when applying the domain wall repulsion principle. Looking at the problem in reciprocal space clarifies things somewhat. Fig. 1a shows the typical situation of a monolayer of rare gas adsorbed on a graphite surface. The fundamental wavevector of the overlayer, $\vec{q}_A(10)$, is close to the $\vec{q}_S(1/3,1/3)$ position (open circle) in substrate reciprocal units, in other words the overlayer is close to adopting a commensurate $\sqrt{3} \times \sqrt{3}R30°$ structure. The wavevector $\vec{q}_A(11)$ is therefore close to $\vec{q}_S(01)$, and the difference vector $\vec{q}_M = \vec{q}_A(11) - \vec{q}_S(10)$ corresponds to moiré fringes with a large spacing.

When the moiré fringes are far apart, atoms near the center of each moiré cell will relax towards substrate potential wells, and the misfit of the adsorbate condenses into the relatively narrow domain walls. This relaxation is what gives physical significance to

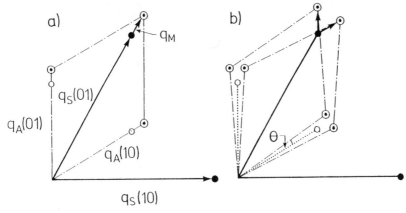

Figure 1. a) Incommensurate unrotated phase of a rare gas on graphite. Filled circles are substrate reflections, dotted circles are adsorbate reflections. The first $(1/3,1/3)$ reflection of the $\sqrt{3} \times \sqrt{3}$ R30° commensurate structure is indicated (open circles), as well as the moiré wavevector $\vec{q}_M$. b) Sketch of an epitaxially rotated phase, rotation angle $\pm\theta$. Note that the moiré wavevector rotates through a much larger angle.

fringes of the moiré pattern. When the moiré unit cell is of nearly the same size as the adsorbate and substrate unit cells, the two structures go in and out of phase so rapidly that there is little energetic advantage in atoms relaxing their positions, and so the moiré pattern loses its physical significance. This is the basis for focusing on the small moiré wavevector $\vec{q}_M$ in Fig.1a, rather than, say, the large difference vector, $\vec{q}_A(10) - \vec{q}_S(10)$.

Of course, it is always possible to find a smaller difference vector if one goes far enough out in reciprocal space. However, reflections with large wavevectors correspond to higher order harmonics of the substrate potential, which will be physically less significant in determining how the adsorbate aligns. There is no general way to decide which moiré wavevector is of most physical significance, one simply looks for a relatively small moiré wavevector produced by relatively low-order adsorbate and substrate reflections.

It might seem that the coincidence site lattice principle contradicts the domain wall repulsion principle, since the former requires, in effect, minimizing the moiré pattern unit cell, while the latter requires maximizing it. It has been pointed out that the two principles apply to different regimes [3]. The domain wall repulsion principle applies if the two crystals have nearly the same structure in the plane of the interface, such that the moiré pattern fringes have a large spacing. If the two structures are very different, the coincidence site lattice principle is usually more appropriate.

From the metallurgists point of view, the two regimes correspond to low angle and high angle grain boundaries. If the orientation of a high angle grain boundary deviates somewhat from a favoured coincidence site lattice, there will result a domain wall structure to accomodate the misfit (secondary misfit dislocation network). The exact coincidence

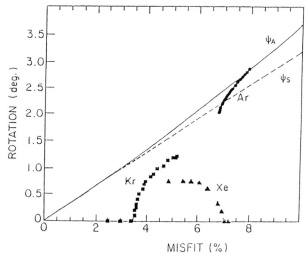

Figure 2. Experimental results for the epitaxial rotation of Ar, Kr and Xe on graphite, from Ref. 4. The adsorbate rotation angle is shown as a function of the misfit, defined as $m=|\vec{q}_A(11)-\vec{q}_S(01)|/|\vec{q}_A(11)|$. The $\psi_A=30°$ and $\psi_S=30°$ solutions are indicated for the case $\vec{q}_A(11)>\vec{q}_S(01)$, which is the case for Ar and Kr but not for Xe.

site lattice therefore represents the case where the separation of these secondary misfit dislocations has been maximized to infinity. Seen from this viewpoint, there is no contradiction with the domain wall repulsion principle.

3. A Symmetry Principle for Epitaxial Rotation

To introduce the symmetry principle for epitaxial rotation, let us look at what actually happens when a rare gas on graphite becomes epitaxially rotated. As the misfit between the adsorbate structure and the commensurate $\sqrt{3} \times \sqrt{3}R30°$ phase increases (by varying the rare gas density), the adsorbate peak splits into two peaks rotated a small angle either side of the substrate high symmetry direction (Fig.1b). In other words, large areas on the surface have rotated either $+\theta$ or $-\theta$ to the high symmetry substrate direction. Now this is curious, because it goes against the domain wall repulsion principle (at fixed adsorbate wavevector, the length of $\vec{q}_M$ increases as the adsorbate is rotated).

Recent experimental results [4] for the epitaxial rotation of several rare gases are summarized in Fig. 2. At about the time this phenomenon was first noticed experimentally [5], Novaco and McTague demonstrated theoretically that such a rotated phase could be energetically favourable [6]. They solved the Hamiltonian for a harmonic rare gas lattice on a stiff substrate, and found that rotation was favoured because the shear modes involved are energetically less costly than the strain modes of the parallel orientation.

Shiba later extended the theory to take explicit account of the interactions of domain walls [7]. The main feature of the Shiba model is the prediction of a finite critical

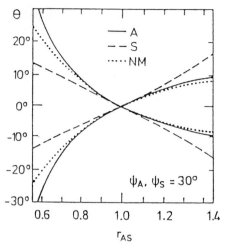

Figure 3. Comparison of the $\psi_A = 30°$ and $\psi_S = 30°$ solutions with the Novaco-McTague result (NM) for a Cauchy lattice.

mismatch up to which the structure is unrotated, and beyond which it begins to rotate. This can be seen in Fig. 2, where Kr begins to rotate at a misfit of 3.5%. (The Novaco-McTague model predicts a roughly linear dependence of rotation angle on mismatch at small mismatch.)

Both the Novaco-McTague and Shiba models were great breakthroughs, and give considerable insight into the origin of epitaxial rotation. However, it should be noted that the Shiba model, with two adjustable parameters, explains quantitatively only the case of Kr. The model would need to be extended substantially to explain the behaviour of Xe on graphite. The risk, as mentioned in the introduction, is that if the theory becomes too elaborate, it may require too many unknowns in order to make quantitative predictions, not to mention that it may become thoroughly inscrutable!

This prospect provides the incentive to find a "rule-of-thumb" for epitaxial rotation which, although it may not give as detailed predictions as more rigorous theories, will provide a starting point for analysing and predicting epitaxial rotation phenomena in a wide variety of systems. The principle we propose is based on the notion that the interface energy may depend not only on the *spacing* of domain walls, but also on their *orientation* relative to both the adsorbate and substrate lattices. In particular, we suggest that the alignment of the domain walls with high symmetry directions of either substrate or adsorbate will correspond to a local, and possibly global, minimum of the interface energy as a function of adsorbate rotation. This, then, is the symmetry principle.

Now of course, the unrotated structure already satisfies this symmetry principle, since the moiré vector in Fig. 1a is aligned with a high symmetry direction of substrate. However, by rotating the adsorbate lattice only a small angle, it is possible to align the moiré vector along *another* high symmetry direction of the substrate. This is due to a well-known property of moiré patterns, namely that they magnify the relative rotations of the lattices which form them. This effect is apparent in Fig. 1b: a small rotation of $\vec{q}_A(10)$

produces a large rotation of $\vec{q}_M$. It is a simple exercise in geometry to derive an expression for the rotation angle θ of the adsorbate required to rotate the moiré wavevector through some large high-symmetry angle, ψ_S. Defining the wavevector ratio $r_{AS}=|\vec{q}_S|/|\vec{q}_A|$ (where $\vec{q}_M=\vec{q}_A-\vec{q}_S$),

$$\cos\theta = r_{AS}\sin^2\psi_S + \cos\psi_S\sqrt{1 - r_{AS}^2\sin^2\psi_S} \tag{1}$$

For an equivalent rotation, ψ_A, relative to the adsorbate, θ is obtained by replacing r_{AS} with $r_{SA}=1/r_{AS}$ in the above equation. This means, of course, that the moiré cannot be parallel to high symmetry directions of both adsorbate and substrate at the same time, except in the trivial case of $\theta=0°$.

Evidently, from the symmetry principle alone, we have no idea which of the two solutions might be energetically favourable, nor which high symmetry angles of the substrate or adsorbate will be preferred. However, for lattices with triangular symmetry, the angles $30°$ and $60°$ are obvious candidates for ψ_A or ψ_S. In Fig. 3, the $\psi_A=30°$ and $\psi_S=30°$ solutions are plotted as a function of r_{AS}, along with the prediction of Novaco and McTague.

The similarity of the results, indeed the near identity at small mismatch, is quite surprising. After all, the Novaco-McTague result is developed in the harmonic lattice approximation, in other words with only a very weak relaxation of the adsorbate into a domain-wall structure. It should be noted, though, that the Novaco-McTague result is plotted for the particular case of a Cauchy adsorbate lattice, which means that the ratio of the longitudinal and transverse sound speeds is $\sqrt{3}$. Most simple solids are Cauchy to a close approximation, but any deviation of the ratio from $\sqrt{3}$ will show up as a change of slope of the Novaco-McTague result near the origin. So the exact overlap of the $\psi_A=30°$ and $\psi_S=30°$ solutions with the Novaco McTague result is in some ways fortuitous. The two solutions are sketched in Fig. 2 for comparison with the experimental results.

According to the symmetry principle, if one could rotate the adsorbate relative to the substrate and measure the interface energy continuously, one might obtain something as sketched in Fig. 4, where there is a minimum for the parallel alignment, and other minima for the $\psi_A=30°$ and $\psi_S=30°$ solutions. Here we can only speculate about the existence and relative depths of these minima. Two such speculations are that:

i) even if the global minimum is for the unrotated phase, there may be local minima, corresponding to metastable states, for the ψ_A or ψ_S solutions (as depicted in Fig. 4).

ii) the relative depth of the minima corresponding to rotated and unrotated states may change as a function of the mismatch, leading to the rotated-unrotated phase transition at a critical mismatch, as described by Shiba, and also to transitions between different rotated phases.

The next two sections elaborate on these two speculations, and compare the predictions of the symmetry principle with experiment.

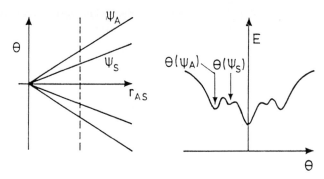

Figure 4. The sketch on the left represents the ψ_A and ψ_S solutions as a function of r_{AS}, the mismatch parameter. The vertical dashed line indicates the position of the cut through the interface energy surface, which is shown on the right. Minima corresponding to the ψ_A and ψ_S solutions are indicated.

4. Stable and Metastable Rotational Phase Transitions

Recent studies by Gibbs et al. [8] of the clean surface reconstruction of Au(100) reveal an interesting case of epitaxial rotation. The situation is that, at temperatures below 1170 K, the uppermost Au layer forms a close-packed phase, modulated by the square substrate ("distorted hexagonal"). The (10) of the overlayer has a length of 1.207 ± 0.01 in units of the substrate (110), and this value shows no temperature dependence. On cooling below 970 K, the structure develops satellites rotated $\pm(0.81° \pm 0.05°)$ to the fundamental peaks of the overlayer (Fig 5).

In order to apply the symmetry principle, one must decide which is the relevant moiré vector to consider. The difference vector between the (10) of the triangular overlayer and the (110) of the square substrate is a possiblity, but the large misfit would produce domain wall separations of only a few lattice constants. Instead, we focus on the fact that 1.207 is very nearly 6/5, in other words the overlayer is on the verge of adopting a commensurate "5" reconstruction along the (110) direction. Thus the moiré vector $\vec{q}_M = \vec{q}_A(10) - (6/5)\vec{q}_S(110)$ is chosen as the physically relevant one, corresponding to a very large domain wall spacing.

Applying Eqn 1 with $\psi_A = 30°$ (the ψ_A and ψ_S solutions degenerate for such small misfit values), one obtains $\theta = 0.20° \pm 0.02°$. For $\psi_A = 60°$, however, the result is $\theta = 0.58° \pm 0.09°$, in fair agreement with the measured value. Furthermore, on occasions where the sample has been quenched from high temperature, extra metastable satellites are seen at $\theta = 0.25° \pm 0.1°$. A possible interpretation is that the $\psi_A = 30°$ solution is metastable for this system, but that the $\psi_A = 60°$ solution is stable. We note that the occurence of *two* epitaxially rotated states, one being metastable, is something quite unexpected in the context of the Novaco-McTague or Shiba models, but follows in a straightforward way from consideration of the symmetry principle. Apparently the energy of the stable rotated phase and the unrotated phase are similar, since they coexist.

Another example of metastable states concerns the epitaxial rotation of Pb clusters on Ge(111) and Si(111), as observed by Grey et al [9]. On these substrates, Pb grows first as a two-dimensional layer, then as three-dimensional clusters (Stranski-Krastanov mode), and the stable cluster orientation is that shown in Fig. 6 (open circle), with Pb(111)//Si(111) and Pb(220)//Si(220). However, room-temperature deposition results in metastable epitaxially rotated clusters growing as well. The proportion of the clusters that are epitaxially rotated can reach over 50%. The clusters can be made to disappear irreversibly by annealing at about 150° C.

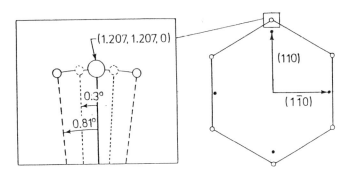

Figure 5. Sketch of reciprocal space of Au(100) surface, based on Ref. 8. The enlarged region shows the epitaxially rotated phase that appears below T=970 K, as well as a metastable rotated state that appears on quenching (dashed circles).

On Si(111) two epitaxial rotation angles are observed, depending on which two-dimensional phase is present. The two-dimensional phase that is produced by room-temperature deposition has the 7×7 periodicity of the clean Si surface [10], and the observed epitaxial rotation for clusters grown on this phase is ±6.1°. If the two-dimensional structure is first annealed, there results a structure slightly incommensurate to $\sqrt{3} \times \sqrt{3}R30°$ (but unrotated), and subsequent deposition of Pb at room-temperature produces ±3.1° rotated clusters.

In Fig. 6, we have plotted these rotated states in reciprocal space (filled circles), along with the 30° and 60° solutions for ψ_A and ψ_S. Here $\vec{q}_M$=Pb(220)−Si(220) has been chosen. Note that in reciprocal space, the solutions of Eqn. 1 as a function of misfit take the form of curves diverging from the commensurate position. In particular, ψ_S solutions appear as straight lines. The measurements suggest that on the 7×7 surface, the ψ_S=60°

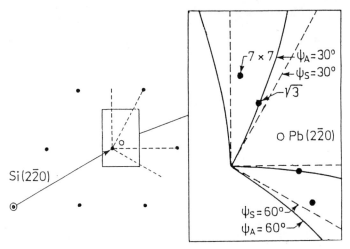

Figure 6. Observed positions in reciprocal space of the (220) reflection of stable (open circle) and metastable (filled circles) Pb clusters on Si(111), based on Ref. 9. The rotation of the clusters depends on the two-dimensional surface structure onto which the clusters are deposited, as discussed in the text.

solution is preferred, while on the $\sqrt{3} \times \sqrt{3}R30°$ surface, the $\psi_A=30°$ or $\psi_S=30°$ rotation is adopted. A possible explanation of this difference is that for the 7×7 surface, the misfit dislocations described by $\vec{q}_M$ tend to line up with the stacking fault boundaries of the 7×7 structure. This is already the case for the unrotated phase, and becomes so again for a 60° degree rotation of the moiré pattern relative to the substrate. The $\psi_A=30°$ solution for the $\sqrt{3} \times \sqrt{3}R30°$ interface is similar to the behaviour of the rare gases on graphite.

5. Rotational Phase Transitions

The similarity of the $\psi_A=30°$ and $\psi_S=30°$ solutions to the Novaco-McTague result is an interesting curiosity. But how can we understand the more complex (and more realistic) behaviour described by Shiba in terms of the symmetry principle? We shall attempt to give a qualitative explanation here.

The dependence of interface energy on rotation angle shown in Fig. 4 can be thought of as representing a slice through an energy surface, the other parameter being misfit. In Fig 7, two such slices are sketched; one at large misfit where the epitaxially rotated phase is a global minimum, and one at small misfit where the global minimum is for the unrotated phase. The bold line in the left part of the figure indicates the equilibrium rotation as a function of r_{AS}. At a certain critical mismatch, the global minimum switches from rotated to unrotated, and there is an abrupt transition to the parallel orientation. This is a simple interpretation of the critical mismatch and rotated-unrotated transition predicted by Shiba.

91

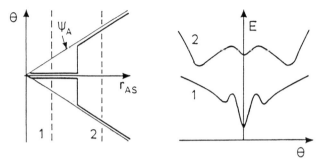

Figure 7. The bold line in the sketch on the left represents an abrupt unrotated to rotated phase transition as a function of the mismatch parameter r_{AS}. The sketch on the right represents two cuts through the interface energy surface, below (1) and above (2) the critical mismatch. After the transition, the epitaxially rotated state is a global minimum of the interface energy.

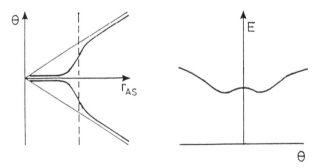

Figure 8. The sketch on the left represents a continuous unrotated to rotated phase transition. The sketch on the right represents a cut through the interface energy surface near the critical mismatch, where the two epitaxially rotated states have nearly merged.

A continuous transition can be envisaged if the minima of the rotated phase are broad, so that at small enough misfit, $+\theta$ and $-\theta$ rotations become unresolved, the net minimum lying at $\theta = 0°$. This is illustrated schematically in Fig. 8. Shiba has derived the minimum energy solution as a function of mismatch, yet it could be interesting to investigate theoretically the complete energy surface, to see whether the qualitative ideas portrayed in Figs. 7 and 8 have any quantitative basis.

Another type of rotational phase transition can be envisaged: if the mismatch can be varied over a large enough range, the physically relevant moiré vector may change. In other words, at a certain mismatch, a new substrate wavevector may more closely match the adsorbate wavevector, so that the adsorbate will tend towards a new commensurate structure. For example, in the case of Xe on graphite, the Xe interatomic distance is too large to adopt the $\sqrt{3} \times \sqrt{3}R30°$ structure, but too small to adopt the less dense 2×2 commensurate structure.

McTague and Novaco noted on the basis of their calculations for Xe on graphite that there appeared to be a second local minimum of the interface energy as a function of adsorbate rotation (see Fig. 9). This occured at a rotation angle of 27°, in other words, 3° rotated relative to the 2×2 commensuration. They predicted a first-order transition to this other epitaxially rotated state if Xe could be expanded sufficiently. The experimental evidence suggests a more complex behaviour (Fig.2). On increasing the misfit to the $\sqrt{3} \times \sqrt{3}R30°$ phase, the Xe actually rotates back towards the parallel orientation. It has been proposed that entropy considerations may have to be introduced to explain this behaviour [4]. The Novaco-McTague and Shiba models do not include thermal effects.

For Ne, a broad, shallow minimum in the interface energy was observed by Novaco and McTague at a rotation angle of about 15°. The Ne interatomic distance is too small to adopt the $\sqrt{3} \times \sqrt{3}R30°$ structure, and a possible interpretation of this minimum is that Ne is undergoing a rotational phase transition to a 1×1 commensuration. The large rotation angle is confirmed by experiment [11]. However, it is difficult to study such rotational phase transitions for the rare gases on graphite, because the range of misfit that can be probed is rather limited.

A system where the adsorbate lattice parameter can be varied over a much wider range is Na/Ru(100) [12]. Here, the adsorbate has been observed to rotate according to the $\psi_A = 30°$ solution over a very wide range of mismatch. (Interestingly, at large mismatch where the $\psi_A = 30°$ solution and the Novaco-McTague result become distinct, the low energy electron diffraction measurement agrees more closely with the former.) The results for Na/Ru(100) are plotted in Fig. 9, which shows rotation angle as a function of mismatch. The upper half of the figure shows the 30° and 60° solutions for ψ_A and ψ_S relative to the 1×1 and 2×2 phases. The lower half shows these solutions relative to the $\sqrt{3} \times \sqrt{3}R30°$ phase. Two solutions exist which connect the $\sqrt{3} \times \sqrt{3}R30°$ phase continuously with the 1×1 phase, and so a *continuous* rotational transition between these two commensurate phases is possible, at least in principle. The experimental data, however, suggests an abrupt transition may occur to the $\psi_A = 30°$ or $\psi_S = 30°$ solution.

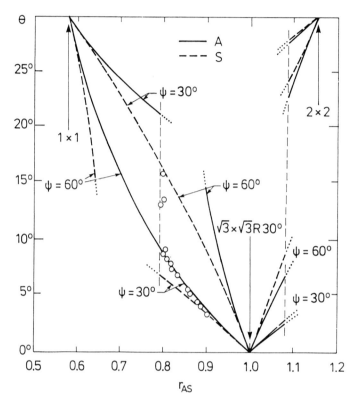

Figure 9. The $\psi_A = 30°, 60°$ and $\psi_S = 30°, 60°$ solutions, as a function of the mismatch parameter r_{AS}, relative to the $\sqrt{3} \times \sqrt{3}$ commensurate phase (bottom half of figure) and to the 1×1 and 2×2 phases (upper half). Note the two solutions connecting the $\sqrt{3}$ and 1×1 phases continuously. For clarity, the other solutions are truncated at the position of equal mismatch to two commensurate phases (vertical dashed lines). Results for Na/Ru(100) are plotted, adapted from Ref. 12.

The veritcal dashed lines in Fig. 9 indicate the position where the structure is equally poorly matched to the two commensurate phases. It should be noted that at these positions, the moiré wavevectors are quite large, corresponding to fringe separations of only a few crystal lattice unit cells. As mentioned in section 2, this lessens the physical significance of the moiré vector, and therefore of the symmetry principle. Nevertheless, it would be most interesting to obtain experimental data for a system such as Na/Ru(100) that spanned the full range of mismatch between two commensurate structures, to see how a rotational phase transition occurs in practice.

6. Conclusions

A virtue of the symmetry principle for epitaxial rotation presented in this article is its ease of use. The experimentalist who stumbles on an instance of epitaxial rotation, or of a rotational phase transiton, in a system far removed from the well-studied case of the rare gases on graphite, may apply the symmetry principle to make quantitative estimates and see whether his observations fit in a simple picture of epitaxial rotation.

We emphasize that the symmetry principle is not simply a geometrical recipe. It is based on a physically very intuitive assumption: a high symmetry state of a system will in general represent a local extremum of the system's energy, in the parameter space that defines the system. In the case of the interface between two crystals, the parameter space is spanned by mismatch and rotation angle. The curves in Fig. 9 represent locii of such high-symmetry states in the parameter space. The prescription given for obtaining these curves, Eqn. 1, does not *explicitly* contain any parameters related to the interface energetics (hence the ease of use). But aspects of the interface energetics are *implicitly* contained in the choice of small q_M, of low-order q_A and q_S, and of high-symmetry values for ψ_A or ψ_S. Without these conditions, of course, the symmetry principle would generate an infinity of solutions, and would be nothing more than geometry.

After completion of this manuscript, it was pointed out to us that several other systems follow very closely solutions obtained from the symmetry principle, notably Li/Ru(100) [13], which follows the $\psi_S = 60°$ solution, and the γ phase of D_2/graphite [14], which follows the $\psi_A = 60°$ solution. Further, a compilation of data for several alkali metals intercalated in graphite [15] follows the $\psi_S = 60°$, between the simple $\sqrt{7} \times \sqrt{7}R19.1°$ and 2×2 structures. Note that these $\psi = 60°$ solutions are very different from the Novaco-McTague result.

Of course, some systems may not follow the predictions of the symmetry principle at all. Xe on graphite appears to be one example. Other factors, such as entropy, the effect of surface steps or the finite size of real systems may have to be taken into account to obtain even a qualtitatively correct picture. It is our hope, however, that the symmetry principle will prove a useful tool in many cases, and that it gives new physical insight into problems involving rotational epitaxy.

We thank Mourits Nielsen and Jens Als-Nielsen for many stimulating discussions which helped to refine the ideas presented here. We thank Doon Gibbs, Ben Ocko, David Zehner and Simon Mochrie for making their fascinating results available to us prior to publication. We thank Renée Diehl, Samuel Fain and Simon Moss for bringing further interesting examples of epitaxial rotation to our attention.

References

[1] Mailing address: Physics Department, Risø National Laboratory, 4000 Roskilde, Denmark

[2] F. Grey and J. Bohr, *A symmetry principle for epitaxial rotation*, to be published.

[3] W. Bollmann in *Crystal Defects and Crystalline Interfaces*, Springer Verlag, (1970)

[4] K.L. D'Amico, J. Bohr, D.E. Moncton and D. Gibbs, Phys. Rev. **B41**, 4368(1990)

[5] C. G. Shaw, S. C. Fain, Jr., and M. D. Chinn, Phys. Rev. Lett. **41**, 955, (1978).

[6] A. D. Novaco and J.P. McTague, Phys. Rev. Lett. **38**, 1286, (1977); J.P McTague and A. D. Novaco, Phys. Rev. **B19**, 5299 (1979).

[7] H. Shiba, Journ. Phys. Soc. Japan **46**, 1852 (1979); H. Shiba, Journ. Phys. Soc. Japan **48**, 211 (1980).

[8] D. Gibbs, B. M. Ocko, D. M. Zehner and S.G.J. Mochrie, *Structure and phases of the Au(001) surface I: In-plane structure*, to be published.

[9] F. Grey, M. Nielsen, R. Feidenhans'l, J. Bohr, J. Skov Pedersen, J.B. Bilde-Sørensen, R. L. Johnson, H. Weitering and T. Hibma, *Epitaxial rotation of metastable clusters of Pb on Ge(111) and Si(111)* to be published.

[10] F. Grey, R. Feidenhans'l, M. Nielsen and R. L. Johnson, Journ. de Phys. (Colloque) **C7** 181 (1989)

[11] S. Calisti, J. Suzanne and J. A. Venables, Surf. Sci. **115** 455 (1982)

[12] D. L. Doering and S. Semancik, Phys. Rev. Lett. **53**, 66 (1984).

[13] D. L. Doering and S. Semancik, Surf. Sci. **175**, L730 (1986).

[14] J. Cui and S. C. Fain, Phys. Rev. **B39**, 8628 (1989).

[15] M. Mori, S. C. Moss, Y. M. Yan and H. Zabel, Phys. Rev. **B25** 1287 (1982)

STRUCTURES AND PHASE TRANSITIONS IN ALKALI METAL OVERLAYERS ON TRANSITION METAL SURFACES

Renee D. Diehl

Department of Physics
104 Davey Laboratory
The Pennsylvania State University
University Park, PA 16802

INTRODUCTION

Alkali metal adsorption has been studied for many years, primarily because of the great number of technological applications which arise from adding small amounts of alkali metal atoms to other materials. These include the promotion of catalytic reactions, increases in electron and ion emission rates of metals, and the enhancement of the oxidation of semiconductors. In recent years, alkali metal films have also been studied with a more fundamental view toward understanding the unusual interactions between adsorbed alkali metal atoms, the wide variety of surface structures which occur, and the phase transitions associated with them. Alkali metal adsorption has been the subject of several recent reviews[1,2,3] and one book[4] and the purpose of this chapter is not to provide a comprehensive review of this subject, but to discuss some alkali metal overlayer structures which display interesting behavior from the point of view of phase transitions.

CHEMICAL BONDING AND INTERACTIONS

A simple model for alkali metal adsorption was proposed many years ago in which the alkali metal adsorbate valence level forms a broad resonance level which shifts upward in energy as the alkali metal atom approaches the surface[5], as shown in Figure 1. Since the ionization potentials of alkali metals are comparable to the work functions of most transition metals (see Table 1) and the states in the resonance level of the adatom are only occupied up to the Fermi energy of the substrate metal, there will be, within this model, a partial transfer of the alkali adatom's s-electron to the metal surface. A result of this charge transfer is the formation of a static dipole at the surface, arising from the adsorbed ion and its image charge in the metal. Other more recent theoretical[6,7] and experimental[8] results suggest that there may be no actual charge transfer to the substrate, but there is still a substantial polarization of the adatom electrons toward the substrate. This polarization also produces a static dipole at the position of the adsorbed atom, and so the net effect on adsorbate structures is very similar to that of the charge-transfer model.

The interactions between alkali adatoms on metal surfaces can be divided into direct interactions, which occur through the vacuum halfspace and indirect interactions which occur through the substrate. Because of the polarization of adatoms, a large component of the direct interaction for low-density overlayers is a dipole-dipole repulsion, and it varies as $p^2 r^{-3}$ where p is the dipole moment at the position of an adatom and r is the separation between adatoms. This interaction energy would increase as r^{-3} as the density of adatoms is increased, except that the dipole moment per adatom decreases due to the depolarization which occurs as the surface dipoles get closer together.

Table 1. Ionization Potentials and Work Functions of Selected Metal Surfaces

Work functions for transition metals (eV)[9]		Ionization potentials for alkali metals (eV)[10]	
Cu (111)	4.98	Li	5.39
Cu (100)	4.59	Na	5.14
Ni (111)	5.35	K	4.34
Ag (111)	4.74	Rb	4.18
W (110)	5.25	Cs	3.89

The indirect interactions in these systems arises from the interaction of adatom electrons via the substrate or via the screening charge in the metal. These interactions are oscillatory (attractive and repulsive) and in general are anisotropic on the surface[1,11]. Calculations of this interaction energy[11,12] are complicated, but in general it is expected to decrease in magnitude slowly as a function of distance and so may dominate the dipole-dipole interaction at large distances. The equilibrium structures formed by alkali metal overlayers depend on the relative strengths of the direct and indirect interactions and the substrate potential corrugation experienced by the adatoms.

The variation in alkali-metal-atom binding energy at different adsorption sites on the substrate surface depends on both the adsorbed alkali metal atom and the substrate surface. The modulation of adsorption energies has not been measured directly for any of the systems discussed here, but in detailed structural studies, alkali metal atoms usually have been found to adsorb preferentially in the high-coordination sites on transition metal surfaces[13,14,15,16,17,18,19].

The density at which the adatoms become a metallic overlayer has been a much-discussed topic in the literature[2] and probably depends on the details of the adsorption system. But in most cases of alkali adsorption below room temperature, the overlayer atoms seem to remain essentially repulsive, due to either dipole repulsion or to Fermi/Coulomb repulsion, as evidenced by the large unit cells and lack of island growth, at submonolayer coverages. Some irreversible "condensation" of alkali adatoms into islands, which has been observed at

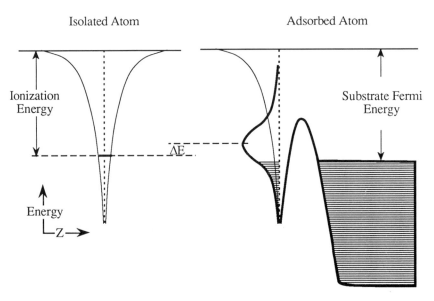

Figure 1. Schematic diagram showing the change in the alkali metal valence level upon adsorption, according to the Gurney model[5]. ΔE denotes the upward shift of the alkali atom valence electron energy level when it approaches the surface.

Table 2.   Some transition metal surface unit cell sizes and alkali metal sizes.

transition metal	surface unit cell length, a (Å)	alkali metal	ionic diameter(Å)	3D solid NN distance (Å)
Ru(001)	2.70	Li	1.46	3.02
Cu(111)	2.56	Na	1.94	3.66
Ni(111)	2.49	K	2.66	4.53
Pt(111)	2.77	Rb	2.94	4.84
Ni(100)	3.52	Cs	3.34	5.24
Cu(100)	3.62			
Rh(100)	3.80			
W(100)	3.16			

higher temperatures for some systems, may be indicative of net attractive interactions under some adsorption conditions.

The fact that adsorbed alkali metal atoms are mutually repulsive at low temperatures and submonolayer coverages for most alkali metal adsorption systems has led to a wide variety of low-density structures which generally are not found in other adsorption systems. On some substrates these low-density phases are solid and on others they appear to be fluid. Density-temperature phase diagrams have been proposed for many alkali metal films on transition metal surfaces, including Na/Ru(001)[20], Cs/Rh(100)[21], K/Pt(111)[22], K and Cs/Ni(111)[23,24], K/Ni(100)[25], Cs/Cu(111)[26], K and Cs/ Cu(110)[27], Cs/W(100)[28], and Na/W(110)[29,30]. In this review, we will discuss a selection of these phase diagrams and draw conclusions about the alkali metal-surface interactions from their similarities and differences. We have grouped these systems into sections having similar substrate symmetry, since this is an important factor in the structures of the overlayer phases.

SUBSTRATES HAVING HEXAGONAL SYMMETRY

Of the substrates mentioned above, Ru(001), Pt(111), Ni(111), and Cu(111) have hexagonal symmetry. The phase diagrams for some of these systems are shown in Figure 2. In general, it can be observed from the phase diagrams that the disordering temperatures for the submonolayer phases increase as the coverage is increased, and they are higher for commensurate phases than incommensurate phases. The coverage at which the second layer begins to grow depends on the relative sizes of the substrate unit cell and the adsorbate atom, since coverage is defined as the ratio of the number of overlayer atoms to the number of substrate atoms in the surface layer. Some substrate unit cell sizes and alkali metal atoms sizes are given in Table 2. This consideration, of course, affects which commensurate structures it is possible to form for any given combination of adsorbate and substrate. At monolayer completion in most alkali metal adsorption systems, the alkali metal atoms are more closely packed than in the closest-packed bulk alkali metal plane. This has often been attributed to the polarization of the alkali metal electrons toward the substrate, but calculations have shown that this compressed arrangement can be a result of the bonding in 2D monolayer itself [31].

Low-Coverage Phases

At low coverages and low temperatures, three of the alkali metal films of Figure 2 form structures which produce isotropic rings in the low-energy electron diffraction (LEED) patterns. (The ring phase is not shown on the Cs/Cu(111) phase diagram; it forms at coverages below 0.07 monolayers[26].) These ring patterns arise from an overlayer which has a reasonably well-defined nearest-neighbor distance, but no long-range rotational order. A LEED photograph of the ring pattern for K/Ni(111) is shown in Figure 3a. As the coverage is increased in this phase, the diameter of the diffraction rings increases, corresponding to a decrease of the mean nearest-neighbor distance. It has been shown explicitly for some systems that this is consistent with a uniform compression of the overlayer as the coverage is increased. The phase diagrams in Figure 2 are all consistent with this phase being fluid, although this cannot be proven directly from the diffraction patterns; the correlation length deduced from the width of the rings in Figure 3 is about 25 Å. Because most studies of alkali metal adsorption systems have not been carried out below 70K, it is not known if lowering the temperature further would cause this phase to solidify into a 2D crystalline solid phase having long-range order, such as the long-period commensurate phase observed for K/Pt(111) (see phase diagram), or if the overlayer simply would freeze to form an amorphous solid phase.

At somewhat higher coverages (but still below 0.25 monolayers), all of these systems produce LEED patterns which can be described as modulated rings(see Figure 3). The momentum transfer of these rings corresponds to a lattice parameter which is incommensurate with the substrate, and the diameter of the rings increases as the coverage is increased due to the compression of the overlayer. It has been shown for some of these systems that over at least part of the phase diagram this modulated ring arises from a fluid having bond-orientational order[23,26] (modulated fluid), and it seems likely that this is the case for all of these systems, at least above 100K. (The modulated ring phase of Na/Ru(001) at high coverages is different from these phases and will be discussed in a later section.) In all of these phases which have been observed so far, the rotational epitaxy of the fluid is the same as

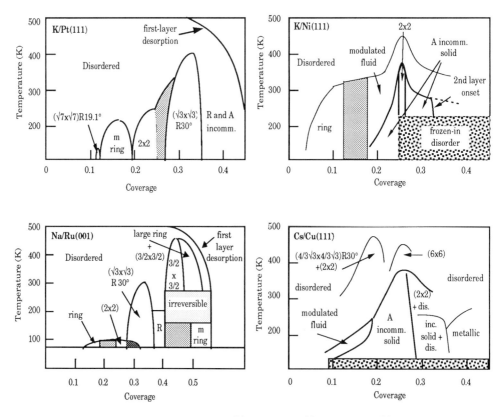

Figure 2.   Phase diagrams for Na/Ru(001)[20], K/Ni(111)[23], Cs/Cu(111)[26], and K/Pt(111)[22]. Coverage is defined as the ratio of the number of first layer atoms to the number of substrate atoms in the surface layer. Shaded areas are regions of phase coexistence or "transition" regions between phases, "m ring" denotes a phase in which the LEED pattern is a modulated ring, R denotes a rotated incommensurate phase and A denotes an aligned incommensurate phase. "Frozen-in disorder" denotes regions of the phase diagrams where the diffusion kinetics were too slow to form equilibrium phases in these experiments.

that of the substrate, i.e. the fluid phase has the rotational symmetry of the substrate, but not its translational symmetry. When the temperature of this phase is reduced or the coverage increased in the K/Ni(111), K/Cu(111) and Cs/Cu(111) systems, the layer has been observed to solidify into an incommensurate solid phase. This phase transition has been studied in some detail and will be discussed in the following section[26,32,33].

Incommensurate Melting

The melting of the low-density incommensurate solid has been studied in detail for K[32] and Cs[26]/Cu(111) and K/Ni(111)[33]. For all of these systems, an apparently continuous transition has been observed between the incommensurate solid and a modulated fluid phase having the same mean density as the solid phase. The transition temperature depends very strongly on the density of the overlayer, which varies by as much as a factor of two in this phase as the coverage is increased. For adsorption on Cu(111), it has been observed from the LEED patterns that the modulated fluid further melts into an isotropic fluid at even higher temperatures.

Detailed LEED spot-profile measurements for K/Ni(111) have shown that the radial and azimuthal diffraction spot widths increase continuously as the temperature is increased in the modulated fluid phase, and that the aspect ratio of the diffraction spots remains constant[33] (see Figures 4 and 5). The melting of incommensurate overlayers has been of general interest in recent years because of the theoretical studies of 2D melting transitions[34,35,36]. In particular, a fluid phase having bond-orientational order has been predicted to occur for 2D melting[34]. It cannot be proven from these experimental results that this alkali-metal modulated fluid is that which has been predicted, but it is interesting to note that similar behavior has been observed for rare gas monolayers[37,38,39,40]. In all of these systems, the

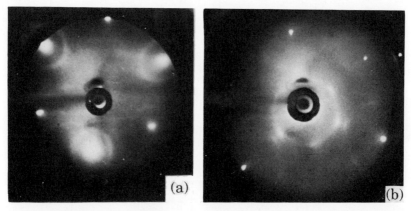

Figure 3.   (a.) LEED pattern from K/Ni(111) at a coverage of 0.11 monolayers and at 120K [23]. The photograph was taken when the primary electron beam energy was 192 eV. The shadow to the left of the center of the pattern is due to the electron drift tube. The specular beam would be just above the center of the pattern if it were visible. The six spots are first-order diffraction spots from the Ni(111) substrate. (The three-fold intensity symmetry arises from the 3-fold symmetry of the Ni(111) surface.) The rings are due to the potassium overlayer. The first-order potassium ring is around the specular beam (not visible at this energy) and the visible rings are from the electrons diffracting once from the overlayer and once from the substrate. The substrate's threefold symmetry is reflected also in the intensity of these rings.
(b.) LEED pattern from K/Ni(111) at a coverage of 0.19 monolayers, T=187K and the primary beam energy is 225 eV[33].The spots are again from the substrate, and the modulated ring around the center is the first-order diffraction from the overlayer.

melting transition is apparently continuous as observed with diffraction (although heat capacity studies of Xe/graphite have shown that transition to be first-order[41]), and the aspect ratio of the diffraction spots is constant in the fluid phase. This last result seems to indicate that these modulated fluid phases arise from a general property of the adsorbed overlayer and are not dependent on the details of the interactions between the adsorbate atoms.

High-Coverage Phases

At coverages approaching 0.25 monolayers and above, it is apparent from the phase

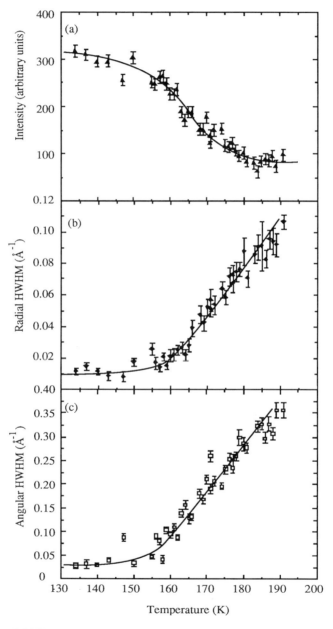

Figure 4.   Fitted LEED spot intensities and radial and angular halfwidths as a function of temperature for K/Ni(111)[33] at a coverage of 0.19 monolayers. The error bars represent the precision of the Lorentzian fit parameters and the solid curves are a guide to the eye.

diagrams that the substrate periodicity has a bigger effect in the Pt and Ru adsorption systems than in the Ni and Cu adsorption systems. The commensurate phases on Pt (111) and Ru(001) exist over larger coverage ranges and there are coexistence regions of commensurate phases, for instance a coexistence of the p(2x2) phase with the ($\sqrt{3}$x$\sqrt{3}$)R30° phase, which indicate that the substrate potential overcomes the mutual repulsion of the alkali metal atoms. This is probably at least partly due to the alkali adsorbate size being smaller relative to the substrate periodicity for Pt and Ru (compared to Ni and Cu). But the strength of the adsorbate-adsorbate interaction relative to the substrate potential modulation will also be related to the dipole moment of the adsorbate atoms, which in turn depends on the relative electronic properties of the overlayer and the substrate.

A particularly interesting aspect of the high-coverage phases for the alkali metal adsorption systems in which the first layer does not saturate until well above a coverage of 1/3 is the observation of aligned and non-aligned rotational epitaxy in the incommensurate layers. Rotated hexagonal incommensurate phases at densities above the ($\sqrt{3}$x$\sqrt{3}$)R30° are observed for both Na[42] and Li[43]/Ru(001), and for Na[44] and K[22]/Pt(111). A schematic drawing of such a rotated phase is shown in Figure 6. Rotated alkali overlayer phases are also observed on substrates having nonhexagonal symmetry, and these will be discussed in the later sections of this chapter.

Non-symmetry rotational epitaxy of incommensurate overlayers has been observed for many years[45], but it has been of particular interest since 1977 when rotated incommensurate overlayers were predicted to occur in physisorption systems by Novaco and McTague (NM)[46] and soon later were observed experimentally[47]. This theory predicted that a rigid incommensurate overlayer on a hexagonal substrate would rotate in order to relieve the stress imposed by the substrate periodicity. The result of the NM theory was extended by Shiba[48] to include density modulations and domain walls, which are very important in adsorption systems in which the substrate potential modulation is not negligible. The alkali metal systems in which these phases have been observed are essentially in agreement with the predictions of NM for rigid overlayers(see Figure 7); they have not been studied yet at very small misfits where Shiba's theory is particularly relevant. In some ways, the alkali metal overlayers provide a better test of the theories than physisorbed overlayers because of the large

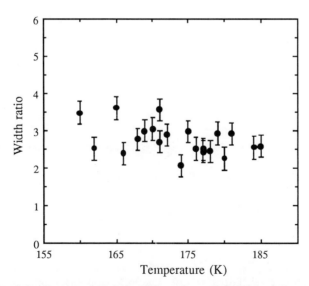

Figure 5.   Ratio of the excess azimuthal width to the radial width of the overlayer first-order diffraction spots as a function of temperature for K/N(111) at a coverage of 0.19 monolayers[33].

density range over which the overlayers can be studied. It is interesting to note that so far there have been no phases observed having non-symmetry rotations at low coverages, e.g. below 0.25 monolayers. This may be because the overlayer is not rigid enough at very low densities to sustain a collective rotation of the overlayer.

An alternative explanation of the rotational epitaxy which has been proposed by Doering[43,49] and by Fusilier et al.[50] is that they arise from a succession of higher-order commensurate phases. Doering has shown that the rotational angle vs. misfit curves for Na and Li/Ru(001) lie along major trajectories of higher-order commensurate phases[43,49]. The

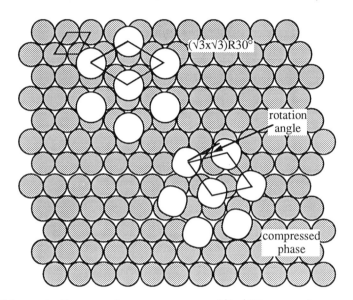

Figure 6. Schematic diagram of a commensurate $(\sqrt{3}x\sqrt{3})R30°$ overlayer structure on a hexagonal substrate and a compressed, rotated structure on the same substrate. The rotation angle shown is the difference in rotational orientation of the two structures.

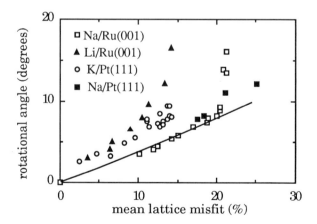

Figure 7. Rotational angles of Li[43] and Na[53]/Ru(001) and Na[44] and K[22]/Pt(111) as a function of mean lattice misfit. The rotational angle and the lattice misfit are calculated relative to a perfect $(\sqrt{3}x\sqrt{3})R30°$ overlayer, where misfit = $(d(\text{overlayer})-d(\sqrt{3}))/d(\sqrt{3})$. The curve is the NM prediction for the speed of sound ratio $C_L/C_T = \sqrt{3}$.

distinction between higher-order commensurate and incommensurate phases is very difficult to make experimentally and cannot be made with the present data.

It will be noted from the phase diagram for K/Pt(111) that the rotated phase coexists with a phase having the same lattice parameter as the rotated phase, but rotationally aligned along the substrate symmetry direction (i.e. 30° from the (√3x√3)R30° phase). Such a coexistence has also been observed for Na/Pt(111)[44]. It is not known if this is due to a degeneracy of energies of the two states or if surface defects such as steps cause the alignment of some regions of the overlayer. The effects of surface steps and defects adsorbed at steps have been studied for rare gases adsorbed on the same surface, Pt(111), and these studies show that the rotational alignment of those overlayers are very sensitive to the nature of surface defects[51].

The modulated ring phase observed near monolayer completion for Na/Ru(001) is not a fluid phase, as the low-coverage modulated-ring phases are, but a solid phase which has domains of at least two orientations on the surface. The resolution of the LEED results is not good enough to determine if this phase is hexagonal or if it has slight distortions from a hexagonal unit cell.   A similar phase having two orientations is present in the Cs/Cu(111) phase diagram and is labeled as a metallic phase.

SUBSTRATES HAVING SQUARE SYMMETRY

Alkali metal overlayers on substrates having square symmetry often have even more ordered phases than those on hexagonal substrates because the square substrate can induce the formation of overlayers having nonhexagonal symmetries.   There have been many studies of alkali metals adsorbed on square-symmetry substrates but few phase diagrams proposed so far. The phase diagram for Cs/Rh(100) is shown in Figure 8. Generally speaking, the lighter transition metals (Ni[52,53,54], Cu[55,56]) appear to provide a smoother surface to the adsorbed alkali metal atoms than the heavier metals (Rh[21], Ir[57], W[28,58]). This was also true for the hexagonal substrates. At very low coverages on Cu(100) and Ni(100), rings are observed in the LEED patterns which indicate that the overlayer is similar to the low-coverage phases observed on hexagonal substrates, while on Rh(100), Ir(100) and W(100) long-period commensurate structures are observed in the same coverage range.

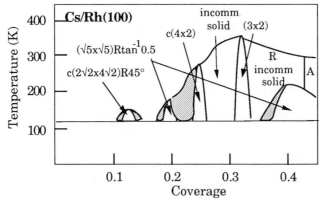

Figure 8.   Phase diagram for Cs/Rh(100)[21].   There are five commensurate phases, as labeled, and three incommensurate phases. The shaded areas are regions of coexistence of two commensurate phases or of a commensurate phase and a disordered phase.   The phase which is labeled incommensurate solid in the coverage range from 0.25 to 0.33 could be interpreted as a coexistence of commensurate phases(see text). The higher coverage incommensurate phases are rotated (R) and aligned (A) with respect to the substrate periodicity.

Commensurate Phases

On the apparently smoother substrates of Ni(100) and Cu(100), commensurate structures are observed only above a coverage of about 0.25, and their symmetry is quasihexagonal. On Rh(100) and Ir(100), however, a succession of commensurate phases are observed over a wide coverage range. The commensurate phases which are observed for Cs/Rh(001)[21] and K/Ir(100)[57] are essentially identical up to a coverage of 1/3; these are labeled on the phase diagram in Figure 8. At low coverages, the observed commensurate structures are consistent with adsorption in the high-coordination 4-fold hollow sites, but the higher coverage phases require some adsorption in other sites, either bridge or on-top sites.

The lowest-coverage commensurate phase observed in both of these systems is the c(2√2x4√2)R45° structure at a coverage of 0.125 monolayers, for which the nearest-neighbor distance is about 12 Å on both surfaces, which is quite large compared to saturation nearest neighbor distances of 4.39Å for Cs/Rh(100) and 3.85Å for K/Ir(100). Although the commensurate structures which are observed have various symmetries, most of them in fact are structures which has a 6-fold or nearly 6-fold coordination in the overlayer. The only exception to this is the (√5x√5)Rtan^{-1}0.5 phase at 0.2 monolayers, which has 4-fold coordination. The preponderance of 6-fold coordination in the commensurate structures is indicative of the strong adsorbate-adsorbate interactions relative to the substrate corrugation potential.

The transitions which have been observed between the commensurate phases have been either clear-cut first-order transitions with a clear coexistence of phases, or transitions in which diffraction spots move continuously and can be interpreted as either mixed phases (coexistence) with very small domain sizes, or a continuously varying lattice parameter between the phases. The "continuous" transition in question is between the c(4x2) and the (3x2) phases, for which proposed structures are shown in Figure 9. It can be seen that a transition between these phases only involves a uniaxial compression of the overlayer. The question of whether the intermediate structure is a mixture of these two structures or a uniaxially incommensurate structure (with or without domain walls) depends on the relative

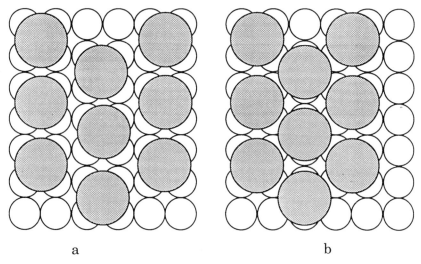

a   b

Figure 9.   Proposed structures for the (a) c(4x2) phase at 0.25 and (b) (3x2) phase at 0.33 coverage for K/Ir(100)[57] and Cs/Rh(100)[21]. The relative sizes of the schematic overlayer atoms to the substrate atoms are not drawn to scale.

interaction energies of the substrate and the overlayer atoms. This question could be resolved by measuring diffracted intensities in the "mixed-phase" region and comparing to model calculations, or by using a kinematic diffraction technique for which the ambiguities which arise from double diffraction in LEED are not present.

Incommensurate and Fluid Phases

Rotated Incommensurate Phases. Rotated incommensurate phases have been observed for Cs/Rh(100)[21] and for K/Cu(100)[59]. In both cases, the rotated hexagonal incommensurate phase is observed at coverages higher than the commensurate (3x2) phase, which has two overlayer atoms per unit cell. In these phases, the overlayer rotation angle relative to the commensurate alignment varies between about 2 and 6 degrees, which is a much smaller range than was observed for overlayers on hexagonal substrates. The origin of this rotation may be the same as the origin of the rotated phases on hexagonal substrates, and in this case also it is impossible from the experimental data to distinguish between a continuous rotation and a succession of closely-spaced higher-order commensurate phases. One interesting finding in the K/Cu(100) study was that when the copper substrate was prepared with a relatively high defect density, the rotated phase was never observed. Instead, an aligned incommensurate phase occurred over the same density range.

Uniaxial Incommensurate Phases. In some alkali metal overlayers on surfaces which have smooth atomic relief, the hexagonal overlayer distorts uniaxially. In all of the observed cases, one lattice vector of the overlayer remains "commensurate" with the substrate over a small coverage range while the overlayer compresses in the perpendicular direction. Uniaxially incommensurate phases have been observed for both K[56] and Cs[55] on Cu(100) and possibly for Cs/Rh(100)[21] and K/Ir(100)[57]. In all of these cases, the lattice parameter of the quasi-hexagonal overlayer is fixed along a substrate [110] symmetry axis, while in the perpendicular direction the average lattice parameter is either expanded or compressed relative to the substrate spacing, as shown in Figure 10.

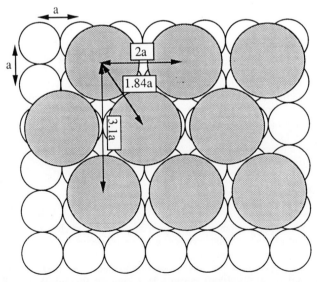

Figure 10. Schematic diagram of a quasihexagonal overlayer on Cu(100). The layer has the commensurate spacing of 2a along the epitaxial (horizontal) direction but is compressed along the perpendicular (vertical) direction. Note that if the layer compresses slightly more along the same direction, it will form a commensurate c(3x2) structure with two atoms per rectangular unit cell.

In the Cs/Cu(100) system, a detailed measurement of the lattice parameters as a function of coverage has not been made, but at 150 K the overlayer forms a hexagonal structure which apparently compresses continuously and uniformly until the lattice parameter along the epitaxial direction (i.e. the direction in which a unit cell vector of the overlayer is parallel to a unit cell vector of the substrate) is twice that of the substrate, and then it continues to contract along the direction which is perpendicular to that. The curious aspect of this result is that experiments at room temperature on the same system show that the monolayer structure has a uniaxially *expanded* structure. Further measurements on this system would be very interesting because it is not obvious how the two results connect on the density-temperature phase diagram.

For the K/Cu(100) system, measurements were made of the LEED spot profiles of the quasi-hexagonal phase as it compressed from one commensurate phase having 3 atoms per unit cell to another commensurate phase having 2 atoms per unit cell. The spacing along the epitaxial direction is the same in both of these phases. It is observed in the LEED spot profile that the overlayer spot remains in a fixed position in the direction corresponding to the epitaxial direction while shifting continuously along the perpendicular direction. The spot width does not change. This is interpreted as a continuous uniaxial compression because the spot does not broaden as might be expected if the intermediate phase were a mixture of the two commensurate phases. However, it should be noted that the spot is always slightly elongated in the perpendicular direction and that its width of 0.06 $Å^{-1}$ corresponds to a real space distance of approximately 100 Å which raises the possibility that the broadening which might occur if there were a mixture of microscopic domains might not be directly observable for domains larger than 100 Å. However, structure factor calculations for such mixtures would be required to consider this possibility in detail.

Another point which must be raised is whether these commensurate phases are "accidentally" commensurate (i.e. a floating solid) with the substrate during the overlayer compression, because one might expect the diffraction spot profiles to change if the commensurate structures are truly locked-in. For instance, LEED spots from locked-in commensurate structures are generally circular rather than elongated. High-resolution spot profiles would go far to help to explain the details of this system, and may also be helpful in identifying the existence of uniaxial domain walls. A detailed study of the melting of these phases would also be very interesting from the point of view of understanding the role of the substrate periodicity in the behavior of these overlayers.

2D Liquid Phases. One phase of K/Cu(100)[60] has been interpreted as being composed of 2D liquid islands on the surface. The experimental evidence which led this interpretation was from a LEED study which showed that at a temperature of 330 K and at coverages between 0.18 and 0.26 monolayers, ring patterns were observed which did *not* increase in diameter as the coverage was increased. This would indicate that the mean density of the observed phase is not changing as a function of coverage. In this interpretation, the 2D liquid phase is isotropic and has a nearest-neighbor distance of 5.6Å compared to a bulk solid nearest-neighbor spacing of 4.53Å. It is argued that at this coverage the induced adsorbate dipoles have depolarized completely and that the interaction between the alkali metal atoms has become attractive. However, the lattice spacing seems rather large for islands having attractive interactions. Further studies as a function of temperature are required for a more complete understanding of this "condensation" phenomenon.

SUBSTRATES HAVING QUASI-HEXAGONAL SYMMETRY

Much of the early adsorption work on alkali metal overlayers was carried out on tungsten and molybdenum surfaces. Since these materials are bcc, their closest-packed (110) faces have a centered rectangular unit cell, giving a quasi-hexagonal array of adsorption sites. Much of the work on the structures and phase transitions of these overlayers was done quite early compared to the work on other transition metal surfaces, and there are several reviews of it[1,61]. The potential relief for alkali metal adsorption on W and Mo surfaces is larger than that on many other transition metal surfaces, and for this reason many commensurate structures are observed even on the smoothest bcc surface, the (110) surface. The disordering transitions of commensurate and incommensurate structures have been studied using LEED, and it has been generally found that commensurate structures display a sharper transition than incommensurate structures. First-order transitions were usually observed between commensurate structures.

The relatively strong site bonding of Na on W(110) has allowed the use of a lattice gas model to describe this system with a reasonable amount of success particularly at low coverages. A recent Monte Carlo simulation[30] of this system has reproduced most of the structures which were experimentally observed by using only a dipole-dipole interaction between adsorbed atoms. Qualitatively, the order-disorder transition temperatures agree with the experimentally observed ones, but it was found that corrections to the dipole potential used were needed out to a distance of at least twice the W lattice constant.

OVERLAYER-SUBSTRATE SPACING

One of the present controversies in alkali metal adsorption is the question of whether alkali metal atoms do indeed form an ionic bond to the substrate at low coverages, which gives rise to a picture of adsorption in which there is an "ionic to covalent transition" as the coverage is increased. This has been discussed in the literature[2,4,6,7,8,64,65], and there is still not a consensus on the model for adsorption at very low coverages. If there *is* an ionic to covalent transition, one might expect this change to manifest itself in the structure as a change in the overlayer-substrate spacing, since the ionic radii of alkali metal atoms are significantly smaller than their covalent radii (see Table 2).

There have been no comprehensive and systematic studies of the overlayer-substrate spacing in an alkali metal adsorption system. However, the measurements that have been made for various alkali metal adsorption systems are shown in Figure 12, plotted as a function of coverage. It can be seen that there is a large amount of scatter in the data, and no general conclusions can be drawn which are consistent with all of the data. Some of these

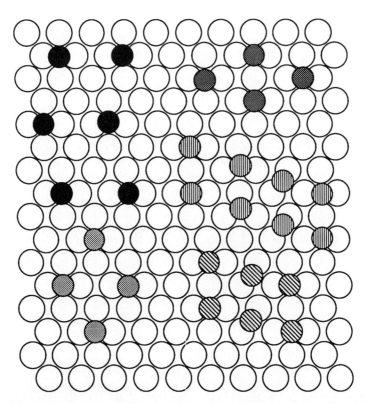

Figure 11. Schematic diagram of the commensurate structures observed for Na/W(110)[29]. These phases span a coverage range from 0.167 monolayers to 0.4 monolayers.

results must be unreliable because they are not consistent with each other. However, for the few systems where there are data points at more than one coverage, the observed trend is for the spacing to be somewhat larger at the higher coverage, and this is expected to be due to the change in bonding of the alkali metal atoms at different coverages. This result would be more convincing if a study were carried out on a system which is either incommensurate or disordered over a wide coverage range. Commensurate structures are more likely to have spacings which are affected by the details of the site symmetry, or may have different adsorption sites, and this will complicate both the analysis of the experiment and the interpretation of the results. The K/Ni(111) LEED study was carried out on incommensurate layers, but the measured overlayer-substrate spacing seems anomalously small, consistent with an ionic potassium radius, and an independent measurement of this system is required before these results can be considered to be reliable. Some of the other measurements using LEED IV techniques have also given results which seem anomalous, and some of the difficulties associated with these measurements are discussed in reference 13.

OTHER ASPECTS OF ALKALI METAL ADSORPTION

High Temperature Commensurate Structures

In some alkali metal adsorption systems there are irreversible transitions into commensurate structures as the temperature is increased, usually above room temperature. This is the case, for instance, for the two high-temperature commensurate structures shown in the Cs/Cu(111) phase diagram in Figure 2. Such behavior has not been described theoretically, but may due to a potential energy minimum which is a result of the substrate potential being stonger than the dipole repulsion. This minimum may not be accessible at low temperatures due to a potential barrier. Whatever the cause of these structures, they suggest the possibility of strong domain walls in overlayers having average densities slightly different from the commensurate density. Such domain wall structures have been proposed for Cs adsorption on Ru(001)[68] on the basis of splitting of spots in the LEED patterns.

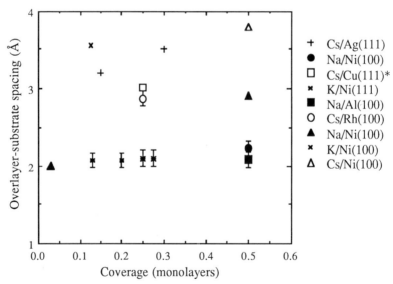

Figure 12. Overlayer-substrate spacing plotted as a function of coverage for a variety of alkali metal adsorption systems, Cs/Ag(111)[64], Na/Ni(100)[13,14,15,66], Cs/Cu(111)[16], K/Ni(111)[67], Na/Al(100)[17], Cs/Rh(100)[19], K/Ni(100)[66], and Cs/Ni(100)[66]. * The Cs/Cu(111) result is rather unusual because it is the only study which has found the adsorbed atoms to be in on-top sites rather than hollow or hollow and bridge sites.

Other interpretations, such as multiple diffraction, cannot be ruled out however, and further studies of these phases using different scattering techniques would be very useful.

Substrate Reconstruction and Surfaces Having a Strong Corrugation

One of the most studied aspects of alkali metal chemisorption in recent years is their ability to cause some surfaces to reconstruct. The most common of these reconstructions is the (1x2) reconstruction of fcc(110) faces, although alkali metal induced reconstruction of other surfaces has also been observed[4]. The structures which have been deduced for the fcc(110) (1x2) reconstruction are generally variations of the missing row model, and the large corrugation of this structure is reflected in the phase diagrams of Cs and K adsorbed on the reconstructed surface of Cu(110)[27]. Many studies of alkali metal chemisorption on surfaces of strong potential relief, including reconstructed surfaces, have been carried out and are the subject of several reviews[1,61].

CONCLUSION AND ACKNOWLEDGMENTS

Alkali metal chemisorption systems are a source of many interesting types of structures and associated phase transitions. The understanding of the interactions in alkali metal adsorption systems has progressed rapidly in the past five years, and it is anticipated that many of the controversial aspects of alkali metal adsorption, such as the adsorbate-substrate bonding and the details of the coverage-dependent interactions which determine the structures, will be resolved in the next few years. It is a pleasure to acknowledge the many interesting and fruitful discussions on alkali metal chemisorption I have had with many people over the past few years, including the members of my research group at Liverpool: Sumant Chandavarkar, David Fisher, and Zi-You Li.

REFERENCES

1. A. G. Naumovets, Sov. Sci. Rev. A Phys. **5** (1984) 443.
2. T. Aruga and Y. Murata, Prog. Surf. Sci. **31** (1989) 61.
3. C. T. Campbell, Annual Rev. Phys. Chem., to be published.
4. *Alkali Adsorption on Metals and Semiconductors*, ed. H. P. Bonzel, A. M. Bradshaw, and G. Ertl (Elsevier, Amsterdam) 1989.
5. R. W. Gurney, Phys. Rev. **47** (1935) 479.
6. E. Wimmer, A. J. Freeman, J. R. Hiskes, and A. M. Karo, Phys. Rev. B **28** (1983) 3074.
7. H. Ishida, Phys. Rev. B **38** (1988) 8006; **39** (1989) 5492; **40** (1989) 1341.
8. D. M Riffe, G. K. Wertheim, and P. H. Citrin, Phys. Rev. Lett. **64** (1990) 571.
9. C. Kittel, *Introduction to Solid State Physics* (Wiley, 6th edition, 1986).
10. N. W. Ashcroft and N. D. Mermin, *Solid State Physics* (Holt, Rinehart and Winston, 1976).
11. O. M. Braun, Sov. Phys. Solid State **23** (1981) 2779.
12. T. L. Einstein, CRC Critical Rev. Solid State Sci. **7** (1978) 261.
13. S. Andersson and J. B. Pendry, Solid State Comm **16** (1975) 563.
14. J. E. Demuth, D. W. Jepsen and P. M. Marcus, J. Phys. C **8** (1975) L25.
15 N. V. Smith, H. H. Farrell, M. M. Traum, D. P. Woodruff, D. Norman, M. S. Woolfson and B. W. Holland, Phys. Rev. B **21** (1980) 3119.
16. S. A. Lindgren, L. Wallden, J. Rundgren, P. Westrin, and J. Neve, Phys. Rev. B **28** (1983) 6707.
17. B. A. Hutchins, T. N. Rhodin, and J. E. Demuth, Surf. Sci. 54 (1976) 419.
18. M. Van Hove, S.Y. Tong, and J. Stoner, Surf. Sci. **54** (1976) 259.
19. C. Von Eggeling, G. Schmidt, G. Besold, L. Hammer, K. Heinz and K Müller, Surf. Sci. **221** (1989) 11.
20. D. L. Doering and S. Semancik, Surf. Sci. **129** (1983) 177.
21. G. Besold, Th. Schaffroth, K. Heinz, G. Schmidt, and L Müller, Surf. Sci. **189/190** (1987) 252.
22. G. Pirug and H. P Bonzel, Surf. Sci. **194** (1988) 159.
23. S. Chandavarkar and R. D. Diehl, Phys. Rev. B **38** (1988) 12112.
24. S. Chandavarkar, R. D. Diehl, A. Faké and J. Jupille, Surf. Sci. **211** (1989) 432.
25. D. Fisher and R. D. Diehl, to be published.
26. W. C. Fan and A. Ignatiev, J. Vac. Sci and Technol. A **6** (1988) 735.

27. W. C. Fan and A. Ignatiev, Phys. Rev. B **38** (1988) 366; W. C. Fan and A. Ignatiev, J. Vac. Sci and Technol. A**7** (1989) 2115.

28. K. Müller, G. Besold, and K Heinz, in ref. 4. It should be noted that W(100) has a reconstructed surface below room temperature, and this naturally affects the Cs structures which occur. The adsorbed Cs also affects the reconstruction by suppressing the disordering temperature of the reconstructed superlattice.

29. V. K. Medvedev, A. G. Naumovets and A. G. Fedorus, Soviet Physics - Solid State **12** (1970) 375.

30. L. D. Roelofs and D. L. Kriebel, J. Phys. C **20** (1987) 2937.

31. E. Wimmer, Surf. Sci. **134** (1983) L487.

32. W. C. Fan and A. Ignatiev, Phys. Rev. B **37** (1988) 5274.

33. S. Chandavarkar and R. D. Diehl, Phys. Rev. B **40** (1989) 4651.

34. J.M. Kosterlitz and D.J. Thouless, J. Phys. C **6**, 1181 (1972); B.I. Halperin and D.R. Nelson, Phys. Rev. Lett. **41**, 121 (1978); Phys. Rev B **19**, 2457 (1979) A.P. Young, Phys Rev B **19**, 1855 (1979).

35. S.T.Chui, Phys. Rev. Lett. **48**, 933 (1982); Phys. Rev. B **28**, 178 (1983)

36. H.Kleinert, Phys. Lett. **95A**, 381 (1983)

37 E. D. Specht, A. Mak, C. Peters, M. Sutton, R. J. Birgeneau, K. L. d'Amico , D. E. Moncton, S. E. Nagler, and P. M. Horn, Z. Phys. B. **69** (1987) 347.

38. S. E. Nagler, P. M. Horn, T. F. Rosenbaum, R. J. Birgeneau, M. Sutton, S. G. J. Mochrie, D. E. Moncton and R. Clarke, Phys. Rev. B **32** (1985) 7373.

39. N. Greiser, G. A. Held, R. Frahm, R. L. Greene, P. M. Horn, and R. M. Suter, Phys. Rev. Lett. **59** (1987) 1625.

40. K. L. D'Amico, J. Bohr, D. E. Moncton, and D. Gibbs, Phys. Rev. B **41** (1990) 4368.

41. A. J. Jin, M. R. Bjurstrom and M. H. W. Chan, Phys. Rev. Lett **62** (1989) 1372.

42. D. L. Doering and S. Semancik, Phys. Rev. Lett. **53** (1984) 66.

43. D. L. Doering and S. Semancik, Surf. Sci. Lett. **175** (1986) L730.

44. J. Cousty and R. Riwan, Surf. Sci. **204** (1988) 45.

45. E. Bauer, Appl. Surf. Sci. **11/12** (1982) 479.

46. A. D. Novaco and J. P. McTague, Phys. Rev. Lett. **38**, 1286-1290 (1977); J. P. McTague and A. D. Novaco, Phys. Rev. B **19** (1979) 5299.

47. C. G. Shaw, S. C. Fain, Jr., and M. D. Chinn, Phys. Rev. Lett. **41** (1978) 955.

48. H. Shiba, J. Phys. Soc. Jpn. **46** (1979) 1852.

49. D. L. Doering, J. Vac. Sci. Technol. A **3** (1985) 809.

50. C. R. Fuselier, J. C. Reich and N. S. Gillis, Surf. Sci. **92** (1980) 667.

51. K. Kern, P. Zeppenfeld, R. David, R. L. Palmer, and G. Comsa Phys. Rev. Lett. **57** (1986) 3187.

52. R. L. Gerlach and T. N. Rhodin, Surf. Sci. **17** (1969) 32.

53. S. Andersson and U. Jostell, Solid State Comm. **13** (1973) 829.

54. S. Andersson and B. Kasemo, Surf. Sci. **32** (1972) 78.

55. J. Cousty, R. Riwan, and P. Soukiassian, Surf. Sci. **152/153** (1985) 297.

56. T. Aruga, H. Tochihara, and Y. Murata, Surf. Sci. **158** (1985) 490.

57. K. Heinz, H. Hertrich, L. Hammer and K. Müller, Surf. Sci. **152/153** (1985) 303.

58. A. U. MacRae, K. Müller, J. J. Lander and J. Morrison, Surf. Sci. **15** (1969) 483.

59. T. Aruga, H. Tochihara, and Y. Murata, Phys. Rev. Lett **52** (1984) 1794.

60. T. Aruga, H. Tochihara, and Y. Murata, Surf. Sci. Lett **175** (1986) L725.

61. A. G. Naumovets, in *Physics of Solid Surfaces 1984* (Elsevier, Amsterdam, 1985) p. 41.

64. G. M. Lamble, R. S. Brooks, D. A. King, and D. Norman, Phys. Rev. Lett. **61** (1988) 1112.

65. S. Å. Lindgren, C. Svensson and L. Walldén, Phys. Rev. B **42** (1990) 1467.

66. S. Andersson and B. Kasemo, Surf. Sci. **32** (1972) 78.

67. D. Fisher and R. D. Diehl, unpublished.

68. R. Duszak and R. H. Prince, Surf. Sci.**226** (1990) 33.

ORIGINAL PROPERTIES OF THIN ADSORBED

FILMS ON AN IONIC SURFACE OF SQUARE SYMMETRY

AND HIGH SURFACE HOMOGENEITY: MgO(100)

J.P. Coulomb

Faculté des Sciences de Luminy , C.R.M.C.[2] , Département de Physique
Case 901 , 13288 Marseille Cedex 9 .

Introduction

Since the pioneering work of A. Thomy and X. Duval [1], who discovered the existence of two-dimensional (2d) phases at the beginning of the seventies , physisorption studies on highly homogeneous surfaces have pointed out several evolution stages . A couple of decades have been devoted on the one hand to the understanding of the specific (2d) matter properties with a particular interest for the (2d) melting transition and on the other hand to the "subtle" substrate influence which is well illustrated by commensurate ⇔ incommensurate (2d) solid transition observations. Three International Congresses have been dedicated to this low dimensionality physics [2][3][4]. After this stage the (2d) physics community interest switched toward thicker films properties analysis. J.G. Dash article's title "Between two and three dimensions" [5] expresses very well the wish to link the (2d) matter behavior to the bulk (3d) one . During this period growth modes of films and wetting phenomena have been extensively analysed . The most recent stage of this physisorption development has been reached by the middle of the eighties when surface phase changes such as the roughening transition and surface premelting have been observed in "bulk " materials . Indeed physisorbed wetting films are well suited for such surface phase transitions.

Another branch of natural evolution of (2d) pure physisorbed films concerns the co-adsorption of two gases . Investigations of the properties of mixed (2d) films present some difficulties both at the experimental and the theoretical point of view, but this field is spreading out [6][7]. In my opinion the more promising physisorption development in the future comes from the possibility of using new substrate types . In the past interesting results have been obtained on square symmetry substrates by B.B. Fisher and W.G. McMillan [8], T. Takaishi and M. Mohri [9] and J.G. Dash et al. [10]. But a novel era in physisorption study on such square symmetry supports began in 1984 when J.P. Coulomb and O.E. Vilches succeeded in preparing MgO powders of high surface homogeneity [11]. This paper summarizes the main results obtained in the thermodynamic and structural analysis of rare-gas (Ar, Kr, Xe) and molecular films (CH_4, N_2, CO) films adsorbed on such MgO powders . It is very difficult to prepare homogeneous MgO single crystal surfaces. The results obtained up to now with air-cleaved MgO single crystals clearly show that the surface presents some heterogeneity [12]. Quite recently surface homogeneity improvement has been observed for ultra-high vacuum cleaved MgO samples, by T. Angot and J. Suzanne [13] .

1) MgO powder characteristics [11]

Our MgO powders are prepared by burning magnesium ribbons in a dry atmosphere (O_2 - Ar or O_2 - N_2 mixtures). O_2 concentration governs in a sense the MgO powder specific surface S_s ; for the mixture O_2 (20%) - Ar (80%), S_s equals 8 m^2/g (crystallites mean size is equal to 2000 $\overset{\circ}{A}$). Larger S_s values are obtained for higher oxygen concentrations . More

Fig. 1. TEM micrograph of MgO smoke obtained by burning Mg ribbons in air . Well shaped micro-cubes of several tenths of a micron can be observed .

details are given elsewhere [11]. MgO powders prepared in such a way are essentially composed of cubic (100) faces. Well shaped micro-cubes are observed by T.E.M., figure 1.

A (100) MgO surface is represented schematically figure 2. It is composed of two types of ions, Mg^{2+} and O^{2-} organized in a square symmetry arrangement. The MgO structure is of the NaCl type [14].

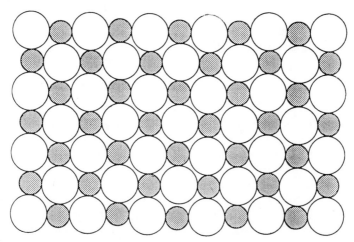

Fig. 2. Schematic representation of the (100) MgO surface. It is composed of two types of ions Mg^{2+} and O^{2-} organized in a square symmetry arrangement (the respective ion size is $d_{Mg^{2+}} = 1.72$ Å (shaded circles), $d_{O^{2-}} = 2.52$ Å (open circles)).

For the first time the signature of a (2d) phase transitions between dense phases was observed in krypton adsorption isotherms measured at 77.4 K on such MgO powders, figure 3c . By comparison Kr adsorption isotherms measured at the same temperature on others kind of MgO powders prepared by different groups working in the same field are also represented figures 3 a,b and d [10][15][16].

With regard to the previous lamellar materials which have allowed (2d) phase transitions observations (the so-called graphite but also the boron nitride and some lamellar halides) the MgO (100) surface presents the following original characteristics :

- the square symmetry of the lattice potential wells ;
- a rather large geometric corrugation $\mathcal{E}_g$ due to the Mg^{2+}, O^{2-} difference ions size
  ($d_{Mg^{2+}} = 1.72$ Å , $d_{O^{2-}} = 2.52$ Å ; $\mathcal{E}_g = 0.40$ Å) [17];
- a strong periodic electrical field $\vec{E}$.

We have investigated the influences of these original surface characteristics on adsorbed films properties . First with rare-gas (Ar, Kr, Xe) or spherical molecule (CH_4) we have studied the influence of the square symmetry of the lattice sites . Indeed it is of prime interest to analyse the energetic competition between adsorbate - substrate interaction $U_{ads.-sub.}$ which favours commensurate square structure and the adsorbate - adsorbate interaction

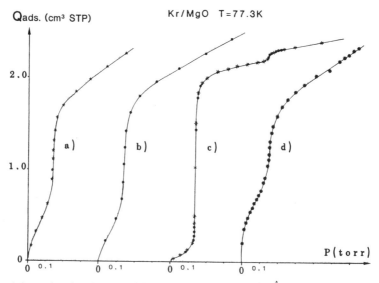

Fig. 3. Adsorption isotherms of Kr measured at T = 77.3 K on several kind of MgO powders prepared in different laboratory ; a) J.G. Dash et al [10] , b) J. Jordan et al [15] , c) J.P. Coulomb and O.E Vilches [11] , d) A.G. Shastry et al [16] . A well developed vertical step is observed in adsorption isotherm (c) , this feature is a proof of high surface homogeneity . Moreover for the first time a substep is also observed which is the "signature" of a (2d) phase transition between two (2d) dense phases .

$U_{ads.-ads.}$ which leads to a close packed arrangement. Secondly , with simple molecules as N_2 and CO which possess permanent multipole moments , we have begun the investigation of the surface electrical field $\vec{E}$ influence . The coupling of $\vec{E}$ with the multipole moments generates extra attractive interactions in addition to the Van der Waals one .

2) Influence of the MgO(100) surface square symmetry on thin adsorbed films properties

As a consequence of its rather large geometric corrugation $\mathcal{E}_g$ it was commonly assumed that adsorbed films on MgO give rise to localised adsorption for most of the adsorbates Ar,Kr,CH$_4$ [18]. Our studies show that the results are not so simple. It is not $\mathcal{E}_g$ but the energetic corrugation $\mathcal{E}_e$ which governs the (2d) film commensurability ($\mathcal{E}_e$ is the variation of the potential energy when the adsorbed molecule is displaced on the surface) . For the MgO (100) surface $\mathcal{E}_e$ depends largely on the adsorbed molecule characteristics (size, morphology and electrical properties) .

2-1) Thermodynamical properties of rare gases (Ar, Kr , Xe) and spherical molecule CH$_4$ monolayers adsorbed on MgO(100) surface [19][20]

All of these gases exhibit stepped adsorption isotherms , therefore for thin films at least, they grow layer-by-layer on the MgO (100) surface. An Ar/MgO adsorption isotherm measured at 60 K is represented for example in figure 4 . From the (2d) condensation pressure value $P^{(1)}$ relative to the saturated vapor value $P^{(\infty)}$, we can deduce by comparison to the same gases physorbed on graphite basal planes that the MgO(100) surface is less attractive than the graphite (00.1) one. The free energy excess ΔF { $\Delta F = RT \log P^{(1)}/P^{(\infty)}$ } due to the substrate is lower for MgO than for graphite, in spite of the induction forces (resulting from the MgO surface electrical field polarisation of the adsorbed atoms) , see figure 5 .

Sets of adsorption isotherms have been measured for each of the considered adsorbates [19][20]. Xe, Kr and Ar rare-gas present a sub-step in their adsorption isotherms in a characteristic temperature range, CH$_4$ does not . Figures 6, 7 and 8b represent Xe, CH$_4$ and Ar adsorption isotherms sets respectively . The sub-step which is a signature of (2d) transition between two dense phases A and B, merges at low temperature with the (2d) transition   dilute phase $\Leftrightarrow$ dense phase A and gives rise to a (2d) triple point T_{2t} . For Ar, T_{2t} is determined by extrapolation because the pressure is too small for standard capacitance gauge measurements ($P \geq 10^{-3}$ torr) , figure 8a . When the temperature increases the vertical step of the adsorption isotherms bends. The (2d) coexistence phases regime disappears giving rise to a (2d) critical point T_{2c} (only one (2d) fluid phase covers the MgO(100) surface above T_{2c} in the sub-monolayer regime). T_{2t} and T_{2c} for each adsorbates are summarized in table 1.

For an incommensurate hexagonal close packed (2d) solid, the ratio T_{2t}/T_{3t} is constant ; $T_{2t}/T_{3t} = 0.61$ [21] . Xe and Kr (2d) triple points are in agreement with this value within the experimental uncertainty . Recently structural studies of Xe and Kr (2d) films have given direct proof of their h.c.p. atomic arrangement [13][22] . But for Ar, T_{2t} has an unusually

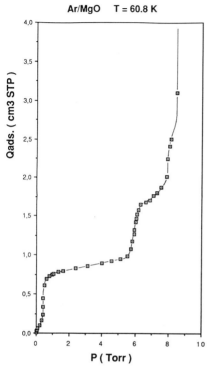

Ar/MgO   T = 60.8 K

Fig. 4. Stepped adsorption isotherm of Ar measured at T = 60.8 K on high homogeneous surface MgO powder .

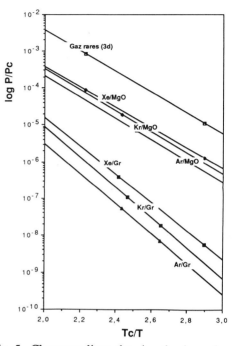

Fig. 5. Clapeyron lines showing the dependence of the condensation pressure P of the first mono-layer on the inverse of the temperature 1/T, for rare-gas (Ar, Kr, Xe) adsorbed on MgO (100) surface and on the graphite basal plane (001) (pressure and temperature are normalised to their critical value for each gas). At a fixed temperature the free energy excess ΔF, due to the substrate attractive forces , is proportional to the distance between the (2d) line and the bulk line . MgO substrate is less " attractive " than the graphite one for rare-gas.

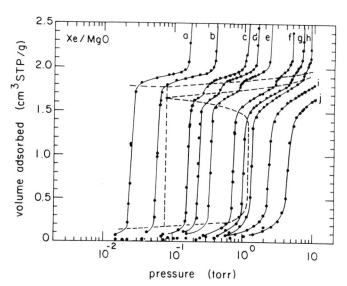

Fig. 6. Xe / MgO adsorption isotherms measured at several temperatures [19] ; a) 96.86K; b) 100.47 K ; c) 106.20 K ; d) 108.44 K ; e) 111.02 K ; f) 116.14 K ; g) 118.72 K ; h) 121.15 K ; i) 126.17 K ; j) 131.19 K . Dashed lines indicate estimated phase boundaries between the (2d) gas , the (2d) liquid and the (2d) solid .

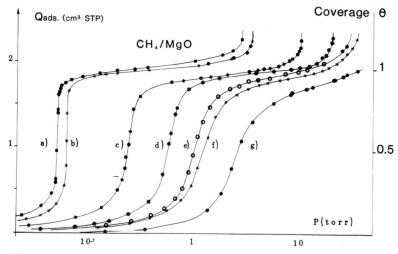

Fig. 7. CH_4 / MgO adsorption isotherms measured at several temperatures [20] ; a) 76.32K ; b) 77.35 K ; c) 83.14 K ; d) 87.31 K ; e) 89.60 K ; f) 91.19 K ; g) 95.0 K.

119

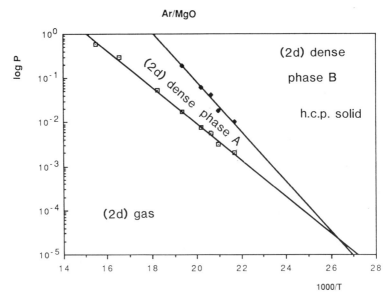

Fig. 8 - a . Clapeyron lines representing the pressure dependence of the two (2d) transitions , (2d) gas ⇔ (2d) dense phase A and (2d) dense phase A ⇔ (2d) dense phase B, versus 1/T . The intersection of the two lines defines the triple point temperature T_{2t} ; $T_{2t} = 38 \pm 2$ K (neutron diffraction results show that the dense phases A and B are respectively a (2d) "liquid cristal" like phase and a h.c.p. (2d) solid) .

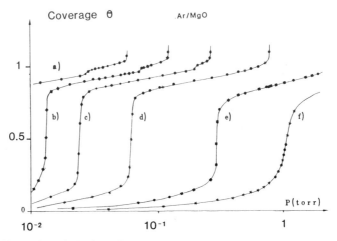

Fig. 8 - b . Examples of Ar / MgO adsorption isotherms [20] ; a) 47.70 K ; b) 49.57 K ; c) 51.65 K ; d) 54.84 K ; e) 60.36 K ; f) 65.50 K .

	T_{2t}	T_{2c}	T_{2t}/T_{3t}	T_{2c}/T_{3c}	T_{2t}/T_{2c}
Xe	101 ± 1	119 ± 2	0.62	0.41	0.85
Kr	67 ± 1	87 ± 2	0.57	0.41	0.77
Ar	38 ± 2	63 ± 2	0.45	0.41	0.60
CH$_4$		80 ± 2		0.42	
Corresponding state law	....	....	0.61	0.39	0.86

low value. In many other cases substrate influence always displaces T_{2t} toward the higher temperature side [21]. Structural analysis by neutron diffraction of the Ar/MgO system has allowed a clear description at the atomic scale of this original melting phenomenon.

The (2d) melting is not observed for CH$_4$ films ; neutron diffraction study has thrown light on this original result. It is interesting to notice that the (2d) critical temperatures T_{2c} are almost constant for the four gases and are in reasonable accord with the expected ratio value $T_{2c}/T_{3c} = 0.39$ [21] which characterizes the (2d) liquid ⇔ (2d) gas critical point.

2-2) Structural properties of Ar and CH$_4$ monolayers adsorbed on MgO(100) surface

Neutron diffraction experiments have been done at the High Flux Reactor of the Institut Laüe Langevin , on D$_{1B}$ diffractometer.

[36]Ar / MgO system [20]

Neutron diffraction spectra of Ar film adsorbed on MgO have been measured for different coverages Θ ; $0.5 \leq Θ \leq 1.0$ layer at 10 K , figure 9 . In addition to the h.c.p. (2d) solid observed for Θ= 1 layer (d= 3.85 Å) a new (2d) Ar atom crystal organization is stable in the sub-monolayer regime. The diffraction peaks of this (2d) solid which should be stabilized by the MgO(100) surface are indexed by the P(2x3) overstructure [20]. The originality of this structure is that all the argon atoms lie along one of the family of channels (rows of Mg^{2+} ions) which characterize the MgO(100) surface, figure 10. Phase coexistence between the two (2d) solids is observed for Θ = 0.91 . We have to note that on a less homogeneous MgO surface (as for the MgO (100) single crystal one) only the P(2x3) structure is observed by LEED in the argon monolayer regime [12][13][23] .

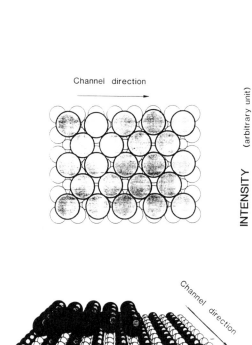

Channel direction

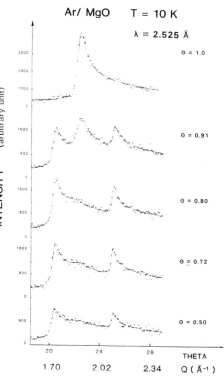

Ar/ MgO   T = 10 K

λ = 2.525 Å

Θ = 1.0

Θ = 0.91

Θ = 0.80

Θ = 0.72

Θ = 0.50

INTENSITY (arbitrary unit)

THETA
Q (Å⁻¹)

Fig. 9. Neutron diffraction spectra of Ar films adsorbed on MgO at T = 10 K in the sub-monolayer and monolayer coverage Θ regime [26] . Different (2d) structures are observed ;
– for Θ = 1 layer , a (2d) hcp structure ;
– for 0.50 ≤ Θ ≤ 0.80 , a (2d) commensurate P(2x3) overstructure ;
– for Θ = 0.91 layer , a (2d) phase coexistence is observed .

Fig. 10. Schematic representation of the commensurate P(2x3) overstructure . All the argon atoms lie in the channels which characterize the MgO(100) surface .

We have investigated the temperature stability of the P(2x3) overstructure. For $\Theta = 0.80$ we have observed that the diffraction peaks widths show quite different temperature dependence, figure 11. The coherence lengths deduced by fitting the diffraction peak by standard methods (Ruland and Tompa analysis for example [24]) are represented figure 12. The (10) diffraction peak located at wave vector q=1.75 Å^{-1} mainly probes the long range order (LRO) along the channel direction (CD); its coherence length $L_{coh.}$ decreases sharply in the temperature range $35 < T < 40$ K. At temperature $T \geq 40$ K only a short range order is observed along CD, with $L_{coh.} \leq 40$ Å. The (11) diffraction peak located at q=2.11 Å^{-1} probes the long range order in the direction perpendicular to the channels ; the LRO persists up to 55 K in that direction with $L_{coh.} = 220$ Å . We can conclude from these results that in the temperature range $35 \leq T \leq 40$ K a new (2d) argon phase (labelled as dense phase A in fig.8-a) appears characterized by two kinds of order , a short range one in the channel direction and a long range one in the perpendicular direction .

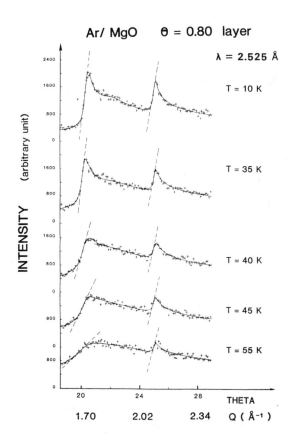

Fig. 11. Neutron diffraction spectra of Ar film adsorbed on MgO in the sub-monolayer coverage regime $\Theta = 0.80$ layer , for different temperatures [26] . The low angle diffraction peak broadens out in the temperature range; $35 < T \leq 40$ K. The second diffraction peak width remains constant in all the temperature range ; $10 < T \leq 55$ K . At $T = 40$ K a new (2d) phase appears, characterized by an anisotropic range order . It looks like a (2d) " liquid crystal " phase .

The "one-directional" disorder which appears between 35 and 40 K is in good agreement with the unusually low value of T_{2t} that we have determined from our thermodynamic study $T_{2t} = 38 \pm 2$ K in comparison with the expected value deduiced from the (2d) corresponding law ($T_{2t} = 50$ K). This (2d) melting is in fact one example of " one-dimensional " melting. Consequently a new (2d) " crystal like " phase has been observed for the first time [20][25][26] . Recent molecular-dynamics simulation of Ar monolayer physisorbed on MgO (100) surface has pointed out also this unusual melting phenomena [27] .

CD$_4$ / MgO system [20][28]

Neutron diffraction spectra of deuterated methane have been measured for different coverages θ ; $0.8 \leq \theta \leq 2$ layers at 10 K , figure 13 . From the main diffraction peak position q= 1.49 $\text{\AA}^{-1}$, we can conclude that the CD$_4$ monolayer crystallizes in a square

O.80 layer Ar/MgO

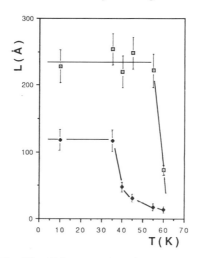

Fig. 12. Coherence lengths L versus temperature of the two diffraction peaks represented fig. 11 and located respectively at the wave vector ;

($\blacklozenge$) q = 1.75 $\text{\AA}^{-1}$ and

($\square$) q = 2.11 $\text{\AA}^{-1}$.

commensurate C(2x2) (2d) solid [20][28] . The C(2x2) overstructure was observed also on MgO(100) single crystal surface by LEED [23] and He atoms scattering [29] . The methane bilayer presents an unusual diffraction profile. It results from the strong modulation of the Bragg-rods which characterizes the bilayer square structure [20]. Diffraction profile calculations for several methane molecule configurations (monopode, dipode, tripode, spherical rotation) compared to the experimental spectrum suggest that the CD$_4$ molecule are in the dipode configuration , figure 14 . Our results are in agreement with a self-consistent fiel (SCF) calculation [30] .

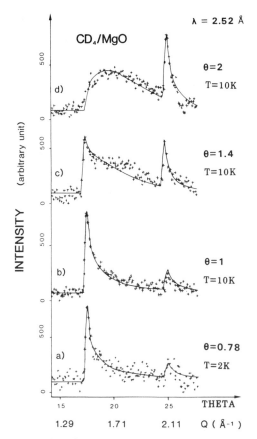

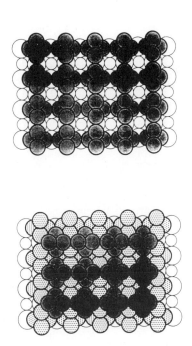

Fig. 13. Neutron diffraction spectra
of CD_4 films adsorbed on MgO at
T =10K in the θ coverage range
1.0 $\leq$ θ $\leq$ 2.0 layers [20] ;
spectra (b) (c) and (d) . Diffraction
spectrum (a) is measured for a CD_4
film coverage θ = 0.78 adsorbed at
T= 2 K on MgO.The methane monolayer film and bilayer film are commensurate C(2x2) structures . Phase
coexistence regime is observed for
spectrum (c) , the MgO surface is
covered respectively by a CD_4 monolayer and a CD_4 bilayer in respective
proportion 0.6 and 0.4 .

Fig. 14. Schematic representation of the
square commensurate C(2x2) CD_4
monolayer structure (on top) and of the
CD_4 bilayer structure (at the bottom).
Relative intensities of the two diffraction
peaks suggest a bipod CD_4 configuration
[20][28] (methane molecules stand up
on the MgO surface on two hydrogen
atoms) .

125

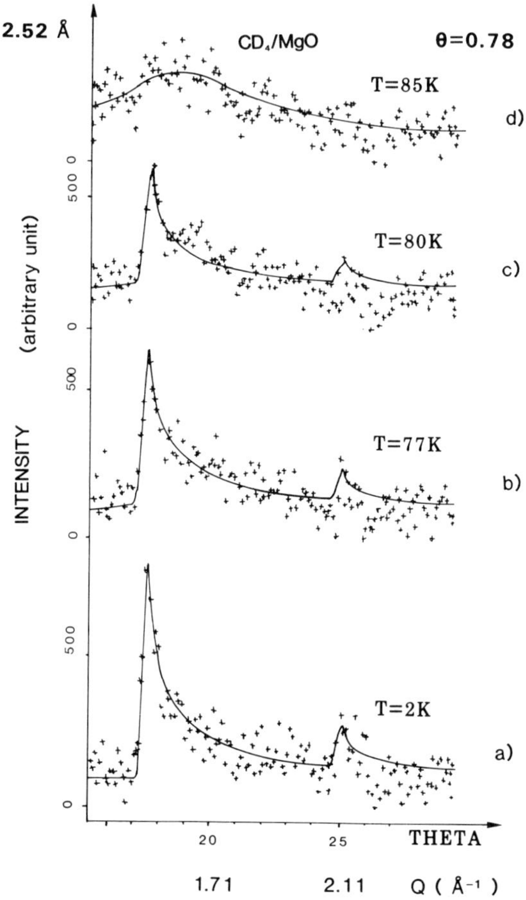

$\lambda = 2.52$ Å

CD$_4$/MgO $\theta = 0.78$

T=85K d)

T=80K c)

T=77K b)

T=2K a)

INTENSITY (arbitrary unit)

500 0

500

500

500

20   25   THETA

1.71   2.11   Q (Å⁻¹)

Fig. 15. Neutron diffraction spectra of CD$_4$ film adsorbed on MgO in the sub-monolayer coverage regime θ $\theta = 0.78$ layer, for different temperatures [20]. The square commensurate C(2x2) (2d) CD$_4$ solid is stable in a large temperature range $2 \leq T \leq 80$ K.

The square C(2x2) methane monolayer is stable in a large temperature range; $2 \leq T \leq 80$ K, figure 15. This result is in accord with our thermodynamical result ($T_{2c} = 80 \pm 2$ K).

The strong influence of the square potential wells arrangement of the MgO(100) surface for methane is probably a consequence of the good parametric agreement between the substrate's second neighbours site distance $d = 4.21$ Å and the methane molecule size $d = 4.17$ Å [31]. The attractive interaction of the CH$_4$ octupole permanent moment with the surface electric field $\vec{E}$ must also favour the commensurability.

With the rare-gas Xe, Kr, Ar atoms and the quasi-spherical CH$_4$ molecule adsorbed on MgO(100) surface we have found examples of all of the possible scenarios of the (2d) films - MgO surface interaction competition :

– For the larger rare-gas as Xe and Kr the MgO(100) energetic corrugation $\mathcal{E}_e$ has little

influence. The (2d) monolayer crystallizes in h.c.p. structure ; $\Delta U_{ads.-sub.} < \Delta U_{ads.-ads.}$
($\Delta U_{ads.-sub.}$ and $\Delta U_{ads.-ads.}$ are respectively the difference of the adsorbate-substrate interaction and of the adsorbate-adsorbate interaction of the considered (2d) structure induced by $\mathcal{E}_e$ in comparison with the h.c.p. (2d) structure) .

– For the methane monolayer , the MgO(100) surface imposes a (2d) film crystallization in square commensurate C(2x2) structure ; $\Delta U_{ads.-sub.} > \Delta U_{ads.-ads.}$;

– Argon presents the more subtle (2d) film - support interaction ;

$\Delta U_{ads.-sub.} \simeq \Delta U_{ads.-ads.}$. Depending on the argon monolayer concentration Θ, two (2d) structures are observed . At sub-monolayer coverage , $\Delta U_{ads.-ads.} < \Delta U_{ads.-sub.}$, the Ar (2d) film presents a commensurate P(2x3) structure and when Θ increases the situation reverses , $\Delta U_{ads.-sub.} < \Delta U_{ads.-ads.}$ and the Ar monolayer forms a incommensurate h.c.p. structure .

3) Influence of the MgO(100) surface electric field $\vec{E}$ on thin adsorbed films properties

3-1) Thermodynamic properties of N_2 and CO monolayers adsorbed on MgO(100) surfaces

With N_2 and CO gases we want to probe the influence of the surface electrical field $\vec{E}$ on adsorbed film properties. Nitrogen and carbon monoxide (2d) films condense on MgO(100) surface at pressure values $P^{(1)}$ far below the saturated pressure one $P^{(\infty)}$ by comparison with the rare gases . This observation qualitatively illustrates the electric field influence . Moreover the free energy excess ΔF due to the MgO substrate is comparable to the graphite one for N_2 but is larger for CO, figure 16. The attractive "coupling" between the electric field $\vec{E}$ and the CO molecules is stronger than that with the $\vec{E}$ - N_2 molecules one . The dipole moment μ and quadrupole moment Q are larger for the CO molecule, see table 2 .

Table 2 . Electrical characteristics of N_2 and CO molecules (polarizability α, dipole moment μ and quadrupole moment Q). The different factors $U_{pol.}$, U_μ and U_Q due to the attractive interaction between the adsorbed molecules and the MgO (100) surface electric field are determined from the Van-Dongen calculation [34].

Gas	α $10^{-24} cm^3$	μ 10^{-18} esu/cm^{-2}	Q 10^{-26} esu/cm^{-2}	$U_{pol.}$ Kcal/mole	U_μ Kcal/mole	U_Q Kcal/mole
N_2	2.4	0	- 1.5	0.12	0	0.74
CO	2.6	0.11	- 2.5	0.13	0.26	1.23

Fig. 16. Clapeyron lines pointing out the dependence of the condensation pressure P of the first monolayer on the inverse of the temperature $1/T$, for N_2 and CO adsorbed on MgO (100) surface and on the graphite basal plane (001) (pressure and temperature are normalized to their critical value for each gas). At a fixed temperature the free energy excess ΔF, due to the substrate attractive forces , is proportional to the distance between the (2d) line and the bulk line . MgO substrate is as "attractive" as the graphite substrate for such molecules characterized by multipole permanent moments (μ and Q). This result illustrates very well the electric field influence of the MgO(100) surface on adsorbed films properties .

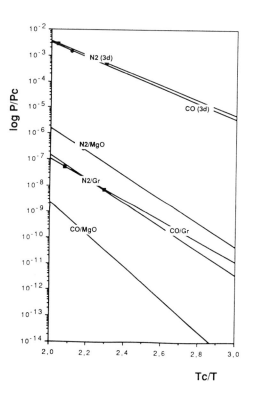

Sets of adsorption isotherms have been measured for the two gases . Their characteristics are similar', for example the isotherms for N_2 are bent, figure 17 . The temperature range where the pressure P is measurable $(P \geq 10^{-3}$ torr) , is far above the (2d) critical temperature T_{2c} . On the other hand second steps are vertical , figure 18 . We have deduced the adsorption heat $Q^{(n)}_{ads.}$ of the first and second layers of N_2 and CO adsorbed on MgO(100) surface', results are indicated in table 3 .

Table 3 . Adsorption heat $Q^{(1)}_{ads.}$ and $Q^{(2)}_{ads.}$ respectively for the first and the second layers of N_2 and CO adsorbed on MgO(100) surface' .

gas	Xe	Kr	Ar	N_2	CO
$Q^{(1)}_{ads.}$ (Kcal/mole)	3.67	2.74	2.00	2.62	3.77
$Q^{(1)}_{ads.}/Q_{sub.}$	0.98	1.05	1.05	1.59	2.11
$Q^{(2)}_{ads.}$ (Kcal/mole)	3.65	2.61	1.96	1.68	1.76
$Q^{(2)}_{ads.}/Q_{sub.}$	0.98	1.00	1.03	1.02	0.98

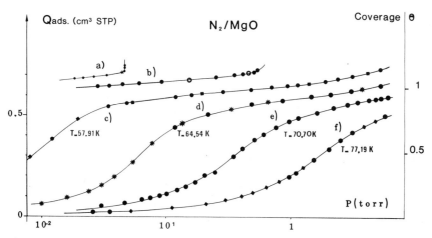

Fig. 17. N_2 / MgO adsorption isotherms measured at several temperatures in the monolayer regime [20] ; a) 42.70 K ; b) 49.70 K ; c) 57.91 K ; d) 64.54 K ; e) 70.70 K ; f) 77.19 K . Isotherms step are strongly bent , the temperature range investigated is apparently far above the (2d) critical temperature T_{2c} ; $T_{2c} << 58$ K . The adsorption heat of the first layer $Q^{(1)}_{ads.}$ has been determined for $Q_{ads.} = 0.325$ cm^3 STP .

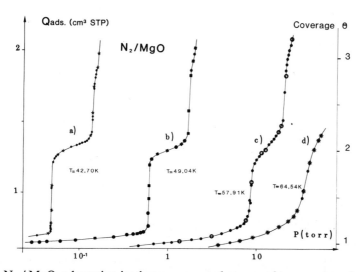

Fig. 18. N_2 / MgO adsorption isotherms measured at several temperatures in the second and third layers regime [20] ; a) 42.70 K ; b) 49.70 K ; c) 57.91 K ; d) 64.54 K . Isotherms steps are vertical in the temperature range $43 \leq$ T < 58 K , above the second layer isotherm step bends. We can deduce its (2d) critical temperature $T^{(2)}_{2c}$; $T^{(2)}_{2c} = 58 \pm 2$ K. The adsorption heat of the second layer $Q^{(2)}_{ads.}$ has been determined for $Q_{ads.} = 0.975$ cm^3 STP .

Our results clearly demonstrate the influence of the surface electrical field $\vec{E}$ on N_2 and CO films by comparison with rare-gas results [32][33] (the ratio $Q^{(n)}_{ads.} / Q_{sub.}$ is also given , $Q_{sub.}$ is the bulk sublimation heat of each adsorbate) .

We can also notice that the electrical field "perturbation" is limited to the first monolayer . This result is in accordance with the Van-Dongen calculation, which predicts a fast decrease of the electrical field absolute value $|E|$ versus the distance z from the surface [34] . For the MgO(100) surface $|E|$ at the point of coordinates x,y,z (expressed in angstom) is :

$$|E(x,y,z)| = 1.1416 \cdot 10^{12} \cdot [\; \cos^2(2 \cdot \pi \, x/a) + \cos^2(2 \cdot \pi \, y/a)\;]^{1/2} \cdot e^{-2.1106 \cdot Z}$$

(a is the MgO lattice parameter , a = 4.21 Å) .

For instance above the cation Mg^{2+} at a height $z = 3$ Å from the MgO surface one finds : $E = 2.9 \cdot 10^9$ V/m . The main attractive term comes from the interaction of $\vec{E}$ with the permanent quadrupole moment of the N_2 and CO molecules. Estimation from point charge calculations of the different factors $U_{pol.}$, $U\mu$ and U_Q are indicated table 2 .

$U_{pol.} = 1/2 \; \alpha \cdot E^2$, $U\mu = \mu \cdot E$ and $U_Q = 1/6 \; Q \cdot \partial E / \partial z$; α , μ and Q are respectively the polarisability and the dipolar and quadrupolar permanent moments . These rough calculations are qualitatively in agreement with our experimental results, more recent calculations are in progress [35][36] .

3-2) Structural properties of N_2 monolayer adsorbed on MgO(100) surface [32]

The nitrogen (2d) film condensed on MgO (100) shows a great variety of (2d) structures. We have observed four (2d) solids S_1 , S_2 , S_3 and S_4 in the coverage Θ range ; $0.8 \leq \Theta \leq 1.2$ layer and temperature T range $10 \leq T \leq 50$ K, figures 19 and 20 . S_1 S_2 and S_3 are (2d) structures of the type P(2x"N") stabilized by the "potential channels" of the (100) MgO surface (as for the P(2x3) argon (2d) solid) . The difference with argon results is that N_2 (2d) film undergoes an unidirectional compression along the channels . S_4 solid observed at higher temperature (T = 50 K) has a hexagonal close packed structure which is probably a consequence of the thermal activation of N_2 molecular rotational motion. The structural parameters (a , b , γ) of the S_1 , S_2 , S_3 and S_4 unit cells are given in table 4 (structural parameters of the dense planes d_{111} of bulk N_2 solid phases α and β are also indicated) .

The three (2d) structures S_1 , S_2 and S_3 of N_2 adsorbed on MgO (100) surface at T= 10 K , are different from the bulk d^{α}_{111} dense planes . On the other hand S_4 structure is close to the h.c.p. d^{β}_{111} dense plane . We can notice that the (2d) structure S_3 is denser than the h.c.p. orientationally ordered bulk d^{α}_{111} plane . In such a plane a quarter of the N_2 molecules stand up , the others make an angle $\alpha \simeq 70°$ with the surface perpendicular direction . During the unidirectional compression of the N_2 (2d) film , when the coverage Θ increases , the nitrogen molecules probably lift out of the plane .

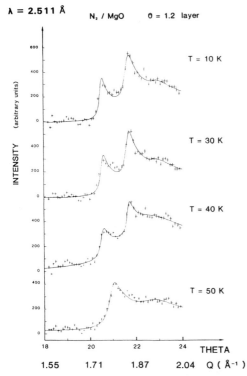

λ = 2.511 Å   N₂ / MgO   θ = 1.2 layer

INTENSITY (arbitrary units)

T = 10 K

T = 30 K

T = 40 K

T = 50 K

18   20   22   24   THETA
1.55   1.71   1.87   2.04   Q (Å⁻¹)

Fig. 19. Neutron diffraction spectra of nitrogen films adsorbed at T = 10 K on MgO , around the monolayer coverage regime [33]. Three (2d) structures S_1 , S_2 and S_3 are observed respectively at the coverage θ value ; 0.80 , 1.0 and 1.2 layer . Only the low angle diffraction peaks are shown (one additional peak is observed at $q = 2.11$ Å⁻¹)

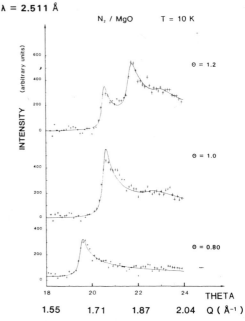

λ = 2.511 Å

N₂ / MgO   T = 10 K

INTENSITY (arbitrary units)

θ = 1.2

θ = 1.0

θ = 0.80

18   20   22   24   THETA
1.55   1.71   1.87   2.04   Q (Å⁻¹)

Fig. 20. Neutron diffraction spectra of nitrogen film adsorbed on MgO around the monolayer coverage regime θ = 1.2 layer for several temperature T [33] . The (2d) S_3 structure is stable in the temperature range ; $10 \leq T \leq 40$ K . Above at T = 50 K a new (2d) structure S_4 appears.

Table 4 . Structural parameters (a , b, γ) of the S_1 , S_2 , S_3 and S_4 unit cells . The commensurability P (2xN) of each (2d) overstructure is also indicated.

parameter	S_1	S_2	S_3	S_4	d^{α}_{111}	d^{β}_{111}
a (Å)	3.84	3.72	3.78	4.06	3.99	4.04
b (Å)	3.84	3.72	3.58	4.06	3.99	4.04
γ (°)	101.6	106.3	108.3	120	120	120
S_M(Å2)	14.45	13.29	12.84	13.79	13.8	14.2
$d_{/\!/}$(Å)	4.86	4.46	4.30			
Θ (layer)	0.8	1.0	1.2	1.2	bulk	bulk
P (2xN)	P(2x"∞")	P(2x3)	P(2x"∞")	h.c.p.		

S_M(Å2) = surface unit cell area ; $d_{/\!/}$(Å) = nearest neighbour molecular distance along the channels direction .

At the present time the surface electric field influence on the structural properties of the N_2 (2d) film is not clear . As for the argon monolayer the channels of the (100) MgO surface have a pronounced influence on the low temperature (2d) structures . The difference comes from the fact that the corrugation along the channels $\mathcal{E}_{/\!/}$ has a strong influence on the Ar atoms (pinning of the Ar monolayer in the commensurate P(2x3) structure) but $\mathcal{E}_{/\!/}$ has no influence on the N_2 molecules (no pinning is observed for the N_2 monolayer along the Channels).

We will continue the N_2 / MgO structural analysis in more detail and we want also to investigate the CO / MgO system by neutron diffraction experiments . In the latter case the electric field influence should be larger . Recently the uniaxial compression of the N_2 monolayer has been confirmed by LEED measurements [13] .

4) Conclusion

Because of the limited space we have restricted our overview mainly to adsorbed (2d) films properties . Interesting results have been also obtained in film's growing mode and premelting surface phenomena studies [37][38] . They are discussed in detail in M. Bienfait's paper .

132

Concerning our purpose the most outstanding results are the following ; new (2d) solid , (2d) phase transition and (2d) phase have been observed for the first time :

- the square (2d) solid , CD_4 / MgO [28] ;
- the " one-dimensional " melting phenomena and the (2d) " liquid crystal " like phase ; Ar / MgO [20] [26] .

Physisorption study on MgO powders of high surface homogeneity is a promising field for future investigation (for instance in the mixed (2d) films properties analysis). Moreover the recent improvement of the MgO single crystal surface homogeneity is interesting. I think that adsorption on MgO surface will be the main development axis in the surface films transition studies .

References

1) A. Thomy and X. Duval , J. Chim. Phys. 67 (1970) 1101 .
2) Proceeding of the Colloque International du C.N.R.S. ; " Phases Bidimensionnelles adsorbées " , J. Physique Fasc. 10 - C4 (Marseille - 1977) .
3) Proceeding of the International Conference on " Ordering in Two Dimensions " , Edit. S.K. Sinha (Lake Geneva - U.S.A. - may 1980) North Holland (1980) .
4) " Phase Transitions in Surface Films " , Edit. J.G. Dash and J. Ruvalds , Nato Advanced Study Institutes Series - B : Physics , Plenum Press (1980) .
5) J.G. Dash, Physics Today, Volume 38 Issue 12, December (1985), 26 .
6) Hoydoo You, S.C. Fain, Sushil Satija and L. Passel, Phys. Rev. Lett. 56 (1986) 244.
7) B. Mutaftschiev, Phys. Rev B, 40 (1989) 779 .
8) B.B. Fisher and W.G. McMillan, J. Chem. Phys. 28 (1958) 549 .
9) T. Takaishi and M. Mohri, Faraday Trans. I 68 (1972) 1921 .
10) J.G. Dash, R. Ecke, J. Stoltenberg, O.E. Viches and O.J. Whittemore, J. Phys. Chem. 82 (1978) 1450 .
11) J.P. Coulomb and O.E. Vilches , J. Phys. (Paris) , 45 (1984) 1381 .
12) T. Meichel, J. Suzanne, C. Girard and C. Girardet , Phys. Rev. B 38 (1988) 3781.
13) T. Angot and J. Suzanne (in press : " The Structure of Surfaces III " , Springer Series in Surface Sciences , Springer - Verlag) .
14) The Oxide Handbook, Editor G.V. Samsonov, Plenum - New York - London , (1973) 23 .
15) J. Jordan, J.P. McTague, J.B. Hasting and L. Passel, Surface Sci. 150 (1985) L82.
16) A.G. Shastri, H.B. Chae, M. Bretz and J. Schwank, J. Phys Chem. 89 (1985) 3761.
17) K.H. Rieder, Surface Sci. 118 (1982) 57 .
18) P.R. Anderson, J. of Colloid and Interface Sci., 43 (1973) 43 .

19) J.P. Coulomb, T.S. Sullivan and O.E. Vilches, Phys. Rev. B $\underline{30}$ (1985) 4753 .

20) K. Madih, Thesis University of Aix - Marseille II (1988) .

21) C. Tessier and Y. Larher , Proc. Int. Conf. " Ordering in Two Dimension " , Edit. S.K. Sinha (North Holland - 1980) p. 163 .

22) D. Degenhardt, H.J. Lauter and R.Frahm , Surface Sci. $\underline{215}$ (1989) 535 .

23) T. Meichel, J. Suzanne et J.M. Gay, C.R. Acad. Sciences Paris $\underline{11}$ (1988) 989 .

24) W. Ruland and H. Tompa, Act. Cryst. A 24 (1968) 93 .

25) D. Degenhardt, Thesis Kiel (1988) .

26) D. Degenhardt, K. Madih, H. Lauter and J.P. Coulomb (to be published)

27) A. Alavi and I.R. McDonald, Mol. Phys. $\underline{69}$ (1990) 703 .

28) J.P. Coulomb, K. Madih, B. Croset and H. Lauter , Phys. Rev. Lett. $\underline{54}$ (1985) 1536 .

29) D.R. Jung, J.Cui, D.R. Frankl, G. Ihm, H.Y. Kim and M.C. Cole, Phys. Rev. B $\underline{40}$ (1989) 11893 .

30) B. Deprick and A. Julg, New J. Chem. $\underline{11}$ (1987) 299 .

31) D.R. Baer, B.A. Fraass, D.H. Riehl and R.O. Simmons, J. Chem. Phys. $\underline{68}$ (1978) 1411 .

32) M. Trabelsi, K. Madih and J.P. Coulomb (in press, Phase Transition Special Issue)

33) M. Trabelsi, K. Madih and J.P. Coulomb (to be published)

34) R.H. Van - Dongen, Thesis Delft (1972) .

35) A. Lakhlifi and Girardet (to be published)

36) A. Alavi and Lynden-Bell (to be published)

37) K. Madih, B. Croset, J.P. Coulomb and H. Lauter, Europhysics Lett. $\underline{8}$ (1989) 459 .

38) J.M. Gay, J. Suzanne and J.P. Coulomb, Phys. Rev. B $\underline{41}$ (1990) 11346 .

NEUTRON SCATTERING STUDIES OF QUANTUM FILMS

H.J.Lauter, H.Godfrin, V.L.P.Frank and P.Leiderer*

Institut Laue-Langevin, BP 156X, F-38042 Grenoble, France
*Institut fur Physik, Universitat Konstanz, D-7750 Konstanz,FRG

INTRODUCTION

The phase diagrams of a monolayer of adsorbed gases or light molecules on flat substrates look similar to the phase diagrams in 3-dimensions (3-D). This means that the usual coexistence regions, the triple point and the critical point are present as well in 2-dimensions (2-D) as in 3-D [1]. The phase diagrams can be studied by adsorption isotherm or heat capacity measurements, which reveal mainly the coexistence regions or the phase boundaries, respectively. However, the substrate can not always be regarded to be ideally flat. In many cases the adsorbate does see the adsorption sites of the substrate and locks into a commensurate phase (C-phase). In the case of graphite as substrate the $(\sqrt{3} \times \sqrt{3})$ R 30° overstructure is seen in many cases. This structure is shown in fig.1. The heavier rare gases and the light molecules (N_2, CD_4) exhibit the C-phase only if the lattice parameter of the dense plane in 3-D is close to the nearest neighbor distance in the C-phase. This is the case because the nearest neighbor distance nearly does not change in the densest plane if the adjacent planes are taken off which means the dimensionality changes from 3-D to 2-D. The quantum gases however all show a C-phase in their 2-D phase diagram despite a much

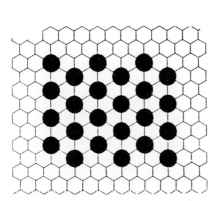

Fig.1: Model of the commensurate $(\sqrt{3} \times \sqrt{3})$ R 30° overstructure on graphite.

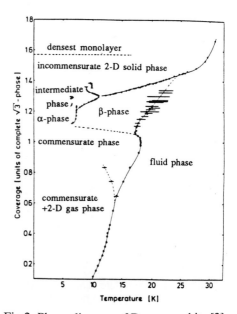

Fig.2: Phase-diagram of D_2 on graphite [2].

Phase Transitions in Surface Films 2
Edited by H. Taub *et al.*, Plenum Press, New York, 1991

denser nearest neighbor distance in 3-D. This is due to the zero-point motion which gives a repulsive contribution to the nearest neighbor interaction and consequently a high compressibility to the system. Thus if as before the dimensionality is changed and many of nearest neighbor atoms are missing the 2-D lattice expands. If in addition the corrugation of the adsorption potential is added, the quantum gases recognize the dilute density structure of the C-phase as the ground state.

The phase diagram of D_2 on graphite is shown in figure 2 as an example of an adsorbed quantum gas. The location of the phase boundaries has been determined by heat capacity measurements [2]. The definite allocation of the different phases to structures is given by scattering techniques. In this case it has been done by neutron diffraction and LEED [3,4].

The inelastic neutron scattering gives additional information of the interaction between the adsorbed particles itself but also between the adsorbate and the substrate. In particular interesting is the search for the phonon gap at the zone center which characterizes the loss of the translational invariance of the adsorbed layer in the C-phase. The transition from the C-phase to the incommensurate higher density phase is characterized by different intermediate phases as seen in figure 1. Theories of the commensutrate-incommensurate transition predict domain walls in the transition region. In these domain wall phases the commensurate phase is still locally present. Thus the study of the phonon gap will reveal interesting features.

EXPERIMENTAL

The experimental set-up is described in Ref.5. The sample consists of a stack of exfoliated graphite sheets (Papyex) with a diameter of 2 cm and a height of about 7 cm. The surface area is in the order of 200 m^2. The graphite is a 2-D powder. Only the axis perpendicular to the basal planes of the graphite shows a certain order of a mosaicity of about 30° FWHM. The coherence length of the adsorbed layer is about 300 Å in the C-phase. The basal planes of the graphite are parallel to the scattering plane of the neutron spectrometer. For the inelastic studies the IN3 spectrometer of the ILL has been used with a fixed final energy of 1THz and a Be-filter on the analyser end. The resolution across the elastic line was 0.03THz. For the elastic studies ZYX-graphite was used with a coherence length of about 2000 Å.

The density of the adsorbate was controlled by adsorption isotherms. But also the highest intensity of the Bragg-peak of the adsorbate in the C-phase as a function of coverage at constant temperature determines the best commensurate phase $\rho=1$. $\rho=1$ means that all adsorption sites in the C-phase are occupied by adsorbed atoms or molecules. The definition of $\rho=1$ with diffraction is within 2-3% identical with the $\rho=1$ coverage defined by the highest melting temperature of the C-phase in figure 2. The small difference may be a real temperature effect.

MEASUREMENTS

I. THE ($\sqrt{3}$ x $\sqrt{3}$) PHASE

The verification of the C-phase has to be done by diffraction. In fig.3 neutron diffraction patterns are shown at various average adsorbate densities [3]. The spectrum A at a slightly overfilled C-phase shows the maximum intensity at a momentum transfer $Q=1.702$ Å^{-1} which corresponds for a triangular lattice to 4.26 Å, the nearest neighbor distance in the C-phase (fig.1). All the spectra shown, the elastic and the inelastic ones, show difference counts. The signal from the cell and substarte without adsorbate has always been subtracted.

The inelastic neutron measurements have been taken at two different momentum transfers Q. The reason for that is depicted in fig.4 which shows the reciprocal space of the triangular lattice of the C-phase. The scan taken with a $Q=1.7$ Å^{-1} collects all excitations with wave vector q along the circle with radius Q with the help of the Bragg-points which are marked by the vector τ. These excitations with wave vector q have mainly transverse polarization and the highest intensity is expected from the zone boundary phonons because of the high density of states. If a phonon gap exists at the zone center a second high density of states is expected at this point. As the 2-D powder averaging crosses also the Γ-points a second peak is expected in energy in the scan with a Q=1.7 $\mathring{A}^{-1}$.

The scan taken with Q=0.85 $\mathring{A}^{-1}$ collects the longitudinal zone boundary phonons due to the 2-D powder averaging (fig.4). So with the correct choice of the momentum transfer different modes can be separated even in a powder-like sample.

The scans with the different momentum transfers are shown in figure 5. The data points in fig.5a (Q=1.7 $\mathring{A}^{-1}$) show clearly a double peak. The one at lower energies represents the phonon gap, whereas the one at higher energies shows the collected transverse zone boundary phonons. The scan with a Q=0.85 $\mathring{A}^{-1}$ (fig.5b) shows only one peak which results from collected longitudinal zone boundary phonons. The fit to the data is a two

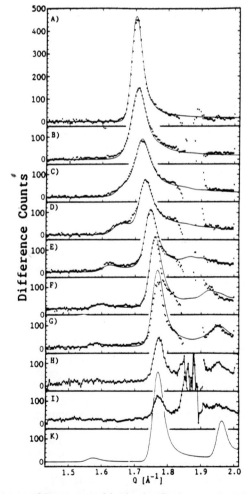

Fig.3: Diffraction pattern of D_2 on graphite in the C-phase and α-phase at various average densities of the adsorbate; A) ρ=1.07, B) ρ=1.10, C) ρ=1.11, D) ρ=1.13, E) ρ=1.16, F) ρ=1.20, G) ρ=1.23, H) ρ=1.24, I) ρ=1.25; T=2K. The solid lines are fits with the model in fig.7 [3].

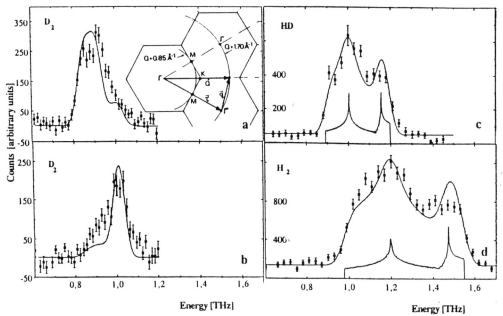

Fig.4 and fig.5: Fig.4 is the inset in fig.5a. Fig.4 shows the reciprocal space of a triangular lattice. The circles are shown on which phonons are collected due to the powder averaging for Q=1.7 Å^{-1} and Q=0.85 Å^{-1}. Q is the total momentum transfer; τ is a reciprocal lattice vector and q is the phonon wave vector [3]. Fig.5: Neutron inelastic data of the hydrogen isotopes adsorbed on graphite in the C-phase [8]. Q=0.85 Å^{-1} in b and Q=1.7 Å^{-1} in a,c and d. In a and b the spectra taken with D$_2$ are shown, which is a coherent scatterer. Whereas in c and d HD and H$_2$ is used, respectively, which are incoherent scatterers for neutrons. So the density of states is measured.

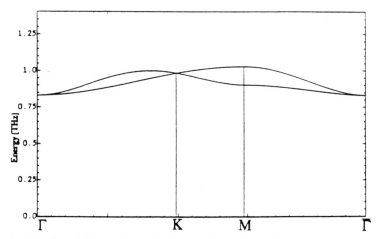

Fig.6: Calculated dispersion relation for D$_2$ on graphite in the C-phase [5].

138

parameter fit. The adsorbate molecules are thought to be connected by each other by a net of springs and each molecule is connected by a spring to the substrate [5]. The model is a very simple one and is not able to reproduce the clear separation between the phonon gap and the transverse zone boundary phonons seen in the data (fig.5a). But it is however interesting to note the difference in the spring constants. α is the spring constant between the adatoms and

Table 1: Parameters obtained from the fits characterizing the density of states for the in-plane modes of the C-phase of the hydrogen atoms adsorbed on graphite. In parenthesis are given the values of Ref.7.. Indicated are the following: z.c.gap is the zone center gap energy, width is the width of the density of states, trans.peak and long.peak are the peaks arising from the transverse and longitudinal phonons in the density of states, respectively. All values are in Kelvin (48 K = 1THz = 4.14 meV). (* The experimental data of fig.5a suggest a value of 43.3K for the transverse peak)

	H_2	HD	D_2
z.c.gap	47.3 (46.6)	43.2	40.0 (36.9)
width	27.5 (42.1)	14.7	9.5 (14.8)
trans.peak	57.9 (64.9)	48.8	43.3*(44.2)
long.peak	71.4 (83.8)	55.8	48.1 (50.3)

Table 2: Parameters that characterize the adsorbate - graphite system in the commensurate phase for Kr^{11}, CD_4^{12}, N_2^{13}, D_2^2, HD and H_2^3 and $^3He^{14}$, the adsorbates are ordered with increasing de Boer parameter.
[a] The de Boer parameter indicates the quantum character of the adsorbate.
[b] This column presents the width of the in-plane phonon density of states (DOS). These values are not always easy to determine with inelastic neutron scattering, since the intensity of the structure factor decreases with increasing energy.
[c] The phase diagrams of Kr, N2 and CD4 present a commensurate phase region that extends to higher temperatures, when the total coverage is slightly higher than the commensurate one. Details of the phase diagrams can be found in refs.15, 16 and 17, respectively.

Adsorbate	Mass [a.u.]	de Boer[a] parameter	Lennard-Jones parameter		Gap energy [K]	Gap ratio $\Delta_{meas}/\Delta_{calc}$	DOS width[b] [K]	Melting Temperature
			ε [K]	σ[Å]				
Kr	83.8	0.10	165.3	3.63	8.7	1,0	-	~125[c]
CD4	20.0	0.23	137.0	3.68	14.5	0,9	48	~55[c]
N2	28.0	0.42	35.6	3.32	19.3	1,7	34	~72[c]
D2	4.0	1.26	35.2	2.95	40.0	1,1	9.5	18.5
HD	3.0	1.43	35.9	2.95	43.2	1,1	14.7	19.4
H2	2.0	1.74	36.7	2.95	47.3	1,0	27.5	20.5
3He	3.0	3.10	10.2	2.56	10.9	0,3	40	3.05

β the in-plane one between the adsorbate molecule and the substrate at the adsorption site. α has been determined to 0.016 N/m and β to 0.182 N/m. The ratio is about 1/10. This small ratio is equally well seen in the rather flat dispersion in fig.6. Thus the molecule exhibits nearly "Einstein behavior" [6], but the influence of adatom-adatom interaction is still visible through the dispersion. Calculations [7] are in good agreement with the value of the energy of the phonon gap, however the adatom-adatom interaction visible by the width of the density of states is still by a factor of 1.5 too small (see table 1). This indicates that probably not the parameters at the minimum of the interaction potential have to be modified but the shape of this potential at a larger intermolecular spacing due to the increased intermolecular spacing in the C-phase.

The effect of the isotope mass can be probed by using in addition to D_2 also HD and H_2 as adsorbate [8]. The spectra are shown in fig.5c and 5d. Both spectra make use of the incoherent scattering cross section of HD and H_2. The calculated density of states is seen as a solid line in the lower part of the figures. The line through the points is the density of states folded with the resolution of the instrument. The characteristic values of the dispersion relations are summarized in table 1. The same model as described for D_2 has been applied to the other isotopes. The theory [7] describes well the isotope shift which is not only due to the different mass but also due to the anharmonicity of the potentials. The isotope effect is seen with decreasing mass as well in the shift of the phonon gap to higher energies as in an increasing width of the density of states. As with D_2 the width of the density of states is wider by the same factor 1.5 (table 1).

The C-phase has also been detected for other adsorbates and a collection of gap energies and widths of the density of states is given in table 2 [9,10]. The gap energy is related to the curvature of the corrugation of the in-plane adsorption potential at the adsorption site. The agreement between theoretical calculations of the gap energy and the measured values is indicated in the 7th column of table 2 [7,9,10,18]. The effective curvature is influenced by the movement of the atoms or molecules in the adsorption potential. All the quantum gases are governed by the zero point movement, so that the gap stays nearly unchanged up to the melting temperature of the C-phase. Kr and CD_4, however, show a lowering of the gap energy by a factor of 2 when the temperature is raised to the melting.temperature due to the enhanced mean-square displacement of the adsorbate and the anharmonicity of the adsorption potential [9,19,20].

Also the width of the density of states exhibits the difference between the quantum gases and the heavier gases. The quantum gases show a smaller width of the density of states, because the nearest neigbor distance in the C-phase is increased with respect to the one in 3-D. This leads to a lower interaction between adsorbate atoms. The heavier gases show more interaction, because the nearest neigbor distance in 3-D matches the one in 2-D. The high value of ^3He should be taken with care. On one side the very high zero point motion may allow for a higher adsorbate-adsorbate interaction, but on the other side we are not absolutely sure about the interpretation of the data [14].

II. THE COMMENSURATE-INCOMMENSURATE TRANSITION

Once the inelastic signals are understood in the C-phase the adjacent phases can be investigated. This will be described first for the α-phase. In fig.3 several scans are shown taken at different coverages and constant temperature. They follow a path from the C-phase into the α-phase (fig.2). The diffraction peak shifts with increasing density to higher Q-values according to the compression. At the same time satellites are moving outwards from the main peak. The feature around 1.88 Å^{-1} is the (002) graphite peak, which is due to an interference phenomena. It is of no importance for this study and cuts unfortunately out a

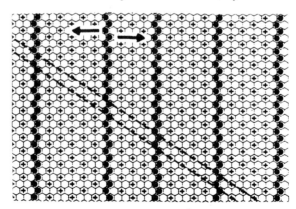

Fig.7: Striped superheavy domain wall model for D_2 on graphite [3].

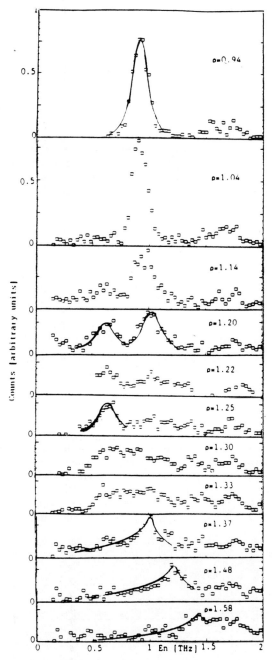

Fig.8: Neutron inelastic data of D_2 on graphite at T=4K for various coverages (Q=1.7 Å^{-1}) [9]. The lines are guides to the eye.

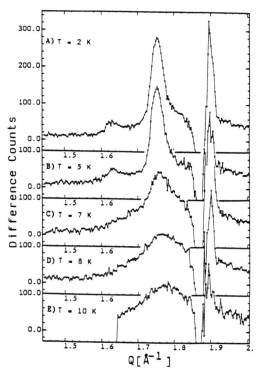

Fig.9: Diffraction pattern of D_2 on graphite at constant coverage $\rho=1.16$ (see fig.2) at various temperatures showing the α-β transition between 5K and 7K[3b].

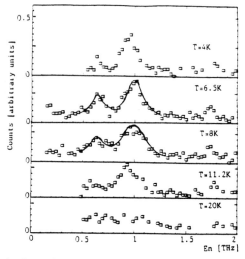

Fig.10: Neutron inelastic data of D_2 on graphite at constant coverage $\rho=1.16$ (see fig.2) at various temperatures across the α-β transition [9]. The lines are guides to the eye.

certain region in Q, where the signal of the adsorbate is not observable. The fit to the data has been done with a model of striped superheavy domain walls [3,4] which is depicted in fig.7. For a more detailed discussion the reader is referred to the Refs.3 and 4. It should only be mentioned that the crosses in fig.7 represent the molecules still in the commensurate position. The filled circles mark the molecules in the domain walls in an ideal position. However they are too close and a certain relaxation indicated by the arrows takes place. In the applied model the distance between the domain walls has a distribution which depends on the domain wall density. The width of the domain walls is fairly small. It extends only across 4 rows of molecules. This is again due to the high compressibilty of the 2-D quantum gases as mentioned in the introduction.

The inelastic study of the domain wall structure is shown in fig.8 [21]. The top two spectra are taken in the C-phase and show the phonon gap and the transverse zone boundary phonon in one peak due to the relaxed instrument resolution (0.06 THz) used for these scans. If the average density is further increased the α-phase is entered (see fig.2) and the signal around 1 THz starts to decrease. This is understood as due to the amount of the molecules in the C-phase decreasing due to the model of the domain walls. At the same time a signal at about 0.6 THz appears which finally at $\rho=1.25$ is the only signal to be seen. This signal attributed to an excitation of the domain walls is still to be calculated. It is best at the densest α-phase ($\rho=1.25$, see fig.2). This is a first example of the usefulness of the inelastic studies to complete and to understand the events in this phase diagram.

The following scans in fig.8 show that in the region between $\rho=1.30$ and $\rho=1.33$ the signals can not be any more resolved probably due to too many different excitations in this phase which was modeled by a hexagonal heavy domain wall structure [3,4]. In the region beyond the density of $\rho=1.33$ the inelastic response changes again and indicates the pure transverse zone boundary phonon of an incommensurate 2-D solid.

The next object to study is the β-phase. In fig.9 diffraction patterns are shown taken at constant coverage $\rho=1.16$ as a function of temperature [3]. At T=2K and 5K the already known satellite structure of the α-phase is seen, which could be modeled by striped superheavy domain walls. For higher temperatures than the $\alpha-\beta$ transition (fig.2) the spectrum looks like a liquid structure factor in particular if the temperature is raised. The structure of this "reentrant liquid" was not known, but got new interest because this phase seems to be separated from the normal 2-D liquid by very broad peaks in specific heat (fig.2). The inelastic neutron measurements are shown in fig.10. They have also been taken with the relaxed resolution (fig.8). So again the excitations of the commensurate parts (phonon gap and transverse zone boundary) are seen and at lower energy the signal from the domain walls. Here no change in the spectrum is seen if the temperature crosses the $\alpha-\beta$ transition at 7.2 K. The consequence is that the domain walls do still exist in the β-phase because the excitation belonging to them are still visible and also the excitations from a commensurate phase. The solution for a structural picture is to introduce patches of domain walls as depicted in fig.11. This model still allows for the inelastic features and the structure factor, seen as inset in fig.11 [3b], fits the data. This is a proof that the β-phase is a disordered domain wall phase. The evolution with temperature can be modeled by shorter and shorter domain walls until melting around 20K. Perhaps the unbinding of domain walls can be treated in the class of Kosterlitz-Thouless transitions [22,23].

The phase diagram of the monolayer of the heliums adsorbed on graphite are similar to the ones of the hydrogens in that .a pronounced C-phase shows up with a transition to the incommensurate phase probably via a striped superheavy domain wall phase [24-26]. We present here the identification of the domain wall (D) phase of ^{3}He as a striped superheavy domain wall phase.

In fig.12 the explored coverage range between 0.05 and 0.35 atom/$Å^2$ of ^{3}He on graphite is shown. The density of the adsorbed layer can be calculated from the diffraction peak position if a homogeneous triangular lattice is assumed. The first five points in fig.11 show all the same ordinate of 0.0636 at/$Å^2$, the density of the C-phase. The abscissa's calibration (the total number of adsorbed atoms on the surface of the sample) is given by the amount of

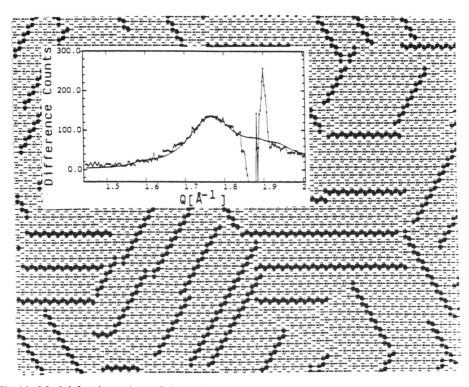

Fig.11: Model for the β-phase. It is made up of patches of the α–phase structure [3b]. The inset shows an overlay of a measured spectrum in the β-phase (spectrum c in fig.9) and calculated one using the structure from the same figure [3b].

³He that produces the highest Bragg-peak intensity within the C-phase. This defines the best C-phase: all adsorption sites of the C-phase on the graphite are occupied by an adsorbate atom. This calibration was first made with D_2 on graphite [3] and defined a value of 14.34 cc (STP) of gas. This agrees with ³He, despite its low scattering power and high neutron absorption cross section.

The straight full line in fig.12 through the best C-phase point and the origin does not match the points in the IC1-phase. This shows clearly that the calibration of the C-phase and of the incommensurate (IC1) phase do not coincide. One possible explanation is that the effective surface area of the C-phase is smaller than the one of the IC1-phase. Another possibility may be that in the IC1-phase atoms are additionally adsorbed on other planes than the basal ones. The calibration of the IC1-phase is shown in fig.12 by the dashed line, which does not pass through the origin. The difference at monolayer completion amounts already to 6% .

The deduced densities in the D-phase (fig.2) show a behavior that is equivalent to the case of D_2 [3,4], which has been identified as a striped superheavy domain wall phase. The much less favorable scattering conditions of ³He with respect to D_2 made it impossible to measure any satellites. However, the characteristic coverage dependence in fig.12 is a proof that this domain wall phase exists as proposed by heat capacity measurements for ⁴He [25]. The density deduced from the diffraction peak position does of course not represent any more the real density of the layer, but serves just for visualization. The densest stage is reached when the superheavy walls are separated by one row of atoms in commensurate position (7) as for D_2 [3,4].

III. THE SECOND ADSORBED LAYER

For the helium isotopes, the second adsorbed layer is less dense than the first one [27] in contrast to the hydrogen isotopes [28], where both layers have the same density. The weaker adsorption potential and the enhanced zero point motion of the helium atoms are clearly visible in the low density of the second layer. Under the pressure of the second layer the density of the first layer increases slightly and finally locks into a 8*8 and 9*9 overstructure for ³He and ⁴He [3b], respectively.

The second layer promotion is evidenced as a sharp knee between the IC1 and IC2-phase in fig.12. The increasing pressure of the second layer on the first one induces a further compression of the first layer, seen by the small slope in the IC2-phase. This slope increases slightly towards the S-phase, changing over to a zero slope at 0.2 at/Å². At this point the density of the first layer (0.1106 at/Å²) corresponds to an 8*8 overstructure with respect to the graphite, having 37 atoms per unit cell. The approach to this clear lock-in is marked in the S-phase by some diffraction peaks, whose line shape is better described by a double peak. This splitting is shown in the insert of fig.12. It can be interpreted as a slight distortion of the unit cell due to a one directional registry with the substrate. This partial registry agrees for two points with the 4*4 lattice spacing and for one point with the 8*8 one, before finally the pure 8*8 overstructure is reached.

The diffraction of the second layer itself could be seen only at higher coverages around a total density of 0.3 at/Å² (fig.12)
Similar features in the bilayer region are seen for ⁴He on graphite in fig.13 [3b]. For intensity reasons the signal from the C-phase could not be seen. Thus only the peak positions from the IC1-phase (see fig.12) appear at low densities. For the same reason we could not calibrate the x-axis on the best C-phase as was done in fig.12. Thus the x-axis is in square root of the amount of the filling. The substrate used is Papyex with a coherence length of about 350 Å. But in addition the data obtained with ZYX-substrate (coherence length about 2000 Å) are shown in fig.13 with a rescaling at the monolayer completion for the x-axis.

Again the second layer promotion is evidenced as a sharp knee at a density of 0.112 at/Å² (³He at 0.106 at/Å² in fig.12) and the increasing pressure of the second layer induces a small further compression of the first layer At a filling of 19 cc³/² the first layer locks into an

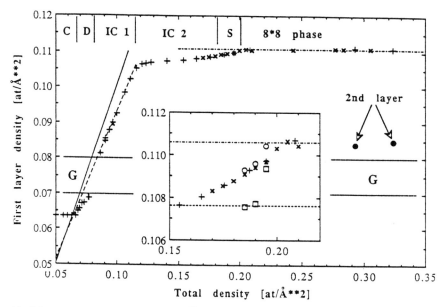

Fig.12: Measured density of the first ^{3}He layer vs. total coverage at T=1.0K (+) and at T=0.06K (x). (□) and (o) indicate the splitting of the diffraction peak in the insert. (•) is the position of the second layer density. The identification of the different phases is described in the text. **G** marks a region disturbed by a graphite background reflection. The full and dashed line depict the monolayer behavior. The dashed-dotted line shows the density of the 8*8 phase. The dashed line (insert) indicates the 4*4 density.

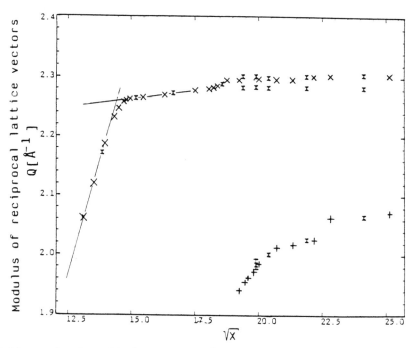

Fig.13: Measured reciprocal lattice vectors for ^{4}He adsorbed on Papyex (x) at T=1.2 K as a function of the square root of the total coverage taken in units of cm^3 STP ($\sqrt{X}$). (x) and (+) are the ZYX data, rescaled to the coverage at the monolayer completion for the Papyex sample.

overstructure with respect to the graphite. This time it is the 9*9 overstructure for the ZYX-substrate. For the Papyex a splitting of the diffraction peak is observed, which indicates a one directional registry with the substrate induced by the small coherence length of this substrate.

The second layer diffraction peaks become visible at a filling of 19 cc$^{3/2}$ and a compression of the second layer density is seen with the third layer promotion in fig.13.[3b]

IV. THE ^{4}HE FILM

A ^{4}He film on graphite is composed of two solid layers adjacent to the substrate (below 2K) and subsequent liquid layers (see e.g.Ref.13). So the liquid ^{4}He film has two boundaries,

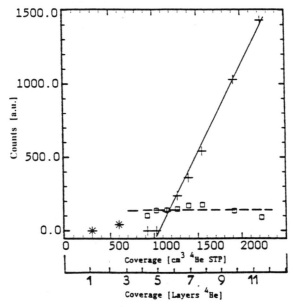

Fig.14: Intensities of the bulk signal (+) and the interface signal (o) at the energy of 0.6 meV as a function of coverage (^{4}He on graphite powder [31]). The signal at a coverage of a monolayer and at 2.6 layers is marked by (*) (it is a constant background !); T=0.8K.

the solid-liquid one and the liquid-gas one. Excitations can propagate along these interfaces, which are the freezing-melting wave [29] and the ripplon [30], respectively. Any excitation with a dispersion like a freezing wave could not yet be measured with neutron scattering. But in addition to the signals arising from the bulk ^{4}He some modes could be detected which have no dispersion (the energy does not change as a function with wave vector). These modes (at 0.4 meV, 0.6 meV....) are localized at the solid-liquid interface because they still exist if the sample cell is completely filled with helium (this means that the liquid-gas interface is suppressed) [31]. On the other hand these modes disappear between a coverage of 2.5 and 4 layers as shown in fig.14. This range has unfortunately not been investigated in more detail. But for coverages beyond 4 the intensity of this mode does not increase as function of coverage in agreement with the explanation that it is bound at the solid-liquid interface. These modes can be taken to explain the high transmission of phonons through the interface between a soilid and liquid helium (Kapitza-resistance). In contrast the bulk signal increases linearly with coverage. It extrapolates to zero at a coverage of 5 layers. This indicates that in addition to the two solid layers three liquid layers do not contribute to superfluidity, if superfluidity is connected to the measurability of the phonon-roton dispersion curve.This statement will be refined for the 4He film on Payex-graphite, which is described in the following.

At the second interface a ripplon should be visible. Indeed it could be measured [31-33] and shows the dispersion expected from theory [30]. The attribution of this mode to be a ripplon was strengthened by the fact that this mode disappears if the liquid-gas interface is suppressed by filling the sample cell completely with helium. This is visualized in the figs. 15 and 16. In the figures the different colors indicate the behavior of the intensity as a function of energy and momentum transfer. It is clearly seen in fig.15 that besides the intensity on the phonon-roton curve there is intensity on an energetically lower lying branch. This branch coincides with the calculated dispersion of the ripplon using the parameter set in Ref.20. The agreement is very good, it seems to be so even up to 1.5A-1. At still higher Q the roton intensity combined with the one of the flat modes becomes to high to distinguish the ripplon signal. This good agreement allows to say that the temperature dependence of the surface tension is really based on an experimentally verified dispersion relation. It still remains to prove the modified parameter set [20] by a theory.

In fig.16 the result of the completely filled sample cell is shown. In the region where the ripplon should show up, the colors are the same as in fig.15. Only near the phonon-roton intensity and the flat bar due to the multiple scattering [31,32] the attribution of the colors to intensity has been modified. Thus this figure shows that no signal of the ripplon intensity is visible in fig.16 the filled cell data, although below 0.7 Å-1 it would have been distinguishable from the overwhelming quasi bulk phonon roton intensity. This disappearance proves that the ripplon signal is really bound to the gas-liquid interface.

The fig.17 shows the evolution of the signal height of the ripplon with coverage. Again the saturation shows that this mode is bound to an interface. It becomes visible above 3 layers (two of them solid !). Like in fig.14 the evolution of the bulk signal is exhibited. The extrapolation is made only with coverages below 5 layers. This shows that the bulk signal disappears at about 3.5 layers. There is no discrepancy with the dependence in fig.14 because here in fig.17 the scale is much finer and in principle a linear tail of the bulk signal height as a function of coverage is seen. (A rough extrapolation from still higher coverages extrapolates to about 4.5 layers in agreements with fig.14.)

In conclusion it has been shown that a lot of information can be drawn from adsorbates even on powder like graphite substrates using elastic and inelastic neutron scattering. These studies give insight into the dynamical behavior of the adsorbates in the commensurate phase itself, but which is also very valuable for the modelling of 2-D phases in the demonstrated case of the commensurate-incommensurate transition. All the adsorbed quantum gases have a commensurate-incommensurate transition, where the striped superheavy domain wall phase appears as intermediate phase. This domain wall phase shows again phase transitions as a function of temperature which implies the creation of shorter pieces of domain walls. The evolution of the first layer structure is demonstrated towards the densest monolayer and under the pressure of the second layer. Due to the second layer pressure the first layer of the helium show overstructures with the substrate. In particular interesting are the interface excitations of an 4He film. At the solid-liquid helium interface localized excitations could be detected and at the liquid-gas helium interface the ripplon could be measured. Thus an overview has been given what can be measured with neutron scattering in the case of adsorbed quantum gases on graphite ranging from a submonolayer to a film.

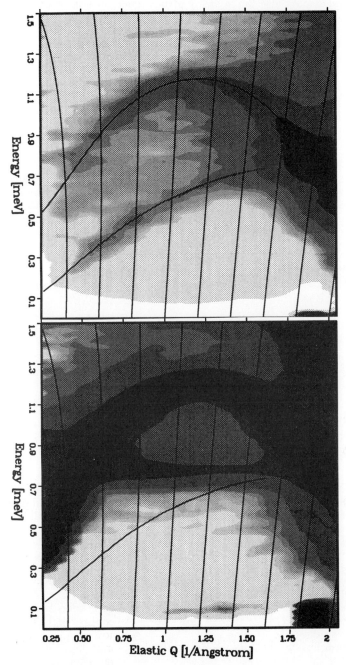

Fig.15 (top)and 16 (bottom): Intensity along the phonon-roton curve and along the ripplon curve in the energy-Q plane. The intensities originately displayed as different colors are visualized by the graduation from white to black,which marks the highest intensity (e.g. at the roton minimum in fig.15). Fig.15 shows the signal from 5.06 adsorbed layers on Papyex and fig.16 the signal from the completely with ^{4}He filled sample cell. In both figures the phonon-roton curve of bulk helium is seen as a black line as well as the ripplon dispersion.

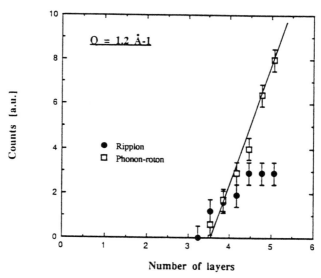

Fig.17: Intensities of the bulk signal (□) and the ripplon (●) as a function of coverage (^{4}He on Papyex-graphite); T=0.6K.

ACKNOWLEDGEMENTS

This work has been partially supported by the Federal Ministry of Research and Technology (BMFT), F.R.G.

REFERENCES

1. Thomy A., Duval X. and Regnier J., Surf.Sci.Rep.1, 1 (!981)
2. Freimuth H. and Wiechert H., Surf.Sci. 178, 716 (1986)
3a.Schildberg H.P., Lauter H.J., Freimuth H., Wiechert H. and Haensel, Jap.J. Appl. Phys.26, 345 (1987),(Proc.18th Int.Conf.on Low Temperature Physics, Kyoto);
3b.Schildberg H.P., Thesis, University of Kiel R.F.A. (1988)
3c. H.J.Lauter, in "Phonons 89", Hunklinger. S., Ludwig W. and Weiss G.edts.(World Scientific, 1990) p.871
4. Cui.J, Fain S.C., Freimuth H., Wiechert H., Schidberg H.P. and Lauter H.J., Phys.Rev.Lett.60, 1848 (1987)
5. Frank V.L.P., Lauter H.J. and Leiderer P., Phys.Rev.Lett.61, 436 (1988)
6. Nielsen M., McTague J.P. and Passell L, Phase Transitions in Surface Films, Plenum Press (1980) p.127
7. Novaco A.D., Phys.Rev.Lett.60, 2058 (1988)
8. Lauter H.J., Frank V.L.P., Leiderer P. and Wiechert H., PhysicaB156&157, 280 (1989)
9. Lauter H.J., Frank V.L.P., Taub H. and Leiderer P.,LT-19, PhysicaB165&166, 611 (1990)
   Frank V.L.P., Lauter H.J., Godfrin H. and Leiderer P., NATO Workshop, Exeter 1990
10 Bruch L.W. same NATO Course
11 Frank V.L.P., Lauter H.J. and Taub H., unpublished data
12 Moeller T., Lauter H.J., Frank V.L.P. and Leiderer P., "Phonons 89", Hunklinger. S., Ludwig W. and Weiss G.edts.(World Scientific, Singapore 1990) p.919
13 Hansen F.Y., Frank V.L.P., Taub H., Bruch L.W., Lauter H.J. and Dennison J.R., Phys.Rev.Lett.64, 764 (1990)
14 Frank V.L.P., Lauter H.J., Godfrin H. and Leiderer P.,"Phonons 89", Hunklinger. S., Ludwig W. and Weiss G.edts.(World Scientific, Singapore 1990) p.1001

15 Butler D.M., Litzinger J.A. and Steward G.A., Phys.Rev.Lett.44, 466 (1980)

16 Chan M.H.W., Migone A.D., Miner K.D. and Li Z.R., Phys.Rev.B.30, 2681 (1984)

17 Kim H.K., Zhang Q.M.and Chan M.H.W., Phys.Rev.B 34, 4699 (1986)

18 Ni X.Z. and Bruch L.W., Phys Rev.B 33, 4584 (1986)

19 Hakim T.M., Glyde H.R. and Chui S.T., Phys.Rev.B 37, 974 (1988)

20 Frank V.L.P., Lauter H.J., Godfrin H. and Leiderer P.,"Phonons 89", (World Scientific, Singapore 1990) p.913

21.Frank V.L.P, Lauter H.J. and Leiderer P., Jap.J.Appl.Phys.26, 347 (1987),(Proc.18th Int.Conf.on Low Temperature Physics, Kyoto)

22.Kosterlitz M. and Thouless D.L., Prog in Low Temp.Phys. VII B, 371 (1987)

23.Halpin-Healy T. and Kardar M., Phys.Rev.B 34, 318 (1986)

24.Hering S.V, Van Sciver S.W and Vilches O.E., J.Low Temp.Phys. 25, 793 (1976)

25.Motteler F.C, Thesis (Univ. of Washington, 1985)

26.Lauter H.J., Godfrin H., Frank V.L.P. and Schildberg H.P., LT-19, Physica B165&166, 597 (1990)

27.Lauter H.J.,Schildberg H.P., Godfrin H., Wiechert H.and Haensel R., Can.J.Phys. 65, 1435 (1987)

28.H.P.Schildberg, H.J.Lauter, H.Freimuth, H.Wiechert and R.Haensel, Jpn.J.Appl. Phys. 26, 343 (1987)

29.Keshishev K.O., Parshin A.Ya. and Babkin A.B., Sov.Phys. JETP 53, 362 (1981)

30.Edwards D.O. and Saam W.F., Prog. in Low Temp.Phys. VII A, 283 (1978)

31.Lauter H.J., Frank V.L.P., Godfrin H. and Leiderer P., Elementary Excitations in Quantum Fluids, Ohbayashi K and Watabe M. Eds.,Springer Series in Solid-State Sciences 79 (1989) p.99
Lauter H.J., Godfrin H. and Wiechert H., in "Phonon Physics",Kollar J.et. al. edts., (World Scientific, Singapore 1985) p.842

32.Godfrin H., Frank V.L.P., Lauter H.J. and Leiderer P.,"Phonons 89", Hunklinger. S., Ludwig W. and Weiss G.edts.(World Scientific, Singapore 1990) p.904

33.Lauter H.J., Godfrin H., Frank V.L.P. and Leiderer P., NATO Workshop, Exeter 1990

A MOLECULAR DYNAMICS STUDY OF THE EFFECT OF STERIC PROPER-

TIES ON THE MELTING OF QUASI TWO-DIMENSIONAL SYSTEMS

Flemming Y. Hansen

Fysisk-Kemisk Institut,The Technical University
of Denmark, DTH 206, DK-2800 Lyngby, Denmark

H. Taub

Department of Physics and Astronomy, University
of Missouri, Columbia, Mo. 65211, USA

ABSTRACT

The effect of molecular steric properties on the melting of quasi two-dimensional solids is investigated by comparing molecular dynamics simulations of the melting of
butane and hexane monolayers adsorbed on the graphite (002)
surface. These molecules differ only in their length, being
members of the n-alkane series $[CH_3(CH_2)_nCH_3]$, where n=2 for
butane and n=4 for hexane and have similar solid monolayer
structures on graphite. The simulations show a qualitatively
different melting behavior for the butane and hexane
monolayers consistent with neutron and x ray scattering
experiments. The melting of the low-temperature herringbone
(hb) phase of the butane monolayer is abrupt and characterized by a simultaneous breakdown of translational order and
orientational order of the molecules about the surface
normal. In contrast, the hexane monolayer exhibits polymorphism in that the solid (hb) phase transforms to a rectangular centered (rc) structure with a short coherence
length in coexistence with a fluid phase. The formation of
gauche molecules is essential for the melting process in the
hexane monolayer but unimportant for butane. A mechanism for
the melting of the (hb) phase is proposed. It is based on
vacancy creation within the monolayer by molecules reducing
their "footprint" on the surface either by a conformation
change, tilting with respect to the surface, or promotion to
a second layer.

1. INTRODUCTION

Interest in two-dimensional (2D) melting has been particularly stimulated by a theory of Kosterlitz, Thouless,[1], Halperin, Nelson [2,3] and Young [4] (KTHNY). In

Phase Transitions in Surface Films 2
Edited by H. Taub *et al.*, Plenum Press, New York, 1991

this theory, melting occurs via two *continuous* transitions as opposed to a single first-order transition in three-dimensional (3D) systems. The first transition in the melting process involves the unbinding of dislocation pairs in the 2D solid and results in a fluid, termed the 'hexatic phase', which has quasi long-range orientational order but short-range translational order. In this context, orientational order refers to the ordering of the directions of vectors between neighboring atoms; short-range correlations are defined to vanish exponentially as exp(-ar) with separation r while quasi long-range correlations decay algebraically as $r^{-\gamma}$. The second transition in the melting process is from the hexatic phase to a normal, isotropic fluid with short-range orientational order. This fundamentally different melting behavior is related to the absence of conventional long-range order in 2D solids [5].

Since the KTHNY theory assumes a strictly 2D system, it can be questioned whether it applies to monolayer films physisorbed on solid substrates. In such films, motion of the adsorbed atoms or molecules perpendicular to the surface is possible; and further, there is a lateral periodic variation of the adatom-surface potential, the so-called corrugation in the holding potential. For monolayers which form structures commensurate with the substrate corrugation, the KTHNY theory does not apply and the melting may be better described by a lattice gas model.

Physisorbed monolayers which are incommensurate with a substrate are most closely approximated be the 2D models. Strandburg [6] has reviewed experimental investigations and computer simulations of the melting of such films. To date, only experiments with ethylene [7,8] and possibly ethane on graphite [9], have exhibited heat capacity signatures indicative of continuous melting; however, details of the transition remain unclear. Extensive computer simulations of the rare gas monolayers xenon, krypton and argon on graphite have been done by Abraham [10-13]. The general conclusion from these studies is that the transitions in the monolayers are consistent with first-order melting and that there is no convincing evidence of the hexatic phase. These results must be interpreted cautiously since the hexatic phase is predicted to be one with long-range correlations and slow dynamics which are difficult to simulate properly.

For some time we have been interested in how the melting of quasi 2D monolayers is affected by the steric properties of the constituent molecules. Our interest in this subject has been heightened by the fact that the most convincing evidence for continuous melting has been found for ethylene and ethane monolayers [7,9]. The systems which we have been investigating are rod-shaped molecules, the *n*-alkanes, adsorbed on the basal plane surfaces of graphite[14]. For these molecules, it is quite plausible that steric effects are more important in the melting of monolayer films than in bulk. Since the molecules physisorb with their long axis parallel to the graphite surface, the energy barrier for molecular rotation about the surface normal could be quite large compared to the steric hindrance encountered in the bulk where the molecules are not confined to a single plane.

One of the main purposes of the simulations presented here has been to assist in the interpretation of neutron and x ray diffraction experiments showing qualitatively different melting behavior for butane and hexane monolayers adsorbed on graphite [14]. Since these rod-shaped molecules are identical in structure except for their length, they provide an excellent opportunity for studying steric effects on melting. There have been no previous simulations of monolayer film melting with flexible rod-shaped molecules. Leggetter and Tildesley [15] have simulated butane and decane on graphite but only in the room-temperature fluid phase.

The n-alkanes on graphite are obviously more complicated to simulate than the rare gases noted above. The Hamiltonians of these flexible molecules contain rotational and intramolecular degrees of freedom in addition to the center-of-mass motion. As Abraham's work on rare gas monolayers containing a very large number of atoms has shown, it is extremely difficult to determine the order of the melting transition. It would be more so for our systems. Instead, the intent of our work is to show how molecular length qualitatively affects melting of the butane and hexane monolayers and the structure of their fluid phase. We shall see that atomic motion perpendicular to the surface plays an essential role in melting so that 2D theories are clearly not applicable. This paper summarizes preliminary results of our simulations. A more detailed account of this work along with a comparison with experiments is in preparation [17].

2. EXPERIMENTAL BACKGROUND

Our approach to studying the effect of molecular steric properties on monolayer melting has been to consider a series of isostructural molecules, the n-alkanes $(CH_3(CH_2)_nCH_3)$, physisorbed on the basal-planes of graphite. The dependence of their monolayer melting behavior on the length of the molecule (n) has been investigated using both neutron and x ray diffraction techniques [14]. In this paper we will focus on the butane (n=2) and hexane (n=4) monolayers with emphasis on hexane whose behavior is more complicated. It is important to bear in mind the flexibility of these molecules. In their lowest-energy trans conformation, the molecules are rod-shaped. However, they may be thermally excited into gauche forms by rotations about the C-C bonds. This gives the molecules a more globular shape.

At low temperature, both monolayers form a rectangular, orientationally-ordered structure in which the molecules lie with their long axis parallel to the surface in a herringbone arrangement as sketched in Fig.1(a). The hexane monolayer is commensurate with the graphite lattice whereas the butane structure is commensurate in only one direction (parallel to the longer edge of the unit cell). Despite the similarity of their solid structure, the two monolayers behave differently upon heating. In the case of butane, there is little change in the intensity of the Bragg peaks of the herringbone phase up to $\approx$ 116K where they vanish abruptly. Above this temperature, there is only a weak, broad peak characteristic of a fluid phase.

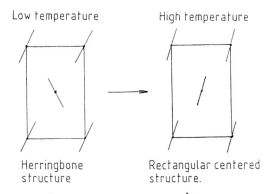

Low temperature High temperature

Herringbone Rectangular centered
structure structure.

a b

Fig.1 a)Sketch of the herringbone (hb) structure. b)
Sketch of the rectangular centered (rc) structure.
The long axis of the molecules are shown.

The solid hexane monolayer shows little change in its
diffraction pattern up to ≈ 170K where the Bragg peaks of
the herringbone phase disappear. However, in contrast to
butane, several broad peaks persist in the diffraction patterns above this temperature indicating a substantial amount
of translational order in the high-temperature phase.
Although they weaken as the temperature is increased further, broad peaks are still visible at room temperature.
This is very unusual behavior for physisorbed monolayers on
graphite, and the interpretation of the data turned out to
be rather difficult.

These results motivated us to initiate molecular
dynamics (MD) simulations of the melting of these
monolayers. Not only were we interested in determining the
structure of the high-temperature hexane phase but also in
understanding the apparently different melting behavior of
the two monolayers. Other questions which we wished to
address were: 1) Is there actually a coexistence of phases
at high temperatures in the hexane monolayer? 2) Is the
flexibility of the molecules important for the melting
process? and 3) Can a driving mechanism be deduced for the
melting transition?

3. THE SIMULATIONS

We shall only briefly summarize the method of the
simulations and refer to [17] for more details. As in
simulations of bulk hydrocarbons [18-20], we have used the
skeletal model for the alkane molecules. That is, the methyl
and methylene groups are replaced by single force centers or
pseudo-atoms at the positions of the carbon atoms. This is
justified because the motion of the light H atoms is
unimportant for the collective behavior being studied. The
interaction between molecules is represented by a sum of

pairwise pseudo-atom interactions, which are taken from bulk simulations to be a simple 6-12 Lennard-Jones potential. The molecule-substrate interaction is also modeled as a sum of atom-atom potentials of the Lennard-Jones type and the method of Steele [21] has been used to calculate the energy.

Simulations of bulk alkane phases have established the importance of retaining the molecular flexibility allowed by variation of the bend and dihedral angles. However, fixing the bond lengths and hence eliminating the faster stretching modes does not affect the reliability of the simulations [22]. Following Weber [19], we have expressed the bend-angle potential as a harmonic in the cosine of the bend-angle. For the dihedral-torsion potential, we have also used Weber's representation [19] which includes a coupling with the bend angles. The parametrization of this potential is based on the most recent experimental [23] and theoretical [24] investigations.

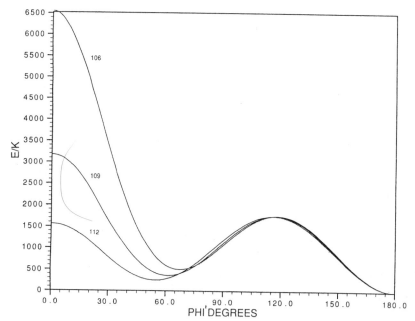

Fig.2 The dihedral potential for different bend angles. The equilibrium bend angle is 109°. A *trans* molecule has a dihedral angle of 180°.

With the bend angles at their equilibrium value the dihedral-torsion potential has the following characteristics:

trans-gauche barrier	:	1770 K
gauche energy	:	360 K
gauche-gauche barrier	:	3180 K

The potential is shown in Fig.2. The *trans-gauche* barrier at $\phi'=\pm120°$ is practically independent of the bend angles, θ', whereas the position and energy of the *gauche* states near $\phi'=\pm60°$ do show a dependence on θ'. The effect of the bend angles on the *gauche-gauche* barrier at $\phi'=0$ is more dramatic. As θ' is increased, it may become even smaller than *trans-gauche* barrier.

Since the nearest-neighbor pseudo-atom distances within the molecules are fixed, we have employed a constraint dynamics scheme to integrate the equations of motion [25]. The simulations are done at a constant temperature corresponding to a canonical sampling of phase space. The time step in the integrations is chosen to be 0.002 ps for the butane system and 0.001 ps for the hexane. This choice is based on a test of the accuracy of the numerical integration by doing a constant energy simulation (microcanonical sampling) in which the temperature fluctuates. In practice a small drift in the total energy of 0.03%/ps was observed in both systems at temperatures near the melting point. This was found to be acceptable.

Periodic boundary conditions were used in two orthogonal directions in the surface plane. This necessitates that the simulation box be commensurate with the graphite lattice. The box dimensions were chosen to be ≃67*67 Å which resulted from a compromise between maximum size and reasonable computing time. This is somewhat smaller than the monolayer cluster size on the polycrystalline graphite substrate used in the neutron diffraction experiments. After deconvolution of the instrumental resolution function, a monolayer coherence length of 105 Å was inferred from the low-temperature diffraction patterns [16]. As discussed below, we do not believe the choice of box size qualitatively influenced the simulations above the melting point of the monolayers .

The results of a simulation are given in terms of a set of thermodynamic functions and structural data. They are determined as an average over time blocks of 10 ps with a sampling every 0.1-0.2 ps. The results from different time blocks are assumed to be statistically independent and used to estimate the variance in the data. The total simulation time varies with the temperature and is largest near the melting point where the systems are allowed to run for 200-350 ps. At other temperatures, a typical run time is 150-200 ps.

The initial configurations of the monolayers are chosen to be the zero-temperature herringbone structures predicted by the intermolecular and molecule-substrate potentials. These structures differ slightly from those inferred from the low-temperature neutron diffraction patterns. In particular, the calculated structures are not commensurate with the graphite lattice. This feature may be of less concern for the butane and hexane monolayers than for rare gases. The reason is that the melting behavior seems to be governed by the length of the molecule and the extent to which it inhibits rotational motion rather than the corrugation of the holding potential, which depends on the molecular conformation.

4. RESULTS

In this summary, we will focus on structural data from

the simulations. A more complete presentation of the simulation results will be given in Ref. 17. The translational order in the monolayers is monitored by calculating the spherically-averaged structure factor $S(\kappa)$ which can be compared with the neutron diffraction patterns from powdered samples. It is also useful to calculate the 2D Fourier transform of the real-space configurations of the monolayer cluster. These Fourier transforms are similar to low-energy electron diffraction (LEED) patterns. Although LEED experiments are not available for comparison with the simulations, we shall see that the Fourier transforms provide a sensitive means of characterizing the film structure. For convenience, we shall refer to them as LEED patterns.

To describe the orientational configuration of the molecules, three angles are of primary interest. They are the polar angle θ between the long molecular axis and the z direction perpendicular to the surface, the rolling angle ψ' about the long axis, and the in-plane azimuthal angle ϕ of the long axis. In these flexible molecules, the long axis is defined to be the principal axis with the smallest moment of inertia.

4.1 Butane monolayer

For the butane monolayer, the simulation box contained 128 molecules. At the melting point, the Bragg peaks in $S(\kappa)$ disappear and are replaced by one broad peak characteristic of a translationally-disordered phase. In the low-temperature herringbone phase, the azimuthal-angle (ϕ) distribution of the butane molecules has two peaks, one for each sublattice. Upon melting this distribution becomes random. The distribution in the rolling angle (ψ') about the long axis of the molecule has a broad peak around 90° in the fluid phase. This indicates that the molecules are not rotationally disordered about their long axis but librate with large amplitude about the orientation for which the molecular plane is parallel to the surface. The distribution in the polar angle θ is rather sharp showing that in the fluid phase the long axis of the molecules is nearly parallel to the surface. Of particular interest, is the dihedral-torsion angle distribution which shows that all the molecules have the *trans* conformation near the melting point. Thus the butane simulations are consistent with the neutron diffraction experiments and indicate that the high-temperature phase is translationally disordered with the molecules in full rotation about the surface normal. The flexibility of the molecules appears not to be of importance for the melting process, since no *gauche* molecules are formed.

4.2 Hexane monolayer

In the hexane simulations, 104 molecules are considered. From the plot of the intermolecular potential energy in Fig.3, one can see that a transition takes place in the temperature range 220K to 225K. The distribution in the azimuthal angle ϕ of the molecules is shown in Fig.4. Below the transition, it has the two peaks characteristic of the herringbone structure, while above the transition a single broader peak is seen. Thus, at high temperatures, a

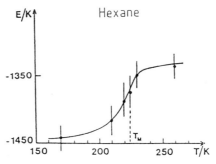

Fig.3 The intermolecular potential energy in the hexane
      monolayer as function of the temperature.

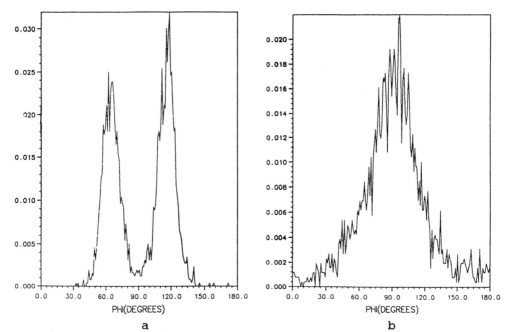

a   b

Fig.4 a) The azimuthal angle distribution in the hexane
      monolayer at 210K, just below the melting point.
      b) Same as a) at 225K, just above the melting
      point.

larger amount of orientational order remains than in the
butane system.

Calculation of the spherically-averaged structure factor $S(\kappa)$ shows that there is also more translational order
in the high-temperature hexane monolayer than for butane.
The Bragg peaks of the hexane (hb) phase indicated by the
arrows (with Miller indices) in Fig.5(a) are replaced at
higher temperatures by a single strong peak in Fig.5(b)
which is narrower than that found for the butane monolayer
above its melting point.

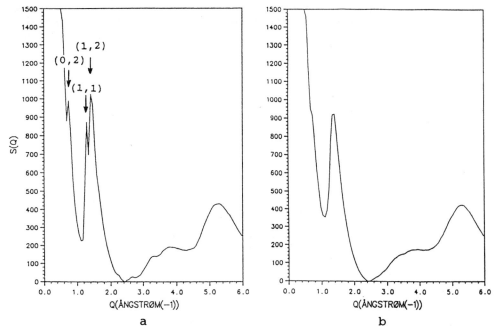

Fig.5 a) The calculated spherically-averaged structure
factor in the hexane system at 210K, just below
the melting point. b) Same as a) at 225K, just
above the melting point.

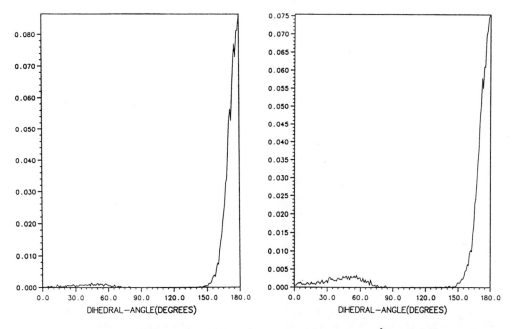

Fig.6 a) The dihedral angle distribution in the hexane
system at 210K, just below the melting point. b)
Same as a) at 225K, just above the melting point.

Another qualitative difference between the results of the hexane and butane simulations is revealed by the dihedral-angle distributions in Fig.6. They show a small number of molecules having a *gauche* conformation below the transition temperature. The butane results showed no evidence of *gauche* molecules below the melting point.

The structure of the hexane monolayer at high temperatures was determined from snapshots of the molecular configurations illustrated in Fig.7 and the LEED patterns calculated from them. We interpret the snapshots as depicting a two-phase system in which patches of a new ordered phase coexist with a disordered phase containing a high concentration of *gauche* molecules. The solid-like patches have a rectangular-centered (rc) structure as sketched in Fig.1(b) consistent with the single peak at $\phi=90°$ seen in the azimuthal angle distribution of Fig.4(b). Their characteristic dimension is 15-20Å which is much smaller than the simulation box size. For this reason, we do not believe the box size selected has qualitatively affected the fluid structure. Further simulations are in progress to test this conclusion.

The calculated LEED patterns support the structure of the high-temperature hexane monolayer inferred from the snapshots. In Fig.8, the pattern for the well-ordered (hb) structure is compared with that of the monolayer just above the melting point. The latter shows both spots and ring formations consistent with coexisting solid and fluid phases. In an (rc) structure, all reflections with an even-odd combination of Miller indices are missing. The disappearance of the (1,2) spot at 225K in Fig.8(b) is then consistent with a transition to an (rc) phase. Note that the (0,2) spot in this pattern appears as a shoulder on the broad central peak of the spherically-averaged structure factor $S(\kappa)$ in Fig.5(b).

The broad peaks observed at high temperatures in the neutron diffraction patterns of hexane monolayers on graphite can be indexed by an (rc) structure [16]. Profile analysis of these patterns reveals that the (rc) phase would be 10-20% denser than the low-temperature (hb) structure. The simulations have not yet provided evidence of (rc) patches with a larger density than that of the (hb) phase. We are now investigating whether such a density increase occurs in simulations on larger systems. Also in progress is an analysis of the relative stability of the (hb) and (rc) structures at zero temperature using the skeletal model to represent the intermolecular interactions [26].

We may summarize the structural features of the hexane monolayer found in these simulations by saying that the hexane system is polymorphic having at least two solid phases. The presence of the (rc) patches at high temperatures gives the hexane monolayer a greater degree of translational and orientational order than in the butane monolayer above its melting point. The hexane monolayer also differs from butane in that *gauche* molecules are formed below the melting point, suggesting that molecular flexibility plays a role in the melting process.

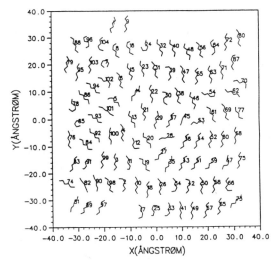

Fig.7 A snapshot of the molecular configuration at a given time of the equilibrated high-temperature phase of the hexane system. The atomic coordinates are projected on the plane of the surface. Patches of disorder are seen to coexist with patches of order with the (rc) structure.

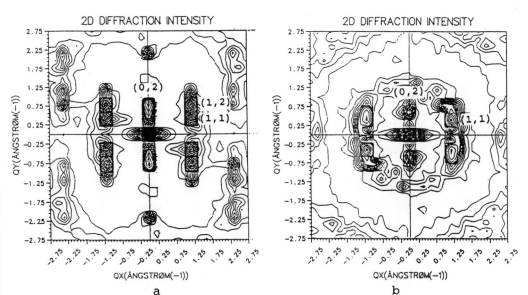

a   b

Fig.8 a) The calculated LEED pattern of the hexane system at 210K, just below the melting point. Main spots are indexed. b) Same as a) at 225K, just above the melting point.

5. A MECHANISM FOR MELTING

We interpret the appearance of the *gauche* molecules in the hexane monolayer as a precondition for melting. The molecules are driven to this conformational change because they are too long in the *trans* state to develop the rotational disorder about the surface normal required for the fluid phase. By colliding with their neighbors, molecules acquire the energy necessary to cross the barrier to the *gauche* state. Since the *gauche* molecules have a smaller footprint on the surface, this conformational change creates vacancies within the monolayer solid, allowing the molecules to rotate more freely and to disorder translationally. Other mechanisms for vacancy formation might involve tilting of molecules so that their long axis is no longer parallel to the surface or promotion of molecules into a second layer. However, it appears that for the longer hexane molecule these mechanisms for vacancy formation are more costly in energy than the conformational change.

In the case of butane, the molecule is short enough that tilting appears to be an energetically more favorable means of vacancy formation within the monolayer. Collisions result in the lifting of one end of the molecule, allowing a neighboring molecule to rotate underneath it. Thus a *trans-gauche* transformation is not required to achieve either orientational disorder of the molecules about the surface normal or translational disorder.

To test this interpretation of the melting process in the butane and hexane monolayers, we have performed simulations where the *trans-gauche* barrier is tripled in magnitude in order to prevent formation of *gauche* molecules. As expected, the melting point of the butane monolayer did not change since tilting was unaffected by use of a rigid molecule. However, the melting point of the hexane monolayer increased by $\simeq 75K$ as shown in the plot of the temperature dependence of the intermolecular potential energy in Fig.9. Because no *gauche* molecules were formed, vacancies were created by tilting of the molecules as for the shorter butane molecule. Evidence of the tilting is seen in the snapshot of Fig.10(a) showing the molecular configurations during a run to achieve equilibration above the melting point. As the running time

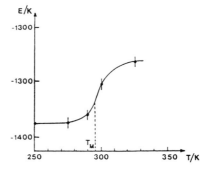

Fig.9 The intermolecular potential energy in a monolayer of "rigid" hexane molecules, where the *transguauche* barrier has been tripled to avoid formation of *gauche* molecules. The melting point is estimated to be 295K.

is increased further, we see that the monolayer structure differs from that of flexible molecules in Fig.7. In Fig.10(b), the *trans* molecules appear to form rafts consisting of a single row of molecules side-by-side. The preferred orientation of the rafts probably reflects the molecular orientations in the initial (hb) structure so that a longer run would result in a more random distribution.

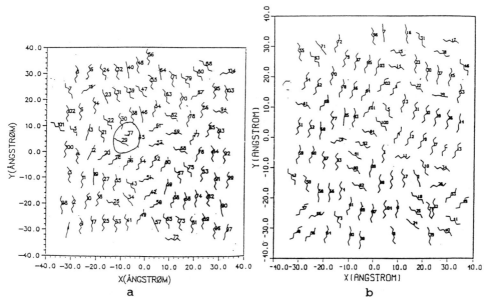

a   b

Fig.10 a) A snapshot of the molecular configuration in the monolayer of "rigid" hexane molecules in an equilibration run at 300K. Notice that molecule 37 stands on end with the long axis of the molecule perpendicular to the surface. b) Same as a) for the equilibrated phase which may be characterized as a 2D liquid crystal phase.

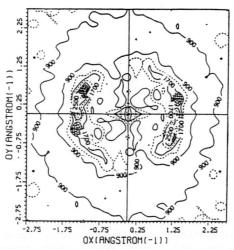

Fig.11 Calculated LEED pattern of the equilibrated phase of a monolayer with "rigid" hexane molecules at 300K just above the melting point.

165

The LEED pattern calculated for this phase is shown in Fig.11. The absence of spots indicates little correlation between the rafts so that the phase may be better described as a liquid crystal.

As a test of the role played by tilting of molecules in the melting of the butane monolayer, we have run simulations in which the magnitude of the molecule-substrate binding energy is tripled. This inhibits tilting and, as shown in Fig.12, results in an increase in the monolayer melting temperature by about 35K.

We believe that our analysis of melting in the butane and hexane monolayers may be applicable to a number of other films with (hb) monolayer phases. As discussed in Ref. 14, the (hb) structure is surprisingly pervasive among monolayers of nonspherical molecules adsorbed on graphite. If, as in the case of butane and hexane, the molecules are sufficiently long that an orientational-disordering transition does not precede melting, then the vacancy formation mechanism which we have discussed here may be operative. The creation of vacancies is required in order for the molecules to develop the orientational disorder about the surface normal and the consequent translational disorder character-istic of the fluid phase. Depending on the particular steric properties of the molecule, vacancy formation can be achieved by tilting, a conformational change of the molecule, or by promotion into a second layer.

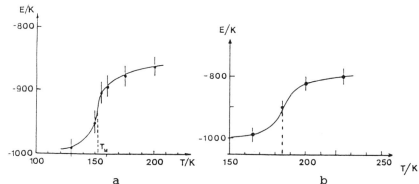

Fig.12   a) The intermolecular potential energy in a system of butane molecules as function of temperature. The melting point is estimated to 152K. b) Same as a) but with a tripling of the molecule-substrate holding potential. The melting point of the "strapped down" butane monolayer is estimated to $\simeq$ 185K.

6. SUMMARY

The focus of this study has been to investigate the structure of the butane and hexane monolayers above their melting point, to relate it to steric properties of the molecules, and to try to formulate a mechanism for melting based on these properties. It was not our aim nor is it possible to conclude from this study whether the melting of these monolayers is first order or continuous. We are now investigating the dynamical state of the films in the vicinity of their melting points both in calculations of molecular translational and rotational diffusion rates from simulations [27] and in quasielastic neutron scattering experiments [28].

The simulations described here have been successful in qualitatively reproducing the different melting behavior observed for the butane and hexane monolayers. However, for both monolayers, the melting temperatures in the simulations are about 30% higher than observed in the neutron and x ray diffraction experiments. This discrepancy could reflect a defect in the skeletal model used to represent the inter-molecular and molecule-substrate interactions, it could result from superheating in the simulation or it might be caused by the relative small cluster size used in the simulation. We note, though, that faulty potentials cannot explain the qualitatively different melting behavior of butane and hexane in the simulations, since the same ones have been used for both monolayers. As for superheating, the qualitative agreement between the simulations and experiment suggests that hysteresis effects may not be that important. Simulations are in progress to test the dependence of the melting point on sample size. Preliminary results on a system of 200 butane molecules ($\approx$50% larger sample than used here) show no change in the melting temperature.

The simulations presented here show the butane monolayer to melt abruptly into a fluid phase in which there is little translational and orientational order. All molecules remain in their *trans* configuration at the melting point so that molecular flexibility does not play a funda-mental role in the melting process. This does not appear to be true for the longer hexane molecule. Depending on rigidity assumed for the molecule, both the melting point of the monolayer and the structure of the high-temperature fluid phase change. At high temperatures, the flexible molecules form solid-like patches of *trans* molecules having an (rc) structure in coexistence with a fluid phase dominated by *gauche* molecules. The rigid molecules, on the other hand, prefer a single liquid-crystal-like phase above the melting point. It would be interesting to perform neutron diffraction studies on more rigid rod-shaped molecules than the alkanes to test these predictions of the simulations.

In conclusion, we find that a strictly 2D theory of melting such as that of KTHNY is not applicable to the monolayers of long molecules which crystallize in the (hb) structure. Instead, we have proposed a general mechanism for melting of these systems which is based on vacancy formation resulting from molecular motion perpendicular to the surface.

7. ACKNOWLEDGEMENT

The authors wish to thank J.M. Phillips for a critical

reading of the manuscript. This work was supported by Danish SNF Grant No. M 11-7015, U.S. NSF Grant Nos. DMR-8704938 and DMR-9011069, and Pittsburgh Supercomputing Center Grant No. DMR-880008P.

REFERENCES

1. J.M. Kosterlitz and D.J. Thouless, J. Phys. C **6**, 1181 (1973).
2. B.I. Halperin and D.R. Nelson, Phys. Rev. Lett. **41**, 121 (1978).
3. D.R. Nelson and B.I. Halperin, Phys. Rev. B **19**, 2457 (1979).
4. A.P. Young, Phys. Rev. B **19**, 1855 (1979).
5. N.D. Mermin and H. Wagner, Phys. Rev. Lett. **17**, 1133 (1966).
6. K.J. Strandburg, Rev. Mod. Phys. **60**, 161 (1988) and references therein.
7. H.K. Kim, Q.M. Zhang and M.H.W. Chan, Phys. Rev. Lett. **56**, 1579 (1986).
8. J.Z. Larese, L. Passell, A.D. Heidemann, D. Richter and J.P. Wickstedt, Phys. Rev. Lett. **61**, 432 (1988).
9. S. Zhang and A.D. Migone, Phys. Rev. B **38**, 12039 (1988).
10. F.F. Abraham, Phys. Rev. B **28**, 7338 (1983).
11. F.F. Abraham, Phys. Rev. Lett. **50**, 978 (1983).
12. F.F. Abraham, Phys. Rev. B **29**, 2606 (1984).
13. S.W. Koch and F.F. Abraham, Phys. Rev. B **27**, 2964 (1983).
14. H. Taub, in *The Time Domain in Surface and Structural Dynamics*, edited by G.J. Long and F. Grandjean, NATO Advanced Study Institute Series C, Vol. **228** (Kluwer, Dordrecht, 1988), p. 467.
15. S. Leggetter and D.J. Tildesley, Berichte der Bunsen Gesellschaft für Physikalische Chemie **94**, 285 (1990).
16. J.C. Newton, Ph.D thesis, University of Missouri-Columbia USA (1989).
17. F.Y. Hansen J.C. Newton and H. Taub, unpublished.
18. J.P. Ryckaert, G. Cicotti, and H.J.C. Berendsen, J. Comp. Phys. **23**, 327 (1977).
19. T.A. Weber, J. Chem. Phys. **69**, 2347 (1978).
20. T.A. Weber, J. Chem. Phys. **70**, 4277 (1979).
21. W.A. Steele, *The interaction of gases with Solid Surfaces* (Pergamon Press, London 1974).
22. M. Pear and J.H. Weiner, J. Chem. Phys. **71**, 212 (1979).
23. H.D. Stidham and J.R. Durig, Spectrochimica Acta **42A**, 105 (1986).
24. K. Raghavachari, J. Chem. Phys., **81**, 1383 (1984).
25. F.Y. Hansen, unpublished.
26. F.Y. Hansen and H. Taub, unpublished.
27. F.Y. Hansen, unpublished.
28. P. Dai, H. Taub, T.O. Brun, F. Trouw, and F.Y. Hansen, unpublished.

PROBING FILM PHASE TRANSITIONS THROUGH MEASUREMENTS OF SLIDING FRICTION

J. Krim

Department of Physics
Northeastern University
Boston, MA 02115, USA

I. Introduction

The sliding friction of a molecularly thin film adsorbed on a solid surface is a topic which is presently quite underdeveloped. One's ability to measure this quantity allows a wealth of new information to become available in a variety of areas such as Darcy flow in dendritic film growth, the "no-slip" boundary condition of fluid hydrodynamics and the study and identification of thin film phase transitions. The purpose of this chapter is to describe how sliding friction measurements can be carried out and in particular be utilized for the identification and characterization of melting and surface melting transitions occurring within thin adsorbed films. At present, these measurements are providing more insight to the microscopic origins of friction than to the actual study of the phase transitions. Nonetheless, as the details of the frictional force laws which govern slippage become more developed, the emphasis should shift back towards characterization of phase transitions.

One might question whether an adsorbed monolayer slips at all, in light of the wide range of experimental situations which support the "no-slip" boundary condition in continuum hydrodynamics. There exists however no compelling theoretical argument which requires this to be so. The topic has attracted the likes of Maxwell,[1] Helmholtz,[2] and J.J. Thompson,[3] who variously argued for and against the validity of the no-slip boundary condition. Many years have passed and yet the topic of fluid flow and slip near solid boundaries, particularly with respect to confined geometries, is still rousing interest.[4] The resistance to sliding of molecularly thin films confined between crossed mica cylinders has been measured,[5-8] but no explicit slipping at the liquid-solid interface was reported. Computer simulations of the contact line separating two immiscible fluids have however called into question the applicability of the no-slip boundary condition,[9,10] and experimental observations of the slippage of physisorbed monolayers have recently been reported.[11-13] An experimental technique for measuring sliding friction of molecularly thin layers will be described in the following sections along with a discussion of its application to the study of surface phase transitions.

II. Sliding Friction of a Block on a Plane

Consider a block with mass m sliding on a plane with velocity v_{rel}, where v_{rel} is the velocity of the block relative to that of the surface of the plane. One could characterize the frictional resistance force F_f at the interface by noting the levels of force required to slide the block along the surface at different speeds. An alternate

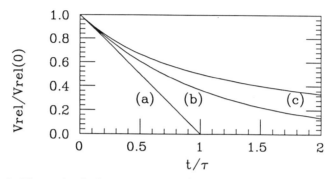

Figure 1. Plots of velocity versus time for an object which slides to a stop from an initial velocity $v_{rel}(0)$ for three different frictional force laws. (a) $F_f = \mu_k F_N$; (b) $F_f = -bv_{rel}$; (c) $F_f = -cv_{rel}^2$.

way to probe the sliding friction is to give the block an initial velocity $v_{rel}(0)$ at time $t = 0$ and then record the time dependence of v_{rel} as it slides to a stop. The motion of the block is described by $F_f = m\dot{v}_{rel}$, and is shown in Figure 1 for three forms of an assumed frictional force law.

(a) Amontons' Law of friction:[14] $F_f = \mu_k F_N$, where F_N is the force normal to the surface and μ_k is the kinetic coefficient of friction. The velocity decays linearly and the characteristic "slip time" τ is that required to reach a complete stop.

$$v_{rel}(t) = v_{rel}(0) - \frac{\mu_k F_N}{m}t; \quad \tau = \frac{mv_{rel}(0)}{\mu_k F_N} \quad (1)$$

(b) Stokes' Law of friction:[15] $F_f = -bv_{rel}$, where the friction coefficient b is a constant. The velocity decays exponentially and the characteristic slip time is that required for the velocity to drop by a factor of e.

$$v_{rel}(t) = v_{rel}(0)e^{-\frac{b}{m}t}; \quad \tau = m/b \quad (2)$$

(c) Newton's Law of friction: [15] $F_f = -cv_{rel}^2$, where the friction coefficient c is a constant. The velocity decays as t^{-1} and the characteristic slip time is that required for the velocity to drop by a factor of 2.

$$v_{rel}(t) = \frac{v_{rel}(0)}{1 + \frac{cv_{rel}(0)}{m}t}; \quad \tau = \frac{m}{cv_{rel}(0)} \quad (3)$$

The case of a macroscopic block can be extended to that of a two-dimensional layer by utilizing the mass per unit area ρ of the film instead of the total film mass. The film's velocity relative to the substrate v_{rel} becomes an average over the microscopic motion of the film particles so that there is no constraint whatsoever requiring the whole film to move in unison. It is typical to utilize Amontons' law for the frictional force of a block sliding on a plane. The choice of a force law for a two-dimensional sliding layer is not however obvious.

Amontons' law is related to an increase in interfacial contact area as the block is pressed to the surface.[16] One probably cannot expect Amontons' law to hold for the case of a sliding monolayer since the entire extent of the layer is in contact with the underlying surface. One might anticipate however that the sliding friction would increase as the "normal force", i.e. the substrate-adsorbate van der Waals interactions became stronger. The second or third forms for friction may potentially be more applicable to the case of a sliding monolayer. Stokes' law is a particularly simple form of an assumed force law and is utilized as a starting point for the analysis of the behavior of a sliding layer.

The characteristic "slip time" τ retains the same meaning for the case of the film as well as the block, keeping in mind that it is the average velocity of the film with respect to the substrate which is decaying. The two-dimensional quantity which is analogous to the frictional damping coefficient is referred to as the "interfacial viscosity" η_2. It is related to the slip time according to $\tau = \rho_2/\eta_2$ for the case of Stokes' Law of friction. The following sections describe how the slip time and interfacial viscosity can be quantitatively measured by means of a quartz crystal microbalance.

III. Acoustic Damping of an Oscillator

If the substrate underlying the adsorbed film oscillates back and forth, then film slippage effects can be probed by the inertial reaction forces which the film presents to the oscillator. Such an experiment can be carried out by studying film adsorption onto metal electrodes which have been evaporated onto the major surfaces of a quartz crystal oscillator vibrating in the transverse shear mode.(Figure 2) If the period of oscillation becomes comparable to the characteristic slip time of the adsorbed film, then slippage effects will become observable. Reaction forces are produced whenever a film adsorbs onto the planar surfaces which are translating back and forth, or if the oscillator is exposed to a three-dimensional gas. Such reaction forces are conventionally described in terms of an acoustic impedance, $Z(\omega) = R(\omega) - iX(\omega)$, where the resistive term $R(\omega)$ is proportional to average energy dissipation and the reactive term $X(\omega)$ is associated with the inertia of the oscillator. Stockbridge has shown that the shifts in the resonant frequency ω_0 and the quality factor Q of a quartz crystal oscillator which is acted on by an acoustic impedance Z are:[17]

$$\delta\left[\frac{1}{Q}\right] = \frac{4R}{\omega_0 \rho_q t_q}, \qquad \delta\omega_0 = \frac{2X}{\rho_q t_q}, \qquad (4)$$

where ρ_q and t_q are the density of quartz and the thickness of the quartz crystal, respectively. (Exposure to one side only will produce shifts which are one-half as large.)

In practice it is convenient to drive the crystal oscillator with a constant level driving force which, through a feedback network, drives the crystal at its series resonant frequency, i.e. $F(t) = F_0 e^{i\omega_0 t}$. The amplitude of vibration of a damped oscillator driven at its own resonant frequency by a constant driving force is given by:

$$A(\omega_0) = \frac{F_0/M}{R\omega_0} = \frac{1}{M\omega_0}\frac{F_0}{R} \tag{5}$$

where M is the total effective mass of the oscillator. The experiments described here are carried out in a regime satisfying the condition $m_{film} << M$, where m_{film} is the

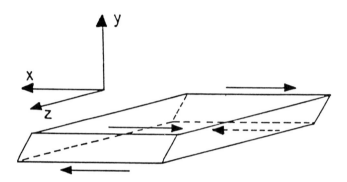

Figure 2. Transverse shear mode of oscillation. Adsorption occurs onto metal films which have been evaporated onto the surfaces parallel to the $x - z$ plane and move antiparallel to each other.

total mass of the adsorbed film. This implies that $\delta\omega << \omega_0$ in the experimental regime of interest. Eq.(5) can therefore be simplified to $A = K\frac{F_0}{R}$ where K is a constant. Since changes in Q^{-1} are directly proportional to changes in R, Eq.(4), it is apparent that:

$$\delta(A^{-1}) \propto \delta(Q^{-1}) \propto \delta R \tag{6}$$

One can therefore choose to record either the quality factor or the amplitude of vibration in order monitor changes in the resistive component of the acoustic impedance.

In order to relate the physical quantities of interest to the frequency and amplitude shifts of the oscillator, it is necessary to derive the acoustic impedance which the film or bulk fluid phase presents to an oscillating planar surface. A planar surface which is driven by a periodic driving force $F(\omega) = F_0 e^{i\omega t}$ in a direction parallel to the plane has velocity $v(\omega) = v_0 e^{i\omega t}$. The ratio $Z_m = F(\omega)/v(\omega)$ is termed the "mechanical impedance". The mechanical impedance of a fluid or film to shear oscillations of a plane is generally expressed as $Z_m(\omega) = Z(\omega)A_s$, where $Z(\omega)$ is the mechanical impedance per unit area, or acoustic impedance and A_s is the area of the plane in contact with the fluid. The acoustic impedance at frequency $f = \omega/2\pi$ of a three-dimensional fluid with mass density ρ_3 and bulk viscosity η_3 is well established: [17] [18]

$$Z_{vapor} = R_{vapor} - iX_{vapor} = (1 - i)\sqrt{\pi\eta_3\rho_3 f} \qquad (7)$$

Equation (7) inadequately describes the impedance of a gas a low pressure, where the time for particles to relax to the equilibrium state after a collision with the oscil-

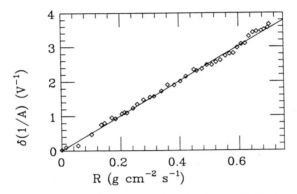

Figure 3. The change in inverse amplitude due to helium gas at 77.4 K is plotted versus the acoustic resistance calculated from the bulk parameters of gas according to Eq.(8a).(From Ref.13) The solid line shows the fit to the data. Its slope determines the proportionality constant between A^{-1} and R in Eq.(6). The initial amplitude of vibration is .125 V.

lator is comparable to the period of oscillation. The real and imaginary components of the acoustic impedance, corrected for this "viscoelastic" effect are as follows: [17]

$$R_{vapor} = \sqrt{\pi\eta_3\rho_3 f} \left[\frac{\omega\tau_r}{1 + (\omega\tau_r)^2} \left(\sqrt{1 + (\omega\tau_r)^{-2}} + 1 \right) \right]^{1/2} \qquad (8a)$$

$$X_{vapor} = \sqrt{\pi\eta_3\rho_3 f} \left[\frac{\omega\tau_r}{1 + (\omega\tau_r)^2} \left(\sqrt{1 + (\omega\tau_r)^{-2}} - 1 \right) \right]^{1/2} \qquad (8b)$$

The time τ_r is that which is required for the excess momentum of a fluid particle to fall to $1/e$ of its initial value after a collision with the plane.

Figure 3 shows a typical quartz oscillator response to a three dimensional gas as the pressure surrounding the oscillator is increased from zero to above 600 torr. The change in inverse amplitude is plotted as a function of the acoustic impedance, Eq.(8a) for helium gas at 77.4 K (film adsorption is negligible at this temperature.) The slope of the line determines the proportionality constant between A^{-1} and R in Eq.(6). This provides a quantitative, *in situ* calibration of the response of the oscillator to the acoustic impedance which acts on it.

In general, an adsorbed film will be present in addition to the vapor phase. If the thickness of the film is much less than the penetration depth of the adsorbed material and the density of the gas phase is significantly less than that of the film phase (as is the case for the films discussed here), then the acoustic impedance is well approximated by: [13]

$$Z(\omega) = Z_{vapor} + Z_{film} = (1 - i)\sqrt{\pi \eta_3 \rho_3 f} + \left[\frac{-i\omega \rho_2}{1 - i\omega\tau} \right] \qquad (9)$$

The first term represents the impedance of the vapor phase, Eq.(7). The second term is the acoustic impedance of the adsorbed film itself, where ρ_2 is the film mass per unit area. The quantity τ is the characteristic slip time discussed in Section II and treated previously in great detail by Krim and Widom from a two-dimensional microscopic viewpoint, assuming Stokes' Law for the form of the frictional force law. [12] The term "slippage" for a film should be treated with caution since, as mentioned earlier, there is no precondition that the whole film be moving in unison. It unreasonable to expect that the atoms in a physisorbed film would all be "slipping" (or "hopping") in the same direction at any moment in time. Nonzero slip time implies only that that *on the average* the film exhibits a finite reaction time to the oscillatory motion of the underlying substrate. It definitely does not imply that the film must have decoupled to any great extent from the motion. It also does not require that the film or the interface be perfectly smooth: in perfect analogy with the block on the plane, one can measure a slip time for both rough and smooth interfaces.

The slip time is determined experimentally by expressing the acoustic impedance of the film, $-i\omega\rho_2/(1 - i\omega\tau)$, in terms of its resistive and reactive components:

$$R_{film} = \frac{\rho_2 \omega^2 \tau}{1 + \omega^2 \tau^2}, \qquad X_{film} = \frac{\rho_2 \omega}{1 + \omega^2 \tau^2} \qquad (10)$$

In practice the measurement goes as follows: An *in situ* calibration such as that shown in Figure 3 is carried out with a gas which does not condense at the temperature of interest. A second gas which does condense is then introduced, and the frequency and amplitude shifts are recorded, generally as a function of pressure. The amplitude shifts are converted to acoustic resistance by means of the calibration. The frequency shifts are converted to acoustic reactance by means of Eq.(4). The acoustic impedance of the gas phase is subtracted from the total impedance, Eq.(9), in order to determine the impedance of the film. The slip time is then deduced from the ratio of the resistive to the reactive component of the film impedance, since $R_{film}/X_{film} = \omega\tau$.

The above analysis is based on the assumption that Stokes' Law is the applicable form of the frictional force law. If the applicable force law is in fact Newton's Law then the deduced slip time will still be quite comparable to the actual value (Figure 1). This is true to a lesser extent for the case of Amontons' Law. The form of the force law can be probed by recording the slip time (determined via the Stokes' Law analysis) as a function of the amplitude of the oscillator. This effectively varies the initial relative velocity, $v_{rel}(0)$, discussed in Section II. If Stokes'

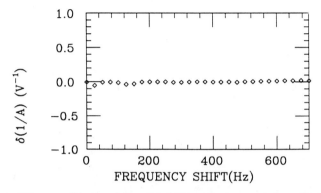

Figure 4. The amplitude of the 7.4 MHz quartz crystal oscillator appears constant for deposition of about ten layers of gold onto the surface electrodes. (From Ref. 13) The characteristic slip time of a gold film on a gold substrate is far below experimental resolution. The amplitude of vibration is .125 V.

Law is indeed applicable, then the slip time will be independent of the amplitude ($\tau = m/b$). Amontons' Law and Newton's Law will result in slip times which are directly ($\tau = [mv_{rel}(0)]/[\mu_k F_N]$), or indirectly ($\tau = m/[cv_{rel}(0)]$) proportional to the amplitude of vibration, respectively.

IV. Slippage of Single-Phase Films

Film dissipative effects due to sliding reach a maximum at $\omega\tau = 1$ when the period of oscillation becomes comparable to the slip time of the film. If the slip time is much less than the period of oscillation, then the film will appear to be rigidly attached whereas if the slip time is much longer than the period, the film motion will become decoupled. This is demonstrated in Figure 4, which shows the changes in inverse amplitude versus frequency shift (proportional to the mass of the added film) for a ten-layer gold film which has been deposited onto a gold substrate. The slip time of the gold film on the gold substrate is evidently much shorter than the period of the oscillator since no damping is detected while the mass load is producing a change in the frequency. This is quite reasonable since to first approximation, the gold film should not slip at all on top of a gold substrate. It raises the question however of the limit of applicability of the technique.

An 8 MHz quartz crystal driven by a conventional Pierce oscillator circuit[19] typically has a quality factor near 10^5. Changes in dissipation of $\delta Q^{-1} \geq 1 \times 10^{-8}$ are easily detectable. A monolayer of liquid nitrogen ($\rho_2 \approx 34.5$ ng/cm^2) adsorbed on both sides of the crystal will produce a frequency shift of 10 Hz ($\delta\omega = 63$ rad/s). Eq.(4) and Eq.(10) can be combined to show that $\delta(Q^{-1}) = 2\tau\delta\omega$, which is used to predict the minimum slip time which will be observable:[20]

$$\tau_{min} = \frac{\delta Q^{-1}_{min}}{2\delta\omega} = \frac{1 \times 10^{-8}}{2(63 \text{rad/s})} = 8 \times 10^{-11} \text{s} \tag{11}$$

Eq.(11) provides a convenient way to estimate whether or not one is likely to detect film slippage effects. In general, however one will not know in advance what the slip time associated with a particular film-substrate interface will be. The case of a gold film deposited on a gold surface is an extreme example of a solid layer whose slippage should be negligible. Physisorbed layers can well be expected to exhibit significantly greater slippage. Such layers are indeed observed to exhibit slip times on the order of ns, well above the threshold of detectability. Figures 5 and 6 show slippage behavior typical of physisorbed liquid films. The slip time as a function of coverage is plotted for water on silver at room temperature and nitrogen on gold at 77.4 K. The coverage in these figures has been determined from the mass per unit area of the adsorbed film, ρ_2, which is quantitatively related to the frequency shift of the oscillator $\delta\omega$ according to Eqs.(1) and (7). The behaviors are similar in several ways. Each film exhibits a high degree of slippage at less than a half-monolayer coverage. This is consistent with the relative freedom that adsorbed particles have in the low coverage regime. The slip times increase as the films thicken, and reach maxima in the two to three layer regime. As the films' thicknesses further increase, the average slip times decrease slowly, indicating that the films become somewhat more pinned as they grow thicker.

It is interesting to note that two very different interfaces, water/silver and nitrogen/gold have very comparable slip times. Similar behavior is observed for liquid rare-gas films as well. As discussed earlier, one might expect the slip time to decrease as the van der waals attraction to the substrate is increased. Although there is evidence that more massive atoms exhibit shorter slip times,[21] the effect is at best, weak. One might also expect that the slip times will decrease as the surface becomes rougher. This effect has indeed been observed. When the gold surface utilized in Fig. 5 was exposed to air for one day, the slip times were markedly reduced. [13] Although the surface did not become geometrically rough on account of the air exposure, the microscopic contamination provided a very effective "pinning" mechanism. This result supports the assumption that the sliding energy dissipation is occurring primarily at the solid-film interface rather than the film-vapor interface.

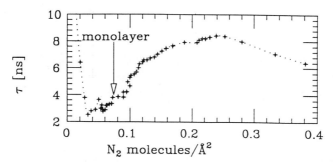

Figure 5. Slip time vs. coverage for nitrogen adsorption on gold at 77.4 K.

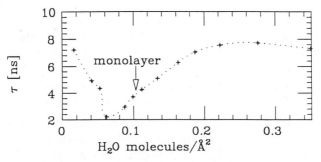

Figure 6. Slip time vs. coverage for water adsorption on silver at room temperature.

Far fewer studies have been carried out on the slipping behavior of incommensurate solid layers. Preliminary evidence indicates that an incommensurate xenon solid layer on a Au(111) surface at 77.4 K has a characteristic slip time which is about twice as long as the liquid layer.[22]

V. The Spreading Diffusion Coefficient

It is reasonable to expect that the slipping behavior of a film is related to the rate at which it will spread on a surface. Widom and Krim[23] have estimated this spreading rate for the case of the liquid layers presented in the previous section. The analysis goes as follows:

Consider N particles adsorbed on an area A_s. Suppose all of the particles were forced to gather at one point on the surface. How long would it take for the particles to spread back over the entire extent of the surface? From the sliding friction viewpoint, the equation of motion of the film is:

$$-\vec{\nabla}\phi = \eta_2 \, v_{rel} \tag{12}$$

where ϕ is the two-dimensional spreading pressure and η_2 is the interfacial viscosity. One can alternatively view the spreading as a diffusive process described by the relation:

$$-D\vec{\nabla}n = n \, v_{rel} \tag{13}$$

where n is the number of film particles per unit area and D is the "spreading diffusion coefficient". The film spreading behavior is determined within the context of two-dimensional thermodynamics:

$$d\phi = sdT + nd\mu \tag{14}$$

where μ is the chemical potential, T the temperature and s the entropy of the system. Eq.(14) can be rewritten as:

$$\vec{\nabla}\phi = n(\frac{\partial\mu}{\partial n})_T\vec{\nabla}n \tag{15}$$

Eqs.(12),(13) and (15) can be combined to obtain an expression for the spreading diffusion coefficient in terms of experimentally measurable quantities:

$$D = \frac{n^2}{\eta_2}(\frac{\partial\mu}{\partial n})_T \tag{16}$$

The spreading diffusion coefficients for the liquid monolayers discussed in the previous section are on the order of 1 cm^2/sec. This is in qualitative agreement with the experimentally observed equilibrium times, and supports the interpretation of D as the rate at which the film can spread into an equilibrium configuration.

VI. Probing Phase Transitions through Sliding Friction Measurements

In the preceding sections, only single-phase films were discussed. Since the slip time, interfacial viscosity and spreading diffusion constant are characteristic of an interface, major changes should be observed in all of these quantities when a system undergoes a phase transition. Moreover, there may be additional peaks in the dissipation associated with the processes involving the phase transition itself.

Figure 7 shows data for frequency and quality factor shifts versus pressure for Kr adsorption at 77.4 K onto a smooth Au surface with no previous air exposure.[24] Kr films grown at this temperature are known to form at least one or two uniform layers at before the onset of bulk crystallites, depending on the uniformity of the substrate.[25] Kr films at 77.4 K are also expected to condense in the form of a liquid which solidifies as the ambient pressure is increased, since 77.4 K is above the two-dimensional melting point of the film (69.5 K, which is .6 of the bulk Kr

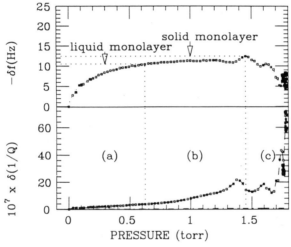

Figure 7. Frequency and quality factor shifts versus pressure for Kr adsorption on gold at 77.4 K. The dash-dot line in the lower figure denotes the vapor contribution to the total shift. This contribution is too negligible to be plotted in the upper figure. See text for a discussion of Regimes (a)-(c).

melting point).[26]   Shifts in inverse quality factor have been obtained according to the calibration method described above. The contribution to the frequency and quality factor shifts due to the gas phase is shown by the dash-dot line in the lower figure, and is too negligible to be plotted in the upper figure. Frequency shifts of 10.4 and 12.5 Hz correspond to film coverages where the particle spacings are equal to the spacing in the bulk liquid and solid, respectively. Regime (a) is labelled as the pressure regime where the film coverage is lower than a complete liquid monolayer. Regime (b) corresponds to an intermediate pressure regime where the complete liquid monolayer is being compressed. The sudden rise in the frequency data near 1.45 torr is characteristic of a liquid-solid transition occurring within the film.

The magnitude of the frequency shift starts to decline in Regime (c), immediately after reaching a value which corresponds to the bulk solid spacing. The first peak in the quality factor data near the beginning of Regime (c) is consistent with what one would expect for a phase transition associated with solidification. The second peak in Regime (c) may well be due a further phase transition, which is as yet unidentified. Analysis of the combined frequency and quality factor shifts according to the manner described above indicates that the film begins to decouple from the motion of the substrate as soon as it reaches the bulk solid spacing: The solid film slides more readily on this surface than the liquid film. The oscillations which are observed in the frequency in Regime (c) are likely to be due to variations in the film coupling to the motion as the amplitude of vibration changes. Condensation of bulk crystallites near 1.8 torr is evidenced in the shift in the inverse quality factor, but not in the frequency shift. This indicates that the crystallites which condense are also substantially decoupled from the motion of the substrate.

Figure 8 shows comparable data for Kr adsorption at 77.4 K on a microscopically rough Ag sample.[24] This surface has been indicated by both liquid nitrogen adsorption measurements[27] and scanning tunnelling microscopy measurements[28] to be a self-affine fractal surface at microscopic length scales. Frequency shifts are greater on this sample due to a higher resonant frequency (8 MHz instead of 5MHz for Figure 7) and because of roughness of the Ag surface (about 1.5 times greater than that of a geometrically flat plane). Phase transitions are smeared out in the frequency data due to the roughness of the substrate. The three pressure regimes of Figure 7 are labelled in Figure 8 for comparison. No decoupling is observed in the frequency shift data, since the frequency shift progresses monotonically upward, and the shifts in inverse quality factor remain low throughout the entire range studied.

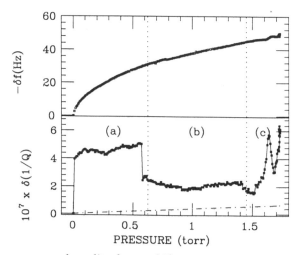

Figure 8. Frequency and quality factor shifts versus pressure for Kr adsorption on a microscopically rough Ag surface. The dash-dot line in the lower figure denotes the vapor contribution to the total shift. Pressure regimes (a)-(c) are placed at the same values as in Figure 7.

In Regime (a), the slippage of the sub-monolayer liquid is quite comparable to that observed on the flat Au surface. A sudden discontinuity is observed close to the beginning of Regime (b) where the liquid monolayer is being compressed, perhaps due to a "locking in" of the film as the monolayer completes. Dissipation peaks are observed in Regime (c), which may be related to the same phase transitions occurring within the solid film phase of Figure 7. This data clearly indicates that, in contrast to the data for the smooth surface, the "solid" phase film slips less on this surface than does the liquid.

Slip times for Kr adsorption in Regime (a) are comparable to those observed for the monolayer regimes of the water and liquid nitrogen films depicted in Figures (5-6). Slip times are more difficult to compute for the the higher pressure regimes since some of the dissipation is attributable to the presence of a phase transition rather than to interfacial slippage. Interfacial viscosities for liquid Kr monolayers in Figures (7-8) are on the order of 40 dynes/cm^2 per cm/s. This implies that 40 dynes/cm^2 of frictional force are present when the film slides along the surface at an average velocity of 1 cm/s. Since the slipping velocity of these liquid films is estimated to be close to .5 cm/sec [24], the estimated frictional force should be of the correct order of magnitude. It is inappropriate to discuss a quantitative sliding force at velocities other than those at which the measurements were carried out without a knowledge of the velocity dependence of the friction. Future work will focus on the velocity dependence of the frictional force law and on the role which surface defects play in determining the degree of sliding which will occur. Such investigations will allow quantitative comparison of these results to theories of interfacial slippage.[29] [30]

Acknowledgements

I am indebted to A. Widom for his continuing interest and collaboration on this work, and to E. Watts, R. Chiarello and V. Panella who have participated in the experiments described here. This work has been supported by the ONR, grant # N00014-89-J-1853 and NSF grant #DMR-86-57211.

REFERENCES

1. J.C. Maxwell, Philos. Trans. R. Soc. London, 157, 49 (1867)

2. H. Helmholtz and X. Piotrowski, Sitz. der k. Akad Wien, 40, (1860)

3. W.C. D. Whetham, read by J.J. Thompson, Phil. Trans. R. Soc. London, 181, 559 (1890)

4. A. Khurana, Physics Today 41 (5), 17 (1988)

5. F.P. Bowden and D. Tabor, *The Friction and Lubrication of Solids*, Part II (Clarendon, Oxford , 1964)

6. J.N. Israelachvilli, Acc. Chem. Res. 20, 415 (1987)

7. J.N. Israelachvilli, P.M. McGuiggan and A.M. Homola, Science 240, 189 (1988)

8. J. Van Alsten and S. Granick. Phys. Rev. Lett., 61, 2570 (1988)

9. J. Koplich, J. Banavar, J. Willemsen, Phys. Rev. Letters, 60, 1282 (1988)

10. P.A. Thompson and M.O. Robbins, Phys. Rev. Lett. 63, 766 (1989); M.O. Robbins and P.A. Thompson, Phys. Rev. A 41, 6830 (1990)

11. J. Krim Bull. Am. Phys. Soc, 33, No 3, 436 (1988)

12. J. Krim and A. Widom, Phys. Rev. B, $\underline{38}$, 12184 (1988); A. Widom and J. Krim, Phys. Rev. B $\underline{34}$, R4 (1986)

13. E.T. Watts, J. Krim and A. Widom, Phys. Rev. B $\underline{41}$, 3466 (1990)

14. Amontons, Mem. Acad. Roy. Soc., 206 (1699)

15. J.B. Marion, *Classical Dynamics* (Academic Press, New York, 1970), p. 53

16. A.W. Adamson, *Physical Chemistry of Surfaces*, (Wiley, New York, 1982)

17. C.D. Stockbridge, *Vacuum Microbalance Techniques*, (Plenum, New York, 1966) Vol. 5

18. E.M. Lifshitz, *Fluid Mechanics*, Vol. 6 of *Course of Theoretical Physics* (Pergamon, London, 1959), pp.88-90.

19. M.E. Frerking, *Crystal Oscillator Design and Temperature Compensation* (Van Nostrand, New York, 1978) pp67-68

20. This is conservative. The actual minimum will depend on oscillator stability.

21. J. Krim, E.T. Watts and J. Digel, J. Vac. Sci. Tech. A, (1990) in press

22. J. Krim and R. Chiarello, to be published

23. A. Widom and J. Krim, to be published

24. J. Krim and R. Chiarello, J. Vac. Sci. Tech. A, in press

25. J. Krim, J. Suzanne and J.G. Dash, Phys. Rev. Lett. $\underline{52}$, 635 (1984)

26. A. Thomy and X. Duval, in *Adsorption at the Gas-Solid and Liquid-Solid Interface*, J. Rouquerol and K.S.W. Sing, Eds. (Elsevier, Amsterdam, 1982)

27. P. Pfeifer, Y.J. Wu, M.W. Cole and J. Krim, Phys. Rev. Lett. $\underline{62}$, 1997 (1989)

28. P. Pfeifer, J. Kenntner, J.L. Wragg, J. West, H.W. White, J. Krim and M.W. Cole, Bull. Am. Phys. Soc. $\underline{34}$, 728 (1989)

29. J.B. Sokoloff, Phys. Rev. B $\underline{42}$, 760 (1990)

30. W. Zhong and D. Tomanek, Phys. Rev. Lett. $\underline{64}$, 3054 (1990)

PHASE TRANSITIONS IN LIPID MONOLAYERS ON WATER: NEW LIGHT ON AN OLD PROBLEM

P. Dutta

Department of Physics and Astronomy
Northwestern University
Evanston, Illinois 60208-3112 USA

I. INTRODUCTION

Long before exfoliated graphite, organic materials provided an easy way to make monolayers. The familiar chains formed by linking $-CH_2-$ groups are hydrophobic, while a variety of organic groups (in particular, polar groups) are hydrophilic; when combined, for example into a saturated single-chain fatty acid $CH_3(CH_2)_n COO^- H^+$, the resulting molecules will usually adopt one of three strategies to minimize their potential energy in an aqueous environment. They may form micelles, which are blobs with the hydrophilic head groups on the outside (in contact with water) and hydrophobic tails kept dry inside; they may form membranes, which are tail-to-tail bilayers with heads on both surfaces; or, if they are at an air-water interface, they will put their heads in the water and tails in the air. This last system is a Langmuir monolayer: you can easily make one in your kitchen sink.

Many reasons are given for studying such films. It has frequently been suggested that they are model membranes; however, membranes have an important degree of freedom that Langmuir monolayers lack--the ability to acquire non-zero curvature. It is of course still true that insights into the behavior of lipids in monolayer form may be useful in understanding bilayer systems. Another reason for studying monolayers is that they can be transferred to solid substrates by simply dipping a substrate in and out of the water; the resulting multilayers ('Langmuir-Blodgett films') are of considerable fundamental and applied importance, in particular because the technique provides a way to build artificial structures. Naturally an understanding of the parent system is important if we are to assemble these structures precisely and reliably, something that is at present possible only in a few simple cases.

However, the real attraction of these monolayers comes from a simple and long-known fact: their pressure-area isotherms are apparently discontinuous. These discontinuities are not due to layer-by-layer growth; since only one layer can have its heads in the water, a multilayer system is not normally stable on the water surface. Rather, the discontinuities indicate phase transitions within the monolayer. These systems are named after Langmuir, in spite of important previous work by Pockels,[1] Rayleigh[2] and others, because in 1917 Langmuir[3] made the first isotherm measurements that revealed these discontinuities.

This paper will be concerned with the obvious questions: what are the phases, and what are the details of the phase transitions? After a discussion of the basics, I will very briefly list some highlights of the substantial body of work performed in this area over the past seventy years.[4] I will then proceed to recent diffraction results which call into question some of the assumptions behind the earlier work.

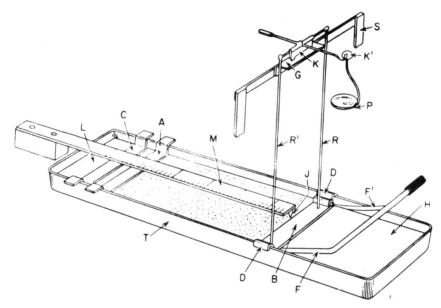

Fig. 1: Langmuir's original apparatus (from Ref. 3). The monolayer (dots) is compressed between the movable barrier (A) and the stationary force-measuring barrier (D). Modern troughs look different (no yardsticks!) but the basic principles are the same.

II. THE BASICS

To make a Langmuir film, you must deliver the molecules to the water surface. When bulk lipid material is placed in contact with the surface, a monolayer will form in equilibrium with it; the pressure and density of this two-dimensional vapor depend on the material but are usually very low. The more common method is to make a dilute solution in a volatile solvent chosen such that the solvent-water surface energy is low (alcohols, hexane, etc. work well). When such a solution is dropped onto the surface, it rapidly spreads; the solvent then evaporates, leaving the monolayer material spread across the surface. A 'barrier' that compresses the monolayer, while allowing water to flow below it, is then used to vary the monolayer density. A second (fixed) barrier dividing regions of plain and monolayer covered water, and attached to a force sensor, can be used to measure the monolayer pressure. Various methods of measuring surface tension are also used to determine the pressure (surface tension is a negative two-dimensional pressure, so that the effect of a positive monolayer pressure is to reduce the measured surface tension; in other words the pressure is the difference between the actual surface tension and the known value for clean water).

The emphasis on 'delivery' implies a metastable system; it is also obvious that attempting to increase the pressure above the equilibrium spreading pressure (i.e. the pressure in the presence of bulk material) must necessarily lead to a metastable state. But so what? The monolayer phases are in two-dimensional equilibrium; whether the monolayer is at a local or global free energy minimum is of no consequence as long as it is confined to this minimum. However, if the monolayer is compressed beyond a certain pressure (or, if any bulk phase is brought into contact with it), it will irreversibly 'collapse' (return to the bulk phase). All single-chain fatty acids and alcohols collapse at about the same area/molecule irrespective of chain length. This means that the molecules stand vertically on the water surface in the phase just preceding collapse, so that the area is determined not by the length but by the cross-section. (Incidentally, the fact that the area/molecule at collapse is roughly that expected for a close-packed monolayer array of vertical chains--not exactly, since the chains may not be close-packed and there may be voids, but certainly to within about 10%--constitutes proof that the film on the surface is a monolayer.[2])

A monolayer isotherm, therefore, covers the region between zero density and the collapse point. The features seen in these isotherms depend on the material, temperature, subphase pH, etc. (for sample isotherms see e.g. Figs. 4 and 9; note that the conventional variable in this field is the area/molecule (A), i.e. the inverse density rather than the density). As the monolayer is compressed, the isotherms in general show 'flat sections' (which may not be quite flat) and/or 'kinks' (which may be rounded changes of slope).

Some precautions are necessary for successful monolayer preparation. First, one must start with clean water. Many organic impurities in water will slowly rise to the surface, and lipid monolayers are extremely sensitive to such contamination. Even highly purified water has appreciable amounts of such impurities, which are removed by letting the water stand and then repeatedly sweeping across the surface with a wiper. Dissolved ions, the other major class of impurities, obviously cannot be removed in this way; they will interact chemically with acid head groups, and their presence presumably also affects the interactions between non-polar hydrophilic head groups. These ions are effectively removed by careful distillation or (preferably) ion exchangers. Some contradictions in early results are now attributed to impure water; even today, experiments performed using commercially purchased distilled water, or in metal or glass 'troughs', will give meaningless results. (Teflon and quartz are good trough materials.)

A related necessity is extremely pure monolayer material. The room-temperature isotherms of commercially purchased pentadecanoic acid, for example, have a rounded section of low but non-zero slope; many creative explanations of this feature have been offered. Such isotherms have also been used to argue that these systems are not good representations of membranes (quite apart from the matter of curvature) because calorimetry shows unmistakable first-order transitions in membranes, while the monolayer isotherms show no flat sections. However, Pallas and Pethica[5] recently found that the rounded regions of pentadecanoic acid isotherms became unambiguously flat when the material was repeatedly recrystallized before use.

III. WHAT IS GOING ON?

Faced with multiple isotherm segments, scientists working in the field came up with the decidedly original idea that they were seeing solid, liquid and gas phases. When some isotherms showed too many phases to be thus labelled, two kinds of order-disorder transitions were postulated[6], involving positional order and orientational order (not to be confused with bond-orientational order). In this scheme there are three condensed phases: a solid, a liquid with molecules all pointed the same way ('Liquid Condensed' or LC), and a liquid that is both positionally and orientationally disordered ('Liquid Expanded' or LE).

This *sounds* reasonable, and as a result the names for the phases have come into widespread use, although of course many other postulates have been put forth. But why postulate? It turns out that, although these are the first monolayer systems to have been scientifically studied, diffraction data did not come along until 1987.[7,8] In retrospect, there were many reasons for this delay. Stacking multiple surfaces (as with, say, exfoliated graphite) is not possible with these systems. Single surfaces were being studied prior to 1987 even with rotating anode sources, but Langmuir monolayers have turned out to be powders in the plane; this naturally extracts a heavy price in intensity. To make matters worse, the substrate is disordered and contributes a high background, whereas a solid substrate would be relatively 'quiet' away from its own diffraction peaks. (Incidentally, another disadvantage of a liquid substrate is that experimental schemes to vary the orientation of the plane normal are not cost-effective; instead, the X-ray beam must be brought down to the horizontal surface.)

Experimental difficulties are by no means limited to diffraction measurements. While the pressure range in which these films are stable and show phase transitions (about 0-50 dynes/cm.) is one that is easily accessible even with 19th-century techniques, the measurement of practically any other property is far more difficult. (Some 'obvious' measurements such as of specific heat have never been performed.) In the rest of this section, I will survey the state of the experimental art prior to the X-ray results. Given the extensive literature, no review can possibly be complete; the present survey will in fact be extremely brief. I will confine myself mainly to providing a flavor of the innovative techniques that have been developed as alternative structural probes.

Mechanical properties

The study of mechanical properties is one way of obtaining macroscopic information that may have microscopic implications; isotherm studies, for example, measure compressibility. Measurements of surface viscosity--many using channel viscometers--outnumber any other class of monolayer measurements except isotherm studies.[4] Although these are time-honored experiments, I will make the obvious observation that one can only measure viscosity in a phase one already knows to be fluid. Although a solid can be distinguished from a fluid by measuring its response at stresses below the elastic limit, these limits are quite possibly low enough for monolayers that they are will not measurably resist being forced through channels; rather, solid films will give spurious 'viscosity' values.

This problem was addressed by Miyano and coworkers,[9] who constructed a variable-strain shear balance useable for static or dynamic measurements. This apparatus distinguishes static shear modulus from viscosity, but the results are still ambiguous[10]: studying a variety of fatty acids and alcohols, they found that the onset of shear modulus (presumably, a freezing transition) coincided in some cases with the highest-pressure discontinuity in the isotherm; however, in other cases this onset did not coincide with *any* of the features of the corresponding isotherms. We recently used a different approach:[11] we measured the monolayer 'pressure' both along and normal to the compression direction, and found that in some isotherm regions the measurements were different, implying shear stresses. The shortest-chain fatty acid studied (pentadecanoic acid) was 'fluid' (within the sensitivity of our measurements) at all pressures, while two others (octadecanoic and tetracosanoic acids) were always 'solid'; only the last material showed changes as a function of pressure that corresponded to changes in the isotherm slope.

Although these techniques are less problematic than flow measurements, the use of macroscopic measurements to determine structure is always subject to some qualifications. A solid supports shear by definition, but the shear modulus may be too small to measure; on the other hand a high viscosity may appear to be a shear modulus over the time scales used in the experiment. Because of low yield strength, or because of slipping between grains in a polycrystalline material, a monolayer may not support shear unless the applied stresses are experimentally indistinguishable from zero.

Optical measurements

Optical techniques have been shown to be powerful probes of structure in organic materials, for example in liquid crystals. There is no fundamental reason why they should not yield equally useful data for Langmuir monolayers, but there are the usual difficulties (small signal from the monolayer, possibly large noise from the water). In the case of multilayers transferred to solid substrates, a variety of techniques such as Fourier Transform Infrared Spectroscopy (FTIR), Brillouin and Raman spectroscopy, polarized second-harmonic generation, surface plasmon generation, etc., have been used as probes of structure (for a review see Ref. 12). However, signal-enhancement techniques such as attenuated total reflection are not easily applicable to the monolayer on water, which therefore remains a more difficult system.

In 1986, Shen and coworkers[13] showed that second-harmonic generation from a Langmuir monolayer could be used to determine the orientation of the molecules (or rather the orientation of the non-linear bonds in the molecule). Unfortunately, the coefficient for second-harmonic generation is small in simple fatty acids, while the air-water interface, being fundamentally non-centrosymmetric, contributes a substantial background. (The use of molecules with built-in chromophores greatly increases the signal, but adds to the complexity of the phase diagram; one would prefer to understand systems of simple molecules first!) However, Shen and coworkers were able to extract orientational information from pentadecanoic acid monolayers that was consistent with the LE-LC picture.

Shen and coworkers also used IR sum-frequency generation,[14] which probes second-order processes involving vibration spectra, and so is sensitive to C-C and C-H bonds (unlike second-harmonic generation, which is thought to occur primarily at the head group). As with second-harmonic generation, the signal is cancelled out in a centrosymmetric system, and thus the technique is sensitive to orientational order. Again, the results (on pentadecanoic acid) were examined in terms of--and found to be consistent with--the LE-LC picture.

A simpler and elegant technique, quasi-elastic light scattering, probes surface waves and thus the *dynamic* longitudinal elasticity and viscosity of the monolayer. Yu and coworkers[15] have used the technique to study both pentadecanoic acid and phosphatidylcholine, finding (among other things) that the technique is a good way to identify a two-phase coexistence region because the elastic constants fluctuate back and forth as 'islands' float in and out of the laser spot.

Fluorescence microscopy

If a small amount of insoluble fluorescent dye is mixed with the monolayer, and if the monolayer is not a single homogeneous phase, one expects that the distribution of the dye will also be inhomogeneous; the impurity is likely to be more soluble in one phase than another. In fact, spectacular experiments by Möhwald and coworkers show---under a microscope[16] ---a variety of fluorescent patterns, from regions with fractal one-dimensional interfaces[17] to spiral-shaped islands[18]. These islands can order into hexagonal superstructures, and they can be controlled with subphase pH[19] or with external fields[20]. This field cannot be adequately reviewed here; from our limited phases-and-phase-transitions perspective, the value of fluorescence microscopy is that it identifies coexistence regions, and thus first-order transitions, although it does not tell us what the structures are. Given the known problems caused by impurities, a sense of unease regarding their deliberate introduction is perfectly justified; however, extremely small amounts are enough, and studies as a function of dye concentration reportedly do not indicate distinct trends.

Summary

Isotherm measurements, surface viscometry, and surface potential measurements (not discussed here) formed the primary tools of the trade before approximately 1980. In the last decade, a variety of new or newly enhanced tools (again, not all discussed here) have been trained on this classic problem, and a substantial amount of progress has been made. Nonetheless, the overall picture is ambiguous; these techniques have told us something about the phases of specific systems, but not unequivocally what the phases are. We next turn to X-ray diffraction.

IV. X-RAY DIFFRACTION: EXPERIMENTAL DETAILS

The basic experimental geometry is one that is familiar to everyone attending this Conference. The incident X-ray beam hits the monolayer at glancing incidence, and diffraction scans are performed above the monolayer plane. Since this plane is always horizontal, the X-ray beam must come down to it. Since the synchrotron beam is also initially horizontal, optical elements (glancing-angle mirrors, tilted diffracting crystals) must be used. Because water is disordered, it is essential that background be minimized by making the incident angle smaller than the critical angle for total reflection (about 0.15°)---this would not be so important if the substrate were solid.

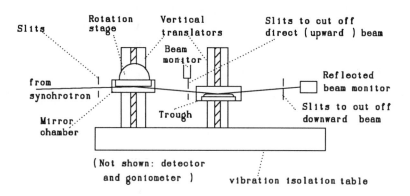

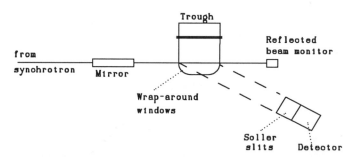

Fig. 2: Schematic of X-ray apparatus, seen from the side (above), and from the top (below). (From B. Lin, Ph.D. thesis, Northwestern University, 1990)

In our own experiments, we use a platinum-coated flat mirror that can be lated vertically and tilted; the trough can be moved up and down to align it in the beam. (Several early experiments[7,8] showed that the monolayers are powders, so that rotating the monolayer about its plane normal is unnecessary). The trough material is Teflon, which is hydrophobic; thus the trough can be over-filled to form an inverted meniscus, allowing the X-rays to reach the water surface without hitting the trough sides. Our trough enclosure has mylar X-ray windows, and channels below and above for cooling/heating water; we control our circulators using a platinum thermometer in the trough, thus holding the

189

trough temperature constant to 0.1°C. The ability to change temperature is crucial to much of our work; although the temperature range to be covered is small (0-100°C at most!), problems with vapor condensation and dripping if the temperature is non-uniform, or is lower than the ambient, have presumably kept other groups doing X-ray work from attempting anything other than near-room-temperature studies. (In fact the problems are unpleasant but hardly insurmountable.)

Vibration isolation is another serious issue that all experiments involving liquid interfaces must address. Two types of solutions exist. Als-Nielsen's apparatus[8] uses flat glass slides under and very close to the surface; capillary waves are damped out in a very thin liquid film. In our apparatus we do not allow anything other than Teflon to touch the subphase, and one cannot form a thin film of water on a hydrophobic material; thus we have chosen to use external vibration isolation. Normal air tables do not hold a constant position with respect to an external index such as the synchrotron beam and are therefore useless. An electronic vibration isolation table (Newport EVIS) turned out to be ideal; our entire system is built on this device.

The final aspect is the X-ray detection system. Since the X-ray beam comes in at a glancing angle, the 'footprint' on the surface is wide in the incident direction (we use a 15 cm. wide trough); and since a horizontally collimated beam, for good 2θ resolution, is necessarily wide, the footprint is wide normal to the incident direction as well. Thus an analyzer crystal, or a non-position-sensitive detector with soller slits, is required. Since the scattered direction cannot be calculated from the position of incidence on a detector unless the scattering region can be approximated by a point, one cannot achieve good horizontal and vertical resolution simultaneously with position-sensitive detectors, unless the footprint is made extremely small (which nullifies the counting-rate advantage gained from a position-sensitive detector). This is a pity; more rapid data collection methods (e.g. *direction*-sensitive detectors) would make a substantial impact in this field.

Note that the monolayer is made up of extended molecules, not 'point particles'; thus, although the diffraction peaks from all monolayers are 'rods', in these systems the intensity varies considerably along the rods. A system of vertical chains will have diffraction rods whose peak intensity is in the monolayer plane; however, if the chains are tilted, the maxima may be in, above or below the plane depending on the direction of the reciprocal lattice vector relative to the tilt direction. (Maxima below the plane obviously cannot be seen.) The vertical width is determined by $2\pi/L$; since L is typically 20-40Å, the widths are small compared to those from monoatomic monolayers, although not as small as the in-plane widths.

Therefore, by locating the diffraction maxima as a function of k_{xy} and k_z, one determines not only the in-plane lattice structure but the tilt direction and magnitude. The monolayer thickness, and thus the average tilt, is also supplied by ellipsometry or X-ray reflectivity, but these techniques are insensitive in the small-tilt limit where $\cos\theta_t \approx 1$. On the other hand, diffraction provides tilt information only in well-ordered systems.

V. X-RAY DIFFRACTION: SOME RECENT RESULTS

There are several notable omissions in this section. Some of the X-ray data published over the past three years have been about monolayers of phospholipids,[8] which are double-chain, complex-head-group molecules, or about monolayers of α-amino acids,[21] also relatively complex molecules. I will not discuss these studies here. Phospholipids are constituents of biological membranes, and studies of phospholipid monolayers are obviously of biological relevance. The α-amino acid studies grew out of an attempt to use a Langmuir monolayer to nucleate three-dimensional crystals, also an important endeavor. However, we are still some distance from fully understanding the phase behavior of simple single-chain amphiphiles, which in the lipids business are the closest we get to krypton on graphite, and I have no idea where more complicated molecules (with complex interactions and conformational possibilities) would fit into the picture I am going to describe below. The final omission is of our own studies of fatty acid monolayers in the presence of metal ions in the subphase.[22] The role of ions in affecting monolayer properties and transfer to substrates is an important one, but in this paper I will concentrate on the phase transition aspects rather than chemical aspects.

In-plane studies

A couple of years ago we[23] and Kjær et al.[24] studied two very similar molecules with almost identical isotherms. Our molecule was heneicosanol $[CH_3(CH_2)_{20}OH]$, while Kjær et al. studied monolayers of arachidic acid $[CH_3(CH_2)_{18}COOH]$. The room temperature isotherms are so simple that I will describe rather than show them here: there is only a single 'kink', so that the isotherm has two sections of non-zero slope with the denser phase naturally having a higher slope.

Both studies consisted of in-plane scans only (no rod scans), so that if chain tilt were to lift the diffraction peaks off the plane and outside the resolution window, the observed peak intensities would drop to zero. In the high pressure phase, both studies reported a single in-plane diffraction peak at about 1.50Å^{-1}. Although the structure cannot be rigorously determined from a single peak, since no other first-order peaks could be seen it is appropriate to assume the simplest structure, namely a hexagonal lattice of vertical molecules.

When the pressure is lowered below the 'kink', the peak intensity begins to drop continuously in both cases. The kink is commonly assumed to be an order-disorder transition, so that we first attempted to explain our results as continuous melting. However, the loss of intensity is not accompanied by an increase in the line width. Kjær et al. proposed that the chains were tilting and moving the peaks up and out of the resolution window; this picture was supported by reflectivity data, and more recently out-of-plane scans[25] have proved that it is the right one.

However, heneicosanol and arachidic acid differ in one important respect. In heneicosanol, the peak intensity goes to zero within a few dynes/cm. below the kink, implying that all three degenerate first-order peaks of the hexagonal structure have moved up off the monolayer plane. This requires a chain tilt direction that is not towards any of the near neighbors, but somewhere in between. In the Kjær et al. data, the peak intensity decreases to a limiting value of one-third the maximum value, even at zero pressure. This implies that the tilt direction is towards a nearest neighbor, which would cause two of the three previously-degenerate peaks to move off the plane, while the third would remain in the plane.

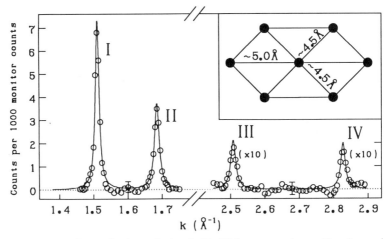

Fig. 3: First and second-order peaks (marked I...IV) from a monolayer of heneicosanoic acid at 5^0C and 35 dynes/cm. Inset: the distorted-hexagonal in-plane structure of what will be referred to as the A phase. (From Ref. 27)

In our heneicosanol study[23] we also varied the temperature while maintaining a high pressures, and found a structural transition along an isobar. This transition is not identifiable when looking at a family of temperature-dependent isotherms. As the temperature is lowered, the 'one-peak phase' (presumably hexagonal) becomes a 'two-peak phase' (presumably distorted hexagonal, i.e. face-centered rectangular with ratio of the two sides of the rectangle $\neq \sqrt{3}$). This transition looks very like the Rotator I to Rotator II transition in paraffins[26], where the hexagonal structure is thought to be made up of chains rotating about their long axis, while the distorted hexagonal (DH) structure is thought to consist of chains that are oscillating around the same axis. (The cross-section of a chain is not circular, so a close-packed structure would be DH. Changes in the DH lattice spacings, seen both in paraffins and in our heneicosanol monolayers, can be explained as due to the chains starting to oscillate around their axis, with increasing amplitudes until they finally begin to rotate freely and form a hexagonal lattice.)

More recently[27] we have seen the second-order peaks of the 'two-peak' phases of heneicosanol and heneicosanoic acid [$CH_3(CH_2)_{19}COOH$], so that it is now established that the structure is DH (see Fig. 3). The second-order peaks are extremely weak, and can only be seen in the highest-pressure phase and at low temperatures ($<10^\circ$C).

Out-of-plane studies

The picture becomes considerably clearer when scans along the diffraction rods are performed, to locate the peak positions in three dimensions. The results described below for heneicosanoic acid in the region 1-8°C are taken from recently published work,[28] but the results at higher temperatures and of heneicosanol are preliminary reports and so are dealt with relatively briefly.

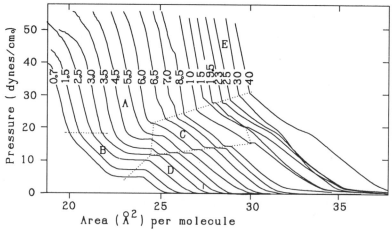

Fig. 4: Isotherms of heneicosanoic acid at the temperatures marked. The isotherms have been shifted for clarity, so that the area axis is correct only for the far left isotherm. The structures determined by X-ray diffraction are shown in Figs. 5 and 8 (From B. Lin, Ph.D. thesis, Northwestern University, 1990)

The isotherms of heneicosanoic acid are shown in Fig. 5; clearly, quite a lot is going on in the phase diagram. For T$\leq$2.5°C, the isotherms show a long plateau plus a rounded change of slope at a higher pressure; for 3°C$\leq$T$\leq$6°C, the single plateau splits into two shorter ones; and for T$\geq$6.5°C, each isotherm contains a plateau and a 'kink'. The kink turns into a small flat section briefly around 20°C and then becomes a kink again, while the lower flat section becomes smaller and vanishes, also around 20°C. The dotted lines in Fig. 4 are guides to the eye connecting similar features; they divide the figure into five regions (labelled A, B, C, D, E for ease of reference).

In Region A, as mentioned before, two first-order in-plane peaks are seen: the positions are ~1.50Å^{-1} and ~1.69Å^{-1}. We will refer to these peaks as Peak I and Peak II. Of the three 'bond lengths' in the distorted-hexagonal lattice, two are ~4.49Å and the third is ~5.01Å; thus Peak I is doubly degenerate.

In Region B, we see the same two peaks, and the magnitudes of the diffraction vectors remain unchanged through the continuous A-B transition. However, Peak I moves continuously off the plane, while Peak II remains in the plane; it can easily be seen that this implies a tilt along the 'long bond' (5Å bond).

In Region C, we again see the same two peaks, but now both peaks move continuously off the plane starting at the A-C transition, in fact Peak II is further off the plane than Peak I. Since Peak I does not split, only one tilt direction is possible: normal to the 5Å bond and thus between near neighbors.

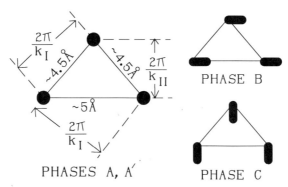

Fig. 5: Schematic showing some of the structures described in the text. Shaded objects are projections of cylinders representing molecules, so that round ≡ vertical and elongated ≡ tilted. (From Ref. 23)

In Region D, we see one in-plane peak at 1.44Å^{-1}, a position that is distinct from that of Peak I or Peak II; thus this is a new lattice structure. The point of greatest interest is that this zero-pressure phase is *not a liquid*; the peak is resolution limited. In Ref. 28, we called it an unidentified 'expanded solid'; more recently, we have observed a second peak off the plane, with both k_{xy} and k_z being strong functions of pressure and temperature. This peak is weak compared to the 1.44Å^{-1} peak, and much broader than the resolution, implying non-isotropic in-plane order. If only peak positions are considered, this structure is qualitatively identical to the B structure (i.e. a distorted-hexagonal lattice with tilt towards a near neighbor) but with different lattice spacings (see Fig. 8).

It is worth noting here that as the chains tilt away from the vertical direction when going from the A phase to the B or C phases, the chain-chain distances (measured normal to the chains) remain constant, which obviously means that the in-plane spacings increase. In the D phase as well, the tilt angle increases as the pressure is decreased, but the chain-chain distances (which are larger than in the A/B/C phases) do not remain constant. On the other hand, while the in-plane spacings are not precisely constant, they change relatively little. We can therefore speculate that the D lattice is formed by the relatively long-range inter-action between polar head groups.

The phase diagram in the 1-8°C region is shown in Fig. 6, with one important addition: the small irregular feature seen in the 6.5-8.5°C isotherms at high pressures turns out to be a first-order phase transition, to a DH structure with very slightly different lattice constants ('bond lengths' ~4.53Å and ~5.01Å). We have labelled this phase A'. The shape of the isotherm at this transition requires

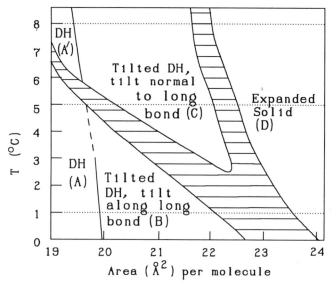

Fig. 6: Qualitative phase diagram (DH=distorted hexagonal) for heneicosanoic acid in the region 0-8°C. The dashed line indicates an unverified section of the A-B phase boundary. (From Ref. 28)

comment; lack of perfect flatness even in some relatively flat sections (cf. Fig. 4) has been a subject of much controversy (it has been suggested, for example, that these are lines of continuous transitions). Since we have observed two-phase coexistence using X-ray diffraction in the transition regions, we can now turn the question around from "The isotherms are not quite flat, are these really first-order transitions?", to "These are first-order transitions, why are the isotherms not flat?" In fact, the familiar mental picture of a first-order transition derives from liquid-gas isotherms; since there is no hydrodynamic equilibrium in a solid-solid transition, there is no reason for the isotherms to be perfectly flat. At the lowest temperatures we see a very flat section (Fig. 4), so impurities are not to blame; presumably, the highest-pressure transitions are particularly rounded or irregular because denser phases are more rigid.

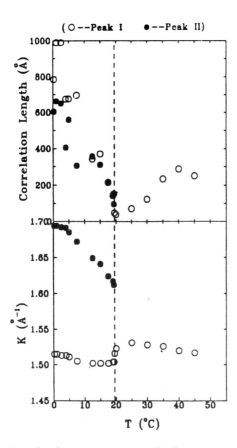

Fig. 7: Correlation length (top) and diffraction peak position (bottom) as a function of temperature for monolayers of heneicosanoic acid along a 35 dynes/cm. isobar. The dashed line marks the transition temperature. Since the peaks are broad at the transition, the peak positions are unreliable there; thus these data do not prove that the transition is first-order. (From B. Lin, Ph.D. thesis, Northwestern University, 1990)

Results from very recent higher-temperature studies are as follows:

(a) Phase E shows only one peak and is presumably hexagonal, as discussed earlier for heneicosanol. The transition occurs exactly at the point where the kink in the isotherm briefly becomes a flat section. The structural transition is accompanied by precursor broadening of Peak I and Peak II, and the single peak in the E phase is always broad (Fig. 7). Because of the considerable peak broadening near the transition, the peak positions are unreliable there; Fig. 8 appears to suggest that the transition is first order, but this may not in fact be so. Since the correlation lengths remain longer than those seen in normal fluids, Phase E is apparently some kind of intermediate structure or 'mesophase'.[29]

(b) The C phase cannot be seen past the temperature at which the A-E transition occurs. At higher temperatures, the single-peak high-pressure phase (E) goes directly to the D phase when the pressure is lowered. Just as at lower temperatures, the in-plane peak in the D phase is resolution limited, while the off-plane peak is broad. Similar features have been reported for arachidic acid in a recent preprint by Kenn et al.[29]

(c) Using unit cell areas determined from the X-ray data, we find that the C-A$'$ transition is indeed continuous; the D-E transition involves a discontinuous change in unit cell area where the isotherm briefly displays a flat section, but the transition becomes continuous as the temperature is increased. In other words, we now have a microscopic 'explanation' for the kink/flat-section/kink region.[30] Heneicosanol isotherms in the same temperature region are shown in Fig. 9. The isotherms are simpler, and our preliminary X-ray results can be quite simply described. The regions marked a, c and e have the same structure as the A, C and E phases of the corresponding acid. We see no B phase or D phase. The absence of a D phase supports our earlier speculation that the D in-plane lattice is determined by the acid head groups: since the electrostatic interactions between alcohol head groups are weaker, they are presumably unable to form the expanded structure.

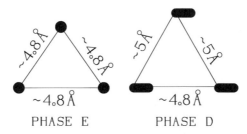

Fig. 8: Schematic showing structures of D and E phases, using the same convention as in Fig. 5.

The discrepancy between the older heneicosanol and arachidic acid in-plane data is now resolved. Arachidic acid shows an D-E transition at room temperature; heneicosanol at room temperature goes from the e phase to the c phase (there is no d phase). The difference is that in the D phase one of the first-order peaks is always in the plane, while in the c phase all peaks move off the plane.

VI. CONCLUDING REMARKS

The most notable feature of the recent results is that at no point do we see traditional fluid phases (those with correlation lengths of the order of a very few intermolecular spacings). In fact, during a quick-and-dirty search at zero pressure, we found the D phase resolution-limited in-plane peak all the way up to $60^\circ C$ (window fogging problems prevent long, careful scans at such high temperatures). Since the bulk phase melting point is above $100^\circ C$, this result is surprising only because of the long-established dogma.

I certainly don't want to suggest that fluid phases are always absent in Langmuir films: other materials, higher temperatures, who knows? It remains true, however, that one cannot determine the structure by looking at the isotherm and then invoking folklore. In retrospect, many studies that 'confirmed' the existence of LE and LC phases were equally consistent with tilted solid structures.

Basic information regarding the structures of phases (in any system) should precede other studies. In the case of Langmuir monolayers, that has not been the historical sequence. We have now obtained 'crystallographic' information; this is

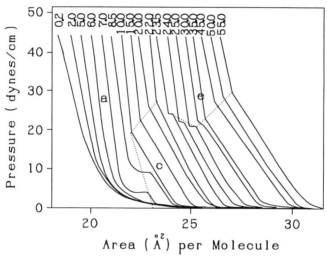

Fig. 9: Isotherms of heneicosanol. The isotherms have been shifted for clarity, so that the area axis is correct only for the far left isotherm. (From B. Lin, Ph.D. thesis, Northwestern University, 1990)

hardly the end of the story, but perhaps it is a new beginning. We hope to be able to extend the characterized region of the phase diagram, and to fully understand the competing roles of the head and tail groups. Meanwhile, we hope both experimentalists and theorists will pay closer attention to this fascinating system in light of the newly available structural data.

VII. ACKNOWLEDGEMENTS

My research in this area was performed at CHESS and at NSLS, in collaboration with B. Lin, M.C. Shih, T. Bohanon, G.E. Ice, J.B. Ketterson, S. Barton, B.N. Thomas, E. Flom and S.A. Rice. This work was supported by the U.S. Department of Energy under grant no. DE-FG02-84ER45125.

REFERENCES

1.   e.g. A. Pockels, *Nature* **43**, 437 (1891)

2.   e.g. Lord Rayleigh, *Phil. Mag.* **48**, 337 (1899)

3.   I. Langmuir, *J. Am. Chem. Soc.* **39**, 354 (1917)

4.   For a review see G.L. Gaines, Jr., Insoluble Monolayers at Liquid Gas Interfaces (Interscience, New York, 1966).

5.   N.R. Pallas and B.A. Pethica, *Langmuir* **1**, 509 (1985)

6.   see, e.g., W.D. Harkins and L.E. Copeland, *J. Chem. Phys.*, **10**, 272 (1942)

7.   P. Dutta, J.B. Peng, B. Lin, J.B. Ketterson. M. Prakash, P. Georgopoulos and S. Ehrlich, *Phys. Rev. Lett.*, **58**, 2228 (1987)

8.   K. Kjær, J. Als-Nielsen, C.A. Helm, L.A. Laxhuber, and H. Möhwald, *Phys. Rev. Lett.* **58**, 2224 (1987)

9.   B.M. Abraham, K. Miyano, S.Q. Xu and J.B. Ketteerson, *Rev. Sci. Instrum.* **54**, 213 (1983)

10.   B.M. Abraham, K. Miyano, J.B. Ketterson and S.Q. Xu, *Phys. Rev. Lett.* **51**, 1975 (1983)

11.   K. Halperin, P. Dutta and J.B. Ketterson, *Langmuir* **5**, 161 (1989)

12.   J.D. Swalen, *J. Mol. Electronics* 2, 155 (1986)

13.   Th. Rasing, Y.R. Shen, M.W. Kim and S. Grubb, *Phys. Rev. Lett.* **55**, 2903 (1985)

14.   P. Guyot-Sionnest, J.R. Hunt and Y.R. Shen, *Phys. Rev. Lett.* **59**, 1597 (1987)

15.   B.B. Sauer, Y.L. Chen, G. Zografi and H. Yu, *Langmuir* 4, 111 (1988)

16.   M. Lösche and H. Möhwald, *Revs. Sci. Instrum.* **55**, 1968 (1984)

17.   A. Miller, W. Knoll and H. Möhwald, *Phys. Rev. Lett.* **56**, 2633 (1986)

18.   W.M. Heckl, M. Lösche, A. Cadenhead and H. Möhwald, *Eur. Biophys. J.* **14**, 11 (1986)

19.   A. Miller, W. Knoll, H. Möhwald and A. Ruaudel-Teixier, *Thin Solid Films* **133**, 83 (1985)

20.   W.M. Heckl, A. Miller and H. Möhwald, *Thin Solid Films* **159**, 125 (1988)

21.   S.G. Wolf, E.M. Landau, M. Lahav, L. Leiserowitz, M. Deutsch, K. Kjær and J. Als-Nielsen, *Thin Solid Films* **159**, 29 (1988); S.Grayer-Wolf, M. Deutsch, E.M. Landau, M. Lahav, L. Leiserowitz, K. Kjær and J. Als-Nielsen, *Science* **242**, 1286 (1988)

22.   B. Lin, T.M. Bohanon, M.C. Shih and P. Dutta, *Langmuir* **6**, 1665 (1990)

23.   S. Barton, B. Thomas, E. Flom, S.A. Rice, B. Lin, J.B. Peng, J.B. Ketterson and P.Dutta, *J. Chem. Phys.* 89, 2257 (1988)

24.   K.Kjær, J. Als-Nielsen, C.A. Helm, P. Tippman-Krayer and H. Möhwald, *J. Phys. Chem.* **93**, 3200 (1989)

25.   J. Als-Nielsen and K. Kjær, in "Phase transitions in soft condensed matter", edited by T. Riste and D. Sherrington (Plenum Press, New York, 1989)

26.   see, e.g. G. Ungar, J. Phys. Chem. 87, 689 (1983)

27. T.M. Bohanon, B. Lin, M.C. Shih, G.E. Ice and P. Dutta, *Phys. Rev. B* **41**, 4846 (1990)

28. B. Lin, M.C. Shih, T.M. Bohanon, G.E. Ice and P. Dutta, *Phys. Rev. Lett.* **65**, 191 (1990)

29. R.M. Kenn, C. Böhm, A.M. Bibo, I.R. Peterson, H. Möhwald, K. Kjær and J. Als-Nielsen, submitted to *J. Phys. Chem.*

30. M. Shih, T.M. Bohanon, J. Mikrut and P. Dutta, in preparation

PHENOMENOLOGY OF SURFACE RECONSTRUCTION

J. Villain, J.L. Rouviere, I. Vilfan*

DRF/MDN, Centre d Etudes Nucleaires de Grenoble
85 X, F-38041 Grenoble Cedex, France
(*) Institut Jozef Stefan, Ljubljana, Jugoslavija

ABSTRACT: *Reconstruction is one of the most fascinating phenomena in surface physics, since it occurs on clean surfaces. In the first part, surfaces and grain boundaries of semi-conductors with diamond structure will be addressed. The various structures which arise from the tendency to preserve four-fold coordination as much as possible will be described and interpreted in terms of dangling bonds and phenomenological potentials. Then the reconstruction of noble metals will be described and related to current theories based on conflicting sp and d electrons and phenomenological potentials. Finally the "deconstruction" phase transition of the (110) face of Au and Pt will be addressed. Its Ising or roughening nature will be discussed in the light of recent experiments, and related to a wetting transition of steps.*

1. INTRODUCTION

Surface reconstruction is the property of many planar crystal surfaces, that their symmetry is lower than the cross-section of the ideal crystal through a plane ("unreconstructed" surface) would have. Examples will be displayed in the present article.

Reconstruction is a very common phenomenon which occurs in materials so different as semi-conductors (Si, Ge) and noble metals (Au, Pt, Ir). The first part of this article will be devoted to semi-conductors and the second part to noble metals. Other elements like W, compound materials and surfaces with adsorbates will not be discussed.

There are so many aspects of reconstruction that the sampling which will be given here can only be arbitrary. The examples are intended to suggest that, although morphologies are often very complicated, it is often possible to get some insight through an appropriate phenomenology. The microscopic mechanisms responsible for reconstruction in Si and Ge will be discussed in the first part of this article, and the second part in mainly devoted to the statistical mechanics of the phase transition (from reconstructed to unreconstructed) of the (110) surface of Au and Pt.

Among other goals, the present review aims at bridging, in a restricted domain, the gap between two branches of theoretical physics in solid state: microscopic (electronic) physics and statistical mechanics. This is only possible if phenomenology is used rather than an *ab initio* approach, and if qualitative insight is seeked rather than quantitative predictions.

A large part of the results which will be presented, in a synthetic and hopefully didactic way, are not ours. However, a few new investigations will also be described.

2. SILICON AND GERMANIUM

These elements crystallize in the cubic system. They have the diamond structure (Fig. 1).

2.1. The (001) face of Si and Ge

This face has, at least with a good approximation, a (1×2) reconstruction at ordinary temperature. This means that the surface structure has a unit cell which is just as large as the unreconstructed unit cell in the (100) direction, and twice as large in the (010) direction (Fig. 1). This superstructure results from a displacement of the

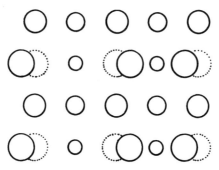

Figure 1. Projection of the Si structure one the (001) plane. The biggest full circles are surface atoms and the dotted circles are the positions they would occupy in the absence of reconstruction. Circles of decreasing radius represent atoms at distance a/4, a/2 and 3a/4 below the surface, where a is the unit cell length.

surface atoms which tend to form pairs or "dimers". The mechanism is simple : while bulk atoms have coordination 4 (4 neigbours), surface atoms have coordination 2. They have two "dangling bonds". This costs a lot of energy. To heal (partly) this unpleasant situation, surface atoms form dimers, thus increasing the electronic overlap and decreasing the energy.

The dimers should be in the plane of the dangling bonds, which is either the [100] or the [010] direction, but which is well defined for a given surface. In the case of Fig. 1b, it is the [010] direction. The dimerisation can occur in two different ways, which implies that, if the surface "deconstructs" at some temperature, the phase transition is in the Ising universality class.

The above picture seems to describe experimental facts in a satisfactory way, but, according to certain calculations (Ihm et al 1983) the bridge formed by each dimer is asymmetric or "buckled", the reconstructed phase should really have a (4×2) superstructure, originating from different orientations of the buckled dimers. More recent calculations (Batra 1990) do not confirm this view.

202

2.2. The (111) face of Si

This face has generally a complicated (7×7) reconstucted superstructure. It is famous because it was one of the tests used by Rohrer and Binnig to demonstrate the power of scanning tunnel microscopy (STM) and to get the Nobel Prize. However the structure proposed by Binnig et al (1983) did not satisfy everybody, and at least 4 other models were suggested. The structure proposed by Takayanagi et al (1984) seems in best agreement with electron microscopy and diffraction, and with STM (Tromp et al 1986). This controversy demonstrates that those methods do not really *see* the individual atoms!

Takayanagi's structure is displayed by Fig. 2. As in the much simpler case of the (001) surface, reconstruction occurs because it decreases the number of dangling bonds. On the unreconstructed surface, all surface atoms (big empty circles) would have coordination 3. Each of them would have, therefore, one dangling bond. This dangling bond would be perpendicular to the surface, so that, in contrast with the (001) face, pairing of dangling bonds is unfavourable. Energy can be lowered by putting additional atoms or "adatoms" (black circles) neighbours to 3 surface atoms. Each adatom has, of course, one dangling bond, but prevents 3 "surface atoms" to have dangling bonds. Those surface atoms preserve their coordination 4, but the tetrahedron is strongly distorted, so that it is not favourable to group all "surface atoms" in triplets with one adatom per triplet. It is preferable to leave one surface atom with a dangling bond among the triplets. This is what happens in one of the two triangles limited by dotted lines in Fig. 2, where the surface atoms occupy approximately the positions they would occupy in the bulk material. The other triangle is shifted. This allows an additional decrease of the number of dangling bonds, thanks to a pairing (or dimerisation) of certain atoms of the layer below surface atoms. Thus, in contrast with the (001) face, the reconstructed (111) surface is organised on several levels.

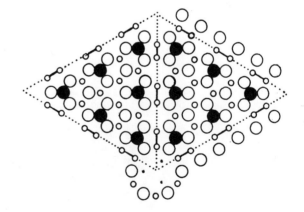

Figure 2. The reconstruction of Si(111). Black circles are "adatoms". Big, empty circles are "surface atoms" (at somewhat lower height). Small empty circles have a lower height yet. Full lines indicate dimerization. Some of the lower atoms are indicated by dots. The dotted lines are domain walls and mark the 7×7 unit cell.

2.3. Microscopic calculations

The band structure of semiconductors is well understood. The electronic mechanisms which determine the surface structure can be put into equations which, in principle, can be solved by a computer. This was done for instance by Batra (1990) in the case of the Si(001)-(2×1) reconstruction. If the structure is too complicated, for instance for the (7×7) reconstruction, present computers are not completely successful. For instance Payne et al (1989) were only able to investigate a 3×3 version of Takayanagi's model. Future computers will presumably do the job. However, not much insight comes from a computer.

Therefore it is interesting to wonder whether simple phenomenological potentials can give valuable information. It is known that pair potentials cannot properly describe hybridized sp3 bonds. An angular term must be included in the potential, for instance by a three body term. The Keating potential (Keating, 1966) in which the energy is expanded in terms of bond-stretching and bond-bending was the first potential that described correctly some silicon bulk properties. But in this potential the tetracoordination of atoms is imposed. Of course, it cannot be used for surface simulations. Recently, to remove this drawback, a lot of empirical potentials have been developed (for a brief review, Baskes, Daw, Dodson and Foiles, 1988). Their parameters have been fitted to some experimental data and also, what is more unusual, to some ab initio calculated energies of simple structures. These potentials, like the Stillinger and Weber one (1985) or the Tersoff one (1988), have been tested on numerous systems ranging from high density phases to system with low symmetry such as surfaces and clusters. These new potentials permit to do dynamical or temperature dependent simulations. However simulations must be compared with experiments. In the next section some results are presented that have been obtained by the association of numerical simulations and experiments. They are not really related to surfaces but to the analogous problem of germanium grain boundaries.

2.4 Germanium (001) tilt grain boundary results

A Grain boundary (GB) is an interface between two semi infinite crystals identical but disoriented from each other (fig. 3). Therefore the complexity of a GB is intermediate between a general crystal-crystal interface and a free surface. Whereas a free surface is defined by its normal, i.e. by two parameters, a GB is characterized by five parameters: for instance its orientation with respect to one of the grains (two parameters) and the rotation that relates the second crystal to the first one (two parameters for the rotation axis and one for the angle).

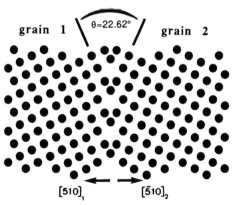

Figure 3 - (001) projection of a grain boundary characterized by the normal $(510)_1$ and the rotation having its axis parallel to the <001> axis and an angle $\theta = 22.62°$.

Few numerical simulations have been performed in semiconductors grain boundaries mainly because of the complexity. No dynamical simulations have been done. Only one ab initio type calculation for one specific relaxed structure has been calculated (Tarnow, Bristowe, Joannopoulos and Payne, 1989). The main reason is the large amount of atoms (typically 60 for the simplest cases) which is generally required for grain boundaries. However semi empirical tight binding methods have been developed for GB simulations (Paxton and Sutton) and they could be applied in surface simulations. This method has the advantage of having a theoretical basis without being too heavy, moreover it gives informations on electronic and atomic structures. The GB results we report here have used empirical potentials (mainly the Tersoff one, (Tersoff (1988)). The simulations have been a tool in the exploitation of the High Resolution Electron Microscopy (HREM) images.

In contrast with free surfaces, only few techniques have access to the structure of the GB interface embedded between two large crystals. HREM is one of these techniques. It permits to "visualize" the atomic structure of crystals along low index axis. However as the two crystals of the GB must be simultaneously imaged and as the interface must be viewed edge on, only GBs with the axis of the rotation perpendicular to the surface normals (tilt GBs) and parallel to a low index axis can be observed. In our case the axis of rotation was parallel to the (001) axis ((001) tilt GBs).

Most of the observed structures were directly determined by high resolution electron images taken along the (001) axis. As shown in figure 4, all the atomic columns of the interface and thus the projection of the GB can be determined. Then some hypotheses on the periodicity along the axis of observation permit to build a three dimensional structure.

In these cases, the use of the simulations is firstly to provide relaxed atomic positions that are used in numerical simulations of HREM images (fig. 4). Secondly, they give an estimate of the energies : a low energy model, in agreement with the HREM images is certainly the correct structure. In some cases, observation in a second direction confirms these structures (Bourret and Rouvière, 1989).

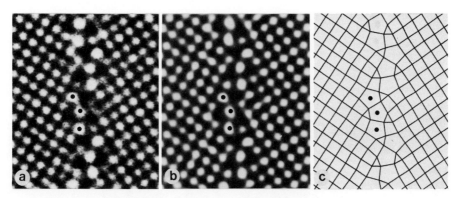

Figure 4 - (a) (001) HREM images of a (001) tilt grain boundary called by the specialist a Σ-13 (510) GB : the interface normal is (510) and the angle of the rotation is θ=22.62°. Atoms are black. (b) Simulation image of the same grain boundary. The correspondance is perfect. (c) (001) projection of this interface. The atomic position have been numerically calculated using a Tersoff potential. The atomic bonds between atoms have been represented in order to permit a better comparison with HREM images : the big white dots in the HREM image (some of them are marked with black spots) correspond to the pentagonal holes of the projection.

However, some interfaces like the one shown in figure 5 could not be resolved directly from HREM images. In these cases, the simulations permit one to reject several high energy models that could be in agreement with the HREM images. These models were generally not tetracoordinated (model I_1 of table I for instance). On the contrary several low energy structures entirely tetracoordinated do not entirely correspond to the HREM images. These structures are slightly asymmetric whereas the HREM images are mainly symmetric (figure 5). However, as some of these structure are symmetrically related and mainly differ by the position of two atoms (figure 6) we proposed a structure composed of a constant part periodically repeated and of a variable core where two atoms can have several stable positions. This model was confirmed by HREM images taken along a direction perpendicular to the <001> axis (Rouvière and Bourret 1989).

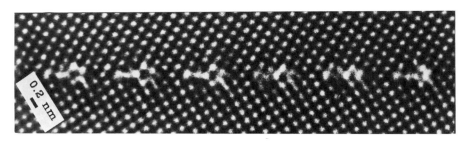

Figure 5 - (001) HREM images of a (001) tilt grain boundary called by the specialist a Σ=13 (510) GB : the interface normal is (510) and the angle of the rotation is θ=22.62°. The structure is different from the one of figure 4 and cannot be directly read out. Atoms are black.

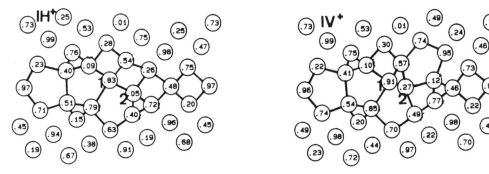

Figure 6 - (001) projections of two models of a (001) tilt grain boundary Σ=13 (510) θ=22.62°. Some bonds between atoms are drawn. Atom heights in units of the lattice parameter a are indicated in units of the lattice parameter a. The structure of figure 5 has been interpreted as a kind of hybrid, IH^+, of these models and symmetrically related structures.

After having recalled the advantage of a collaboration between simulations and experiments we want to summarize the main GB results that can be compared to surface results.

(i) Tetracoordination : the GB problem is simular to the surface one. The bonds of the two crystal surfaces must be reconstructed in order to lower the energy. However, the solution is different. The GB interface is dense enough and the bonds are soft enough to allow the tetracoordination of atoms in all the cases. No dangling bond is observed.

(ii) Periodicity : In a free surface, the elemental periodicities are given by the unreconstructed surface which is the intersection of the ideal crystal by the surface. In the same manner, the elemental periodicities of a GB can be defined by the intersection of the two unreconstructed surfaces of the GB. Except for the structure with a disordered core which is not strictly periodic, all the observed (001) tilt GBs have these elemental periodicities. However, we know that some

Table 1. Tersoff energies (J/m^2) of different models of the same interface. I_1 is a non-tetracoordinated model of rather high energy, which can therefore be discarded although it would fit the experimental data. In the tetracoordinated IH^{-+} model (which offers the best fit with experimental data) the atoms 1 and 2 occupy alternatively the positions they have in the IH^+ and in the IH^- models.

models	IH^+	IH^-	IV^+	IV^-	IH^{-+}	I_1
Tersoff energy	0.326	0.326	0.366	0.366	0.333	0.545

superstructures can exist in (011) GBs (Bourret and Bacmann, 1985)

(iii) multiplicity of structures : a same interface can have several completely different stable structures. Figures 4 and 5 give for instance two structures of a same interface. The possibility of a transition between these two or more structures is an open question.

3. RECONSTRUCTION OF NOBLE METAL SURFACES

3.1. Structures

Gold will mainly be addressed in this review, but Pt and Ir have also a strong tendency to reconstruction. All these materials are face centred cubic.

All simple surfaces of Au, (111), (001) and (110) reconstruct at low temperature. The (110) face has usually a simple (2×1) structure (Fig. 7). This structure is obtained from the unreconstructed surface by removing every second compact atomic row from the surface.

The reconstruction of the (111) and (001) surfaces of Au are more complicated (Fig. 8). A common feature is that, in both cases, *the outer layer is denser* than bulk layers.

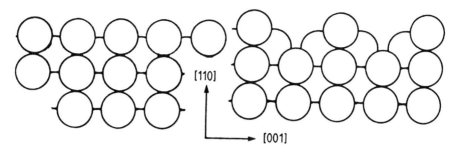

Figure 7. Cross section of a gold crystal perpendicular to the compact [1̄10] atomic rows, showing the missing row reconstruction of the (110) surface (b) compared to the unreconstructed structure of Ag or Cu (a).

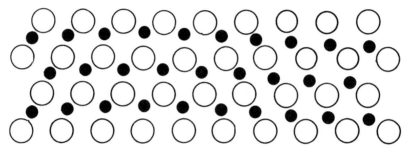

Figure 8. Top view of a (111) gold surface showing its reconstruction. The true reconstructed structure is 23×1. This is a fictive, (10×1) model which aims at giving an idea of the real geometry.

3.2. Microscopic mechanism

The simple (1×2) reconstruction of the (110) face of Au and Pt is simple enough to be accessible by *ab initio* calculations (Ho and Bohnen 1987). From the mysterious output of the computer (Table 2) one can extract a qualitative feature which is not unexpected: *relaxation plays an important part* in that reconstruction. In other words, as in the (111) and (001) faces, surface atoms like to have an interatomic distance which is not the same as in the bulk.

It is seen from table 2 that even without relaxation the unreconstructed structure would be unstable. It would be of interest to know whether the (1×2) structure itself would be stable with respect to other (1×n) structures, in particular with respect to the formation of macroscopic (1̄11) facets.

Table 2. Energy of the unreconstructed and reconstructed (110) surface of Au calculated by Ho and Bohnen (1987) with and without relaxation.

Structure	(1×1)		(1×2)	
	rigid lattice	relaxed	rigid lattice	relaxed
total energy	1.43	1.38	1.40	1.31

The large relaxation in all faces of Au is accounted for by the qualitative theory of Heine and Marks (1986). The main contribution to the cohesion energy in gold comes from 5d and 6sp electrons. It turns out that 5d electrons favour a large atomic distance while 6sp electrons would like this distance to be shorter (Fig. 9). The existence of a surface allows the sp electrons to "flow" to places where there are few d electrons, and the interatomic distance at these places shrinks.

A more quantitative approach has been successfully used by Ercolessi et al (1986) and we shall try to relate it to the work of Heine and Marks. According to Heine and Marks the 5d electrons can be represented by pair potentials (as can be deduced from the tight-binding approximation when the d band is full, as is the case in noble metals). On the other hand the 6sp band is not full, and is broad. It follows that the 6sp electrons can flow to the most favourable places as said above. Therefore, if we want to represent their effect by an effective interaction between atoms, this interaction is *not* pairwise: if the atomic distance is too short at some place, the electrons will go somewhere else, however not too far from the atom i to which they are bound by the Coulomb attraction. This suggests the following potential energy (Ercolessi et al 1986) to be added to a pair interaction.

$$\mathcal{H}_{glue} = \sum_i U(n_i) \qquad (3.1)$$

where

$$n_i = \sum_{j \neq i} \rho(r_{ij}) \qquad (3.2)$$

U and ρ are phenomenological functions and r_{ij} is the distance between atoms i and j. i and j are intended to be neighbours, so that $\rho(r)$ should vanish for large r (Fig. 10).

Ercolessi et al have been able to fit the parameters of their model so as to get a good agreement with experimental data, especially the reconstructed structures.

Another theory of the reconstruction of gold has been given by Guillopé and Legrand (1989).

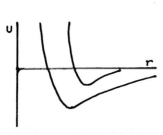

Figure 9. The 5d and 6sp pseudopo-
tential in Au according to Heine
and Marks.

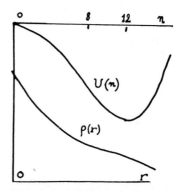

Figure 10. The phenomenological
functions U and ρ according to
Ercolessi et al.

209

4. THE DECONSTRUCTION TRANSITION OF Au(110) AND Pt(001)

4.1. Au(110): a phase transition of a new universality class ?

Although the present contribution is a part of a book about phase transitions on surfaces, we did not really discuss this topic so far. Indeed, the amount of literature devoted to phase transitions related to reconstruction is limited. One reason is that experimental data are usually very sensitive to impurities and other defects, and therefore not easily reproducible. The case which has been most often studied seems to be the missing row reconstruction of Au(110) and Pt(110), which will be addressed in this Section. An obvious advantage is that this (1×2) reconstruction is simple. But in addition, it will be seen to raise fundamental and, to our knowledge, unsolved problems.

The (110) surface of Au and Pt is reconstructed below some temperature T_c and unreconstructed above. The reconstructed ground state has a two-fold degeneracy since the missing rows (Fig. 7) can be chosen in two different ways odd rows or even rows). Since this two-fold degeneracy is also a property of the Ising model, the transition might be expected to be in the universality class of the two-dimensional Ising model (Bak 1979). On the other hand, the disappearance of the Ising order (asymetry between odd and even rows) may well coincide with a *roughening* transition. If it does not, then a roughening transition may be expected to take place at some temperature $T_R > T_c$. Indeed, most of (110) surfaces roughen well below the melting temperature. In the case of Pt(110), such a transition has never been observed. However, this does not mean that it does not exist, because the roughening transition is usually extremely difficult to observe : no observable specific heat anomaly, no characteristic line in the diffraction spectrum; just a characteristic shape (Villain et al 1985, Conrad et al 1985) which can be observed only if inelastic scattering is carefully subtracted.

The roughening transition is known (Knops 1977, Van Beijeren and Nolden 1987) to have the same critical behaviour as the phase transition of the two-dimensional XY model studied by Kosterlitz and Thouless (1973) and Kosterlitz (1974). These properties were derived for a model (the so-called S.O.S. model) which has an important difference with Au(110) : the smooth state is not reconstructed.

Is the universality class modified by reconstruction ? This is basically the question raised in a recent paper by Robinson et al (1989). These authors reported an observation suggesting that the surface does roughen at T_c although the exponents are those of the two-dimensional Ising model, as first noticed by Campuzano et al (1985). The crucial observation of Robinson et al was that the superlattice diffuse scattering line observed by X-ray diffraction at grazing incidence is shifted, and the ratio of the shift to the linewidth is approximately constant when the temperature varies. It is rather difficult to explain the reason of this shift in a limited space, we wish only to stress that this shift is related to the existence of an Ising order below T_c and is not present in other materials (see Kern 1990 for details).

It can be shown (Vilfan and Villain 1990) that the disordered (110) gold surface would exhibit a shifted diffuse scattering peak even it were flat, provided certain conditions are satisfied. This remark weakens the argument of Robinson et al in favour of a roughening transition. However, certain features (e.g. the constant ratio of the shift to the linewidth) are better explained in Robinson's picture. In the remainder of this Section, we address the following questions on purely theoretical grounds. i) Can a roughening transition be in a universality class different from that of the usual SOS model, namely the Kosterlitz-Thouless class? ii) Can the type of transition (roughening or Ising) be related to microscopic data? Certain appropriate concepts will first be introduced.

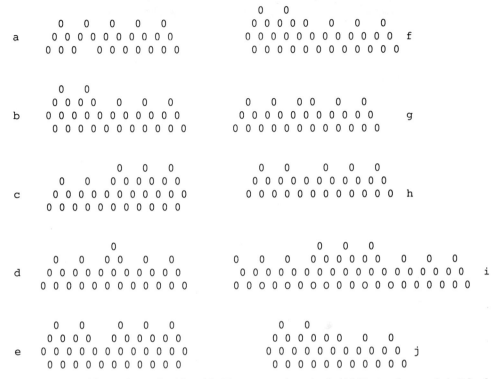

Fig. 11. Side view of the (1x2) reconstructed (110) surface. (a) Ideal surface without defects, (b,c) elementary excitations, characteristic of a roughening transition, (d,e) Ising excitations, (f,j) higher energy roughening excitation, and (g,h,i) higher energy Ising excitation.

4.2. Domain walls, steps and kinks

In an ordinary Ising model, order is destroyed by domain walls. This property was successfully used by Müller-Hartmann and Zittarz (1977) who calculated the transition temperature of the two-dimensional Ising model by writing that the line tension of a domain wall vanishes. Similarly, in the S.O.S. model, roughness is created by steps. In the case of Au(110) where both Ising order and roughness can be present, both domain walls and steps might be expected to be relevant. However, it will be seen that domain walls are made of two steps, so that it is in principle sufficient to consider steps.

This is easily seen at zero temperature. Of course, walls and steps are not present at T=0 at equilibrium, but if they are introduced, they are straight and parallel to the [1$\bar{1}$0] direction (hereafter called y). Their structure is in that case defined by a cross-section of the sample perpendicular to y.

Fig. 11 shows a schematic cross section a) of the ideal, reconstructed surface (the same as Fig. 7 b). b and c) of two steps. d and e) of two domain walls. It is clear from these figures that domain walls are made of two steps. Fig. 11 f shows another type of step and Fig. g and h shows other types of domain walls. These were the only defects considered by Villain and Vilfan (1988) when they argued that a transition of the Ising kind should occur first, and then a roughening transition at a higher temperature. Since this time it was realized that defects (f,g,h) have a higher energy than (b,c,d,e), so that the defects to be considered are (b,c,d,e). This results from the observation by Mochrie (1987) on Au(110) of a (1×3) structure, which is a dense array of "defects" of type (d). The same observation was made on Pt(110) by Fery et al (1988). These observations are in agreement with calculations by Garofalo et al (1987) based on the glue model.

The narrow walls displayed by Fig. 11 d and e are not the only possible wall structures. Fig. 11 i displays a thick wall. If even rows are missing on the left hand side, odd rows are missing on the right hand side and conversely. The parity of missing rows may be regarded as an

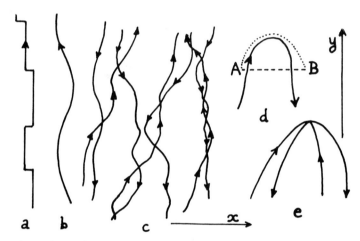

Fig. 12. a) Schematic representation of a step as a line with kinks. The up arrow indicates that the surface level is higher on the right hand side. b) More schematic representation at smaller scale. Kinks are not explicitly shown. c) Typical step configuration in the model of roughening transition investigated in subsection 4.2, with $\alpha = v = 1/2$. d) A configuration forbidden in the model and in Au(110). e) A configuration forbidden in the model, but not in Au(110).

Ising variable, just as the spin in a conventional Ising model. Quite generally, if one crosses a step of type (c) ("up" step) and then a step of type (b) ("down" step) or vice-versa, the Ising variable is changed. This description holds at T=0, but extension to finite temperatures is straightforward. The only change is that walls and steps are not straight but may have kinks. Kinks are shown on Fig. 12 a, where a step is schematically represented by a line, with an arrow which distinguishes up and down steps. More schematically yet, steps can be viewed on long scales as smooth lines (Fig. 12 b) where the various kinks are not explicitly represented. Thus a typical state of a rough surface is represented by Fig. 12 c. Two steps of identical sign cannot cross. Two steps of different sign can cross and each crossing has a free energy U.

212

4.3. A roughening transition with α=ν=1/2

In Subsection 4.1 it was seen that the roughening transition of the S.O.S. model, which has been extensively studied, is in the Koster-litz-Thouless universality class, which corresponds to a specific heat critical exponent $\alpha=-\infty$ and a correlation length critical exponent $\nu=\infty$. In this subsection a roughening transition will be investigated for a particular model, which has different critical exponents. Although these exponents are not the experimentally observed ones ($\beta=1/8$, etc.) in Au(110) it will be argued that the model might apply to surfaces with the same symmetry as Au(110) but different energy parameters.

The starting point of the model is the following property of the missing row reconstruction. The configuration of Fig. 12 d ("overhang") is forbidden. Indeed if one goes from A to B through the dashed line, the parity of missing rows should change, why it would not change along the dotted line. Configuration e is allowed, but, in some sense, unlikely if U is not sufficiently negative. In Au(110) it may be argued that U *is* rather strongly negative. However, one may consider a hypothetical model where U would be positive. Then, configurations of Fig. 12 e can probably be neglected.

We are now led to a model of lines oscillating around the y direction without overhang (Fig. 12 c) and crossing the whole sample from the bottom to the top. There are two different types of lines, denoted by up and down arrows. Lines of the same sign do not cross and lines of opposite sign can cross with some crossing energy U as explained above. The free energy of an isolated step per atomic distance ("line tension") vanishes at some temperature T_R calculated in the next subsection. If U is positive or not strongly negative, a roughening transition occurs at T_R. Only this case will be addressed in the present subsection.

This model can be solved exactly. It is similar to the model solved by Gruber and Mullins (1967), De Gennes (1968) and Pokrovskii and Talapov (1979). In those models, there are no arrows on the lines. However, it turns out that arrows do not change the results, at least for $U=\infty$ and for $U=0$, and presumably for any U. The detailed proof will not be given, but only the principle of the method and the main results will be outlined. Since the lines of Fig. 12 c have no overhang, they are similar to the trajectories of one-dimensional particles, if the y coordinate is replaced by time. Since the lines can have two different arrow orientations, these particles have spin 1/2. Since two lines of the same kind are not allowed to cross, the particles are fermions. Since there is a crossing energy U, these fermions are subject to a (one-dimensional) Hubbard Hamiltonian. The equivalence of the above defined model with a system of one-dimensional fermions can be derived exactly by the transfer matrix formalism. This is a straightforward extension of the spinless case (Villain 1980) which is left as an exercise to the reader.

The one-dimensional Hubbard model is exactly solvable (Lieb 1968). However, since the exact solution is complicated, we shall just consider the simple cases $U=\infty$ and $U=0$. It turns out that small values of U can be treated perturbatively (Villain and Vilfan 1990) and do not change the results basically.

In the case $U=0$ the system of Fig. 12 c is the superposition of two non-interacting systems : lines with up arrows and lines with down arrows. For each system the arrows can be ignored, and therefore the problem is exactly that solved by Gruber and Mullins. The specific heat exponent above the transition is $\alpha=1/2$ and the average distance ℓ between steps diverges as $(T-T_R)^{-\nu}$, with $\nu=1/2$ (Pokrovskii and Talapov 1979, Villain 1980). The mean square height difference is

$$\left\langle [z(\vec{r}+\vec{R})-z(\vec{r})]^2 \right\rangle = \left\langle (n^+(\vec{R})-n^-(\vec{R}))^2 \right\rangle$$

where $n^+(\vec{R})$ (resp $n^-(\vec{R})$) is the number of down (resp up) steps crossing the vector $\vec{R}$. Defining $\delta n^{\pm} = n^{\pm} - \langle n^{\pm} \rangle$, one finds

$$\left\langle [z(\vec{r}+\vec{R}) - z(\vec{r})]^2 \right\rangle = 2\left\langle (\delta n^+(\vec{R}))^2 \right\rangle \sim \ln R \qquad (4.2)$$

The case $U = \infty$ can be solved as follows. Let the arrows on Fig. 12 be first ignored. Then one can distribute arrows randomly among the lines. To each configuration without arrows correspond 2^p configurations with arrows. The partition function with arrows is equal to p ln 2 plus the partition function without arrows (calculated by Gruber and Mullins, etc). For a sample of large size N, p is proportional to $\sqrt{N}$ and negligible. Thus, $\alpha = 1/2$ again. ν is also equal to 1/2. Formula (4.2) can also be derived for $\vec{R}$ parallel to y. However in the other directions the situation is different. For instance, if one walks on a distance R in the x (i.e. [001]) direction one crosses on the average (R/ℓ) lines. The mean square height difference is therefore

$$\left\langle [z(\vec{r}+\vec{R}) - z(\vec{r})]^2 \right\rangle = R/\ell \qquad (4.3)$$

However this case is presumably exceptional. For all finite values of U, the roughness should be logarithmic as predicted for (4.2), and as in the S.O.S. model.

The present model is completely uninteresting below the transition. The transition occurs when the free energy of a single, isolated step changes its sign. Below the transition the energy of a step is positive, and equal to $+\infty$ for an infinite sample. Therefore there are no step at all. Therefore, in particular, $\beta = 0$.

The basic difference between the present model and the SOS model is that the configurations of Fig. 12 d and e are forbidden in the present model. One can speculate that, if configuration d is forbidden and configuration e is allowed (as is the case in gold) then the experimental values of the exponents might be found for appropriate values of U. This is a challenge to theorists (of conformal invariance, in particular). A first attempt has been done by Den Nijs (1990) who treated numerically by finite size scaling the case (which he calls "non-chiral") where steps of type b, c and f on Fig. 11 have the same energy, as well as walls of type d, e, g and h. He found that the exponents (e.g. β, α and ν) have the Ising value even when roughening occurs at the deconstruction transition. This very interesting result would be in agreement with the experimental observation of Robinson et al (1990) on Pt. However, the experimental system is undoubtfully "chiral". Thus, further theoretical investigation is necessary.

4.4. Is deconstruction of Au(110) an Ising or a roughening transition ?

If the deconstruction transition is assumed to be of the Ising type, the transition temperature T_c can be calculated with a good approximation as follows. One calculates the partial partition function $Z_I(1)$ of states with one domain wall, assuming no overhang. The transition occurs when $Z_I(1) = Z(0)$, where $Z(0)$ is the partition function of the ground state, hereafter taken to be equal to 1 (thanks to an appropriate choice of the energy origin). Similarly, if the deconstruction transition is assumed to be of the Ising type, the transition temperature T_R can be calculated as follows. One calculates the partial partition function $Z_R(1)$ of states with one step, assuming no overhangs. The transition occurs when $Z_R(1) = Z(0)$. Obviously, only the lowest of the two temperatures makes sense. If $T_R < T_c$ there is a roughening transition at T_R, and the Ising order also disappears at this temperature. If $T_c < T_R$ there is an Ising transition at T_c, *and a roughening transition should occur at some higher temperature.* In the latter case, the present argument does not exclude roughening to occur only at or very close to the melting point, but this is not very likely if one compares with other materials like Cu (Lapujou-lade 1990, Kern 1990) where the (110) face roughens well before melting.

Since an Ising wall is made of 2 steps one expects the wall energy η (at T=0) to be about twice the step energy ϵ, or at least $\eta > \epsilon$. This favours the roughening transition. However, thermal fluctuation should obviously be taken into account, and they imply excited states. Presumably the energy η' of an excited wall (Fig. 11 g,h) is smaller than the energy ϵ' of an excited step (Fig. i). This might favour an Ising transition. Thus it is not easy to answer the question "Ising or roughening transition?"

Villain and Vilfan (1988) tried to do that (and the answer was : Ising!") but they took into account the defects of Fig. 11 f,g,h. These are now known to have higher energy than those of Fig. 11 b,c,d,e. Therefore the 1988-theory must be updated. This is not too complicated if the possibility of a long distance between walls is neglected. In that case it is found after a long but straightforward calculation (to be published) that T_R is given by $\theta_R = 1$ and T_c by $\theta_I = 1$, with

$$\theta_R = \frac{A+B}{2} + \sqrt{\left(\frac{A-B}{2}\right)^2 + 4C^2} \quad \text{and} \quad \theta_I = \frac{A'+B'}{2} + \sqrt{\left(\frac{A'-B'}{2}\right)^2 + C^2}$$

where $A=\exp(-\epsilon/T)$, $B=\exp(-\epsilon'/T)$, $C=\exp(-W_o/T)$, W_o is the energy of a kink, $A'=\exp(-\eta/T)$ and $B'=\exp(-\eta'/T)+2\exp(-W_o/T)$. These formulae allow, in principle, the comparison between T_c and T_R . The discussion will not presented here, and the result is unfortunately unclear in the absence of knowledge about ϵ, ϵ', η and η'.

Furthermore, even if the transition is of the Ising kind, it follows from the experiments of Robinson et al (1989) that walls are very broad. Otherwise the shift of the diffuse scattering peak could not be understood (Vilfan and Villain 1990). This means that the calculation of $Z_I(1)$ should take into account wall broadening (Fig. 11 i). In fact, what should be calculated is the partition function $Z_R(2)$ of two steps with opposite sign. If $Z_R(2) < Z_R(1)$ at the transition, this means that the two steps form a bound state. This bound state should be identified with a domain wall, so that $Z_R(2)=Z_I(1)$ when there is a bound state. A system of two opposite steps may undergo an unbinding transition or *wetting tran-sition*. The explicit calculation of $Z_R(2)$ when the configurations of Fig. 11 i are taken into account will not be given here. In principle it is similar to the calculation given by Abraham (1980) and Rujan et al (1986). It would be of interest to observe explicitly the unbinding transition of a system of two walls. Unfortunately the experiment is hardly feasible, since it it would not be easy to force two halves of the surface to be in different Ising states .

5. CONCLUSION

The above examples, arbitrarily taken from the Physics of reconstruction, are intended to show the advantage of phenomenological concepts such as defects (domain walls, steps, kinks) and effective potentials. In spite of the growing, and possibly unlimited possibilities of computers which make ab initio calculations possible, we believe that phenomenology is most able to cast insight into complicated phenomena. Phenomenomenological arguments certainly have to be complemented by microscopic calculations. It would be of interest to use these calculations to determine the parameters (e.g. defect energies) which enter the phenomenological theories. Presently, the possibilities of microscopic calculations are still very restricted, and the importance of phenomenology is correspondingly high.

REFERENCES

ABRAHAM, D.B. (1980) Phys. Rev. Lett. **44**, 1165.

BAK, P. (1979) Solid State Commun. **32**, 581.

BASKES, M., DAW, M., DODSON, B.W., FOILES, S. (1988) MRS Bulletin fev, 28

BATRA, I.P. (1990) Phys. Rev. B **41**, 5048.

BINNIG, G. ROHRER, H. GERBER, Ch., WEIBEL, E. (1983) Phys. Rev. Lett. **50**, 120.

BOURRET, A. and BACMANN, J.J. (1985) Surface Sci. **162**, 495.

BOURRET, A. and ROUVIERE, J.L. (1989) in *Polycrystalline semiconductors,* Möller, H.P., Strunk, H.P., Werner, J.H. (Ed.). Springer-Verlag p.8.

CAMPUZANO, J.C., LAHEE, A.M., and JENNINGS, G., (1985) Surface Sci. **152/153** 68.

CAMPUZANO, J.C., JENNINGS, G., and WILLIS, R.F. (1985) Surface Sci. **162**, 484.

CONRAD, E.H., ATEN, R.M., KAUFMAN, S.D, ALLEN, R.L., ENGEL, T., den NIJS, M., RIEDEL, E.K. (1986) J. Chem. Phys. **84**, 1015.

DEN NIJS, M. (1990) This Conference and Phys. Rev. Lett. **64**, 435.

ERCOLESSI, F., TOSATTI, E., PARRINELLO, M. (1986) Phys. Rev. Lett. **57**,719.

FERY, P., MORITZ, W. and WOLF, D. Phys. Rev. B **38** (1988) 7275.

GAROFALO, M., TOSATTI, E., ERCOLESSI, F. (1987) Surface Sci. **188**, 321.

de GENNES, P.G. (1968) Solid State Comm. **6**, 163.

GRUBER, E.E., MULLINS, W.W. (1967) J. Phys. Chem. Solids **28**, 875.

GUILLOPE, M. and LEGRAND, Surface Sci. **215** (1989) 577.

HEINE, V., MARKS, L.D. (1986) Surface Sci. **165**, 65.

HO,K. M., and BOHNEN, K.P. (1987) Phys. Rev. Lett. **59**, 1833.

IHM, J., LEE, D.H., JOANNOPOULOS, J.D., XIONG, J.J. (1983) Phys. Rev. Lett. **51**, 1872.

KEATING, P.N. (1966) Phys. Rev. **145**, 637.

KERN, K. (1990) This Conference.

KNOPS, H.J.F. (1977) Phys. Rev. Lett. **39**, 766.

LAPUJOULADE, J., this conference.

LIEB, E. (1968) Phys. Rev. Lett. **20**, 1445.

MOCHRIE, S.G.J., Phys. Rev. Lett. <u>59</u> (1987) 304.

MÜLLER-HARTMANN, E. and ZITTARTZ, J., Z. Phys. B <u>27</u> (1977) 261.

PAYNE, M.C., ROBERTS, N., NEEDS, R.J., NEEDELS, M., JOANNOPOULOS, J.D. (1989) Surface Sci. **211/212**, 1.

POKROVSKII, V.L., TALAPOV, A.L. (1979) Phys. Rev. Lett. **42,** 65.

ROBINSON, I.K., VLIEG, E. and KERN K. (1989), Phys. Rev. Lett. **63**, 2578.

ROUVIERE, J.L. and BOURRET, A. (1989) in *Polycrystalline semiconductors*, Möller, H.P., Strunk, H.P., Werner, J.H. (Ed.). Springer-Verlag p.19.

RUJAN, P., SELKE, W., UIMIN, G. (1986) Z. Phys. **65**, 235.

STILLINGER, F. and WEBER, T. (1985) Phys. Rev. **B31**, 5262.

TAKAYANASHI, K., TANISHIRO, Y., TAKAHASHI, M., TAKAHASHI, S. (1985) J. Vac. Sci. Technol. **A 3**, 1502.

TARNOW, BRISTOWE, P.D., JOANNOPOULOS, J.D. and PAYNE, M.C. (1989) J. Phys. : Condens. Matter **1**, 327.

TERSOFF, J. (1988) Phys. Rev. **B37**, 6991.

TROMP, R.M., HAMERS, R.J., DEMUTH, J.E. (1986) Phys. Rev. **B15**, 1986.

VAN BEIJEREN, H., NOLDEN, I. (1987) in *Structure and Dynamics of Surfaces II*, W. Schommers and Von Blanckenhagen (eds) Springer, Heidelberg.

VILFAN, I., VILLAIN, J. (1990) Phys. Rev. Lett., submitted.

VILLAIN, J., GREMPEL, D., LAPUJOULADE, J. (1985) J. Phys. F **15**, 809.

VILLAIN, J. (1980) in *Ordering in Strongly Fluctuating Condensed Matter Systems*, Ed. T. Riste (Plenum, New York) p. 221.

VILLAIN, J. and VILFAN, I.(1988) Surface Sci. **199**, 165.

VILLAIN, J. and VILFAN, I. (1990) Europhysics Lett. to be published.

J.D. WEEKS (1980) in *Ordering in Strongly Fluctuating Condensed Matter Systems*, Ed. T. Riste (Plenum, New York) p. 293.

THE ROUGHENING TRANSITION ON SURFACES

Jean Lapujoulade and Bernard Salanon

Centre d'Etudes Nucléaires de Saclay
DPhG/PAS Bat.462
91191 Gif-sur-Yvette Cedex
France

INTRODUCTION

The concept of a roughening transition on surfaces has been introduced
in 1951 by Burton,Cabrera and Frank[1] in order to explain why the growth of
crystals was observed to proceed in two very different regimes: layer by
layer or tridimensional. They have shown that a surface layer of a simple
cubic crystal in which every site can be either occupied or empty, is
formally equivalent to an assembly of up and down spins known as the 2-d
Ising model. From the Onsager solution of this model[2] they have predicted a
transition from a smooth phase at low temperature to a rough phase at higher
temperature. However there is no reason to restrict the roughening to the
outermost layer (empty or occupied) and it was recognized later that the
removal of this restriction shifts the transition in an other universality
class first described by Kosterlitz and Thouless[3]. The first experimental
evidence of the existence of the roughening transition, given in 1974 by
Pawlowska and Nenow[4], has strongly stimulated researches in this field:
theory,simulation and experiment. The nature of the roughening transition is
now well understood and realistic models have been solved either
analytically or by simulation. They can be compared to an increasing number
of accurate experimental data. The roughening of surfaces is a very
interesting problem in two respects:

i/ It is a good system for testing phase transition theories since
exact calculated data are not too difficult to obtain from models.
ii/ As mentionned above it is very relevant to crystal growth
mechanisms. Below the roughening transition nucleation is needed to
create a new layer and the growth proceeds layer by layer at a low
rate. On the other hand above this transition 3-d crystallites are
created without activation energy and the growth rate is high.

In spite of their technical importance we shall not discuss here
kinetics problems and we shall restrict ourselves to surfaces in internal
thermodynamical equilibrium. We shall first expose the theoretical
background of the roughening transition. Then we shall describe the
experimental techniques allowing its study. Finally we shall discuss some
recent results obtained on metal surfaces.

Phase Transitions in Surface Films 2
Edited by H. Taub *et al.*, Plenum Press, New York, 1991

THEORETICAL BACKGROUND OF THE ROUGHENING TRANSITION

In this section only an introductory basis of the theory is presented. For further details the reader may refer to previous review articles[4,5].

Definition

The roughening transition is related to the free energy of a step on the surface. Creating a step on a surface costs energy since a step atom is less bound to the crystal than a surface atom. But on the other hand, for a given energy the step can assume a large number of meandering shapes which introduces a configuration entropy term. When the temperature is increased this term can compensate the increase of internal energy and thus the free energy of a step may vanish at some temperature T_R. This temperature is called by definition the roughening temperature. Above T_R steps are spontaneously created.

We shall explain later that below T_R crystal facets are present in the equilibrium shape of a crystal. Above T_R the crystal is completely rounded, so the disappearance of a facet in the equilibrium shape can also be used to characterize the roughening transition.

One can define the correlation function of the height fluctuations from a plane surface as:

$$G(d) = \langle [z(d)-z(0)]^2 \rangle \tag{1}$$

d being the distance parallel to the surface plane and z the height

It will be shown later that for $T < T_R$ the correlation function $G(d)$ remains finite when $d \to \infty$, while for $T \geqslant T_R$, $G(d) \to \infty$ when $d \to \infty$. This divergence of $G(d)$ above T_R is also used to define the roughening transition. We must emphasize that the three definitions: vanishing of the step free energy, disappearance of flat facets or long distance divergence of the height correlation function are physically equivalent.

Two approaches are commonly used to derive the roughening transition:

i/The thermodynamic approach using the concept of surface tension. The atomic structure is not explicitly introduced. It is contained in the dependence of the surface tension upon orientation.

ii/The statistical approach in which the solid and its surface are modelled by an assembly of interacting atoms. All the thermodynamic properties can in principle be derived from the knowledge of the partition function.

THERMODYNAMIC APPROACH

Crystal equilibrium shape

This subject is well developped [6-10], so we just present here a general outlook.

The surface of a solid is characterized by its free energy per unit area or surface tension: Υ. The crystal anisotropy due to its atomic nature is taken into account here through the dependence of Υ upon the orientation of the surface with respect to crystal axis defined by its polar angle θ and its azimuthal angle φ. A polar plot of $\Upsilon(\theta,\varphi)$ is called the Υ-plot. Then at equilibrium a solid with a constant volume V will assume the shape which minimizes the total surface free energy. For a liquid Υ is independant from the orientation and the equilibrium shape is obviously a sphere in the

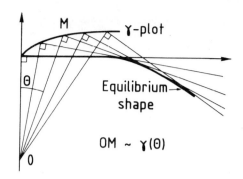

Fig.1 The Wulff construction

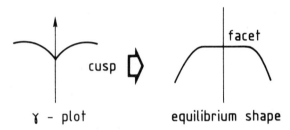

Fig.2 A facet on the crystal is due to a cusp in the γ-plot

absence of gravity. In the general case the minimization leads to the well known Wulff construction: at every point M of the γ-plot one defines a plane which is perpendicular to the line OM (O origin of the γ-plot). The equilibrium shape is defined by the inner envelope of these planes. This is schematically represented in Fig.1 for a cylindrical crystal.

The appearance of facets on the crystal is connected with the existence of non-analyticities ("cusps") in the γ-plot (Fig.2). A cusp in the γ-plot is a direct consequence of a non zero free energy for the step formation. This is very easy to show for a cylindrical geometry. Fig.3 shows the struture of the surface at an atomic scale in the vicinity of a high index face. The surface curvature is due to steps of height a (Fig.3).

The surface free energy can be written in the form:

$$\gamma(\theta) = \gamma_0 + \beta n + E_{int} \qquad (2)$$

where:

γ_0 : free energy of the facet
β : free energy of a step
n : step density $n = 1/(a\, ctg\theta)$
E_{int}: Interaction energy between steps

The second term is proportional to $|\theta|$ while the third is of the order of $|\theta|^3$. Thus if $\beta \neq 0$ a cusp appears in $\gamma(\theta)$. So the disappearance of the

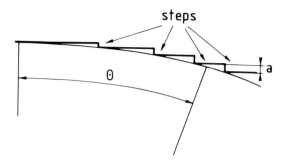

Fig.3 The surface in the vicinity of a facet

Fig.4 Evolution of the crystal shape with temperature

flat facet corresponds to the vanishing of the step free energy. Of course the step free energy depends upon the facet orientation and to every different stable facet corresponds a different roughening temperature. As a crystal is heated up the different facets disappear sequentially according to their respective T_R as shown in Fig.4. When the last facet has disappeared the crystal becomes quasi-spherical.

<u>The continuum model for a plane surface</u>

This model was originally exposed by Ohta and Kawasaki[11] and then developed by Gallet and Nozières[12].

<u>Fluctuations of a free surface on continuous solid</u>

The surface energy per unit area is γ_0. The effect of thermal fluctuations is simply to increase the surface area. The total surface energy is then given by a first order expansion (grad z $\ll$ 1):

$$E = \gamma_0 S + \frac{1}{2} \int_S d^2\mathbf{R} \; \tilde{\gamma} \; (\text{grad } z)^2 = \gamma_0 S + \Delta E \qquad (3)$$

$\tilde{\gamma} = \gamma_0 + \dfrac{d^2\gamma}{d\theta^2}$ is the so called surface stiffness and $S = L^2$.

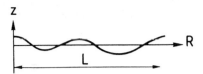

Fig.5 Fluctuations of a free surface

A Fourier expansion of the fluctuations $z = \sum_K z_K e^{iK \cdot R}$ leads to:

$$\Delta E = \sum_K \frac{1}{2} \tilde{\gamma} L^2 K^2 \left\langle z_K^2 \right\rangle \tag{4}$$

For this free fluctuation regime the modes are not coupled and each of them contributes $k_B T/2$ to the total energy, so that:

$$\left\langle z_K^2 \right\rangle = \frac{k_B T}{\tilde{\gamma} L^2 K^2} \tag{5}$$

and the mean square deviation is easily obtained:

$$\langle z^2 \rangle = \sum_K \left\langle z_K^2 \right\rangle = \frac{k_B T}{\tilde{\gamma} L^2} \sum_K \frac{1}{K^2} \tag{6}$$

For large L the discrete sum over modes can be transformed into an integral:

$$\langle z^2 \rangle = \frac{k_B T}{4\pi^2 \tilde{\gamma}} \int_{\pi/L}^{\pi/\xi} \frac{2\pi K dK}{K^2} = \frac{k_B T}{2\pi \tilde{\gamma}} \ln \frac{L}{\xi} \tag{7}$$

ξ is a length scale for the "width" of a step on the surface which is of the order of an interatomic distance. This parameter is needed to avoid the infrared divergence of the integral. It is left undetermined in this theory. The height correlation function is obtained in the same way:

$$G(d) = \langle [z(R+d) - z(R)]^2 \rangle = \frac{k_B T}{\pi \tilde{\gamma}} \ln (d/\xi) \tag{8}$$

We get the important results:

$$G(d) \to \infty \text{ as } d \to \infty$$

Thus the surface is rough at any non zero temperature.

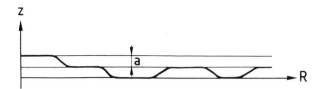

Z

a

R

Fig.6 The surface pinned on parallel equidistant planes

Pinning of the surface by a periodic potential

In order to simulate the crystallographic structure we introduce a periodic potential $V(z)$ which forces the surface to be preferentially located on equidistant parallel planes (Fig.6).

The surface free energy becomes now:

$$E[z(\mathbf{R})] = \int_S d^2\mathbf{R}\left[\gamma_0 + \frac{1}{2}\tilde{\gamma}\,(grad\ z)^2 + V(z)\right] \tag{9}$$

Only the first Fourier component of $V(z)$ is relevant :

$$V(z) = V\,(1 - \cos\,(2\pi\,\frac{z}{a})) \tag{10}$$

a is the distance between planes

Then :

$$E(\{z_K\}) = \sum_K \frac{1}{2}\,L^2\,\tilde{\gamma}\,\mathbf{K}^2\,|z_K|^2 + V(\{z_K\}) \tag{11}$$

The potential introduces a coupling between the modes which makes the calculation more difficult. One has to calculate the partition function :

$$\mathbb{Z} = \int \ldots \int_K e^{-E(z_K)/k_B T} \tag{12}$$

The integration limits are again : $\frac{\pi}{\xi} = \Lambda_0 < |\mathbf{K}| < \Lambda = \frac{\pi}{L}$

The renormalization procedure can be applied to the calculation of $\mathbb{Z}$. One defines a new cut-off $\bar{\Lambda}$ such that: $\Lambda < \bar{\Lambda} < \Lambda_0$. The integral is then divided into two parts:

short wave length modes z_p: $\bar{\Lambda} < |\mathbf{K}| < \Lambda_0$
long wave length modes z_k: $\Lambda < |\mathbf{K}| < \bar{\Lambda}$

One integrates now over short wave length modes:

$$\int_{\bar{\Lambda}}^{\Lambda_0} dz_p\ e^{-E(z_p,z_k)/k_B T} \equiv e^{-\bar{E}(z_k)/k_B T} \tag{13}$$

222

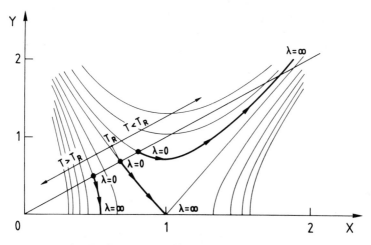

Fig.7 The renormalization trajectories

and from (12) it remains:

$$\mathbb{Z} = \int \cdots \int_{\Lambda}^{\bar{\Lambda}} dz_K \; e^{-\bar{E}(z_K)/k_B T} \tag{14}$$

Then it can be shown by an expansion in power of V up to second order that $\bar{E}(z_K)$ retains its initial form (11) provided that $\tilde{\gamma}$ and V are replaced by renormalized parameters $\bar{\tilde{\gamma}}$ and $\bar{V}$ which include the effect of short wave length fluctuations. The validity of this expansion is of course limited to small values of V ($V \approx k_B T$). This procedure is then iterated by defining differentially a new cut-off $\bar{\Lambda} + d\bar{\Lambda}$ and calculating $\bar{\tilde{\gamma}}(\bar{\Lambda}+d\bar{\Lambda})$ and $\bar{V}(\bar{\Lambda}+d\bar{\Lambda})$ in terms of $\bar{\tilde{\gamma}}(\bar{\Lambda})$ and $\bar{V}(\bar{\Lambda})$. We define dimensionless variables :

$$\lambda = \ln(\Lambda_0 / \bar{\Lambda}) \;\;,\;\; x = 2\bar{\tilde{\gamma}}a^2/\pi k_B T \;\;,\;\; y = 4\pi\bar{V}/k_B T\Lambda^2$$

and after some algebra one finally gets the set of renormalization equations first obtained by Kosterlitz and Thouless in another context[3]:

$$\left.\begin{array}{l} dy/d\lambda = 2y(1 - 1/x) \\[2mm] dx/d\lambda = (y^2/2x) \; A(2/x) \end{array}\right\} \tag{15}$$

$A(2/x)$ is a slowly varying function which has been calculated exactly by Nozières and Gallet[12]: $A(2/x) = 0.4$ in the vicinity of T_R. The set of equations (15) defines an array of quasi-hyperbolas as shown in Fig.7.

Each quasi-hyperbola corresponds to a different temperature. The curves are generated by the variation of the scale parameter λ. They start at $\lambda = 0$ (atomic scale). All the starting points lie on a straight line which corresponds to the initial unrenormalized values of the parameters $\tilde{\gamma}$ and V. When λ increases to infinity there are clearly two kinds of possible behaviour according to whether T is lower or higher than a temperature T_R characterized from (15) by $x = 1$.

These two regimes have the following properties :

$$\boxed{T < T_R}$$

There is always a scale $1/\bar{\Lambda}$ above which the renormalized potential goes to infinity. The fluctuations are then pinned on the crystallographic planes. The surface is thus smooth at large scale.

$$\boxed{T \geq T_R}$$

The renormalized pinning potential $\bar{V}$ always vanishes at large scale . Thus the surface behaves like a free surface: it is rough at large scale. The height correlation function is still given by Eq.(8) provided that the surface stiffness $\tilde{\gamma}$ is replaced by its renormalized value $\tilde{\gamma}$ for λ infinite :

$$G(d) = \frac{k_B T}{\pi \tilde{\tilde{\gamma}}_\infty} \ln (d/\xi) + \text{cte} \tag{16}$$

For the special case where $T = T_R$ one gets the universal relation:

$$x = x_R = \frac{2}{\pi} \frac{\tilde{\tilde{\gamma}}_R a^2}{k_B T_R} = 1 \tag{17}$$

and:

$$G(d) = (2a^2/\pi^2) \ln (d/\xi) + \text{cte} \tag{18}$$

Remarks:

i/ This calculation gives very easily the universal properties of the roughening transition. However it is not so easy to derive the non-universal quantities like $\tilde{\tilde{\gamma}}_\infty$.

ii/ The validity of the calculation relies upon an expansion in powers of the pinning potential which is only valid for small values of V. This approximation is probably not true in real crystal except perhaps for Helium crystals. However it is reasonable to think that the renormalization trajectories in the vicinity of the fixed point are still correct.

iii/ In any case the approach fails completely to describe the detailed behaviour below T_R since $\bar{V}$ becomes then infinite.

These remarks emphasize the need for a more microscopic approach of the roughening transition which will be developped next.

The statistical approach

The SOS models

This is the original approach developped by Burton,Cabrera and Frank[1]. It is based on a very simple model of the crystal known as the Kossel crystal depicted in Fig.8.

The crystal is composed of elementary cubic blocks representing the unit cells. The cohesive energy of the crystal is a sum of pairwise potentials between adjacent cube faces. Moreover in the so-called Solid-On-Solid (SOS) model, overhangs are forbidden as shown in Fig.8. This

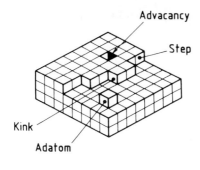

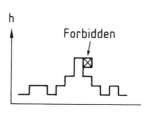

The Kossel crystal   The SOS model

Fig 8. The Kossel crystal and the SOS model

is certainly a good assumption for the gas-solid interfaces . The ground
state of a crystal surface is a compact arrangement of the cubes having one
face lying in the same plane. The various possible excitations are adatoms,
advacancies, steps, kinks (Fig.8). The excess energy E associated with these
excitations is assumed to depend on height differences between neighboring
sites. So for the SOS model the configurational energy can be cast into:

$$E = J/2 \sum_{j,\delta} f(h_j - h_{j+\delta}) \tag{19}$$

where δ labels one of the four columns adjacent to the jth one and J is
the interaction energy between two adjacent cubes. Then it is possible to
define the so-called Absolute SOS (ASOS) or Discrete Gaussian SOS (DGSOS):

$$ASOS: \quad f(x) = |x| \; , \quad DGSOS: \quad f(x) = (x)^2 \tag{20}$$

In order to calculate the equilibrium properties of the surface at
any tmperature one has to calculate the partition function:

$$\mathbb{Z} = \sum_{\{h_j\}} e^{-E(j,\delta)/2k_B T} \tag{21}$$

The sum is extended over all possible configurations $\{h_j\}$. This
partition function has not yet been calculated analytically. However it has
been shown by Chui and Weeks[13] that the DGSOS model is isomorphic to the
Coulomb lattice which in turn is isomorphic to the planar-XY Heisenberg
model[14]. This last system was already known to undergo a Kosterlitz-Thouless
transition[2]. Note that the low temperature phase of the DGSOS model
corresponds to the high temperature one of the two others and vice versa.
Simulations have confirmed the existence of the transition for the DGSOS
model and for the ASOS also[15]. In addition they give the relation between T_R
and J:

$$ASOS: \quad k_B T_R/J = 1.24 \; , \quad DGSOS: \quad k_B T_R/J = 1.48$$

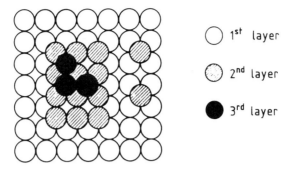

Fig.9 The BCSOS model

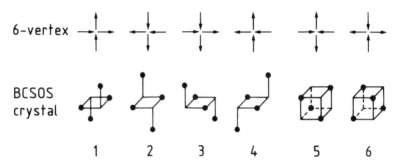

Fig.10 The correspondence between the BCSOS and the 6-vertex model

The BCSOS model[16]

The easiest way to understand this model is to consider the crystal as made of hard spheres instead of hard cubes (Fig.9).

The crystal is made of layers of spheres forming a compact square array. Layers are arranged in such a way that every sphere is always in contact with four spheres of the lower adjacent layer. The upper layers can be incomplete provided that this condition is still fullfilled. Consequently the columns of spheres are divided into two shifted sublattices: the height difference between two neighbouring columns belonging to a different sublattice is either $-1/2$ or $+1/2$ if 1 is the distance between neighbouring spheres in the same column, while the height difference between neighbouring columns belonging to the same sublattice is either -1 or 0 or $+1$. The needed condition of forced contact between spheres belonging to adjacent layers is achieved if one assumes that their attractive interaction is infinite. The model is now developed by assigning to nearest neighbouring spheres within a layer, an interaction energy J. This model is an improvement on the SOS models insofar that it takes into account the real crystal structure and moreover, as pointed out by van Beijeren[16], it has the great advantage of being isomorphic to the 6-vertex model for which an analytic solution is already known[17]. This model which has been developped in the context of ice or ferroelectric crystal is known to present a Kosterlitz-Thouless type transition. The correspondence between the BCSOS and the 6-vertex model is shown in Fig.10, see also ref.(16).

226

The above defined rules for the height difference between columns correspond in the 6-vertex model to the so-called ice-rule: at every vertex there must be two arrows in and two arrows out. The ground state of the surface corresponds to only type 5 and 6 vertices. The excitations are represented by type 1,2,3 or 4 vertices.

The BCSOS model is also applicable to fcc(100) surfaces and can be extended to fcc(110) by introducing an anisotropy in the interactions ($J_x \neq J_y$)[18,35]. J_x represents the interaction between atoms along a row while J_y is the interaction between atoms belonging to adjacent rows. One can remark that excitations on fcc(100) surfaces corresponds to the formation of an elementary (111) facet. Similarly on fcc(110) the excitations correspond to (111) or (100) facets according to their orientation[18]. Table I shows the correspondance between the vertex energy in the 6-vertex model and the excitation energy in the BCSOS model.

Table I Correspondence between the vertex energies in the 6-vertex model and the excitation energies in the BCSOS model.
A is the unit cell area of the corresponding basic face
θ_1 and θ_3 are the angles between the excited facets and the basic plane
σ_{xxx} is the surface tension of the (xxx) face

fcc(100)	fcc(110)
$e_1=e_2=e_3=e_4=J$ $e_5=e_6=0$	$e_1=e_2=J_x$ $e_3=e_4=J_y$ $e_5=e_6=0$
$e_1=e_2=e_3=e_4=A\sigma_{111}/\cos\theta_1$ $e_5=e_6= A\ \sigma_{100}$	$e_1=e_2=A\sigma_{111}/\cos\theta_1$ $e_3=e_4=A\sigma_{100}/\cos\theta_3$ $e_5=e_6=A\sigma_{110}$

Note that only the energy differences between the excited states and the ground state are relevant.

For fcc(100) surfaces the relation between T_R and J is:

$$k_B T_R /J = 1/\ln 2 \simeq 1.44$$

For fcc(110) surfaces the following equation has to be solved numerically:

$$e^{-J_x/k_B T} + e^{-J_y/k_B T} = 1 \qquad (22)$$

The expressions giving the free energy can be found in Ref.17 from which the fraction of the surface sites which are occupied by the various excitations is easily obtained. For instance we have represented (full curve) in Fig.11 the fraction of sites occupied by an excitation irrespective of its orientation (1,2,3 or 4) for the isotropic case : fcc(100).

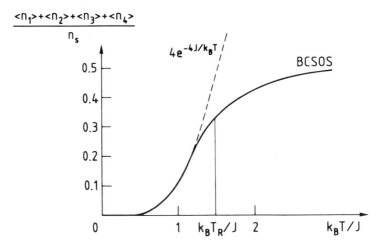

Fig.11 Fraction of excited sites vs. reduced temperature $k_B T/J$ for a fcc(100) surface

This curve deserves some comment:

i/ At low temperatures, the point defects which are adatoms and advacancies are the predominant surface defects. Insofar as they do not form clusters each of them costs $e_1 + e_2 + e_3 + e_4 = 4J$. Thus the curve is expected to behave like $\exp(-4J/k_B T)$ at low temperature (dashed curve) which is well predicted by the 6-vertex model. Thus at low temperature the curve gives four times the number of point defects.

ii/ As T is increased point defects become adjacent; they form domains delimited by steps and the exact curve departs from the exponential shape. Nothing special happens at T_R since the transition is very smooth (infinite order). The divergence of the correlation function is a long distance property which is not directly related to the defect concentration. It must be emphasized that excitations appear well below T_R. The threshold is located around $T_R/2$. This property is characteristic of the roughening transition and does not depend very much upon the particular model.

iii/ Above T_R the full curve saturates at 0.5. This is peculiar to the BCSOS model: the ice-rule limits the total number of possible excitations. This would not be the case with ordinary SOS models. As the roughening temperature is generally not much lower than the bulk melting temperature or the surface melting temperature if any, the problem of the high temperature behaviour of the model is rather academic when the lattice site reference becomes meaningless.

Further developments of the BCSOS model have been recently proposed that we will not discuss in detail since they are exposed elsewhere in this volume :

i/ It can be shown that the BCSOS model is also able to predict the missing row reconstruction and deconstruction of fcc(110) surfaces provided that an interaction energy between vertices is introduced.

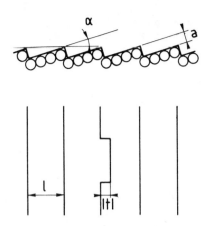

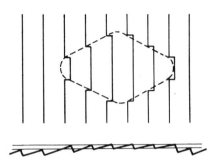

Fig.12 Vicinal surface   Fig.13 Domain structure

ii/ In the BCSOS model only nearest neighbour interactions are
taken into account. In their Restricted SOS model (RSOS) Rommelse and
Den Nijs[19] have introduced next nearest neighbour interaction. They
have predicted a richer phase diagram with new transitions
(preroughening, four-clock step model) which are still awaiting
experimental evidence.

The roughening of vicinal surfaces

A vicinal surface is a surface making a small angle α with a low index
plane. It displays a periodic array of monoatomic steps delimiting low index
terraces (Fig.12).

On close-packed surfaces the elementary excitation at low temperature
is a pair of adatom advacancy which costs four broken bonds. On vicinal
surfaces the formation of a pair of kinks costs only two broken bonds hence
one can expect that the roughening of step edges by kink proliferation will
occur at a lower temperature than the terrace roughening.

It is important to point out that an isolated step is always unstable
with respect to kink formation. Indeed assuming that the kink creation
energy is W_0 the probability to form a kink at any lattice site along the
step is: $\epsilon = \exp(- W_0/k_B T)$ and the displacement u of the step from its
straight position is the solution of a simple random walk problem for which
the correlation function is readily obtained:

$$\langle [u(y+d)-u(y)]^2 \rangle = \frac{2\epsilon}{1+2\epsilon} \frac{d}{a} |t|^2 \tag{23}$$

Thus the correlation function of the one-dimensional step diverges
linearly at any non-zero temperature (note the difference with the two
dimensional surface case). The roughening temperature is always zero even if
the displacements are discrete. However the metastability of vicinal
surfaces is a well known experimental fact. This implies the existence of
repulsive interactions between steps. There are two main classes of
interaction:

i/ statistical interaction: topologically steps cannot intersect
each other.

ii/ direct interaction: The crystal lattice is distorted in the vicinity of a step. When two steps come sufficiently close together these distortion fields interact and create repulsive forces. This is the so-called elastic interaction. Electrostatic dipole-dipole repulsion may also occur but it is generally thought to be weaker than the elastic one.

Under the action of these repulsive forces kinks have a tendancy to form domains (Fig.13) with the same structure as the perfect vicinal surface . These domains are shifted by an integer multiple of a unit vector. Domain boundaries are lines of kinks which can be called "secondary steps". The proliferation of these secondary steps leads to a new mechanism ending to roughening on vicinal surfaces.

Step roughening

The same procedure as previously will be used here: we shall expose first a thermodynamic approach and then then come to the statistical one.

Thermodynamic approach

We follow here the model developed by Villain et al.[20] where the total interaction between steps is explicitly introduced in contrast with another model by Den Nijs et al.[21] which takes implicitly into account only the statistical interaction. The direct interaction has revealed itself necessary in order to fit the experimental data. The array of steps is depicted as an array of strings in mutual interaction (Fig.14).

This two-dimensional array is characterized by two excess surface stiffnesses: $\tilde{\gamma}_x$ and $\tilde{\gamma}_y$. $\tilde{\gamma}_y$ and $\tilde{\gamma}_x$ reflects respectively the step rigidity against kink creation and the step-step interaction. They are in general not equal and $\tilde{\gamma}_y$ is expected to be larger than $\tilde{\gamma}_x$. A periodic potential V(x) is introduced in order to localize the steps on crystallographic positions like in the case of the close packed surface. The total surface excess energy is then:

$$E = \frac{1}{2} \sum_q (\tilde{\gamma}_x q_x^2 \ell^2 + \tilde{\gamma}_y a_y^2 q_y^2) \, |u_q|^2 + V \sum_{m,y} \left[1 - \cos\left(\frac{2\pi}{|t|} u_m(y)\right) \right] \tag{24}$$

where : t is a vector defining the unit step displacement parallel to the terrace
a_y is the lattice constant along the step
ℓ is the step-step distance
q_x, q_y are the parallel components in the vicinal surface coordinates of the momentum exchange vector $q[\mathbf{Q}(q_x, q_y), q_z]$

and :

$$u_m(y) = (1/N^{1/2}) \sum_{q_x, q_y} u_q \, e^{-(iq_x m\ell + iq_y y)} \tag{25}$$

where $N = L^2/a_y \ell$ (L defines the size of the surface)
The renormalization procedure can be applied in the same way as previously. The transition falls into the Kosterlitz class and we only

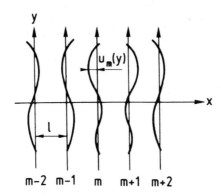

Fig.14 The array of interacting step edges

present the main results as follows :

for $\boxed{T < T_R}$

$\overline{V}$ becomes infinite at large scale and the steps are pinned on their crystallographic positions. The surface is smooth.

for $\boxed{T \geq T_R}$

$\overline{V}$ vanishes at large scale and the array of step behaves like a free surface. The correlation function diverges logarithmically. Note the difference with the isolated step behaviour: the step-step interaction has moved the problem from one to two dimensions. One finds:

$$\left\langle \left[u_m(y) \; u_m(0) \right]^2 \right\rangle = A(T) \; \ln \rho \qquad (26)$$

with:

$$\rho^2 = (y/a_y)^2 \; \left(\tilde{\tilde{\gamma}}_y / \tilde{\tilde{\gamma}}_x \right)^{1/2} + m^2 \; \left(\tilde{\tilde{\gamma}}_y / \tilde{\tilde{\gamma}}_x \right)^{1/2} \qquad (27)$$

and:

$$A(T) = k_B T \; |t|^2 / \pi a_y \ell \left(\tilde{\tilde{\gamma}}_x \tilde{\tilde{\gamma}}_y \right)^{1/2} \qquad (28)$$

—   We note that the distance ρ is renormalized by the anisotropic surface stiffnesses.

At $T = T_R$ one obtains the universal relation :

$$k_B T_R \; / \; a_y \ell \left(\tilde{\tilde{\gamma}}_x \tilde{\tilde{\gamma}}_y \right)^{1/2} = 2/\pi \qquad (29)$$

Statistical approach

The step roughening can also be described by an SOS model as shown in Fig.15.

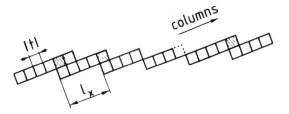

Fig.15 The SOS model of step roughening

Instead of considering columns normal to the terraces we describe the step edge as the extremity of columns lying parallel to the terraces. The energy can be written as:

$$E = \sum_{m,n} f_y(u_{m,n+1} - u_{m,n}) + \sum_{m,n} f_x(u_{m+1,n} - u_{m,n}) \qquad (30)$$

Along the step the ASOS : $f_y(u) \equiv W_0|u|$ or the DGSOS : $f_y(u) \equiv W_0(u)^2$ (cf.Eq.(20)) conditions may be used; W_0 is the kink energy creation. Villain et al. have used also the DGSOS condition for the step-step interaction $f_x(u) \equiv (w_1/2)(u)^2$; w_1 is the energy spent per unit length when a step is displaced between its two neighbours by one unit displacement a_x. Provided that $w_1 \ll k_B T \ll W_0$ and $T > T_R$ they have found an approximate relation between the parameters of the statistical and the thermodynamical model :

$$\tilde{\bar{\gamma}}_x = \frac{1}{2} k_B T \, e^{W_0/k_B T} \quad \text{and} \quad \tilde{\bar{\gamma}}_y = w_1 \qquad (31)$$

The validity range of these relations (32) has never been checked carefully. For vicinal surfaces with short terraces (like fcc(113) or (115)) the condition $w_1 \ll W_0$ is not fulfilled and (32) is not applicable.

For the step-step interaction a more realistic assumption is given by:

$$f_x(\Delta u) = \begin{cases} 0 & \text{for } \Delta u \geq 0 \\ w_p & \text{for } \Delta u = -p \ , \quad -1 < p < -p_{max} \\ \infty & \text{for } \Delta u < -p_{max} \end{cases} \qquad (32)$$

$$\text{where} : p_{max} = (\ell \cos \alpha) / |t|$$

The first line of the right hand term indicates that the inter   n becomes vanishingly small when the step-step distance becomes large   n its equilibrium value ℓ. In the second one w_p increases as the ste   ) distance becomes smaller than ℓ. The third one forbids step crossi   e shall call this model the TLK model. It has been first studied by ..... in et al.[20] and numerically investigated by Selke and Szpilka[22] with only one term p = 1 as applied to fcc(113) surfaces. There is no known analytic solution for this model. The numerical results obtained by simulation are in agreement with the thermodynamic approach and will be discussed in connection with experimental data.

The kink creation energy W_0 is not expected to depend upon the step-step distance ℓ at least in a first approximation. On the other hand w_p strongly depends upon ℓ since both statistical and direct interactions have a R^{-2} dependance. Thus T_R is expected to decrease when ℓ increases.

General remark

The physical significance of roughening which is the basic concept of these models is often misunderstood. As a critical phenomenon it is connected with the behavior of the surface at large scale. It would be perhaps appropriate to use the term "thermodynamical roughening" for the transition itself in order to distinguish it from the "atomic roughness"[23]. Indeed at a macroscopic scale, below T_R the surface is smooth in the sense that the equilibrium shape exhibits a facet. Nonetheless defects (adatoms, steps) are present in this phase and we use the expression "atomic scale roughness" to describe this microscopic morphology. Conversely, even when the surface is thermodynamically rough, flat regions of small size(a few atomic distances) still exist.

So it is now clear that atomic roughness appears well below the critical temperature T_R and that its temperature dependence always scales with T_R. This is why the knowledge of T_R is important to understand surface properties like the chemical reactivity even if they are more connected with the atomic roughness than with the thermodynamical roughening.

EXPERIMENTAL TECHNIQUES

The experimental techniques used to study the roughening transition may be classified in a similar way as the thoretical models:

    i/ macroscopic techniques based on crystal shape measurements.
    ii/ microscopic techniques relying upon the determination of the height correlation function.

MACROSCOPIC MEASUREMENTS

Equilibrium shapes

The measurement of the crystal equilibrium shape gives the most direct information on the roughening transition. The temperature at which a facet disappears is precisely its roughening temperature. Moreover the shape of the rough part in the vicinity of a facet contains information about the step-step interaction (see Eq.2). However transport mechanisms make the true thermodynamic equilibrium very difficult to achieve in a reasonable time except for very small crystals. Both "positive" and "negative" crystals have been used as depicted in Fig.16.

In the former method the crystal shape can be observed directly under equilibrium condition while in the latter the crystal has to be quenched before observation. The positive technique has been applied to α-Ag_2S crystals[24] and more recently to Pb crystals[25]. The negative technique has been used for organic crystal (diphenyl, naphtalen, tetrabrommethan)[26].

However the most attractive case is that of the equilibrium of a helium crystal with its superfluid phase. Superfluidity makes the equilibrium very easy to achieve even for cystals as large as a few centimeters. Moreover the continuum model is expected to be well applicable to this case since the pinning potential is supposed to be small enough. Indeed this system has been extensively studied and the theoretical predictions were found to be essentially correct. We shall not discuss this special case since a good review by Balibar and Castaing has already been published[27].

Growth shapes

In most cases the thermodynamic equilibrium cannot be achieved, nevertheless the roughening temperature can still be deduced from the study of the crystal shape during growth. As pointed out in our introduction we know that the growth is slow on smooth facets (where $T_R > T$) and fast on rough parts (where $T_R < T$). Thus during a growth at constant temperature only the facets for which $T_R > T$ survive. This method has been extensively developped by Heyraud and Métois[28] for the study of metals (Au,Pb,In).

MICROSCOPIC MEASUREMENTS

They are based on the measurement of the height correlation function $G(d)$ which diverges at large d above T_R and saturates to a constant level below T_R as indicated schematically in Fig.17.

The correlation function might be obtained from a surface image such as given by a scanning tunnel microsocope. This very promising technique has not yet been used and all the known data come from diffraction experiments which give more or less directly the Fourier transform of the correlation function.

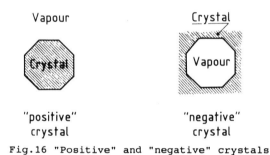

Vapour

Crystal

"positive" crystal

"negative" crystal

Fig.16 "Positive" and "negative" crystals

Diffraction technique analysis

Three different diffraction techniques have been used to date in order to study surface roughening. Thermal energy atom scattering (TEAS) came first then grazing X-ray diffraction and more recently high resolution low energy electron diffraction (HRLEED). In all cases one has to solve the problem of the scattering of a plane wave by a rough surface in order to analyse the diffraction data. In general this is a very difficult calculation but it is greatly simplified if a single scattering approximation can be assumed. This approximation corresponds to the so-called kinematical theory in LEED and X-ray diffraction, to the eikonal approximation in TEAS and to the Helmholtz approximation in optics (Huyghens principle). Then it can be shown that the expression for the scattered intensity factorizes into two terms:

$$I(\mathbf{Q}) = I_0(\mathbf{Q}) \sum_{\mathbf{R},d} e^{i\mathbf{Q}\cdot\mathbf{R}} \langle e^{i\mathbf{q}\cdot\mathbf{t}[n(\mathbf{R}+d)-n(\mathbf{R})]} \rangle \qquad (33)$$

where:

Q and q_z are respectively the tangential and the normal component of the momentum exchange with respect to the surface.

n(**R**) is an integer which defines the local displacement of the surface in terms of a unit vector **t**. For a close packed surface **t** is normal to the surface (|**t**| = a). For a vicinal surface **t** is parallel to the terraces and normal to the steps.

I_0(**Q**) is the so-called form factor (also called the structure factor in some X-ray papers). It depends upon the atomic positions within the unit cell of the perfectly ordered surface but is insensitive to roughness. It governs the intensities of the Bragg peaks which form the diffraction spectrum of the perfect surface. This form factor is also dependent upon the diffraction technique which is used.

The second term is called the structure factor (also called the interference function). It only depends upon the roughness statistics. It is independent of the diffraction technique.

However the single scattering approximation is really only suitable for X-ray scattering under the usual incidence conditions. It becomes

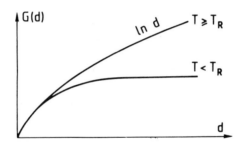

Fig.17 The height correlation function

questionable for total reflection grazing X-ray scattering, certainly not very good for LEED and quite bad for TEAS. Fortunately there are situations where formula (33) can still be used in presence of multiple scattering. When the disorder consists in large shifted domains as depicted in Fig.13, multiple scattering from one domain to another may be neglected and Armand and Salanon[29] have proved that the intensity is the sum of two terms. The first one which can be called <u>domain scattering</u>, corresponds to scattering by these large domains for which formula (33) is recovered with all multiple scattering effects contained in the form factor. The second one is related to the scattering by large domains boundaries and small domains such as point defects; it is expected to be very diffuse and will be called <u>diffuse scattering</u>. Moreover, although a rigorous proof has not been given, this result is presumably still valid for the scattering by the long wave length Fourier components of any type of roughness. This means that (33) remains valid in the general situation for Q not too large. This is exactly what is needed to study the roughening transition since it is a long range property. We shall now distinguish the two cases below and above T_R.

Scattering below the roughening transition

The existence of a stationary part in the correlation function leads to δ-functions in the scattering pattern. They correspond to the Bragg peaks of the perfect surface but their intensity is reduced, the remaining part of the intensity going into a more or less broad component which corresponds to the non-stationary part of the correlation function. The δ-function and the broad component of the domain scattering will be called respectively coherent scattering and incoherent scattering. The terms coherent and incoherent are related to the surface structure and have nothing to do with the test beam.

As pointed out previously when $T \ll T_R$ point defects (adatoms and advacancies) dominate at the surface. Then the intensity of coherent peaks can be written as[30]:

$$I_{coh} = I_{coh}^0 \left(1 - \sigma \frac{n}{n_s} \right) \qquad (34)$$

where:

I_{coh}^0 is the intensity of a Bragg peak on the perfect surface
σ is a cross section for diffuse scattering by the point defect.
There is no incoherent scattering in this case.
n is the number of point defect and n_s the number of site.

For X-ray diffraction or LEED σ is expected to be of the order of a unit cell area while it is much larger for TEAS[31]: $\sigma \simeq 50-100$ Å^2. Thus TEAS is especially attractive to study the early stage of atomic roughness which necesarily preceeds thermodynamic roughening.

Scattering above the roughening transition

The thermal average in (33) is difficult to calculate exactly but can be easily obtained in the so-called Gaussian approximation which becomes exact if the height probabilities obey a Gaussian law. One obtains:

$$\langle e^{i\mathbf{q}.\mathbf{t}[n(\mathbf{R}+\mathbf{d}) - n(\mathbf{R})]} \rangle = e^{- G(d) \, \varphi(\mathbf{q}.\mathbf{t})} \qquad (35)$$

φ is a periodic function of the argument $\mathbf{q}.\mathbf{t}$ where $\mathbf{t}$ is a vector defining the unit surface displacement. Then (33) is splitted into a series of more or less broad peaks centered on Bragg positions. The peak shapes are the Fourier transform of the structure factor given by (35) which only depends upon the correlation function G(d).

When $\mathbf{q}.\mathbf{t}$ is an even multiple of π (in-phase scattering) the waves scattered from various parts of the rough surface interfer constructively and the corresponding peak is a δ-function. The peak shape is not sensitive to the roughening.

When $\mathbf{q}.\mathbf{t}$ is an odd multiple of π (anti-phase condition) the interference is destructive, and the peak has a maximum broadening; this peak is the most sensitive to the roughening phenomena.

Note that the simple observation of the peak width oscillation as a function of $\mathbf{q}$ is a necessary (but not sufficient) condition to prove that the surface is above its roughening transition.

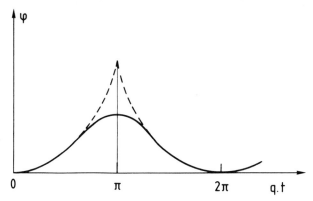

Fig.18 The function φ vs. **q.t**, full line : exact numerical calculation
dashed line : Gaussian approximation

We also emphasize that above T_R the domain scattering is completely
incoherent except in the strict in-phase condition where it becomes
completely coherent. In any case diffuse scattering is always superimposed.

This is all quite well, however, the validity of the Gaussian
approximation can be questioned. If it is strictly valid one gets:

$$\varphi(\mathbf{q.t}) = (1/2)[(\mathbf{q.t}\ \text{mod}\ 2\pi)^2] \tag{36}$$

To check this simulations have been done for various SOS models[32]. The
result is shown in Fig.18.

It is shown that (35) is a very good approximation except in the vicinity of
the anti-phase condition. As measurements are generally carried out in
anti-phase, neglecting this effect leads to an overestimation of the
roughening transition temperature.

For a Kosterlitz-Thouless transition $G(d) \approx \ln d$. Then from (32) the
incoherent scattering in the isotropic case is given by :

$$I(\mathbf{Q}) \approx (a\mathbf{Q})^{2-\tau} \tag{36}$$

with:

$$\tau = (T/\pi a^2\ \bar{\bar{\gamma}})\ \varphi(\mathbf{q.t}) \tag{37}$$

At $T = T_R$ and for anti-phase conditions for the isotropic case:

$$\bar{\bar{\gamma}} = (\pi/2a^2)T_R \quad \text{and} \quad \varphi(\mathbf{q.t}) = (\pi^2/2).(1-\epsilon) \quad \text{thus:}\ \tau = 1-\epsilon \tag{38}$$

where ϵ is a correction to the Gaussian approximation.

For the anisotropic case (vicinal surfaces) one has:

$$I(\mathbf{Q}) \approx \left[\left(\bar{\bar{\gamma}}_y/\bar{\bar{\gamma}}_x\right)^{1/2} a_y^2 q_y^2 + \left(\bar{\bar{\gamma}}_x/\bar{\bar{\gamma}}_y\right)^{1/2} \ell^2 q_x^2\right]^{1-\tau/2} \tag{39}$$

with:

$$\tau = T / \pi a_y \ell \left(\bar{\bar{\gamma}}_x \bar{\bar{\gamma}}_y \right)^{1/2} \varphi(\mathbf{q} \cdot \mathbf{t}) \qquad (40)$$

For $T = T_R$ and anti-phase condition we have again : $\tau = 1 - \epsilon$. We point out that formula (39) shows that the peak shape is anisotropic in the x and y directions and two parameters are now needed for its characterization : the exponent and the ratio of the two surface stiffnesses.

Survey of diffraction data analysis

In the vicinity of in-phase condition the diffraction pattern always displays a δ-function coherent peak superimposed to a diffuse scattering. As T is increased the amplitude of the coherent peak decreases and the diffuse scattering increases. This is due to the effect of point defects which dominates at low T and to the scattering from domain boundaries at high T. Away from in-phase condition an incoherent component is also present which increases with T and contributes to enhance the fall of the coherent component.

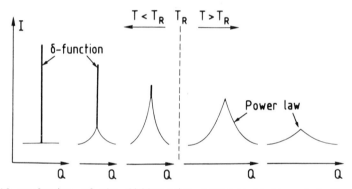

Fig.19 Evolution of the diffraction pattern in the vicinity of an anti-phase condition

In the vicinity of anti-phase condition the evolution of the diffraction pattern vs. temperature is described in Fig.19. At low T the diffraction pattern is similar to the in-phase case a δ-function coherent peak superimposed to a diffuse background. When T increases the coherent peak amplitude decreases until it vanishes at T_R. In the same time diffuse and incoherent scattering both increase. The shape of the incoherent component is presumably Lorentzian at the beginning and becomes more and more like a power law as T_R is approached. Above T_R the coherent part disappears. The incoherent component decreases away from the Bragg position as a power law with an exponent which increases with T.

The evolution of the amplitude of the coherent peak and of the power law exponent are shown in Fig.20. At $T = T_R$, I_{coh} vanishes and τ reaches its universal value with a vertical tangent. However instrument finite resolution must be taken into account for data analysis. For instance the limited angular resolution of the detector can make a δ-function

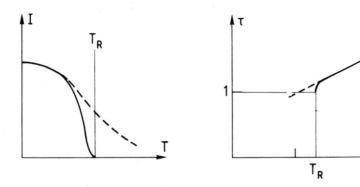

Fig.20 Coherent peak intensity and power law exponent vs. T

undistinguishable from a strongly peaked incoherent component. In these conditions the intensity measured in Bragg position does not vanish at T_R as shown in Fig.20. For the same reason the vertical tangent in the τ vs.T curve is generally not observed and the data can be fitted by a power law even somewhat below T_R. The exponent thus obtained below T_R has not a clear physical significance. Note that the effect of the Gaussian approximation and of finite instrument resolution on the determination of T_R are opposite so that they have a tendency to compensate each other. It is clear in Fig.20 that it is rather delicate to determine T_R from intensity measurements alone. It is safer to deduce it from τ vs.T dependence obtained from peak profile measurements.

EXPERIMENTAL RESULTS

Many types of surfaces have been investigated: noble gases, metals, inorganic and organic compounds. The most extensively studied cases to date are Helium crystals and metals. As the Helium case has been carefully discussed in Ref.27 we shall concentrate on the roughening of fcc metal surfaces which do not reconstruct. The growth shape of In and Pb have been studied by Heyraud and Métois[28]. TEAS has been applied to Cu(100)[33], Cu(110)[34], Cu(113)[35], Cu(115)[36], Cu(1,1,11)[37], Cu(311)[38], Cu(310)[38], Ni(113)[39], Ni(115)[40]; X-ray diffraction to Cu(110)[41], Cu(113)[42], Ni(113)[43], Ag(110)[44] and HRLEED to Pb(110)[45], Ni(100)[46] and Ni(110)[46].

Table II The roughening transition temperatures of fcc(110) surfaces

Surface	T_R (K)	T_m (K)	T_R/T_m	Technique
In(110)	290	420	0.69	Growth shape
Pb(110)	390(1)-415(2)	600	0.65-0.69	growth shape(1) and HRLEED(2)
Ag(110)	720	1234	0.58	X-Ray diffaction
Cu(110)	1000	1355	0.73	TEAS (extrapolated)
Ni(110)	1300	1720	0.76	HRLEED (extrapolated)

Close packed surfaces: fcc(111) and (100)

Growth shapes and diffraction results are in agreement and demonstrate that these surfaces do not undergo a roughening transition below the melting point. A very recent experiment[25] seems to indicate that on Pb the (100) facet disappears 4 K below T_m. However it is not clear whether this is due to asurface roughening or melting..

fcc(110) surfaces

These faces are found to undergo a roughening transition below T_m, the critical temperatures are listed in Table II. For Cu and Ni, T_R is estimated from an extrapolation of measurements carried out below T_R. T_R scales approximately with the melting point T_m: $T_R/T_m = 0.68 \pm 0.1$.

fcc vicinal surfaces

Roughening transitions of the steps have been observed on every studied vicinal surface: Cu and Ni. As mentioned in the theoretical part there are now two characteristic parameters: the exponent and the peak anisotropy.

Temperature of the step roughening transition

The measured T_R are reported in Table III.

Remarks:

i/ For Cu(113) the TEAS value has been obtained from the analysis of the peak shape above T_R while only peak intensity has been analysed in the X-ray experiment. Thus the second value is certainly less reliable than the first.

ii/ Freezing of the surface precludes the determination of too low T_R. At low T the mobility is not high enough to allow the surface to reach its thermodynamic equilibrium. For copper freezing has been found to occur around 300 K[47] and consequently Cu(1,1,11) and Cu(310) remains frozen in their rough state below this temperature.

iii/ As expected, for a given metal T_R decreases when the step-step distance increases.

Table III The roughening transition temperature of vicinal surfaces

Surface	T_R (K)	Technique
Cu(113)	720(1)-620(2)	TEAS(1)-X-ray diffraction(2)
Cu(115)	380	TEAS
Cu(1,1,11)	<300	TEAS
Cu(331)	650	TEAS
Cu(310)	<300	TEAS
Ni(113)	750(1)-770(2)	TEAS(1)-X-ray diffraction(2)
Ni(115)	450	TEAS

iv/ As previously, for the same face of different metals T_R scales approximately with T_m: $T_R/T_m = 0.48 \pm 0.05$ for (113) and 0.27 ± 0.01 for (115).

Peak anisotropy

Eq.(39) shows that even in the absence of energetic anisotropy there is a geometric anisotropy due to the anisotropy of the step structure ($a_y < \ell$). From the geometric anisotropy one expects that peaks will be broader in the y-direction (parallel to the steps) than in the x-direction (perpendicular to the the steps). As $\bar{\bar{\gamma}}_y$ is expected to be larger than $\bar{\bar{\gamma}}_x$ the energetic anisotropy acts in the opposite sense of the geometric anisotropy. In fact in almost all cases the experimental peaks have been found to be broader in the x-direction than in the y-direction which indicates a strong energetic anisotropy and rules out the Den Nijs et al.[21] model which does not account for it. TEAS measurement for Ni(113) where the peak displays only the geometrical anisotropy within experimental accuracy are in strong disagreement with a subsequent X-ray diffraction experiment.

The energetic anisotropy is expected to increase when the step-step distance ℓ increases and so does the experimental peak anisotropy.

Atomic energetic parameters: W_0 and w_1

The analysis of the data by fitting a statistical model has only been done for Cu(113) and Cu(115). Ni(113) data have not been analyzed with an exact calculation but only with approximate formulas somtimes clearly out of their validity range : this leads to inconsistent results. Thus we shall only discuss the copper results. The most complete analysis is obtained for Cu(113) were there is accurate data below and above T_R.

The (113) face has an interesting aspect: in one hand it is a stepped surface which can be described by a TLK model but on the other hand it looks very much like the (110) and thus may be also quite accurately represented by a BCSOS model. The numerical simulations have confirmed this view: a fit of the experimental data with any of these two models gives practically the same values for W_0 and w_1 remembering that J_x and J_y in the BCSOS model corresponds respectively to $w_1/2$ and W_0 in the TLK model.

The exponent τ and the anisotropy $\bar{\bar{\gamma}}_y / \bar{\bar{\gamma}}_x$ have been obtained from the experimental peak shapes above T_R by a fit onto an anisotropic power law (Eq.(38)). Then the parameters of the BCSOS or TLK model are adjusted to reproduce the exponent and the anisotropy as well as their temperature dependance. Analytical formula are available for the BCSOS model and a numerical Monte-Carlo simulation is needed for the TLK. The result is:

$$Cu(113): J_y = W_0 = 800 \text{ K and } J_x = w_1/2 = 280 \text{ K}$$

The intensity of the specular peak below T_R has also been analysed[48]. The experimental intensities are plotted in Fig.21 vs. T for in-phase and anti-phase conditions. The intensities have been carefully corrected for the Debye-Waller factor and normalized to unity at T = 0. We recall that the in-phase intensity decrease is due to the direct scattering by the domain boundaries (here lines of kinks) while the additional decrease in anti-phase comes from the interference between domains. These anti-phase intensities do not drop to zero at T_R due to finite intrument resolution. The low

temperature part of the in-phase curve is expected to be described by
Eq.(33) where σ is now the cross section for incoherent scattering from an
isolated kink. σ is not known precisely but is expected to be of the order
of 50-100 Å^2. The number of kinks is calculated from the BCSOS model with
the above determined values of J_x and J_y. A good fit of the experimental
data is obtain with σ = 70 Å^2. The roughening of Cu(113) is thus completely
described by the BCSOS model in the whole temperature range.

For Cu(115) the terraces become larger and the TLK model must be used.
There are now three parameters: W_0, w_1, w_2. As it expected that $w_2 > w_1$ one

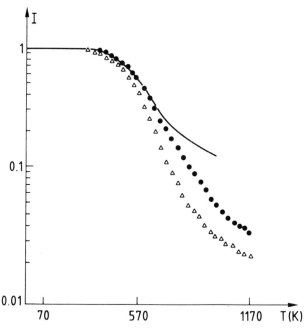

Fig.21 Specular intensity vs. T for Cu(113)

experimental data : • in-phase condition

△ anti-phase condition

BCSOS calculation : full line

can take $w_2 = \infty$ as a first approximation valid if the temperature is not
much higher than T_R. Then the same peak shape analysis as for Cu(113) gives:

Cu(115): W_0 =850 K and w_1 = 120 K

Freezing effect precludes the analysis below T_R on this surface.

The step-step interaction decreases from the (113) to the (115) face as
expected. It is also very satisfactory to find the same value of the kink
energy creation for the two faces. This value of W_0 is much smaller than a

broken bond estimation (W_0 = 3400 K) or a rough band structure estimation
(W_0 = 2300 K)[20]. A more realistic calculation[49] gives W_0 = 1200 K at T = 0
K. Indeed W_0 is the free energy of creation of a kink and it should contain
a vibrational entropy contribution lowering W_0 at finite T. The order of
magnitude of this contribution can be estimated from the calculation of the
creation entropy of a vacancy by Wynnblatt[50]: $S/k_B \simeq 2$. The experimental
value W_0 is thus quite reasonable.

The behaviour of low index faces below T_R

The BCSOS model is expected to describe (100) and (110) faces fairly
well. However available diffraction data on these surface are always taken
below T_R thus the determination of the energetic parameters by the peak

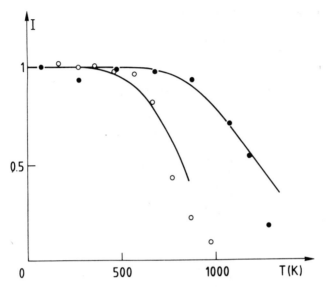

Fig.22 Specular peak intensity vs. T for Cu(100) and Cu(110)
experimental data : • Cu(100)
o Cu(110)
BCSOS calculation : full line

shape analysis above T_R is not possible. Nevertheless a fit of the low
temperature coherent peak intensity is still possible[48]. For the (100) face
one has to adjust two parameters: J , σ and three for the (110): J_x, J_y, σ.
The result of this adjustment for Cu(100) and Cu(110) is shown in Fig.22.

The intensity of the specular peak, corrected for the Debye-Waller
factor and normalized to unity at T = 0 K, is plotted vs.T. A good fit is
obtained with the BCSOS model with the following parameters:

Cu(100): J = 1200 K and σ = 80 Å^2 which predicts: T_R = 1800 K

Cu(110): J_x = 500 K, J_y = 1000 K and σ = 90 Å^2 which predicts: T_R = 1000 K
These results are quite consistent with the previous ones for Cu(113)
from which we conclude that close packed fcc surfaces are very well

described by the BCSOS model. This model predicts that T_R is larger than T_m for (100) faces as generally expected and that $T_R = 1000$ K for Cu(110), which is compatible with the conclusions of Zeppenfeld et al.[34].

However Zeppenfeld et al. have concluded from their energy resolved measurements that for Cu(110) the intensity decrease is not due to incoherent scattering onto defects but to inelastic scattering due to enhanced surface anharmonicity. The two interpretations are apparently conflicting and the problem is not yet completely understood. Let us just point out the following remarks:

i/ The existence of defects well below T_R is inherent to the very nature of the roughening transition. Indeed the results of Zeppenfeld et al. show that the ratio of diffuse and incoherent elastic scattering with respect to coherent elastic scattering increases with T in agreement with the present model.

ii/ The cross section σ is possibly overestimated in the fit and part of the observed intensity decrease may be due to inelastic scattering.

iii/ The defects may contribute to the inelastic scattering more than the ordered parts of the surface. This is supported by a recent molecular dynamic calculation on Cu(110) which shows that defects and enhanced anharmonicity appear approximately at the same temperature[51].

This debate emphasizes the importance of carrying out energy resolved scattering experiments when dealing with high temperature transitions in order to clearly discriminate between diffuse and incoherent elastic scattering, and inelastic scattering.

CONCLUSION

The roughening transition on unreconstructed surfaces is now quite well understood from the theoretical point of view. Its existence in real surfaces has been demonstrated for different systems by means of various accurate experimental techniques. So roughening is a very general phenomenon and it should probably occur on every kind of surfaces. Careful comparisons between experimental data and simulations on realistic models give access to important physical quantities like creation energy of surface defects or interaction energy between defects. This is certainly an axis for future developments. Ab initio calculations of these quantities are very scarce and further efforts are strongly welcome.

The state of theory and experiment for reconstructed surface is less achieved and is an active field today. It is reviewed elsewere in this volume. The influence of impurities or adsorbates upon the roughening transition has not been considered to date and is also a research field to be considered for the future.

The roughening transition is basically a pure statistical mechanics equilibrium lattice concept. Lattice vibrations are not taken into account in its premises. Lattice vibrations are through the Lindeman criterion the driving mechnism for surface melting while lattice defects characteristic of the surface are the elementary excitations which permit surface roughening. How do the two mechanisms interplay? This is a fully open field. The following question have to be answered:

i/ Is there any connection between surface melting and surface roughening ? If so does roughening always preceeds surface melting ?

ii/ Does the separation between the roughening transition and melting retain any meaning at high temperature where vibrational amplitudes are large?

The roughening transition is defined at thermodynamic equilibrium and very little is known in non-equilibrium situations. For instance:

i/ What kinetics governs the achievement of the equilibrium state?

ii/ What is the influence of the roughening transition upon diffusion mechanisms?

ACNOWLEDGEMENTS

We are very indebted to J.Villain for introducing us in the sometimes esoteric field of phase transition theory. The part relating to the continuous model is greatly inspired by the enlightening lectures of P.Nozières at the College de France in Paris. We thank also all those who have contributed to this review: G.Armand, H.J.Ernst, F.Fabre, D.Gorse, B.Loisel, J.R.Manson, V.Pontikis. We especially thank A.Khater for critically reading of the manuscript.

REFERENCES

1. W.K.Burton, N.Cabrera and F.C.Frank, Phil.Trans.Roy.Soc. (London) 243A:299(1951).
2. L.Onsager, Phys.Rev. 65:117(1944)
3. J.M.Kosterlitz and D.J.Thouless, J.Phys. C6:1181(1973)
   J.M.Kosterlitz, J.Phys. C7:1046(1974)
4. J.D.Weeks, in : Ordering of strongly fluctuating condensed matter, T.Riste, Ed., Plenum, New York (1980)
5. H.van Beijeren and I.Nolden, in : Structures and dynamics of surfaces, W.Schommers and P.von Blanckenhagen, Ed., Springer, Heidelberg (1987)
6. N.Cabrera, Surf.Sci. 2:320(1964)
7. A.F.Andreev, Sov.Phys.JETP, 53:1063(1981)
8. C.Rottman and M.Wortis, Phys.Rep. 103:59(1984)
9. C.Jayaprakash and W.F.Saam, Phys.Rev.B, 30:3916(1984)
10. H.J.Schultz, J.Physique(Paris) 46:257(1985)
11. T.Ohta and K.Kawasaki, Prog.Theor.Phys. 60:365(1978)
12. P.Nozières and F.Gallet, J.Physique(Paris) 48:353(1987)
13. S.T.Chui and J.D.Weeks, Phys.Rev.B 14:4978(1976)
14. J.V.Jose, L.P.Kadanoff, S.Kirkpatrick and D.R.Nelson, Phys.Rev.B 16:1217(1977)
15. W.J.Shugard,J.D.Weeks and G.H.Gilmer,Phys.Rev.Lett.41:1399(1978)
16. H.van Beijeren, Phys.Rev.Lett. 38:993(1977)
17. E.H.Lieb and F.Y.Wu, in : Phase transitions and critical phenomena, vol.1, C.Domb and M.S.Green, Ed., Academic Press, New York (1972)
18. A.Trayanov, A.C.Levi and E.Tosatti, Europhys.Lett. 8:657(1989)
19. K.Rommelsee and M.den Nijs, Phys.Rev.Lett. 59:2578(1987)
20. J.Villain, D.Grempel and J.Lapujoulade, J.Phys.F:Metal Phys. 15:809(1985)
21. M.den Nijs, E.K.Riedel, E.H.Conrad and T.Engel, Phys.Rev.Lett. 55:1689(1985) and Erratum, 57:1279(1986)
22. W.Selke and A.M.Szpilka, Z.Phys.B 62:381(1986)
23. This remark is mainly due to M.Wortis
24. T.Ohachi and I.Taniguchi, J.Cryst.Growth, 65:84(1983)

25. A.Pavlowska, D.Dobrev, K.Faulian and E.Bauer, Symposium 3S90, La Plagne, France (March 11-17 1990), to be published.
26. A.Pavlowska and D.Nenow, J.Cryst.Growth, $\underline{8}$:209(1971), $\underline{12}$:9(1972), $\underline{39}$:346(1977)
27. S.Balibar and B.Castaing, Surf.Sci.Rep. $\underline{5}$:87(1985)
28. J.C.Heyraud and J.J.Métois, J.Crys.Growth, $\underline{82}$:269(1987)
29. G.Armand and B.Salanon, Surf.Sci. $\underline{217}$:317(1989)
30. G.E.Tommei, A.C.Levi And R.Spadacini, Surf.Sci. $\underline{125}$:312(1981) . G.Armand and B.Salanon, Surf.Sci. $\underline{217}$:341(1989)
31. G.Comsa and B.Poelsema, Appl.Phys.B $\underline{38}$:153(1985)
32. B.Salanon and J.Lapujoulade, Vacuum in press.
33. D.Gorse and J.Lapujoulade, Surf.Sci. $\underline{162}$:847(1985)
34. P.Zeppenfeld, K.Kern, R.David and G.Comsa, Phys.Rev.Lett. $\underline{62}$:63(1989)
35. B.Salanon, F.Fabre, J.Lapujoulade and W.Selke, Phys.Rev.B $\underline{38}$:7385(1988)
36. F.Fabre, D.Gorse, B.Salanon and J.Lapujoulade, J.Physique(Paris), $\underline{48}$:1017(1987)
37. F.Fabre, B.Salanon and J.Lapujoulade, Sol.Stat.Com. $\underline{64}$:1125(1987)
38. B.Loisel, Thesis, University Paris VII (1989)
39. E.H.Conrad, L.R.Allen, D.L.Blanchard and T.Engel, Surf.Sci. $\underline{187}$:265(1987)
40. E.H.Conrad, R.M.Aten, D.S.Kaufman, L.R.Allen, T.Engel, M.den Nijs and E.K.Riedel, J.Chem.Phys. $\underline{84}$:1015(1986) and Erratum, $\underline{85}$:4856(1986)
41. S.G.J.Mochrie, Phys.Rev.Lett. $\underline{59}$:304(1987)
42. K.S.Liang, E.B.Sirota, K.L.D'Amico, G.J.Hughes and S.K.Sinha, Phys.Rev.Lett. $\underline{59}$:2447(1987)
43. I.K.Robinson, E.H.Conrad and D.S.Reed, J.Physique(Paris) $\underline{51}$:103(1990)
44. G.A.Held, J.L.Jordan-Sweet, P.M.Horn, A.Mak and R.J.Birgeneau, Phys.Rev.Lett. $\underline{59}$:2075(1987)
45. H.N.Yang, T.M.Lu and G.C.Wang, Phys.Rev.Lett. $\underline{63}$:1621(1989)
46. Y.Cao and E.Conrad, Phys.Rev.Lett. $\underline{64}$:447(1990).
47. F.Fabre, D.Gorse, B.Salanon and J.Lapujoulade, Surf.Sci.Lett. $\underline{175}$:L693(1986)
48. B.Salanon, H.J.Ernst, F.Fabre and J.Lapujoulade, Proc. of the 3rd ICSOS meeting, Milwaukee USA, July 1990, in press
49. B.Loisel, D.Gorse, V.Pontikis and J.Lapujoulade, Surf.Sci. $\underline{221}$:365(1989)
50. P.Wynnblatt, Phys.Stat.Sol. $\underline{36}$:797(1969)
51. B.Loisel, V.Pontikis and J.Lapujoulade to be published

246

INTERPLAY BETWEEN SURFACE
ROUGHENING, PREROUGHENING, AND RECONSTRUCTION

Marcel den Nijs

Department of Physics, FM-15
University of Washington, Seattle, Washington 98195

ABSTRACT

Solid-on-solid models provide a comprehensive description of surface roughening, preroughening, and reconstruction. The origin and properties of the preroughening transition and disordered flat (DOF) phase are reviewed. The present experimental evidence for realizations of DOF phases in (110) facets of face-centered cubic (FCC) metals and in (111) facets of noble gas crystals is discussed. A chiral 4-state clock-step model is introduced to describe the deconstruction and roughening of missing row reconstructed (110) facets of FCC crystals. It is shown that a reconstructed rough phase is not possible. Instead roughening induces a simultaneous deconstruction transition, with central charge $c=1.5$ and both Ising and conventional roughening critical exponents. This, together with the experimental evidence suggests that at deconstruction Pt(110) becomes rough and simultaneously undergoes an incommensurate melting transition with respect to its reconstruction degrees of freedom.

1. INTRODUCTION

The interplay between surface reconstruction, roughening, and surface melting is important. The nature of each transition is well established theoretically as long as each takes place at a well separated temperature. In many experimental systems they are not well separated. From a theoretical perspective this leads to interesting new types of critical behaviour. I want to discuss two specific topics. I review in section 2 the properties and origin of disordered flat (DOF) phases and the preroughening (PR) phase transition [1,2,3]. I discuss in section 3 the roughening of missing row (MR) reconstructed structures, of the type observed in Au(110) and Pt(110) [4,5]. These sections represent the situation before the meeting. A discussion of the developments during the meeting is presented in section 4.

We discovered preroughening and DOF phases during a study of the restricted solid on solid (RSOS) model [1]. Preroughening appeared as a necessity; to reconcile the ferromagnetic (F) and anti-ferromagnetic (AF) sides of the phase diagram. Fig. 1 shows the phase diagram of the RSOS model. It is interesting to compare Fig. 1 with the behaviour of the (110) facets of the noble face-centered cubic (FCC) metals: Ni

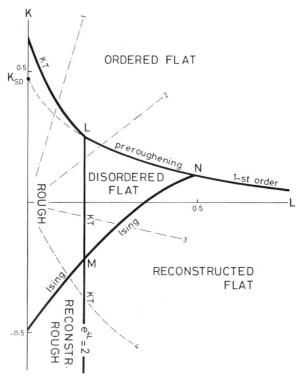

Fig. 1. Schematic phase diagram of the RSOS model. K and L are the nearest and next-nearest neighbour interactions. Both are in units of $k_B T$. Infinite temperature corresponds to $(K, L) = (0, 0)$. The $- \cdot - \cdot -$ lines represent characteristic experimental paths [3].

and Cu, Pd and Ag, and Pt and Au. Fig. 1 gives a comprehensive description of the interplay between surface roughening, preroughening, and reconstruction. Lines of constant K/L represent specific experimental systems.

The structure of Fig. 1 is quite generic. Many aspects of the model do not matter. For one, the RSOS model in section 2 describes (100) facets of simple cubic crystals, not (110) facets of FCC crystals. However, as discussed in section 3, only for $K \ll 0$ (along paths of type 4 in Fig. 1) will the structure of the phase diagram be different.

Secondly, the reconstructed phase of the RSOS model in section 2 has the uncommon checkerboard type structure. However, the models can be modified to replace this by the familiar MR reconstruction without changing the structure of the phase diagram [6]. Finally, and most seriously, the interactions in the model are overly simplistic compared to those in real metals. However, the energy differences between the ordered flat and reconstructed structures, ΔE, have been calculated recently for the noble metals [7]. In my RSOS model this energy difference is equal to $\Delta E = 2K k_B T$. These microscopic calculations allow us to identify experimental systems with specific paths in Fig. 1. At this stage only a qualitative comparison is possible. A quantitative comparison requires additional microscopic information, e.g. step energies, to locate the multicritical points L, N, and M. In the model they vary with the precise choice of the next nearest neighbour interactions [1,2].

Path 1 represents systems with $\Delta E >> 0$. The low temperature phase is ordered flat (OF), and roughens directly by a conventional Kosterlitz-Thouless (KT) transition into the rough phase. According to the microscopic calculations ΔE is largest for Ni(110) and Cu(110) [7]. Path 1 applies most likely to them (assuming that bulk melting and/or surface melting do not interfere), but the experimental evidence is confusing. Initially roughening was reported to take place in Cu(110) at $T/T_m = 0.64$ [8] (T_m is the bulk melting temperature). More recently this signal has been attributed to the onset of large anharmonic vibrations along the surface [9]. In Ni(110) the same has been observed at $T/T_m = 0.52$, and evidence of roughening appears only at a much higher temperature, $T/T_m = 0.75$ [10]. It seems plausible that these anharmonic vibrations are associated with a tendency towards reconstruction. Au and Pt reconstruct into the MR structure at almost the same reduced temperature, at $T/T_m = 0.49$ in Au [4] and at $T/T_m = 0.53$ in Pt [5]. I wonder whether Ni follows a path of type 2. A softening of the lateral vibrations at the preroughening transition seems consistent with the nature of the DOF phase.

Path 2 represents the preroughening scenario. ΔE is still positive, but small. First the ordered flat phase preroughens into the DOF phase, followed at higher temperatures by a conventional KT type roughening transition.

In the DOF phase the surface contains an array of steps with positional disorder but long range up-down-up-down order. The surface remains flat on average. Compare Fig. 2b and Fig. 2c. In the rough phase the surface resembles a terraced mountain landscape with positionally disordered steps. Likewise, in the DOF phase the surface contains steps that are disordered positionally, but up and down steps alternate. At the PR transition the free energy of creating steps vanishes, although the free energy associated with tilting the surface does not vanish.

Ag(110) and Pd(110) are candidates to follow paths of type 2. According to the microscopic calculations ΔE is positive yet quite small [7]. These surfaces have been studied extensively [11], but their roughening seems unexplored. Another possibility is to study surfaces in the presence of an adsorbate. Low coverages of alkali metal induce the MR type reconstruction in e.g. Ni(110) and Pd(110) [12]. By fine-tuning the impurity density it might be possible to switch among paths of type 1-4.

Path 3 represents the conventional scenario where a reconstructed surface deconstructs well below its roughening temperature. Deconstruction is in the Ising universality class because the checkerboard (as well as the MR) structure is two-fold degenerate. Experiments [4,5] and microscopic energy calculations [7] agree that Au(110) and Pt(110) reconstruct into the MR structure. The experimental results for Au(110) have been interpreted in terms of a path of type 3 [4].

The average height of the surface is shifted by one-half in the reconstructed phase (be it checkerboard or MR) with respect to the ordered flat phase. At the deconstruction transition the MR order parameter vanishes without changing the average height. In the DOF phase it is still shifted by $\frac{1}{2}$. Everywhere in Fig. 1 below the line K_{SD}-L-N, the average surface height is shifted by $\frac{1}{2}$. In the rough phase the distinction is meaningless since the interface width diverges with the system size.

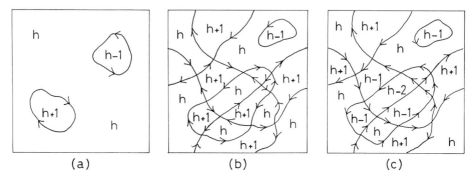

Fig. 2. Typical configurations in the ordered flat (OF) phase (a), the disordered flat (DOF) phase (b), and the rough phase (c). The height of the surface is indicated by h, and the arrows indicate whether a step is up or down (see also eq. (2.1)).

This clarifies the origin of the PR transition along paths of type 2. At preroughening the average height of the surface switches spontaneously from the value in the OF phase to the value in the nearby reconstructed state. The DOF phase can be viewed as an entropy-stabilized manifestation of an energetically nearby reconstructed state. The reconstructed state is its condensate.

It is possible to imagine different types of DOF phases, again with positionally disordered steps, but e.g. with different types of long range up-down step order. The likelihood of finding such a generalized DOF phase depends on how nearby in energy its corresponding surface reconstructed state is. DOF phases can thus be identified and classified according to their parent reconstructed state.

DOF phases are of interest also in a different context. They are equivalent to valence bond state (VBS) type phases in one dimensional quantum spin chains [2]. VBS phases describe quantum fluids with long range topological order, e.g. in the theory of the fractional quantum Hall effect.

Paths of type 4 represent surfaces that roughen before they reconstruct. Fig. 1 has a reconstructed rough phase. In section 2 we will see that in the (100) facets of simple cubic crystals the reconstruction and roughening type order parameters decouple in this part of the phase diagram. This allows the Ising reconstruction line and the KT roughening line to simply cross each other, and results in the presence of the reconstructed rough phase.

The noble metals form FCC crystals and their (110) facets have a anisotropic body-centered structure. Their ordered flat (OF) and MR structures have an additional checkerboard type corrugation, Fig. 3, which is ignored in the RSOS model description. At first this omission looks irrelevant. Indeed it is irrelevant along paths of type 1-3, but not along paths of type 4. In section 3 I introduce a chiral 4-state clock-step model to describe the local region around multicritical point M, and find that along paths of type 4 roughening induces a simultaneous deconstruction transition.

2. PREROUGHENING AND DISORDERED FLAT PHASES

In this section I review the nature and origin of disordered flat (DOF) phases and the preroughening (PR) transition. The discussion applies to the (100) facets of simple cubic crystals. Consider the restricted solid-on-solid (RSOS) model [1,2,3], on a square lattice, with nearest (K) and next nearest neighbour (L) interactions

$$\mathcal{H}_{RSOS} = K \sum_{\langle \mathbf{r}, \mathbf{r}' \rangle} (h_{\mathbf{r}} - h_{\mathbf{r}'})^2 + \sum_{(\mathbf{r}, \mathbf{r}'')} \{ L_1 \delta(|h_{\mathbf{r}} - h_{\mathbf{r}''}| - 1) + L_2 \delta(|h_{\mathbf{r}} - h_{\mathbf{r}''}| - 2) \}. \quad (2.1)$$

The heights are integers, $h_{\mathbf{r}} = 0, \pm 1, \pm 2, \ldots$. The step height between nearest neighbour (NN) columns is restricted to $\delta h = 0, \pm 1$. Fig. 1 corresponds to the special case $L_2 = 4L$ and $L_1 = L$.

It is advantageous to characterize the configurations of the crystal surface by steps instead of column heights. The steps form closed loops. Associate a step variable $S_{\mathbf{r}, \mathbf{r}'} = h_{\mathbf{r}} - h_{\mathbf{r}'} = 0, \pm 1$ with each bond. It is important to distinguish between up and down steps, because $S_{\mathbf{r}, \mathbf{r}'} = -S_{\mathbf{r}', \mathbf{r}}$. Place arrows along the steps, as shown in Fig. 2. In the direction along the arrow the height to the left of the step is (one unit) lower.

From a ferromagnetic perspective, the DOF phase is stabilized by a combination of step entropy and step interactions. Consider the OF phase close to the PR transition. The surface contains thermodynamically excited terraces where the surface is higher or lower (by one unit). Nearest neighbour (NN) interactions are blind to the arrows. They contribute only to the step energy. Further-than-NN interactions look across more than one step, and represent short range interactions between the steps. Steps with parallel arrows repel each other. Steps with opposite arrows attract each other. Consider the extreme case where the repulsion is infinitely strong, $L_2 \to \infty$, and attraction is absent, $L_1 = 0$. Steps with parallel arrows are forbidden to approach each other closer than the interaction range, while steps with anti-parallel arrows can approach each other at will. The steps have more meander entropy in configurations where the steps have alternating up-down-up order than in configurations where neighbouring steps have parallel arrows. This illustrates that the DOF structure is favoured by a combination of entropy and further than nearest neighbour interactions.

This is a quite general argument, independent of the SOS model description, but is not conclusive because it does not tell whether the effect is strong enough to stabilize the long-range AF step order of the DOF phase. This is where surface reconstruction comes into play. It is important to know whether a reconstructed state is energetically close to the OF phase. In the case of eq.(2.1) this state is the checkerboard structure.

The DOF phase just described is only the simplest prototype. The long range arrow order can be of a more complex type. Suppose you increase the range of the interactions in the RSOS model further, and fine-tune the signs of these interactions, such that a more complicated type of reconstructed structure is almost stable, e.g. a structure with a larger unit-cell and a more complex up-down step structure. Then, in the flat phase at short distances, steps prefer this more complex up-down structure. This, combined with the above entropy argument then leads to stabilization of the corresponding more complex type of DOF phase.

The DOF phase has simultaneously order and disorder. The degrees of freedom associated with the positional step disorder and the ones associated with the long range step-up step-down order can be separated by rewriting the RSOS model as an Ising model coupled to a 6-vertex (6V) model [1,2]. Assign an Ising spin to each column height, $\sigma_{\mathbf{r}} = \exp(i\pi h_{\mathbf{r}}) = \pm 1$. It represents the parity of the column height. An Ising Bloch wall indicates the presence of a step between two nearest neighbour columns. Introduce a 6V model on the lattice formed by the Ising Bloch walls to represent whether a step is up or down. The model can be viewed as a modified Ising model or as a bond-diluted 6V model. From the point of view of the Ising model the Boltzmann weight of each configuration is modified with the partition function of a 6V model; i.e. multiplied with trace over all arrow configurations, with the arrows placed on the Ising Bloch walls such that at each Ising Bloch wall intersection the flux of the arrows is equal to zero,

$$Z_{RSOS} = \sum_{\{\sigma_{\mathbf{r}}\}} \exp[-H_I(K, L_1)] \ Z_{6V}(\{\sigma_{\mathbf{r}}\}, L_2). \tag{2.2}$$

H_I is the conventional Ising Hamiltonian with nearest (K) and next nearest neighbour (L_1) interactions. From the point of view of the 6V model eq.(2.2) represents a bond diluted 6V model; a 6V model on a lattice that is not rigid but has annealed fluctuations in its shape and number of bonds.

This formulation of the model clarifies the general structure of the phase diagram. Fig. 1. The coupling constants K, and L_1, govern the Ising type order. They control the density and positional order of the steps. L_2 governs the 6V type order, and controls the long-range AF order in the steps. The Ising spins are AF ordered in the reconstructed phase. In the limit $K \rightarrow -\infty$ each bond contains an Ising Bloch wall, and the model reduces to the exactly soluble Body-Centered (BC) SOS model [14]. At $\exp(L_2) = 2$ the reconstructed flat (RF) phase roughens via KT transition into the reconstructed rough phase.

At finite values of $K << 0$ the Bloch walls form a square array, but with missing bonds (closed loops) at length scales smaller than the Ising correlation length. These loops melt at the deconstruction transition, the Ising critical line in Fig. 1. In the DOF phase and the rough phase, the Ising Bloch walls form a disordered array. For increasing K this array contains an increasing number of disconnected pieces, but it always includes one infinitely large connected backbone cluster. Each finite cluster acts as an independent 6V model. Its arrows are disordered because finite systems can not maintain long-range order. The 6V model defined on the fluctuating backbone cluster is solely responsible for the long-range order in the DOF phase; it undergoes a conventional roughening transition. Its rough phase (at small values of L_2) represents the rough phase (Fig. 2c). Its flat phase (at large values of L_2) represents the DOF phase (Fig. 2b).

This establishes the long-range order of the DOF phase: the surface contains a disordered arrays of steps, but remains flat on average because the height fluctuations are limited by the long-range AF arrow order on the backbone. The backbone disintegrates at preroughening. In the OF phase the Ising spins have long-range ferromagnetic order. Only finite Ising Bloch wall lattices (disconnected terraces) remain.

The formulation of eq.(2.2) also clarifies how to define order parameters to distinguish between the OF, DOF, and RF phase [2,3]. Consider the spin-spin correlation function of the Ising degrees of freedom (the parity of the column heights),

$$G_H(\mathbf{r}_n - \mathbf{r}_0) = \langle \sigma_{\mathbf{r}_n} \sigma_{\mathbf{r}_0} \rangle = \langle \exp[i\pi(h_{\mathbf{r}_n} - h_{\mathbf{r}_0})] \rangle. \tag{2.3}$$

The Ising spins are disordered in the DOF but ordered in the OF phase. In the DOF phase G_H decays exponentially to zero but in the OF flat phase exponentially to the square of the Ising magnetization,

$$\rho = \langle \exp[i\pi h_{\mathbf{r}}] \rangle. \tag{2.4}$$

G_H decays to the square of the staggered Ising magnetization, ρ_s, in both reconstructed phases (the flat and the rough one). Consider the step-step correlation function

$$\begin{aligned} G_S(\mathbf{r}_n - \mathbf{r}_0) &= \langle (h_{\mathbf{r}_n} - h_{\mathbf{r}'_n}) \exp[i\pi(h_{\mathbf{r}_n} - h_{\mathbf{r}_0})] (h_{\mathbf{r}_0} - h_{\mathbf{r}'_0}) \rangle \\ &= \langle \sin(\frac{1}{2}\pi(h_{\mathbf{r}_n} + h_{\mathbf{r}'_n} - h_{\mathbf{r}_0} - h_{\mathbf{r}'_0})) \rangle \end{aligned} \tag{2.5}$$

with $\mathbf{r}'_n$ a nearest neighbour site to the left (above) site $\mathbf{r}_n$ on the square lattice. The phase factor contributes a plus (minus) sign to the correlation function when the height difference between sites $\mathbf{r}_n$ and $\mathbf{r}_0$ is even (odd). In the DOF phase, and also in the RF phase, the steps have long-range AF arrow order. Their arrows are predominantly parallel (anti-parallel) if this height difference is even (odd). G_S decays exponentially to zero in the OF phase but exponentially to the square of the order parameter

$$\psi = \langle (h_{\mathbf{r}'} - h_{\mathbf{r}}) \exp(i\pi h_{\mathbf{r}}) \rangle = \langle \sin(\frac{1}{2}\pi(h_{\mathbf{r}'} + h_{\mathbf{r}})) \rangle \tag{2.6}$$

in the DOF and RF phase. To summarize: in the OF phase only ρ is non-zero; in the DOF phase only ψ is non-zero; and in the RF phase both ψ and ρ_s are nonzero. Conjugate to these three order parameters are three types of interfaces which we studied to determine the phase diagram numerically [1,2].

In the rough phase the surface resembles a terraced mountain-like landscape. At large length scales the discreteness of the step heights has become irrelevant, and the surface can be described by the Gaussian model. The height-height correlations diverge logarithmically

$$\langle (h_{\mathbf{r}+\mathbf{r}_0} - h_{\mathbf{r}})^2 \rangle \simeq \frac{1}{\pi K_G} \log(r). \tag{2.7}$$

The amplitude K_G^{-1} is known as the roughness parameter and fully characterizes all correlations in the rough surface. The Kosterlitz-Thouless (KT) theory is based on a stability analysis of the rough phase with respect to the discreteness of the step height. A sine-Gordon (SG) model is constructed which interpolates between the SOS model and the Gaussian model [15]. Critical fluctuations are large length scale phenomena. Therefore it is sufficient to consider the continuum version of this SG model,

$$\mathcal{H} = \int d\mathbf{r} \{ \frac{1}{2} K_G (\nabla \phi_{\mathbf{r}})^2 - u_2 \cos(2\pi \phi_{\mathbf{r}}) - u_4 \cos(4\pi \phi_{\mathbf{r}}) \}. \tag{2.8}$$

This model is equivalent to a 2 dimensional Coulomb gas [16,17]. Each $\exp(i\pi q)$ excitation represents an electric charge. Eq.(2.8) contains $q = \pm 2$ charges, with fugacity $u_2/2$, and $q = \pm 4$ charges with fugacity $u_4/2$. At the KT transition the $q = \pm 2$ charges unbind; the rough phase becomes unstable with respect to the discreteness of the step height.

The variable $\phi_{\mathbf{r}}$ in eq.(2.8) does not represent an individual column height, but merely the local average height. Therefore u_2 is not infinitely large, but finite and temperature dependent. There must be a line in Fig. 1 where $u_2 = 0$, because in the OF phase the average height is an integer, i.e., $u_2 > 0$, but in the DOF and RF phase the average height is an half-integer, i.e., $u_2 < 0$. In Fig. 1 the $u_2 = 0$ line starts at the self dual point K_{SD} [3], and its continuation beyond point L must be the PR line. Along this line $q = \pm 2$ charges are absent and the rough phase remains stable until the $q = \pm 4$ charges unbind.

The Coulomb gas representation is identical to that of the Ashkin-Teller model. Point N in Fig. 1 can be identified with its 4-state Potts point, the PR line with its Baxter line, and the rough phase to the left of point L with its critical fan [18].

The roughness parameter K_G determines all scaling properties in the rough phase and along the PR line. The charge correlation functions decay as

$$G_q(\mathbf{r}_n - \mathbf{r}_0) = \langle \exp[iq(h_{\mathbf{r}} - h_{\mathbf{r}'})] \rangle \simeq |\mathbf{r} - \mathbf{r}'|^{-2x_q} \tag{2.9}$$

with $x_q = q^2/(4\pi K_G)$ [15]. Inside the rough phase $K_G < \frac{1}{2}\pi$. At the roughening transition $K_G = \frac{1}{2}\pi$ (the $q = \pm 2$ charges unbind; $x_{2\pi} = 2$). Along the PR line K_G increases from $\frac{1}{2}\pi$ at point L to 2π at point N. At point N the $q = \pm 4$ charges unbind ($x_{4\pi} = 2$). The specific heat diverges only close to point N. The specific heat exponent,

$\alpha = 2 - 2/y_T$, varies as $-\infty \leq \alpha \leq \frac{2}{3}$ between point L and N ($\cos(2\pi\phi)$ is the energy operator; $y_T = 2 - x_{2\pi} = 2 - \pi/K_G$). The order parameters ρ and ψ vanish at opposite sides of the PR transition. Their critical exponent $\beta = x_\pi/y_T$ is the same and varies as $\frac{1}{12} \leq \beta \leq \infty$ between points N and L.

The analysis presented in this section applies to the most elementary type of DOF phase, the one with the checkerboard type reconstruction as parent state. The generalization to the missing row type DOF phase is straightforward. It is the uniaxial version of the same up-down step order [6]. More general types of reconstructions lead to more complex types of DOF phases, and can be handled by the same type of analysis.

A different type of generalization of the DOF phase arises from the inclusion of the corner interactions introduced by Rys [19]; an energy difference between inside and outside corners of terraces. This interaction represents the lack of valley-hill symmetry in real systems, and originates from multi-particle interactions between particles. This must be important especially in metals.

Originally, Rys [19] suggested that his corner interaction destroys the roughening transition and the rough phase. This turns out not to be the case; the rough phase is stable with respect to these corner interactions and the KT nature of the roughening transition is unchanged [3]. However, the corner interaction has interesting implications for preroughening. The distinction between OF and DOF phases starts blur. The average height of the surface becomes temperature dependent, $\langle h \rangle = n + \theta(T)$, because the corner interaction gives up- and down-terraces a different energy. This leads to the notion of a θ-DOF phase [3]; $\theta = 0$ represents the OF phase, and $\theta = \frac{1}{2}$ the DOF phase of Fig. 1. Theoretically the θ-DOF phase is characterized by the generalized parity order parameter $\rho(\theta) = \langle \cos(\pi(\phi_\mathbf{r} - \theta)) \rangle$. Its lattice formulation varies with θ; for $\theta = \frac{1}{2}$ it is given by ψ in eq.(2.6).

The PR transition is washed-out by these corner interactions [3]. By switching the corner interactions on and off you can circumvent the PR critical point and deform the OF phase continuously into the $\theta = \frac{1}{2}$ DOF phase without a phase transition. As function of temperature, at fixed strength $H \neq 0$ of the corner interaction, the PR transition (at $H = 0$) shows-up as a rapid change in θ from almost zero to almost $\frac{1}{2}$, at $T \simeq T_{PR}$. See ref.3 for more details.

How can we distinguish OF and DOF phases experimentally? The line shapes in diffraction experiments, e.g. atomic beam diffraction [20,21], are determined by the correlation functions G_q. In the rough phase they decay as power laws, see eq.(2.9), with exponents proportional to the roughness parameter K_G. The present experimental resolution is good enough to extract the value of the roughness parameter K_G from the line shape [21].

In a flat phase the G_q correlations decay exponentially to a constant, $|m(q)|^2$; i.e. the line shapes are superpositions of Lorentzians and δ-peaks. At the PR transition the order parameter $\rho = m(\pi)$ vanishes (the average height switches from $\theta = 0$ to $\theta = \frac{1}{2}$). The δ-peak at the anti-Bragg angle, is non-zero in the OF phase, but vanishes in the DOF phase.

Only the $m(q)$ at the anti-Bragg angle vanishes. Consider the generalized θ-DOF phase. At very low temperatures only the heights $h_r = n$ and $n + 1$ are exposed and $m(q)$ is easy to evaluate.

$$|m(q)|^2 = 1 - 2(1 - \theta)\theta(1 - \cos(q)) \tag{2.10}$$

has a minimum at the anti-Bragg angle, $q = \pi$. The location of the minimum does not vary with θ, but the intensity at the minimum vanishes at $\theta = \frac{1}{2}$. Be aware that at the PR transition all $m(q)$ vanish momentarily. All line shapes are power-laws because the correlation length diverges. The $m(q)$ scale as $m(q) \sim |T - T_{PR}|^{\beta(q)}$ with $\beta(q) = x_q/y_T$, see eq.(2.9).

The characteristic aspect of the PR transition is the shift in the average height. In layer-by-layer growth of adsorbed films one observes sharp (first-order transition) steps in the isotherms associated with completion of layers. The location of these steps in the phase diagram should switch abruptly at the PR transition temperature by one-half layer.

The above suggested experiments probe the change in average surface height. It would be better to observe the long-range step up-down order more directly. One obvious choice is scanning tunneling microscopy (STM). This will require freezing the DOF structure by a temperature quench because the DOF phase is stabilized by entropy. STM can handle only static structures.

3. ROUGHENING INDUCED DECONSTRUCTION

In section 2 we found that (100) facets of simple cubic crystals that follow a path of type 4 enter a reconstructed rough phase. They deconstruct after they roughen. In this section I demonstrate that such a reconstructed rough phase does not exist for missing row (MR) reconstructed (110) facets of FCC crystals. Roughening induces a simultaneous deconstruction transition. I discuss the nature of this transition.

The MR structure in (110) facets of FCC crystals, see Fig. 3, is characterized by the four possible locations of the top rows, $n_t = 1, 2, 3, 4$ (mod 4). Define an angle variable $\theta = \frac{1}{2}\pi n_t$. The domain walls shown in Fig. 3 are characterized by $(d\theta, dh)$. $d\theta$ denotes the character of the wall with respect to the MR order, and dh represents the step height. At every $d\theta = \pm\frac{1}{2}\pi$ type interface the step height is odd, while at every $d\theta = \pi$ type interface the step height is even (including $dh = 0$). This a consequence of the topology, independent of whether the microscopic structures shown in Fig. 3 are indeed the energetically most likely realizations of these $(d\theta, dh)$ walls; the column heights are restricted to be even at one and odd at the other checkerboard type sublattice. For the same reason the MR structure is actually only two-fold degenerate. At a specific surface height only $\theta = 0, \pi$ or only $\theta = \pm\frac{1}{2}\pi$ can be realized.

Along a path of type 3 the $\theta = \pi$ walls are energetically most favourable. They play the same role as the Ising walls in the RSOS model of section 2. When they melt the surface loses its Ising type order parameter (the spontaneous symmetry breaking

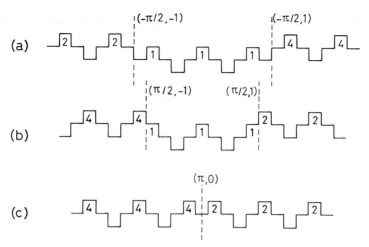

Fig. 3 Typical domain walls in the missing row (MR) reconstructed (110) facets of FCC crystals [23]: (a) the anti-clockwise walls, $(d\theta, dh) = (\frac{1}{2}\pi, \pm 1)$; (b) the clockwise walls, $(-\frac{1}{2}\pi, \pm 1)$; and (c) the Ising walls, $(\pi, 0)$.

between θ and $\theta + \pi$), and enters the DOF phase. The surface remains flat until the roughening temperature where the $\theta = \pm\frac{1}{2}\pi$ walls melt.

Along paths of type 4 the $\theta = \pm\frac{1}{2}\pi$ walls are energetically most favourable. They change both the height of the surface and the reconstruction order. When the $\theta = \pm\frac{1}{2}\pi$ walls melt the surface roughens and looses its reconstruction order simultaneously.

The situation is fundamentally different from that in the (100) facets of simple cubic crystals. In section 2 we represented the reconstruction order by the AF Ising (parity type) order parameter ρ_s, see eq.(2.4), not by the position of the top rows, $\theta = 0, \pi$. This difference amounts to switching between $\theta = 0$ and $\theta = \pi$ each time the height changes by one. The reconstructed phase in the RSOS model has two types of domain walls; in the θ formulation: the Ising type $(d\theta, dh) = (\pi, 0)$ walls, and the $(\pi, \pm 1)$ steps. Along paths of type 4 the $(\pi, \pm 1)$ steps melt first. In the θ formulation they seem to change the reconstruction order, but in the more proper AF Ising spin formulation they do not; in eq.(2.2) the Ising and 6-vertex degrees of freedom decouple. Therefore, along paths of type 4, when the $(\pi, \pm 1)$ walls melt, the surface becomes rough but maintains its reconstruction order. For the (110) FCC facets it is impossible to disentangle the order parameters (except to some extend in the limit of strong chirality); there is no way of redefining the reconstruction order parameter, by switching between θ labels at different heights, such that the $(d\theta, dh) = (\pi, \pm 1)$ do not couple to the reconstruction order. This will be explained more clearly at the end of this section. It means that the reconstructed rough phase does not exist.

The appropriate model to replace the RSOS model is an anisotropic BCSOS model [14] with an increased interaction range. I did not map out its phase diagram numerically, but am quite sure that paths of type 1, 2, and 3 are not affected [22]. Only the behaviour along paths of type 4 changes.

Consider instead the following cell-spin model, to describe the phase diagram locally around multicritical point M. Construct a rectangular lattice oriented along the grooves of the missing rows. Each cell, i.e. the unit cell of this lattice, is large compared to the missing row unit cell, but small compared to the correlation length (which is large, because we are close to point M). Associate with each cell a 4-state clock model variable $\theta_{n,m} = 0, \pm\frac{1}{2}\pi, \pi$, to represent the 4 equivalent MR reconstructed states. Consider the partition function

$$Z = \sum_{\{\theta_{n,m}\}} \exp\{\sum_{n,m} [K_m \cos(\theta_{n,m} - \theta_{n,m+1} - \Delta) + Q_m \cos(2\theta_{n,m} - 2\theta_{n,m+1})$$

$$+ K_n \cos(\theta_{n,m} - \theta_{n+1,m}) + Q_n \cos(2\theta_{n,m} - 2\theta_{n+1,m})]\} Z_{6V}(\{d\theta = \pm\frac{1}{2}\pi\}, L). \quad (3.1)$$

This is a 4-state clock model, with chiral symmetry breaking, and with every configuration weighted by a 6-vertex (6V) model partition function defined on the lattice formed by the $d\theta = \frac{1}{2}\pi$ domain walls. Only the $d\theta = \pm\frac{1}{2}\pi$ walls carry arrows; the $d\theta = \pi$ walls in Fig. 3 do not change the height. Notice the similarity with eq.(2.2). Instead of a 6V model defined on an annealed fluctuating lattice formed by Ising Bloch walls, we have now a 6V model defined on domain walls of the 4-state clock model. I call this the chiral 4-state clock-step model.

The coupling constants, K_m, Q_m, and Δ, represent the domain wall energies. The wall energies couple only to the clock variables because the microscopic structure of the walls is the same whether the step is up or down, $dh = \pm 1$ (the left-right symmetry in Fig. 3). The clockwise walls, $d\theta = \frac{1}{2}\pi$, and anti-clockwise walls, $d\theta = -\frac{1}{2}\pi$, have a different microscopic structure. The corresponding energy difference is represented by the chirality Δ. The coupling constants K_n and Q_n represent kink energies. Finally, domain walls repel or attract each other at short distances. L represents a wall interaction that couples to the step heights. Interactions that do not couple to the step heights give rise to next nearest neighbour and many-body interactions between the θ spins.

This model is quite complex. The chiral 4-state clock model without these novel height degrees of freedom is familiar from the theory of commensurate-incommensurate (C-IC) phase transitions in adsorbed monolayers. Its phase diagram is quite well understood at $\Delta = 0$ and also at large chirality. Questions about the nature of incommensurate melting at intermediate values of Δ still linger [13].

I decided to start with a numerical study of the $\Delta = 0$ case, a study of the finite size scaling (FSS) behaviour of semi-infinite strips at $\Delta = 0$, $L = 0$, and isotropic interactions $K_n = K_m$ and $Q_n = Q_m$. The maximum accessible strip width, $N = 7$, is quite small. It is large enough to establish the structure of phase diagram and the order of the phase transitions, but universal FSS amplitudes are accurate at best up to a few percent, and delicate crossover scaling behaviour is difficult to resolve.

Interface free energies are defined as $\eta(\phi, k) = N(f(\phi, k) - f(0, 0))$, with $f(\phi, k)$ the free energy for the boundary condition: $\theta(n + N, m) = \theta(n, m) - \phi$ and $h(n + N, m) = h(n, m) + k$. The only possible combinations are: $(\phi, k) = (0, 2n)$, $(\pi, 2n)$, and $(\pm \frac{1}{2}\pi, 2n + 1)$. The vanishing of $\eta(\pi, 0)$ indicates the disappearance of the MR reconstruction order parameter. The vanishing of $\eta(0, 2)$ indicates surface roughening.

Fig. 4 shows the phase diagram. To the right of point M the $(d\theta, dh) = (\pi, 0)$ walls are more favourable. Indeed, $\eta(\pi, 0)$ vanishes first. $N\eta(\pi, 0)$ scales to zero everywhere above the deconstruction temperature. The location of the deconstruction

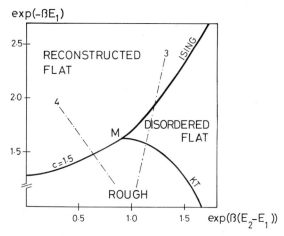

Fig. 4. phase diagram of the 4-state clock-step model at zero chirality. The flat phase represents the MR reconstructed phase. E_2 is the energy of an Ising wall, $(\pi, 0)$; E_1 is the energy of a $(\pm \frac{1}{2}\pi, \pm 1)$ wall, $\beta = 1/(k_B T)$. The $-\cdot-\cdot-$ lines represent paths of type 3 and 4.

line follows from the crossing points where the $N\eta(\pi, 0)$ curves at successive values of N cross. At the transition $\eta(\pi, 0)$ scales with the correct Ising universal FSS amplitude $N\eta(\pi, 0) = 2\pi x_H$ ($x_H = \frac{1}{8}$). At an higher temperature $\eta(\frac{1}{2}\pi, 1)$ and $\eta(0, 2)$ vanish, indicating surface roughening. In the rough phase they scale as powerlaws: $\eta(\frac{1}{2}\pi, 1) = \frac{1}{4} K_G/N$ and $\eta(0, 2) = K_G/N$, with K_G the roughness parameter defined in eq.(2.7). I determined the location of the roughening line from the condition that K_G is universal, $K_G = \frac{1}{2}\pi$, at a KT transition.

Towards the left in the phase diagram, Fig. 4, the $d\theta = \pi$ walls become less favourable than the $d\theta = \pm \frac{1}{2}\pi$ walls. As expected the Ising and KT line meet, and merge. Surprisingly, the simultaneous roughening-deconstruction transition is not first-order. I expected to see this numerically as follows. At the crossing points of $N\eta(\pi,0)$ the value of K_G, as determined from $N\eta(0,2)$, must then be smaller than the universal KT value $K_G = \frac{1}{2}\pi$, because the first-order transition would preempt the KT mechanism. This is not the case. The points $N\eta(0,2) = \pi$ remain at the high temperature side of the crossing points of $N\eta(\pi,0)$, at each value of N, while converging to the same estimate for the critical temperature. Further to the left of point M (Fig. 4) they cross, but never significantly.

Ising and roughening critical behaviour seem to be superimposed. Not only at the multicritical point M, but along the entire roughening driven deconstruction line. I find evidence for simple Ising exponents in quantities that couple to the reconstruction ($N\eta(\pi,0) = \frac{1}{4}\pi$) and pure KT exponents in quantities that couple to the roughening ($N\eta(0,2) = \pi$) along the entire line. The central charge (i.e., the FSS scaling amplitude of the free energy $f(0,0)$) behaves consistent with superimposed Ising and roughening behaviour. Inside the rough phase c takes the correct value $c = 1$; and close to the transition line it swings-up to the value $c = \frac{3}{2}$. In case of a first-order transition the numerical value of c should decrease significantly while moving away from point M; it does not. Only mixed quantities that couple simultaneously to both the reconstruction and roughening degrees of freedom behave unusually. For example, $\eta(\frac{1}{2}\pi,1)$, the free energy of a $(\frac{1}{2}\pi,1)$ type interface, scales as $N\eta(\frac{1}{2}\pi,1) = 1.005 \pm 0.001 \simeq \pi/3$ (the numerical accuracy is high because at $Q = 0$ the corrections to scaling in this quantity appear to be virtually absent).

I tried a powerlaw fit, $\eta \simeq |T - T_c|^{y_T}$, to the temperature dependence of all the interface free energies η and find consistently $y_T = 1.0 \pm 0.1$, i.e. an Ising singularity. So the specific heat diverges with an Ising singularity, $\alpha = 0$. Recall that in a conventional KT transition the correlation length and specific heat have an essential singularity. The convoluted Ising-KT character of our transition implies probably a more complicated scaling form than a simple powerlaw, but the present numerical data do not allow a more detailed analysis, and the pure powerlaw gives an adequate fit.

We need analytical insight into the values of the scaling dimensions, especially the critical exponents of the operators that couple the Ising and roughening degrees of freedom simultaneously. The central charge value $c = \frac{3}{2}$ is associated typically with conformal field theories with super symmetry [24]. In its simplest form super symmetry is a continuous fermion-boson exchange invariance in a field theory with decoupled massless free fermions and bosons [24]. This is related to our problem because at criticality the Ising model is equivalent to a massless free fermion theory and the Gaussian model, describing the rough phase, is a boson theory.

The deconstruction transition (the Ising critical line) inside the rough phase in Fig. 1, is a realization of such a $c = \frac{3}{2}$ super symmetric conformal field theory; the Ising and roughening order parameters effectively decouple. The situation in Fig. 4 is remarkable close to this. Apart from the fact that mixed operators have modified scaling dimensions the only difference seems to be that in Fig. 4 the Ising line is required to coincide with the KT transition line.

The following transformation elucidates this point. Rewrite each θ variable in terms of two coupled Ising spins, $(S, T) = (+,+), (+,-), (-,-), (-,+)$. Eq.(3.1) then takes the form

$$Z = \sum_{\{S_{\mathbf{r}}, T_{\mathbf{r}}\}} \exp\{ \sum_{\langle \mathbf{r}, \mathbf{r}' \rangle} [\frac{1}{2} K(T_{\mathbf{r}} T_{\mathbf{r}'} + S_{\mathbf{r}} S_{\mathbf{r}'}) + Q S_{\mathbf{r}} S_{\mathbf{r}'} T_{\mathbf{r}} T_{\mathbf{r}'}] \} Z_{6V}(\{S_{\mathbf{r}}, T_{\mathbf{r}}\}, L) \quad (3.2)$$

with $\mathbf{r}$ and $\mathbf{r}'$ nearest neighbour sites, and $\Delta = 0$. In the MR reconstructed phase the magnetizations $\langle S \rangle = \langle T \rangle$ and the polarization $\langle ST \rangle$ are non-zero. In the DOF phase only the polarization $\langle ST \rangle$ remains non-zero. In this formulation the 6V model is defined on the joint lattice formed by the Bloch walls of both Ising models, but bonds where their Bloch walls overlap are excluded. When the magnetizations $\langle S \rangle = \langle T \rangle$ vanish the surface deconstructs. It roughens simultaneously, because the Bloch walls of the S and T spins form immediately an infinite large backbone lattice for the 6-vertex arrows, and we know that at small values of L the 6V model is rough (compare also with section 2).

Redefine the spins such that the 6V model arrows live on the Bloch walls of one Ising model only. This can be achieved in two ways: define $\sigma = ST$ and keep S or keep T; I call this S-T invariance. In the (σ, T) representation eq.(3.2) takes the form

$$Z = \sum_{\{\sigma_{\mathbf{r}}, T_{\mathbf{r}}\}} \exp\{ \sum_{\langle \mathbf{r}, \mathbf{r}' \rangle} [\frac{1}{2} K T_{\mathbf{r}} T_{\mathbf{r}'} + (\frac{1}{2} K T_{\mathbf{r}} T_{\mathbf{r}'} + Q) \sigma_{\mathbf{r}} \sigma_{\mathbf{r}'}] \} Z_{6V}(\{\sigma_{\mathbf{r}}\}, L). \quad (3.3)$$

This can be rewritten as a RSOS model coupled to an Ising model, using eqs.(2.1)-(2.2),

$$Z = \sum_{\{h_{\mathbf{r}}, T_{\mathbf{r}}\}} \exp\{ \sum_{\langle \mathbf{r}, \mathbf{r}' \rangle} [\frac{1}{2} K_t T_{\mathbf{r}} T_{\mathbf{r}'} + (\frac{1}{2} K_s T_{\mathbf{r}} T_{\mathbf{r}'} + Q)(1 - 2(h_{\mathbf{r}} - h_{\mathbf{r}'})^2)] \} \quad (3.4)$$

with $K_s = K_t = K$, and for simplicity $L = 0$. Finally by performing a duality transformation on the RSOS degrees of freedom eq.(3.4) maps into an XY-model coupled to an Ising model, which is essentially the same model studied in recent years in the context of the fully frustrated XY model [24]. Monte Carlo results for these models are consistent with my numerical results; in particular it confirms the second-order nature of the transition.

The S-T invariance, associated with the alternate (σ, S) representation, implies that $K_s = K_t$ and is a fundamental property of the walls in Fig. 3. Assume for the moment that K_s and K_t can be different. At $K_s = 0$ and $K_t \neq 0$ the Ising and roughening degrees of freedom decouple. There the model is a simple realization of supersymmetry (at the Ising critical points inside the rough phase). This remains true at small values of K_t, because at $K_t = 0$ the K_t interaction has an irrelevant scaling index, $y_{K_t} = 3$. The S-T transformation implies the same type of behaviour in the opposite $K_s = 0$ limit. Our line $K_s = K_t$ is the separatrix between these two regions.

Consider the model of eq. (3.4.) at $K_t = 0$, or the RSOS model of section 2 along the Ising line inside the rough phase in Fig. 1. Supersymmetry means that an action is invariant under a continuous transformation that mixes fermion and boson degrees of freedom. It requires that the fermions and bosons are both massless. In both cases this invariance is not exact. It is valid only asymptotically in the scaling limit, at large length scales, when the Ising spins are critical and the RSOS model is in the rough phase. Moreover, supersymmetry does not impose a connection between the Ising critical temperature and the roughness parameter K_G, and therefore does not exclude the existence of a reconstructed rough phase. If the roughening and Ising degrees of freedom decouple we can treat in the rough phase the column heights h_r as continuous variables and change K_G freely by a rescaling of the h_r.

Our S-T invariance has the character of a fermion-boson exchange too. On the one hand it is weaker; only a discrete Z_2 type invariance. On the other hand it is stronger; an exact invariance of the model which holds also away from criticality. The S-T symmetry imposes the link between the roughening and deconstruction temperature. From eq.(3.2) it is clear that the two transitions coincide. The roughening induces a simultaneous deconstruction transition.

4. DISCUSSION

This section has been added after the meeting. I like to present the progress made during our stay in Erici. My talk was in the second week of the meeting, which gave me the opportunity to include this material in my talk.

Experimental results presented at this meeting suggest the presence of disordered flat (DOF) phases and preroughening in two different types of solids: the (111) facets of noble gas solids, and the (110) facet of Ni.

Youn and Hess [25] report reentrant first-order layering behaviour in argon films adsorbed on graphite (see Hess' contribution to these proceedings). They use ellipsometry and single crystal graphite. They observe layer-by-layer growth steps in the isotherms until $T \simeq 69K$ where the steps in the isotherms momentarily disappear. They reappear for $69K < T < 80K$, but the location of the steps has shifted by $\frac{1}{2}$ a unit in pressure, suggesting strongly that the layer-by layer growth takes place in this temperature interval between half-filled layers, i.e., $3\frac{1}{2}$, $4\frac{1}{2}$, $5\frac{1}{2}$, $6\frac{1}{2}$, etc. Larese and Zhang [26] have confirmed this reentrant behaviour, using volumetric isotherms for Ar on graphite foam. Moreover, Hess reported that krypton on graphite exhibit the same behaviour.

Preroughening provides an explanation for this reentrant behaviour. In the PR interpretation the (111) facet of Ar undergoes a preroughening transition at $T \simeq 69K$, and a roughening transition not until $T \simeq 80K$. The specific heat peaks observed by Zhu and Dash [27] in Ar on graphite at about $70K$ are attributed then to preroughening instead of roughening. Fig. 5 shows a schematic chemical potential versus temperature phase diagram [28].

A preroughening transition is the only available explanation for the observed reentrant behaviour, but more explicit experimental evidence is needed to distinguish it from other potential explanations. Preroughening is a property of the facet and must be independent of the substrate. Do noble gases adsorbed on MgO show the same reentrant behaviour? The shift in the average height of the top layer is an essential feature of preroughening. So a more precise and independent confirmation of the $\frac{1}{2}$ coverage of the top layer is important. How general is the effect? Do He and H_2 adsorbed on graphite show the same type of reentrant behaviour?

It might be feasible to measure the critical exponents of the PR transition. The thermal exponent, y_T, can be obtained from the specific heat singularity in the thick-film limit (using the crossover scaling behaviour between the asymptotic Ising type layer-by-layer growth exponent and the infinitely thick-film PR value); or, from the universal asymptotic shape of the line of Ising critical points into the PR transition point, see Fig. 5, $T - T_{PR} \sim (\mu - \mu_0)^{y_T}$.

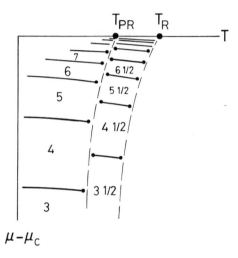

Fig. 5. Schematic chemical potential versus temperature phase diagram for layer-by-layer growth of a system with a preroughening transition. In the DOF phase, between T_{PR} and T_R, the top layer is half-filled. The endpoints of the first-order layering transition lines are Ising critical points [29].

Another issue is the mobility of particles inside the top-layer. In the DOF phase the top layer is half-filled, and behaves fluid-like. Within the above theoretical description of the DOF phase this must be a lattice gas type fluid (particles retain the solid-on-solid restriction), and the dynamics must be characterized by fluctuating steps instead of independent particles (the step density is still low and the correlation length is large). Close to the PR transition these conditions must be satisfied. At higher temperatures the top layer might start to behave more like a 2D continuum particle-fluid instead as a lattice domain wall fluid. This could have interesting repercussions for the roughening transition and the interplay with surface melting. Does the nature of the roughening transition change if the top layer behaves like a non-lattice gas type fluid? Neutron studies of diffusion in the top layer of thick films adsorbed on MgO or graphite, the type of experiments presented at this meeting by Bienfait [30], might give a clue.

The suggestion that the (111) facet of noble-gas solids preroughens is actually quite surprising to me. It suggests that for (111) facets of Lennard-Jones type solids we can identify a reconstructed state which is energetically quite close to the ordered flat surface configuration; see the discussion in section 1. This yet unknown reconstructed state plays the role of parent state for the DOF phase. In this reconstructed state the average height must be shifted by an amount consistent with the observed $\frac{1}{2}$ shift. This is an interesting theoretical microscopic question.

Ed Conrad presented in his poster at this meeting (see also Fig. 4 in Cao and Conrad [10]) experimental evidence that Ni(110) preroughens (see also the discussion in section 1). Earlier I suggested [3] that Pd(110) and Ag(110) are good candidates to observe preroughening in metal surfaces, assuming that Ni(110) and Cu(110) still follow a path of type 1. This result makes a study of Pd(110) and Ag(110) even more interesting.

Cao and Conrad observe in Ni(110) a rapid decrease of the (Gaussian) Bragg-peak contribution of the diffraction peak at the anti-Bragg angle, precisely as suggested above in section 2 [3]. I like to note that the Gaussian (Bragg) contribution does not need to vanish completely at PR. Many-body interactions are likely to be important in these metals, and give rise to a breaking of the valley-hill symmetry. They generate the corner interaction of Rys [3]. As mentioned above (section 2 see also [3]) this washes-out the PR transition. It leaves however a rapid change in the average height $< h >= n + \theta$, and correspondingly a rapid drop in in $m(\pi)$, see eq.(2.10), from a value of θ almost equal to zero to almost equal to $\frac{1}{2}$. The data of Cao and Conrad suggest that PR takes place at about $T = 1300K$ and roughening not until $T > 1400K$. Further experiments are needed to confirm this behaviour in Ni(110) and also in Cu(110).

During the meeting we made progress in understanding the deconstruction and roughening of Pt(110) and Au(110) as well. Jacques Villain describes in his contribution the theory he developed with Vilfan. I like to explain how their work fits into my chiral 4-state clock step model of section 3. Villain and Vilfan (VV) assume that only one type of step is realized, the type shown in Fig. 3b. The walls shown in Fig. 3c are not included, but $(\pi, 0)$ walls reappear in their theory as bound states of two Fig. 3b type steps. On the length scale of the cell spin model, eq.(3.1), this microscopic structure distinction does not figure anymore. The binding energy U in the model of VV is the same as my parameter Q. VV do not include anti-clockwise $d\theta = -\frac{1}{2}\pi$ walls, the ones shown in Fig. 3a. Therefore their theory describes the limit of strong chirality. Finally, they neglect dislocations, i.e., that four clockwise walls can annihilate each other. What is left is the Hubbard model, which has as virtue that it is exactly soluble. The phase transition described by VV is a Pokrovsky-Talapov (PT) type transition from the reconstructed flat phase into a phase that is rough, but incommensurate (IC) floating solid type with respect to the reconstruction order.

It is interesting to speculate about the structure of the full phase diagram of the chiral 4-state clock-step model. See Fig. 6. Some reasonable assumptions lead to good agreement with the experimental data. The structure at zero chirality is known from my analysis in section 3. At non-zero chirality, Δ, the phase diagram is probably complicated, but let's speculate that for $Q < 0$, it looks like the phase diagram of the chiral 3-state clock model (the one studied extensively in the context

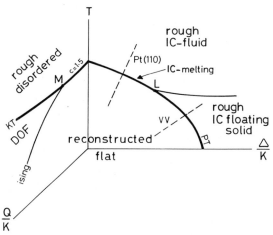

Fig. 6. Proposed partial phase diagram for the chiral 4-state clock-step model, combining Fig. 4 and the phase diagram of the chiral 3-state clock model. The model of Villain and Vilfan correspond to the limit of large chirality, large Δ. I propose that Pt(110) and Au(110) follow a path through an IC melting transition point at intermediate values of chirality Δ. The surface is rough everywhere above the heavy drawn critical line.

of C-IC phase transitions in adsorbed monolayers [13]). This seems reasonable in the light of the simple scaling behaviour of the simultaneous deconstruction/roughening transition line at $\Delta = 0$ (section 3). The PT transition at large chirality is the limit discussed by·VV; there the surface becomes rough and floating IC with respect to the reconstruction degrees of freedom.

At intermediate values of Δ I expect an IC melting transition, similar as in the chiral 3-state clock model, and I propose that the deconstruction transition in Pt(110) and Au(110) is an experimental realization of this. The surface simultaneously roughens and undergoes an IC melting transition with respect to the reconstruction degrees of freedom. Like the chiral 3-state clock model this transition most likely has the same critical exponents as the transition at zero chirality (the Ising transition convoluted with KT roughening as discussed in section 3). The only difference is that the rough disordered phase above T_c has the character of an IC fluid in terms of the reconstruction degrees of freedom. The fluid contains more clock-wise than anti-clockwise walls (Fig. 3b and Fig. 3a); it is a modulated fluid. The modulation wavelength causes a shift, δq, of the diffraction peaks. This shift disappears at T_c. All this agrees nicely with the observations of Robinson et al. [5] for Pt(110). They observe Ising type of exponents and such a shift in the diffraction peaks.

Theoretically we characterize the scaling behaviour the misfit δq by a critical exponent x_q; $\delta q \sim \xi^{-x_q}$, with ξ the correlation length [13]. For reasons still not completely understood, we find numerically and experimentally in the monolayer case (the chiral 3-state clock model) the value $x_q = 1$ suggested by Huse and Fisher (see discussion in [13]). Let's assume that this remains true in our model. Assume that the critical exponents are indeed the same as at $\Delta = 0$, i.e., that $y_T = 1$ (Ising like). This implies a linear vanishing of the misfit $\delta q \sim |T - T_c|^{x_q/y_T} \sim |T - T_c|$. A linear vanishing of the misfit is indeed what Robinson et al observed (Fig. 4 in ref 5).

The assumptions in this argument are compelling, and their implications agree remarkably well with experiment, but they need confirmation. A detailed numerical study of the complete phase diagram of the chiral 4-state clock-step model is needed. Additional experimental evidence is needed too, in particular for Au(110), which probably follows a path of type 4 as well instead of the earlier suggested path of type 3.

ACKNOWLEDGEMENTS

During the meeting I had numerous stimulating discussions with many of the participants. I like to thank in particular Jacques Villain, Ed Conrad, George Hess, Klaus Kern, and John Larese. I like to thank the organizing committee, in particular Haskell Taub, for organizing the meeting, and Sam Fain for critically reading the manuscript. This research is supported by the National Science Foundation Grant No. DMR 88-13083.

REFERENCES

1. K. Rommelse and M. den Nijs, Phys. Rev. Lett. **59**, 2578 (1987).

2. M. den Nijs and K. Rommelse, Phys. Rev. **B 40**, 4709 (1989).

3. M. den Nijs, Phys. Rev. Lett.. **64**, 435 (1990).

4. J. C. Campuzano *et al.*, Surf. Sci. **162**, 484 (1985); H. Derks *et al.*, Surf. Sci. **188**, L685 (1987); E. C. Sowa *et al.*, Surf. Sci. **199**, 174 (1988).

5. I. K. Robinson, E. Vlieg, and K. Kern, Phys. Rev. Lett.**63**, 2578 (1989).

6. Modify the RSOS model as follows: Choose K directional dependent, $K_n \neq K_m$. Let only K_n change sign. Choose the next nearest neighbour interaction L between sites (n, m) and $(n + 2, m)$. The transfer matrix leads to the same spin-1 quantum chain Hamiltonian as the RSOS model, eq.(2.1), in the so-called "time continuum limit" and therefore to the same phase diagram; see also ref.[2].

7. S. M. Foiles, Surf. Sci. **191**, L779 (1987).

8. B. M. Ocko and S. G. J. Monchrie, Phys. Rev. B **38**, 7378 (1988).

9. P. Zeppenfeld, K. Kern, R. David, and G. Comsa, Phys. Rev. Lett. **62**, 63 (1989).

10. Y. Cao and E. Conrad, Phys. Rev. Lett. **64**, 447 (1990).

11. E. Holub-Krappe *et al.*, Surf. Sci. **188**, 335 (1987).

12. C. J. Barnes *et al.*, Surf. Sci. **201**, 108 (1988).

13. For a review see M. den Nijs, in *Phase Transitions and Critical Phenomena*, edited by C. Domb and J. Lebowitz (Academic, London, 1987), Vol. 12.

14. H. van Beijeren, Phys. Rev. Lett. **38**, 993 (1977).

15. For a review see J.D. Weeks, in *Ordering in Strongly Fluctuating Condensed Matter Systems*, T. Riste, ed, p.293. Plenum Press, New. York (1980).

16. M. den Nijs, Phys. Rev. **B 27**, (1983) 1674.

17. For a review see B. Nienhuis, in *Phase Transitions and Critical Phenomena*, edited by C. Domb and J. Lebowitz (Academic, London, 1987), Vol. 11.

18. M. Kohmoto *et al.*, Phys. Rev. **B 24**, 5229 (1981).

19. F. S. Rys, Phys. Rev. Lett. **56**, 624 (1986).

20. J. Villain, D.R. Grempel and J. Lapujoulade, J. Phys. F **15**, 809 (1985).

21. M.P.M. den Nijs, E.K. Riedel, E.H. Conrad and T. Engel, Phys. Rev. Lett. **55**, 1689 (1985) and **57**, 1279 (1986).

22. The BCSOS model is equivalent to the spin-$\frac{1}{2}$ quantum spin chain. As explained in section VI of ref.[2] this model has a DOF phase with the same type of properties as the RSOS model. The PR transition belongs to the same universality class.

23. J. Villain and I. Vilfan, Surf. Sci. **199**, L165 (1988).

24. see e.g. O. Foda, Nucl. Phys. **B 300**, 611 (1988).

25. H.S. Youn and G.B. Hess, Phys. Rev. Lett. **64**, 918 (1990).

26. J.Z. Larese and Q.M. Zhang private communication.

27. Da-ming Zhu and J.G. Dash, Phys. Rev. Lett. **57**, 2959 (1986).

28. It is possible, but less likely that the intermediate phase is reconstructed instead of DOF, i.e., to enter the reconstructed phase via a first-order transition (the phase boundary to the right of point N in Fig. 1) and from there to follow a path of type 3.

29. Fig. 5 is the simplest theoretical phase diagram. The Ising critical points could just as well be triple points with first-order boundaries between them; and even more elaborate phase diagram diagrams with e.g. critical endpoints are theoretically allowed too. However, the second-order nature of the PR transition and also the experimentally observed complete wash-out of the steps in the isotherms make Fig. 5 more likely.

30. P. Zepenfeld *et al.* J. Physique **51** (1990)

THERMAL DYNAMICS OF NONVICINAL METAL SURFACES

Klaus Kern

Institut für Grenzflächenforschung und Vakuumphysik
Forschungszentrum Jülich, Postfach 1913, D-5170 Jülich, FRG

1. INTRODUCTION

The atoms in the surface of a crystal are missing part of their
nearest neighbors which gives rise to a charge redistribution in the
selvedge. This changed force field is responsible for noticeable interlayer
relaxations in the near surface region. Intuitively the inward relaxation
of the outermost surface layer can be explained by the tendency of the
valence electrons to spill over the surface in order to create a lateral
smoothing of the electronic charge density[1]. The new electron distribution
causes electrostatic forces on the ion cores of the surface atoms,
resulting in a contraction of the first interlayer spacing ($d_{12} < d_b$). This
relaxation is most pronounced for open, loosely packed, surfaces. In
addition, the changes in the force field can also favor lateral atomic
rearrangements in the surface plane. The surface "reconstructs" into a
phase with new symmetry. These reconstructive surface phase transition can
either occur spontaneously or be activated by temperature or by small
amounts of adsorbates [2].

So far we have neglected the temperature of the system. As the
temperature rises, however, the lattice vibrational amplitude increases and
the anharmonic terms in the interaction potential gain importance. Due to
the reduced number of nearest neighbors in the surface (a maximum of 9 at
the surface of a fcc-crystal with respect to the 12 nearest neighbors in
the bulk of this crystal) the mean-square amplitude of the surface atoms is
much larger than in the bulk. While in bulk Cu, for example, anharmonicity
is negligible below 70-80% of the melting temperature, anharmonicity on the
Cu(110) surface becomes important at temperatures above 40% of the melting

temperature (the bulk Debye temperature of copper is Θ_B = 343 K, and the melting temperature is T_M = 1356 K).

At the high temperature end it has been demonstrated recently that on a variety of surfaces a disordered quasi-liquid layer wets the surface well below the bulk melting temperature, i.e. the melting of a crystal starts from the surface layer [3]. In view of the Lindemann criterion of melting [4], which states that melting occurs when the mean-square displacement of the atoms surpasses a critical value (~ 10% of the interatomic equilibrium distance), the important role that surfaces play in the melting phase transition is not surprising. As already discussed the mean vibrational amplitude is substantially enhanced at the surface and the Lindemann-criterion predicts a surface instability around 0.75 T_M.

The picture developed above is based on a perfect defect free surface, which is, however, only at zero temperature the stable equilibrium state. At elevated temperatures a certain amount of defects like isolated adatoms and vacancies as well as clusters of those can be thermally excited. Both adatom islands as well as vacancy holes are bordered by steps. Frenkel [5] studied the structure of such steps and argued that they should contain a large number of kinks at finite temperatures. Thus due to thermal fluctuations every crystal surface with steps, should have a certain microscopic roughness at nonzero temperature, the surface remains flat on macroscopic length scales however. Burton and coworkers, demonstrated that the thermal excitation of adatom and vacancy islands and thus the excitation of steps is negligible at low and medium temperatures but gave evidence for the presence of a roughening transition, at a temperature close to the bulk melting temperature, where the surface becomes macroscopically rough[6,7]. The critical temperature of this transition has been termed the roughening temperature, T_R. Burton et al. suggested that at the roughening temperature the free energy associated with the creation of a step vanishes. This was confirmed later by Swendsen in a detailed calculation [8]. One of the fundamental consequences of the existence of a roughening temperature for a certain crystallographic face below the melting temperature is that this face can occur on an equilibrium crystal only at temperatures below T_R.

Let us consider a surface which at T = 0 K is perfectly flat. Upon in-creasing the temperature, thermal fluctuations give rise to vacancies, ad-atoms atoms and steps in the surface layer. The number of these "defects" increases until, at the roughening temperature, the long-range order of the surface disappeares. Long-range order is confined here to the "height-correlation function" and not to the positional correlation function (parallel to the surface plane). Indeed, even above the roughening temperature, the surface atoms populate in average regular lattice sites. It is the fluctuation of the height h(r) which diverges for temperatures T > T_R [9].

$$<[(h(\vec{r}') + \vec{r}) - h(\vec{r})]^2> \infty C(T)\ln(\vec{r}) \tag{1}$$

where C is a temperature-dependent constant and $\vec{r}$ a two-dimensional vector in a plane parallel to the surface. This divergence is very weak. At the roughening temperature $C(T_R) = 2/\pi^2$; the height fluctuation is one lattice spacing for a distance of 139 lattice spacings.

The first direct experimental evidence for a roughening transition was reported in 1979. Several groups have studied the thermal behavior of the basal plane of a hexagonal close-packed ^{4}He crystal [10]. In a beautiful experiment Balibar and Casting obtained for this surface a roughening temperature of $T_R \approx 1.2$ K.

More recently, the question of thermal roughening has also been addressed in the study of metal surfaces. Detailed He diffraction studies from the high Miller index (11n) with n=3,5,7 surfaces of Cu [11] and Ni [12] proved the existence of a roughening transition on these surface.

The microscopic mechanism which leads to the roughening of a low and a high Miller index surface is expected to be different. Indeed, as already mentioned, a low indexed surface – which at T = 0 K is perfectly flat – fulfills the roughening condition, Eq. (1), when the free energy for the creation of a step becomes zero. In contrast, on a high indexed surface – which at T = 0 K is already stepped – Eq. (1) can be fulfilled also without the creation of new steps. It appears that the proliferation of kinks is sufficient to roughen the vicinal surface. Indeed, the ensuing meandering of the step rows, in conjunction with the mutual repulsion between these rows, leads also to the divergence of the "height-correlation function". Thus, the roughening temperature of high indexed surfaces is substantially lower than that of the low indexed ones. In the case of Cu and Ni(11n) the values for T_R range from 300 K to 720 K (Cu) and 450-750 K (Ni) for n = 3,5,7.

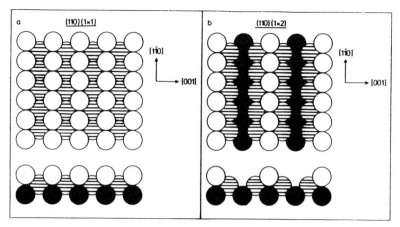

Fig. 1.   Structure of the unreconstructed and reconstructed (110) surface
          of face centered cubic metals.

While the existence of a roughening transition on stepped vicinal surfaces is undisputed, the basic question whether T_R of low indexed surfaces
is lower than the crystal melting temperature T_M or not is still matter of
controversial discussion. The magnitude of T_R is governed by the bond
strength at the surface. Thus, close packed surfaces with more neighbors
and stronger bonds have a higher roughening temperature than more open
surfaces.

Particularly attractive for roughening studies of low index metal surfaces are (110) surfaces of fcc-crystals. Firstly, the (110) surface has
the most open structure of the three densest fcc-faces, (111), (100) and
(110); resembling to some extent the topography of the (113) surface. The
second aspect is surface reconstruction. The (110) surfaces of transition
metals with face centered cubic (fcc) symmetry belong to two different
classes. The first class, including the 3d-elements Cu, Ni and the 4d-
elements Rh, Pd and Ag, have a nonreconstructed (1x1) ground state for the
clean surface, i.e. they keep the bulk termination (they exhibit however
large oscillatory interlayer-relaxations). The second class of fcc metals,
including the 5d-elements Ir, Pt and Au, exhibits a reconstructed (1x2)
ground state. The nature of the (1x2) reconstruction has been studied
extensively by a number of different experimental techniques and there is a
general agreement now that the (1x2) phase of all three 5d-metals is a
missing row geometry [13] with every second close packed [1$\bar{1}$0] row missing
(see fig. 1).

It was suggested that reconstruction and roughening in these systems are indeed related /14/. As pointed out by Garofalo et al. the energies of the relaxed unreconstructed (1x1) surface and the energies for all possible missing-row states (1x2, 1x3, ...,1x4), are all energetically close to one another [15]. Locally the (1xn) reconstructions represent microscopic (111) facets and are expected to be easily excitable at elevated temperatures. Trayanov et al. [14] speculate, that whatever the low-temperature ground state configuration (unreconstructed or reconstructed) it might roughen into a high temperature disordered phase, with a mixture of (1xn) configurations.

2. DETECTION OF STEPS

Very attractive techniques to detect monatomic steps on surfaces are direct imaging methods such as scanning tunneling microscopy (STM). In a typical constant-current topography scan a surface step is easily identified as a vertical movement of the tip by one lattice spacing. The step structure (kinks) and orientation are easily determined. The main drawback of STM in the determination of step densities and distributions is its limited scan range and the narrow temperature range for stable operation (in particular high temperature measurements have not been reported so far).

The classic technqiue for the analysis of defect structures on surfaces is certainly diffraction (LEED, X-ray, He-diffraction). In a first order approximation the diffracted intensity from a perfect single crystal surface drops exponentially with temperature, according to the Debye-Waller formula

$$I \sim \exp(-2W) = \exp[-\langle (Q \cdot u)^2 \rangle] \qquad (2)$$

Thermal excitation of adatoms, vacancies, roughening or lateral disordering but also anharmonicity all give rise to anamalous thermal behavior associated with a more rapid decrease of the diffracted intensity. Thus a simple intensity analysis does not allow to distinguish between these channels.

Step proliferation (i.e. roughening) however can be unambigously detected in a diffraction experiment by the width or line-shape analysis of the diffraction peak under well defined kinematical conditions. To illustrate that we consider a plane wave which is scattered from two adjacent terraces separated by a monatomic step. Beams diffracted from the upper and the lower terraces will interfere. If the interference is constructive or in-phase (i.e. the phase between the two waves is an even multiple of π), the diffraction peak is identical to that of the ideal step free surface. For the case of destructive interference or anti-phase diffraction (i.e. the phase between the two waves is an odd multiple of π), however, all diffraction peaks will be broadened with a maximum halfwidth proportional to the average step density. Thus, in an anti-phase kinematics a diffraction experiment is very sensitive to the presence of steps. At the roughening temperature T_R the anti-phase Bragg-peak is broadened by $\sim$ 60%.

The diffracted intensity which is removed from the peak center is distributed into its wings, i.e. the line shape of the peak changes due to the presence of steps. At the roughening temperature the line shape of the diffraction peak is described by a power law [16]:

$$I(Q_\parallel) \propto Q_\parallel^{-(2-\tau)} \qquad (3)$$

$Q_\parallel$ being the parallel momentum transfer and τ the so called roughness exponent

$$\tau = \frac{\pi}{2} K(T) f(p) \qquad (4)$$

$K(T)$ is the so called roughness parameter. Everywhere in the low temperature phase the effective $K(T)$ is equal to zero (no powerlaw line shape); at the roughening transition it jumps to the universal value $K(T_R)$ = $2/\pi$ (see e.g. ref. 16), and increases with temperature continuously in the rough phase. The function $f(p)$ describes the scattering kinematics and varies frm $f(p)=0$ for in-phase diffraction to $f(p)=1$ in anti-phase geometry. Theoretically $f(p)$ varies with the model and also depends on finite size effects, from a (periodically repeated) square to a simple cosine function of the phase p. Thus, at the roughening temperature in a perfect anti-phase diffraction experiment we have $\tau_R=1$.

In some favorable situations monatomic steps can also give rise to shifted diffraction peaks, and the step density can be extracted directly from the peak shift [17]. Such a case is discussed in chapter 5.

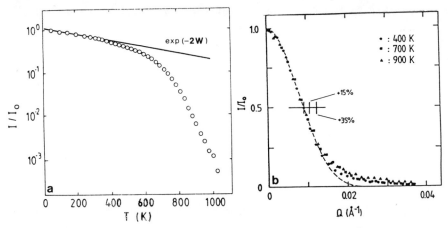

Fig. 2. a) Thermal dependence of the He specular peak height from Cu(110); E_{He} = 18.3 meV, $\vartheta_i = \vartheta_f = 45°$ [21]. b) Specular He-peak profiles normalized at their maximum (same data as fig. 2 in ref. 21). The verticals bars (+ 15% and + 35%) indicate model based theoretically predicted broadenings at the roughening temperature (square and cosine variation of the phase function f(p) in eq. (4)). The dashed line indicates the gaussian resolution function of the instrument.

3. ANHARMONICITY OR ADATOMS

More than ten years ago it had been noticed that the intensities in the photoemission spectra taken from Cu(110) decrease dramatically with temperature above ~ 500 K [18]. Similar effects have been seen recently in low-energy ion scattering [19], in X-ray diffraction [20] and in thermal He scattering [21,22]. The dramatic intensity decrease observed in all cases above 450-500 K could not be accounted for by simple Debye-Waller effects. While Lapujoulade et al. [22] and Fauster et al. [19] proposed as explanation either anharmonic effects or some kind of disorder, Mochrie [20] concluded categorically – without qualitative additional evidence – that he was observing the roughening transition. He even tentatively identified the temperature at which "the intensity has fallen essentially to zero" (870 K) with T_R. A He specular intensity measurement on Cu(110) versus temperature performed in our laboratory shows (Fig. 2a) that also above 870 K the intensity continues to drop (at 1000 K it is one order of magnitude lower) and that there is no sign of saturation even above 1000 K. Whether the intensity becomes "essentially zero" appears to depend on the dynamical range of the instrument, and is not a criterion for the choice of value of T_R. Zeppenfeld et al. [21] have analyzed in detail the energy and angular distribution of the scattered He atoms in the whole temperature range up to

1000 K. In fig. 2b the specular He-peak profiles taken at three different temperatures are shown. By inspection of this figure it is evident that the strong intensity decay can not be attributed to surface roughening; the specular He-diffraction peak measured in near antiphase-scattering geometry does not show any broadening up to 900 K, clearly demonstrating the abscence of step-proliferation. This is also consistent with the behavior of the diffuse elastically scattered He-intensity which has been found to drop continuously between 100 and 900 K, while at T_R a substantial increase of the diffuse scattering is expected [23].

Lapujoulade and Salanon [24] suggested that the anomalous thermal behavior of the specular He-intensity might be exclusively attributed to the thermal excitation of adatoms and vacancies. In order to explain the He-data quantitatively, however, astronomically high equilibrium defect concentrations would be necessary. With the assumption of a random distribution of adatoms and vacancies the attenuation of the specular He-beam I/I_0 is given by $I/I_0 = (1-\theta)^{n_s \cdot \Sigma}$, here n_s is the number of lattice sites, Σ the cross section for diffuse scattering and θ the defect concentration. Taking a diffuse He-scattering cross section of 70 $\mathring{A}^2$ for a Cu-adatom or vacancy, 3.6% defects would be present already at 500 K increasing to 12% at 700 K and eventually to more than 30% at 900 K. These defect concentrations are without any doubt unreasonable. Using low energy ion scattering Fauster et al. [19] have determined upper limits for the defect concentration of Cu(110); at 500 K the adatom and vacancy density is found to be below 1% and at 800 K it is still below 5%. These values agree with LEED measurements of Cao and Conrad[25] for the homologeous Ni(110) surface, demonstrating that adatom and vacancy creation is negligible below 60% of the melting temperature (0.60 T_M = 813 K for Cu). At temperatures of 0.7 T_M they estimate the defects concentration to range between 5 and 15%. This picture is further supported by recent molecular dynamics calculations for the (110) surfaces of Al, Ni and Cu [26]. These simulations show that the thermal excitation of adatoms and vacancies becomes significant only above ~ 0.7-0.8 T_M.

It is generally accepted now that the sharp decrease in coherently scattered intensity above 0.35 T_M is ascribed to an anomalous large increase of the mean-square displacement of the surface atoms $\langle u^2 \rangle$ due to a large anharmonicity in the metal potential at the surface [21,25,27]. This is demonstrated in fig. 3, where the mean square displacement $\langle u^2 \rangle$ at the Cu(110) surface (perpendicular to the surface), as deduced from vastly

276

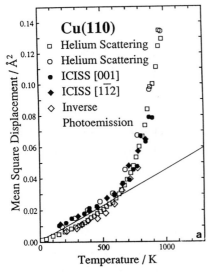

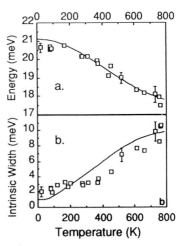

Fig. 3. a) Mean square displacements of surface atoms versus temperature
deduced from ion scattering, He-scattering and inverse photoemission [27]. b) Energy and intrinsic width of the Cu(110) MS_7
resonance phonon at the $\bar{\Gamma}$-point as a function of temperature[51].

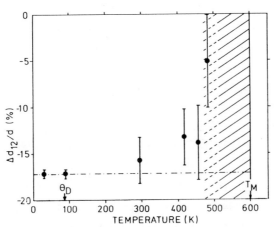

Fig. 4. Surface relaxation of Pb(110) versus temperature [28].

different experimental techniques, is plotted versus temperature. All three techniques agree nicely. A very recent experimental study of the temperature dependent surface phonons has been used to quantify the excess surface anharmonicity of Cu(110)[51]. Above 400 K the MS_7 resonance phonon shows a significant decrease of the phonon frequency accompanied by a substantial broadening of its intrinsic width. A simple anharmonic model reproduces the observed temperature effects and a direct comparison with bulk phonons reveal a surface anharmonicity enhanced about 5 times over that of the bulk.

An enhanced surface anharmonicity on the open (110) surface of fcc metal crystals has been deduced also from theoretical as well as experimental studies of the thermal surface expansion coefficient [28,29]. Nonreconstructed fcc(110) surfaces are strongly relaxed and the interlayer distance between the first and second plane of atoms d_{12} is contracted between 5 and 15% with respect to the bulk value d_b. This relaxation was found to vanish rapidly above ~ 0.4 T_M (i.e. $d_{12}/d_b \to 1$) and can only be ascribed to a dramatic increase of the thermal surface expansion coefficient driven by a strong surface anharmonicity. The corresponding experimental graph for Pb(110) [28] is given in fig. 4.

4. ROUGHENING OF THE UNRECONSTRUCTED fcc(110) SURFACES

The observation of a surface roughening transition on the (110) faces of Ni and Pb was reported recently by Cao and Conrad [25] and by Yang et al.[34], respectively. Using high resolution LEED these authors showed the onset of step proliferation at ~ 0.75 T_M. The roughening transition is preceeded by two stages, a large increase of the mean square displacements of the surface atoms due to excess surface anharmonicity starting at ~ 0.45 T_M followed by adatom and vacancy creation above ~ 0.7 T_M.

While in the case of Pb(110) the logarithmic divergence of the height-height correlation function was demonstrated, the roughening of Ni(110) has been deduced from the broadening of the anti-phase Bragg peak.

A detailed analysis of the Ni(110) LEED angular diffraction profiles by Cao and Conrad reveal a Gaussian central peak superimposed on a broad Lorentzian, which is interpreted in a two-level model. In this model the ratio of the Gaussian to the Lorentzian intensity is proportional to the adatom

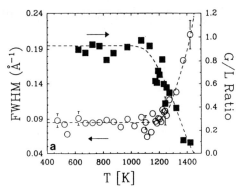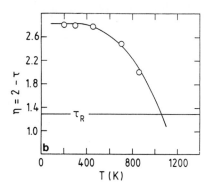

Fig. 5. a) The ratio of the Gaussian to Lorentzian intensity (■) and the
Lorentzian FWHM (o) of the specular anti-phase LEED diffraction
peak from Ni(110) vs temperature [25]. b) The roughening exponent τ
for Cu(110) obtained from log-log plots of the tail part of the
purely elastic specular He-beam.

concentration, the width of the Lorentzian component characterizes the
average step density. In Fig. 5a the temperature dependence of these
quantities as measured by Cao and Conrad is shown. The onset of adatom
creation around 1150 K and the proliferation of steps around 1300 K (= 0.75
T_M) are evident. This behavior of the diffraction line-shape may however
also be consistent with a preroughening-transition of the type discussed by
den Nijs[14,49]. In the preroughening scenario steps are created
spontaneously at the transition, but each step up is followed by a step
down and vice versa and the surface remains flat on a macroscopic length
scale.

From a detailed analysis of the elastic He-diffraction profiles we can
estimate that Cu(110) may indeed roughen at similar relativ temperatures
(with respect to the bulk melting temperature). In fig. 5b we plot the
roughness exponent τ deduced by Zeppenfeld et al. [21] from log-log plots of
the elastic contribution to the polar profiles of the specular peak. At the
roughening temperature τ_R = 1 in ideal anti-phase scattering geometry. Cor-
recting for the actual scattering conditions in the experiment of
Zeppenfeld et al. gives a roughening exponent τ_R = 0.71 (assuming a cosine-
like behavior of the phase factor f(p), which is indeed observed in all
thermally roughened surfaces [11,12]). Extrapolating of the data in fig. 5b
to this value we can estimate the roughening temperature of Cu(110) to be
$T_R \simeq$ 1070 K = 0.79 T_M.

That this value might indeed be the correct roughening temperature of Cu(110) is supported by diffusion measurements of Bonzel and coworkers. These authors studied the surface self-diffusion of various fcc(110) surfaces by monitoring the decay of a periodic surface profile. The profile with periodicities of a few μm is prepared by a photoresist masking techniques and subsequent Argon RF-sputtering and the analysis of the profile decay at elevated temperatures is done by laser diffraction [30-32]. The results for Ni(110) and Cu(110) are plotted in fig. 6. Below the roughening temperature the macroscopic diffusion is expected to proceed by single adatom diffusion while above T_R the macroscopic mass transport should be dominated by meandering steps. Surface diffusion of adatoms on the (110) surface of fcc-crystals is expected to be anisotropic because of the twofold symmetry of the surface. This expectation is confirmed by the data in fig. 6. At low temperatures the activation energy for diffusion along the close packed channels, i.e. along the [110] direction, is found to be only 40% of the barrier for across channel diffusion, i.e. diffusion along the [100] azimuth. For Ni as well as Cu(110), however, this anisotropy vanishes at ~ 0.78 T_M. Above this temperature the mass transport is isotropic, consistent with two-dimensional step diffusion. We thus conclude that the diffusion data in fig. 6 support a roughening transition of Ni and Cu(110) around 78% of the melting temperature.

Rather unusual evidence for the roughening of Cu(110) around 80% of the melting temperature comes from an optical microscopy study of Freyer [33]. In order to cross-check his diffusion measurements he studied the morphology of sinusoidaly grooved Cu(110) surfaces by means of an interference microscope. The samples were first annealed at high temperatures followed by a rapid quench to room temperature. He always observed a strong morphology change at annealing temperatures between 1050 K and 1100 K; but unfortunately missed the significance of the data. Fig. 7 shows two of his interference micrographs after annealing to 1030 K and 1150 K, respectively, clearly demonstrating the roughening above 1100 K (note that the roughness also appears on the right part of fig. 7b, where the Cu(110) surface is not grooved).

Evidence for the roughening of the (110) surface has also been presented recently for the metals In, Ag and Pd. While the roughening of the In(110) [35] surface is generally accepted the experimental results for Ag(110) [36] and in particular Pd(110) [37] are disputed. For palladium Francis

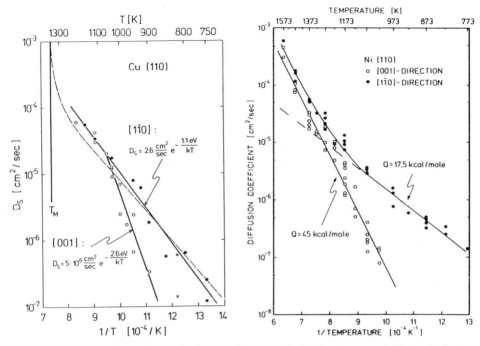

Fig. 6. Arrhenius plot of the surface self-diffusion coefficient for Ni(110) and Cu(110) [30-32]

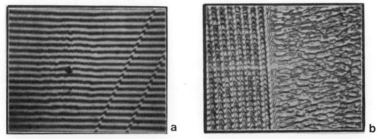

Fig. 7. Optical interference-microscope picture of a grooved Cu(110) surface (average distance between grooves ~ 6μm) upon heating to a) T=1030 K and b) T=1150 K, followed by a rapid quench to room temperature [33].

and Richardson [37] reported an order-disorder transition to occur around 250 K. This transition was, however, not be detected in a series of subsequent experiments [38], and today is believed to be an artifact due to the presence of impurities in the experiments of Francis and Richardson. Ag(110) is an interesting case. This surface was studied by Held et al. with synchrotron x-ray diffraction [36]. Based on a diffraction peak shape analysis they deduced the relatively low roughening temperature of 0.58 T_M. This low transition temperature has been connected with the small stabilization energy with respect to the missing row reconstruction [14] but can also be due to a preroughening transition of the type discussed by den Nijs [14]. In addition Robinson [39] suggested recently, that the data analysis might be influenced by the interference with the theraml diffuse scattering of a bulk diffraction peak. The roughening temperatures of nonreconstructed fcc(110) surfaces are summarized in table I.

Table 1. The roughening temperatures of fcc(110)(1x1) metal surfaces

Surface	T_R [K]	T_R/T_M	References
Ag(110)	720	0.58	36
In(110)	290	0.69	35
Pb(110)	420	0.70	34
Ni(110)	1300	0.76	25
Cu(110)	1070	0.79	21,32,33

5. DECONSTRUCTION OF THE RECONSTRUCTED fcc(110)(1x2) SURFACES

In the case of Au(110), the missing row (1x2) phase has been found to be stable only in a limited temperature range. Upon heating, the half order superlattice LEED-spot was seen to change shape with temperature and eventually disappeared at a critical temperature $T_c \simeq 0.49\ T_M \simeq 650$ K, indicating a continuous phase transition from an ordered (1x2) state into a disordered (1x1) phase [40]. Campuzano et al. have analyzed this phase transition in terms of a two dimensional order-disorder transition and determined critical exponents consistent with the predictions of the 2D-Ising model, which, due to the p2mm symmetry of the Au(110)(1x2) surface, is indeed the appropriate universality class [41].

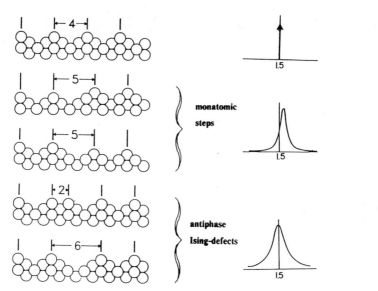

Fig. 8. Thermal excitations of the fcc(110)(1x2) missing row reconstructed
surface; Ising-like antiphase defects and monatomic steps, i.e.
(111) micro facets. Also shown schematically is the profile of the
half-order diffraction peak with and without defects.

A considerable amount of disorder, however, has also been observed to
be present in the low temperature missing row phase of all three metals Ir,
Pt and Au [13,42]. While the coherence along the [110]direction (parallel to
the close packed rows) extends over several hundred Å,the coherence length
along the [001] direction (perpendicular to the rows) hardly surpassed 100-
200Å. Scanning tunneling microscopy [42] assigned this intrinsic disorder to
the presence of some (1x3) and (1x4) reconstructed regions, which are induced by a micro (111) facetting. In theoretical studies it has been shown
that the (1x2) missing row configuration is indeed only marginally stable
with respect to the "higher" missing row states (1x3, 1x4, ..., 1xn). The
energy difference between any of the (1xn) phases of Au(110) has been
calculated to be less than 10meV per atom [15]. Based on this ground it has
been argued by several authors that the missing row configuration should be
thermally unstable with respect to the formation of (111) microfacets,
giving rise to a "rough" surface at elevated temperatures. While Villain
and Vilfan [43] have predicted a succesion of two transitions, an Ising-like
order-disorder transition at ~ 0.50 T_M (spontaneous proliferation of
antiphase Ising-defects, fig. 8) followed by roughening transition at ~
0.57 T_M (onset of (111) micro facetting generating single height steps,
fig. 8), Levi and Touzani [44] have found no evidence for an Ising-like
transition but predicted a direct roughening transition.

In a recent x-ray diffraction experiment Robinson, Vlieg and Kern have studied the thermal behavior of the reconstructed Pt(110) surface [45,46]. The experimentally observed half order diffraction peaks have two characteristics: they are broad in the [001] direction but sharp in the orthogonal [1$\bar{1}$0] direction and always displaced slightly from the exact half order position along [001]. The uniaxial broadening and shift implies disorder in one direction only, i.e. must be associated with line defects oriented perpendicular to the [001] direction. An identical pattern of uni-axially shifted and broadened half order diffraction peaks was observed earlier by Robinson et al. [17] for the Au(110) surface and can be explained

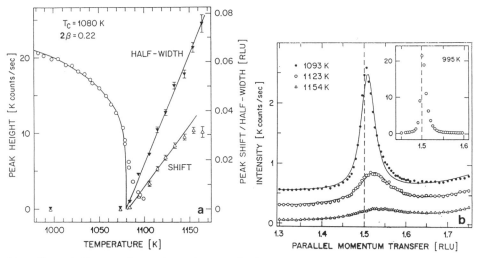

Fig. 9. a) Temperature dependence of the halforder diffraction peak (h, 0.06,0.06) obtained by scanning h [45,46]. b) Temperature dependence of the extracted peak height, width and shift [45,46].

conclusively in terms of randomly distributed single height steps on the surface. It was further demonstrated that the peak shift of the half order spots is exclusively related to the density of these monatomic steps [17,45], while Ising-like defects would only result in a symmetric peak broadening. Indeed, (111) micro facets are also the predominant defects seen in scanning tunneling microscopy images of Au(110) and Pt(110) [42,47].

The temperature dependence of the half-order diffraction profile was measured and found to be have reversibly. The data are summarized in fig. 9, cleary demonstrating a phase transition at a critical temperature of T_C = 0.53 T_M = 1080 ± 1 K. The peak height, I(T), is fully compatible with a theoretical curve $I(T) = I_0|t|^{2\beta'}$ where $t=T/T_c-1$ and $\beta' = 0.11 \pm 0.01$. Above T_c the half-width diverges linearly with the reduced temperature. Both of these aspects are exactly in accord with the predictions of the 2D-Ising model, which has $\beta' = 1/8$ and $\nu = 1$ (correlation length exponent), and agree well with the LEED data for the analogous phase transition of Au(110)(1x2).

Notwithstanding this apparent agreement between Pt(110), Au(110) and the 2D Ising model, we now turn to the behavior of the diffraction peak shift in fig. 9b. Above T_c the peak shifts substantially and completely reversibly. This result is in contrast to the Ising classification because it implies that an equilibrium density of steps appears spontaneously above T_c. This immediately implicates some roughening character. The slopes of the half-width and peak shift versus T in fig. 9b allow us to quantifiy the line defect density in units of probability per lattice site for the monatomic steps $\alpha = 6.6t$ and antiphase Ising defects $\beta = 2.8\,t$; i.e. thermally induced steps are 2-3 times more common.

Two solutions have been proposed to escape from this paradox. Villain and Vilfan [48] suggest that the steps formed above the transition are bound together in pairs. The imposition of paired steps leads necessarily to a phase transition model in the Ising-universality class due to the twofold degenerated ground state. This model forbids any height divergence and the surface is never rough. Villain and Vilfan suggest a step pair unbinding transition at higher temperatures $T_{c'}>T_c$ which eventually roughens the surface. den Nijs [49], however, suggests a transition with real roughening character but Ising-criticality. In the framework of a 4-states chiral clock step model, den Nijs demonstrated that for negligible chirality the reconstructed (1x2)(110) surface deconstructs and roughens in one single transition which is characterized by Ising exponents. This transition has the character of an incommensurate melting transition with respect to the reconstruction degrees of freedom, explaining the peak shift and the linear vanishing of it at T_c. Zero chirality, however, requires that step defects with a phase shift of 3 half-cells (see fig. 8) have also to be present on the surface, but are rarely observed [42] and are expected to be energetically unfavorable [50].

What is needed now is another experiment to test these predictions. It should be possible to measure the temperature dependence of the integer order diffraction peak profiles which are sensitive to roughness. While in the Villain and Vilfan model no change should be seen; the roughening scenario of den Nijs would result in a significant peak broadening.

REFERENCES

1. M.W. Finnis and V.J. Heine; J. Phys. F4, L37 (1974)
2. J.E. Englesfield; Prog. Surf. Sci. 20, 105 (1985)
3. J.F. van der Veen, B. Pluis, and A.W. Denier van der Gon; in "Physics and Chemistry at Solid Surfaces VII", (Springer, Berlin, 1988), p. 455
4. F.A. Lindemann; Z. Phys. 14, 609 (1910)
5. J. Frenkel; J. Phys. USSR 9, 392 (1945)
6. W.K. Burton and N. Cabrera; Disc. Faraday. Soc. 5, 33 (1949)
7. W.K. Burton, N. Cabrera and F.C. Frank; Philos. Trans. Roy. Soc. London 243 A, 299 (1951)
8. R.W. Swendson; Phys. Rev. B17, 3710 (1978)
9. H. van Beijeren and I. Nolden; in Structure and Dynamics of Surfaces II, (Springer, Berlin, 1986), p. 259; and references therein
10. S. Balibar and B. Castaing; Surf. Sci. Rep. 5, 87 (1985)
11. F. Fabre, D. Gorse, B. Salanon, and J. Lapujoulade; J. Physique 48, 1017 (1987)
12. E.H. Conrad, L.R. Allen, D.L. Blanchard, T. Engel; Surf. Sci. 187, 265 (1987)
13. P. Fery, W. Moritz, D. Wolf; Phys. Rev. B38, 7275 (1988)
14. A. Trayanov, A.C. Levi, and E. Tosatti; Europhys. Lett. 8, 657 (1989); M. den Nijs; Phys. Rev. Lett. 64, 435 (1990)
15. M. Garofalo, E. Tosatti, and F. Ercolessi; Surf. Sci. 188, 321 (1987)
16. J.D. Weeks, in Ordering in Strongly Fluctuating Condensed Matter Systems, Ed. T. Riste, (Plenum, New York, 1980) p. 293
    J. Villain, D.R. Grempel, and J. Lapujoulade; J. Phys. F 15, 809 (1985)
17. I.K. Robinson, Y. Kuk and L.C. Feldman; Phys. Rev. B29, 4762 (1984)
18. R.S. Williams, P.S. Wehner, J. Stöhr and D.A. Shirley; Phys. Rev. Lett. 39, 302 (1977)
19. Th. Fauster, R. Schneider, H. Dürr, G. Engelmann, and E. Taglauer, Surf. Sci. 189/190, 610 (1987)
20. S.G.J. Mochrie; Phys. Rev. Lett. 62, 63 (1987)
21. P. Zeppenfeld, K. Kern, R. David, and G. Comsa; Phys. Rev. Lett. 62,63 (1989)
22. J. Lapujoulade, J. Perreau and A. Kara; Surf. Sci. 129, 59 (1983)
23. E.H. Conrad, L.R. Allen, D.L. Blachard, and T. Engel; Surf. Sci. 198,207 (1988)
24. J. Lapujoulade and B. Salanon, this proceeding
25. Y. Cao and E.H. Conrad; Phys. Rev. Lett. 64, 447 (1990)
26. E.T. Chen, R.N. Barnett and U. Landman; Phys. Rev. B41, 439 (1990); P. Stoltze, J. Norskov, and U. Landmann; Surf. Sci. 220, L693 (1989).
27. H. Dürr, R. Schneider and Th. Fauster; Vacuum 41, 376 (1990)
28. J.W.M. Frenken, F. Huussen and J.F. van der Veen; Phys. Rev. Lett. 58, 401 (1987)
29. C.S. Jayanthi, E. Tosatti, and L. Pietronero; Phys. Rev. B 31, 3456 (1985)
30. H.P. Bonzel and E. Latta; Surf. Sci. 76, 275 (1978)
31. H.P. Bonzel, in "Surface Mobilities on Solid Materials", Ed. Vu Thien Binh, (Plenum, New York, 1988), p. 195
32. H.P. Bonzel, N. Freyer, and E. Preuss; Phys. Rev. Lett. 57, 1024 (1986)
33. N. Freyer; Ph.D. thesis, Technische Hochschule Aachen (1985)

34. H.N. Yang, T.M. Lu, and G.C. Wang; Phys. Rev. Lett. 63, 1621 (1989) see also A. Pavlovska and E. Bauer; Appl. Phys. A 51, 172 (1990)
35. J.C. Heyraud and J.J. Metois; J. Cryst. Growth 82, 269 (1987)
36. G.A. Held, J.L. Jordan-Sweet, P.M. Horn, A. Mak, and R.J. Birgenau; Phys. Rev. Lett. 59, 2075 (1987)
37. S.M. Francis and N.V. Richardson; Phys. Rev. B 33, 662 (1986)
38. A.M. Lahee, J.P. Toennies, and Ch. Wöll; Surf. Sci. 191, 529 (1987); K.H. Rieder, private communication.
39. I.K. Robinson; private communication
40. J.C. Campuzano, M.S. Foster, G. Jennings, R.F. Willis and W. Unertl; Phys. Rev. Lett. 54, 2684 (1985)
41. P. Bak; Solid State Commun. 32, 581 (1979)
42. G. Binning, H. Rohrer, Ch. Gerber and E. Weibel; Surf. Sci. 131, L379 (1983)
43. J. Villain and I. Vilfan; Surf. Sci. 199, 165 (1988)
44. A.C. Levi and M. Touzani; Surf. Sci. 218, 223 (1989)
45. I.K. Robinson, E. Vlieg and K. Kern; Phys. Rev. Lett. 63, 2578 (1989)
46. K. Kern, I.K. Robinson, and E. Vlieg; Vacuum 41, 318 (1990)
47. T. Gritsch, D. Coulman, R.J. Behm and G. Ertl; Phys. Rev. Lett. 63, 1068 (1989)
48. J. Villain and I. Vilfan; Phys. Rev. Lett. 65, 1830 (1990)
49. M. den Nijs; this proceeding;
50. L.D. Roelofs, S.M. Foiles, M.S. Daw, and M. Baskes; Surf. Sci. 234, 63 (1990)
51. A.P. Baddorf and E.W. Plummer; J. Electr. Spectr. Related Phenom. 54,541 (1990)

SURFACE-INDUCED MELTING OF SOLIDS

J.F. van der Veen

FOM-Institute for Atomic and Molecular Physics
Amsterdam, The Netherlands

The surface plays a key role in initiating the melting of a solid. At temperatures below the bulk melting point the surface may undergo a continuous and reversible order-disorder transition. It therefore provides the solid with a natural starting point for melting. Recent observations of surface melting generally confirm the predictions of thermodynamics but a number of issues have still remained unsolved.

1. INTRODUCTION

The melting of a solid is one of the most commonly observed phase transitions in our civilization and one would intuitively think that the melting phenomenon is well-understood. Indeed much is known about melting, but our present-day knowledge is largely based on empirical facts. Bulk melting is a first-order phase transition in that the first derivatives of the Gibbs free energy with respect to temperature and pressure are discontinuous across the transition. Upon melting, the system absorbs latent heat while the temperature remains constant. There are abrupt changes in, $e.g.$, entropy, density, heat capacity, electrical and thermal conductivity, diffusivity and elastic properties. Although all these parameters are well-documented for most solids [1,2], a microscopic picture of the mechanism by which a solid melts is still lacking.

Various atomic-scale models of the melting process have been developed since the end of the 19th century. These models viewed the melting as a massive thermally induced instability of the crystal lattice. Sutherland [3] and later Born [4] considered the vanishing of the shear resistance at the melting point T_m to be the cause of melting. However, subsequent experimental investigations have refuted this hypothesis [5]. In fact, the temperature at which such a lattice instability is expected to occur generally lies well above T_m. In other models the lattice instability was attributed to the presence of strong lattice vibrations [6] or to a spontaneous generation, at T_m, of a high concentration of crystal defects such as vacancies [7], interstitials [8], or dislocations [9]. Experiments, however, showed neither a softening of bulk phonons nor an anomalously large production of bulk imperfections in crystals close to T_m [9-11].

To be published in :"*Phase transitions in surface films*",NATO ASI series. Ed. H. Traub (Plenum, New York)

The lattice instability theories might be termed *homogeneous* theories of melting in that they require the breakdown of the lattice to occur uniformly throughout the extent of the crystal at its melting point. However, there is ample evidence that melting is a *heterogeneous* process; it involves the nucleation of liquid at preferred sites in the solid, followed by the growth of the nuclei. The lattice instability models will only predict heterogeneous nucleation of the melt if the crystal defects are nonuniformly distributed so that an accumulation of vacancies, interstitials or dislocations at a specific site in the crystal can generate a liquid nucleus. The above models describe only *how* a solid may melt but essentially fall short of elucidating at which specific sites the liquid nucleates in the solid.

Here we argue that the *surface* is the most likely site from where the solid starts to melt. We discuss a number of recent observations which unambiguously demonstrate the presence of a thin disordered film on the crystal surface already at temperatures below the bulk melting transition.

A premature disordering of the surface, if it indeed commonly occurs, offers an elegant explanation of an intriguing characteristic of melting. On the basis of classical nucleation theory one expects that, upon melting and freezing, hysteresis effects occur such as undercooling of the liquid or superheating of the solid [11]. However, under normal conditions, superheating of solids above T_m is not observed, whereas undercooling of liquids is. This is indicative of a general absence of an energetic barrier for the nucleation of melting, while such a barrier does exist for solidification. Apparently, the disordered surface film paves the way for further melting and precludes superheating.

Tammann [12] was the first to point out that surfaces may play an important role in initiating the melting of a solid. He considered melting as a dissolution of the crystal into its own melt and viewed melting as an inward movement of a melt front that starts at the surface. That melting may commence at the surface can be inferred from an empirical criterion formulated by Lindemann for bulk melting. This criterion states that melting occurs if the root-mean-square thermal vibration amplitude of an atom in a crystal exceeds a certain critical fraction (typically 1/8th) of the interatomic distance [6]. On the basis of this criterion one may argue that the outermost atomic layer of the crystal should disorder far below the bulk melting point. The loosely bonded surface atoms have a reduced number of neighbors and hence, a higher thermal vibration amplitude than in the bulk. Consequently, at the surface the Lindemann criterion is satisfied much earlier.

The first speculations that the melting of a solid starts at the surface have led to many searches for the effect. However, much of the earlier work is inconclusive in that the disordered surface film was not observed directly. The surface melt usually is a few monolayers thick and it is only through the use of advanced surface-science techniques that one is able to detect such a layer. Furthermore, an intrinsic surface melting effect can only be demonstrated if a trivial lowering of the melting point at the surface is prevented. Such a lowering can be caused by adsorption or segregation of impurities. Therefore, the experiments have to be performed in a clean environment and on crystals of high purity. Unfortunately, in many of the earlier studies these requirements were not met.

Recent observations of surface melting on a microscopic scale [13-15] have led to a surge of renewed interest among condensed-matter physicists. Our current understanding of the phenomenon is briefly summarized as follows. Surface melting involves the formation of a thin disordered layer in the surface region of the solid at temperatures below T_m (fig. 1). The thickness $\bar{l}$ of the layer increases with temperature and either diverges as T_m is approached or attains a finite value. The disordered film is *not* a true liquid; a surface film of finite thickness feels the presence of crystalline order in the solid underneath and hence there will always remain some crystalline order in the film. Therefore, a surface melt of finite thickness is usually referred to as a 'quasiliquid'.

Surface melting as implied here has nothing to do with the trivial melting caused by temperature gradients. Therefore, it is not to be confused with familiar phenomena such as the melting of an ice cube floating in a glass of water or the melting of a pat of butter on a stove. In these cases the surface melts simply because the outside is hotter than the inside. Another point of confusion concerns the melting temperature of the surface itself. One is inclined to equate the melting point of the surface to the temperature at which the surface begins to disorder. However, the *true* melting point of the surface, if it is defined as the temperature at which the surface becomes fully liquid, is no different from the melting point of the bulk. For it is only at T_m that the quasiliquid skin may become sufficiently thick that the properties of the quasiliquid approach those of the liquid. It is only at T_m that the liquid and solid phases can be at coexistence.

Thermodynamically, surface melting can be regarded as a case of wetting, namely a wetting of the solid by its own melt. As in the well-known case of adsorption of a gas against a hard wall (see contributions by Chan and Dietrich to this volume), one may have *complete* wetting or *incomplete* wetting, depending on wether the quasiliquid layer thickness diverges or remains finite as $T \to T_m$. By the same analogy, the case of *non*wetting corresponds to a complete absence of surface melting; the surface then remains 'dry' up to T_m. All three cases have been reported in the recent literature on surface melting.

The paper is organized as follows. Section 2 treats the thermodynamics of surface melting. Section 3 reviews recent observations and section 4 briefly identifies some of the questions that remain to be answered.

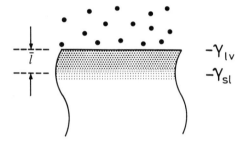

Fig. 1. Wetting of a solid surface by its own melt. In the limit
of macroscopic melt thickness $\bar{l}$, close to T_m, the
interfacial free energies equal those for the solid-liquid
interface (γ_{sl}) and the liquid-vapor interface (γ_{lv}).

2. BASIC THERMODYNAMIC PRINCIPLES

We assume that an increase of crystal temperature in our studies of surface melting corresponds to moving along the solid-vapor coexistence curve in the (P, T) phase diagram (fig. 2). Bulk melting then occurs at the triple point. For the system to move along this trajectory we require the crystal to be in equilibrium with its own vapor. Experimentally this is achieved by putting the crystal in a closed box kept at the same temperature, so that as many atoms condense on the surface as there are leaving. Enclosure in a box is particularly important for materials that have a high equilibrium vapor pressure at T_m (e.g., Ar, Cu, CH_4). This of course makes it difficult to observe the thermal behavior of their surfaces. It is more convenient to investigate materials such as Pb, Al, In, Bi, and Ga, which have such a low vapor pressure at T_m (in the 10^{-7} Pa range) that the evaporation rate is only a few monolayers per hour. Then one can just as well remove the box and place the sample in an ultrahigh-vacuum (UHV) setup. It is true that the evaporating atoms will not return to the sample (they condense on the walls of the vacuum chamber or are pumped away) but per unit time they are so few in number that the melting under such conditions is expected to proceed in essence identically to melting in equilibrium with the vapor. The use of a UHV environment enables us to study atomically clean surfaces free from impurities. An additional advantage is that for the detection of the melt layer surface-science techniques may be used such as medium-energy ion scattering (MEIS) [16], low-energy electron diffraction (LEED) [17], photoelectron diffraction [18] and He atom scattering [19].

What makes the surface to become wet or to remain dry? Suppose a single crystal is cleaved along its {hkl} plane and let us define the specific free energy $\gamma_{sv}^{\{hkl\}}$, which is the work needed to create unit area of the dry surface (the subscript 'sv' refers to the solid-vapor interface). The free energy of a surface of orientation (hkl) that at T_m is covered with a macroscopically thick melt layer, is given by $\gamma_{sl}^{\{hkl\}} + \gamma_{lv}$, where the indices 'sl' and 'lv' refer to the solid-liquid and liquid-

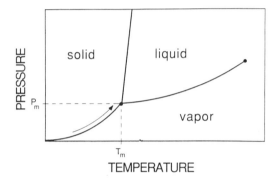

Fig. 2. Schematic (P,T) bulk phase diagram. T_m and P_m are the temperature and pressure at which the solid, liquid and vapor phases are at coexistence. The path along the solid-vapor coexistence curve, indicated by the arrow, corresponds to the trajectory followed in the surface melting experiments.

vapor interfaces, respectively (fig. 1). Surface melting will only occur if there is a gain in free energy, that is, if

$$\gamma_{sv}^{\{hkl\}} - \gamma_{sl}^{\{hkl\}} - \gamma_{lv} \equiv \Delta\gamma^{\{hkl\}} > 0 . \tag{1}$$

Here, $\Delta\gamma^{\{hkl\}}$ is the free energy that the dry surface has in excess of a surface completely wetted by a macroscopically thick melt. On the other hand, if $\Delta\gamma^{\{hkl\}} < 0$, then the surface will remain dry up to T_m.

We note that it is only in the limit $T \to T_m$ that the melt layer attains macroscopic thickness. For temperatures $T < T_m$, at which the surface is wetted by a *quasi*liquid layer of finite microscopic thickness, the free energies $\gamma_{sv}^{\{hkl\}}$, $\gamma_{sl}^{\{hkl\}}$ and γ_{lv} can *not* be equilibrium quantities since the interfaces to which they refer are unstable at these temperatures. This makes it hard to determine for a given crystal face whether or not it will exhibit surface melting. Nonetheless, semiempirical estimates of the interfacial free energies can be made for many materials [16].

292

The sign and magnitude of $\Delta\gamma^{\{hkl\}}$ depend not only on the material but also on the surface orientation. In general, the most open crystal faces (e.g., the {110} face of an FCC crystal) are most likely to exhibit surface melting. In section 3 some examples will be given.

For a system exhibiting complete wetting there is a unique relation between the temperature T and the equilibrium thickness of the quasiliquid layer $\bar{l}$. The fact that there *is* a definite, finite, equilibrium thickness at temperature T is the result of a balance between two opposing thermodynamic forces. On the one hand does the quasiliquid become more liquidlike for increasing layer thickness, which results in a gain in free energy. This corresponds to an effectively *repulsive* force between the interfaces at either side of the melt layer. On the other hand there is a price to pay, namely the energy associated with undercooling the layer. This yields an *attractive* force between the interfaces. For a layer of thickness l the undercooling energy per unit area amounts to $L(1-T/T_m)l$, where L is the latent heat of melting per unit volume. Using Landau mean-field theory one finds for the total free energy $F(l)$ of the surface covered with a melt layer of thickness l [20,21]:

$$F(l) = \gamma_{sl}+\gamma_{lv} + \Delta\gamma e^{-2l/\xi_b}+ L(1-T/T_m)l . \tag{2}$$

The third term describes the repulsive interaction between the interfaces, which here is assumed to be of short range. The parameter ξ_b is a characteristic length scale over which the crystalline order, as measured from the crystal-quasiliquid interface, decays within the layer. The equilibrium thickness $\bar{l}$ is the value of l for which $F(l)$ is minimal (i.e., $dF(l)/dl = 0$):

$$\bar{l}(T) = \frac{\xi_b}{2} \ln\left[\frac{2\Delta\gamma T_m}{L(T_m-T)\xi_b}\right] . \tag{3}$$

We conclude that for a system governed by short-range, exponentially decaying, interactions the melt thickness increases logarithmically with temperature, provided the condition for complete wetting (eq. (1)) is met. But as T approaches T_m very closely, the long-range forces, which are *always* present, must eventually dominate the melting behavior. The short-range forces will have damped out and one is left with van der Waals type dispersion forces. These decay with the interface separation as W/l^2, where W is the Hamaker constant [20,22]. We then obtain the following asymptotic expression for the free energy :

$$F(l) = \gamma_{sl}+\gamma_{lv} + W/l^2 + L(1-T/T_m)l , \tag{4}$$

where, for simplicity, we have neglected the short-range forces altogether. Minimizing $F(l)$ with respect to l yields

$$\bar{l}(T) = \left[\frac{(T_m-T)L}{2T_mW}\right]^{-1/3} . \tag{5}$$

Here it is implied that W is positive, which is the case if the liquid is less dense than the solid. We see that, asymptotically, the layer thickness increases with a power law. A change from logarithmic to power-law growth has been seen on Pb{110}, see below.

3. OBSERVATIONS OF SURFACE MELTING

An obvious way to detect the surface melt is by inspection with the naked eye. A liquid has a dielectric constant different from that of a solid. One therefore expects melting of the surface to be accompanied by a measurable change in optical constants, provided the melt layer is thick enough. For Cu [23] and Pb [24] a change in emissivity has indeed been observed close to T_m, but it is difficult to relate that to structural changes at the surface and to determine the thickness over which the optical constants change. On surfaces of ice ellipsometric measurements have been performed which show evidence for the formation of a watery surface layer as T approaches 0 ^{0}C [25, 26].

The first direct evidence for a disordering of the surface due to melting has come from a MEIS study of the Pb{110} surface in 1985 [13]. Since then, a variety of crystal surfaces has been investigated using a wide range of experimental techniques. Here we do not present an exhaustive review but limit ourselves to discussing some of the recent results obtained on Al, Pb, and Ge crystals using MEIS [15, 24, 27-29], LEED[17, 30], X-ray reflectivity [31] and electron microscopy [32-36]. Calorimetric measurements of surface melting in adsorbed rare gas films have been extensively reviewed by Dash [37]. A survey of recent neutron scattering work has been made by Chiarello and Krim [38].

MEIS and LEED both probe the loss of crystal order associated with surface melting, but they differ in the length scale over which the order is measured (short and long range, respectively). X-ray reflectivity measurements yield the depth profile of the average atomic density and electron microscopy is useful in studies of surface melting on small particles. Obviously, an important parameter in any melting procress is the diffusivity, which should be enhanced to a liquidlike value in the melted layer. Measurements of diffusivity by quasielastic neutron [39] and helium atom [19] scattering form the subject of the review by Bienfait in this volume and will not be further discussed.

3.1 Loss of short-range order

The MEIS technique makes it possible to accurately determine the depth over which the crystalline order is lost or diminished. Let us briefly explain the method of measurement by first considering the case of a *well-ordered surface* (no melt layer).

If a parallel beam of medium-energy (~100 keV) protons is aligned with atom rows in the crystal (fig. 3a), then the deeper lying atoms in each row are shadowed by the surface atom.If, in addition, the detection direction is chosen to coincide with an atom row, then the protons scattered from atoms below the top layer -- the shadowing is not perfect in a thermally vibrating crystal -- are blocked on their way out. The combined shadowing/blocking effect substantially reduces the backscattering yield from subsurface layers relative to the yield from the outermost layer. This gives rise to a distinct 'surface' peak in the energy spectrum of backscattered ions, see fig. 3a [40]. One can directly calibrate the area of the surface peak into the number of 'visible' monolayers or, alternatively, the number of 'visible' atoms per unit area. If the surface is covered with a *disordered melt layer*, then the surface peak area increases as shown in fig. 3b. The disordered atoms do not contribute to the shadowing and blocking effect and therefore are fully visible for ion beam and detector.The measured number of disordered atoms per unit area N can be converted into a layer thickness $\bar{l}$ through the use of $\bar{l}=N/\rho$, with ρ being the average particle density.

Figure 4a shows the temperature dependence of the surface peak area measured on Al(110) [28]. In this experiment a beam of 100 keV protons was aligned with the [$\bar{1}$01] direction and backscattered protons were detected which emerged from the crystal along the [011] direction (see inset). Up to a temperature of ~800 K the number of visible atoms per unit area increases smoothly as expected for a well-ordered crystal (the solid curve is a computer simulation for a dry crystal surface with thermal vibrations). Above ~800 K the yield increases dramatically beyond that expected for an ordered surface. The number of disordered atoms per unit area, as derived from this additional increase, is seen to diverge logarithmically as the temperature is raised to 0.5 K below T_m = 933.5 K (fig. 4b). The Al{110} surface clearly represents a case of complete wetting. On the other hand, experiments performed on Al{111} showed that this close-packed surface remains dry up to T_m. This marked crystal-face dependence has its origin in the fact that $\Delta\gamma^{\{110\}} > 0$ and $\Delta\gamma^{\{111\}} < 0$ [28].

Earlier MEIS investigations on various crystal faces of Pb show essentially the same thermal behavior: the {110} face melts, whereas the {111} and vicinal crystal faces remain thermally

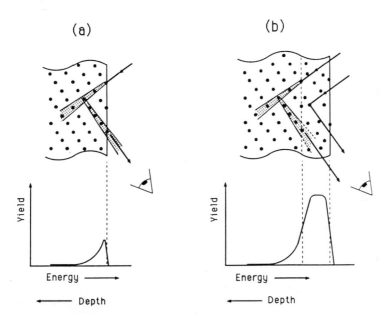

Fig. 3. Schematic representation of 'surface peaks' in backscattering energy spectra, obtained in shadowing-blocking geometry for (a) a well-ordered crystal surface and (b) for a crystal with a disordered surface layer on top. The shadowing and blocking effects are indicated as shaded cones.

stable [15]. The temperature dependence of the disordered layer thickness on Pb{110} could be measured up to 20 mK below T_m = 600.70 K [27]. The thickness first grows logarithmically (regime I in fig. 5a), then with a power law with exponent -1/3 (regime II in figs.5a and b), precisely as described in section 2. We note that for Al{110} the bulk melting point could not be approached sufficiently close that the transition to power-law growth was seen.

A fit of the logarithmic growth law (eq. (3)) to the measured T-dependence of the disordered layer thickness (solid lines in fig.4b and 5a) yields values for the parameters $\Delta\gamma$ and ξ. We find for Pb{110} $\Delta\gamma$=21.2 mJ/m^2 and ξ_b=6.2 Å [27] and for Al{110} $\Delta\gamma$=29.2 mJ/m^2 and ξ_b=4.9 Å [28]. The values for the excess free energy $\Delta\gamma$ are consistent with semiempirical estimates [16, 21]. The correlation length ξ_b/D, normalized to the atomic diameter D, is about the same for Pb and Al: ξ_b/D=1.5-1.6.This value is about equal to the correlation length of ξ_b/D=1.6 calculated for a bulk liquid of hard spheres [41].

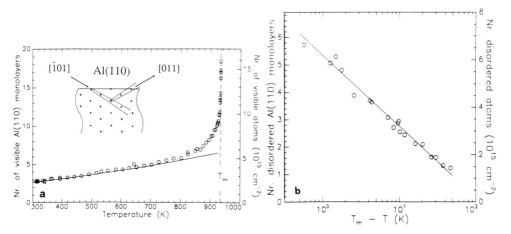

Fig. 4. (a) Surface peak area as a function of temperature for the Al(110) surface, as obtained in the scattering geometry shown in the inset. The solid curve represents the yield expected for an ordered Al(110) surface, as calculated in a Monte Carlo simulation of the experiment. The difference between the experimental data and the simulated yield equals the number of disordered atoms per unit area due to surface melting. From ref. [28].
(b) The number of disordered atoms per unit area on the Al(110) surface due to surface melting,plotted on a logarithmic temperature scale. The solid line represents the best fit of eq. (4) to the data, as discussed in the text. From ref. [28].

The Hamaker constant appearing in the asymptotic power-law dependence of the quasiliquid layer thickness (eq. (5)) has been determined for Pb{110} by fitting to the data of fig. 5b. The measured value of W=(0.40 ± 0.05) x 10^{-21} J compares well with the value of (0.31 ± 0.05) x 10^{-21} J recently calculated by Chen et al. [42] from the known dielectric functions for Pb.

The simple thermodynamic model outlined in section 2 predicts either complete wetting or non-wetting but does not account for the occurrence of *in*complete wetting. Recent temperature-dependent LEED [43] and MEIS [44] measurements on the Ge{111} surface show that above

~1050 K a very thin disordered layer is formed. The MEIS experiment shows that its thickness does *not* diverge but remain constant at a level of $(0.7 - 1.1) \times 10^{15}$ atoms/cm^2, which is equivalent to 1 to 1.5 Ge{111} monolayers (fig. 6). The amount of disorder measured increases with increasing energy of the incident proton beam, i.e., decreasing radius of the shadow cone. This indicates the presence of a significant amount of residual lattice order in the film, which is not surprising given the fact that the layer is extremely thin.

We believe the non-divergence of the melted layer thickness on Ge{111} to be due to layering [44]. For Ge{111} (and Si{111}) a pronounced layering effect is expected, since the diamond-type lattice naturally provides for a layered stacking of atomic planes along the <111> directions. Evidence for a layered disordering in these materials has also come from molecular dynamics simulations [45]. A recent theory of surface melting by Chernov and Mikheev [46], which includes layering effects, derives the conditions for which incomplete melting occurs.

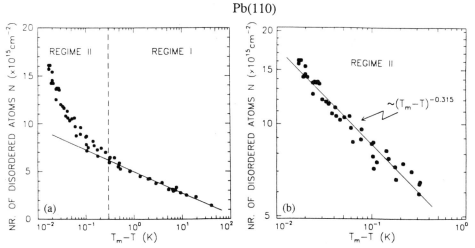

Pb(110)

Fig. 5. The number of disordered atoms per unit area on the Pb(110) surface, as derived from MEIS, as a function of temperature. (a) Semi-log plot, with solid line representing the best fit of eq. (3) to the data in regime I, as discussed in the text. (b) The data of regime II, reproduced on a double-log plot. The solid line represents the best fit of eq. (5) to the data. From ref. [27].

We note that, even if layering were to be absent, the wetting of Ge{111} will be incomplete; the Hamaker constant for Ge is negative (liquid Ge is denser than solid Ge) and the melting will therefore be blocked once the quasiliquid layer has become thick enough that long-range interactions dominate. However, the thickness for which this is expected to occur is much larger than is actually observed on Ge{111}. Hence, we attribute the incomplete melting of Ge{111} to a strong layering effect.

Ge(111)

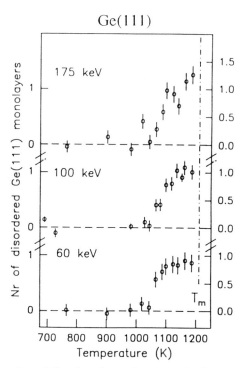

Fig. 6. The number of disordered monolayers on the Ge(111) surface measured by MEIS as a function of temperature. Protons of either 60, 100 or 175 keV energy are incident along the [00$\bar{1}$] axis and emerge from the crystal along the [11$\bar{1}$] direction. From ref. [44].

3.2 Loss of long-range order

For the detection of changes in the long-range order, diffraction is the technique by choice. Ideally suited for this purpose is X-ray diffraction, but very grazing angles are required for high surface sensitivity and this places extreme demands on the flatness of the surface [47]. LEED experiments are much easier to perform, but the interpretation of electron diffraction data from single crystals is complicated by dynamical scattering effects. Nonetheless, the effect of disordering on LEED intensities is generally thought to be describable within the framework of a kinematical scattering theory.

For the derivation of diffracted intensities we define an order parameter field $M(x,z)$ in the surface region and expand it in a Fourier series [48]:

$$M(x,z) = \sum_{G_j} M_{G_j}(z) \, e^{iG_j \cdot x} , \qquad (6)$$

where x and z are coordinates parallel and perpendicular to the surface plane and the summation is performed over the reciprocal lattice vectors in the surface plane. The component $M_0(z)$ for $G=0$ represents the in-plane averaged particle density at depth z: $M_0(z) = \rho(z)$. Associated with each order parameter component $M_{G_j}(z)$ is a characteristic decay length ('correlation length') ξ_j in the z-direction. From Landau mean-field theory it follows that for a system governed by short-range interactions each order parameter component has an exponentially decaying tail into the surface region:

$$M_{G_j}(z) \propto e^{-(\bar{l}-z)/\xi_j}. \qquad (7)$$

At the very surface we obtain

$$M_{G_j}(z=0) \propto e^{-\bar{l}/\xi_j}. \qquad (8)$$

In general, ξ_j decreases for increasing $|G_j|$: $\xi_0 > \xi_1 > \xi_2$, etc. [48,49]. The decay length ξ_0 equals the bulk correlation length ξ_b within the liquid. Since $\xi_0 = \xi_b$ is the largest decay length, it essentially determines the disordered layer thickness $\bar{l}$ in the asymptotic limit. Inserting eq. (3) into (8) we readily find

$$M_{G_j}(z=0) \propto (1 - T/T_m)^{\xi_b/2\xi_j} . \qquad (9)$$

As $T \to T_m$, each surface order parameter component goes continuously to zero according to a power law. In a diffraction experiment one measures the intensities of the various reflections (hk):

$$I_{hk} = |M_{G_{hk}}(z=0)|^2$$

$$\propto (1 - T/T_m)^{\xi_b/\xi_{hk}} , \qquad (10)$$

where $G_{hk} = hg_1 + kg_2$, with g_1 and g_2 base reciprocal lattice vectors in the surface plane. Here it is for simplicity assumed that the experiment only probes the very surface. In a typical LEED (or X-ray diffraction) experiment this will not be the case, which greatly complicates the analysis [17]. Nonetheless, the model describes the trends in a qualititative way. Figure 7 shows temperature dependent LEED measurements of the diffracted (10) and (01) beams for Pb{110} [17]. The intensities shown are corrected for background and have been divided by the Debye-Waller function.

The reciprocal lattice of the {110} surface is rectangular and the Miller-index labeling is such that $|G_{01}|=|G_{10}|/\sqrt{2}$. Above ~500 K the intensities I_{10} and I_{01} start to decrease but I_{10} decreases faster than I_{01}. The latter indicates that ξ_{10} is smaller than ξ_{01}. This is essentially what is expected on theoretical grounds, since ξ_{10} is associated with the larger reciprocal lattice vector. The physical significance of having different decay lengths ξ_{10} and ξ_{01} is that the disordering is anisotropic in the surface plane; the amount of disorder at a given temperature is larger along the close-packed $[1\bar{1}0]$ direction in the surface plane than along the [001] direction [17].

It is interesting to note that the onset of disordering at $T \approx 500$ K, as measured by LEED [17, 30, 50], is close to the onset measured by MEIS [24, 27, 51]. This is also the case for

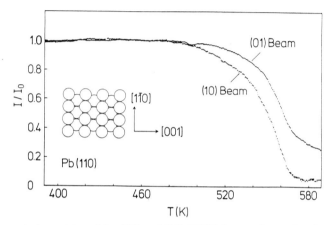

Fig. 7. Decrease in intensities of the (10) and (01) LEED beams due to the disordering of the Pb(110) surface. Measured intensities minus background have been divided by the Debye-Waller function and plotted against temperature. From ref. [17].

Ge{111}. McRae and Malic [43] in their LEED experiment find the surface to disorder from ~1050 K onwards and so do Denier van der Gon et al. [44]. We conclude that, at least for Pb{110} and Ge{111}, the initial stage of disordering proceeds both on long and short length scales.

3.3 Density profile

If the quasiliquid layer is close to being a liquid, rather than being a disordered solid, then it should also have a liquidlike density. For most elements the liquid density is somewhat lower than the solid density at T_m (for Pb, the difference is 3 %). The average density $\rho(z)$ is therefore expected to drop across the solid-quasiliquid interface (fig. 8). The density change, if it exists, can be measured by specular X-ray reflectivity. Since the refractive index for X-rays is slightly different for media of different density (*smaller* for *denser* media), the solid-quasiliquid and quasiliquid-vapor interfaces will contribute both to the reflectivity. There will be constructive and destructive interference between the two scattering contributions, depending on the angle θ or the momentum transfer Q which relates to the angle θ through $Q = (4\pi\sin\theta)/\lambda$, with λ the X-ray wavelength. The interference results in oscillations in the momentum transfer dependence of the reflected intensity R(Q), of which the period is inversely proportional to the separation of the two interfaces.

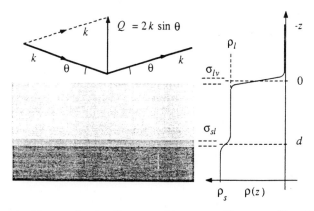

Fig. 8. Specular X-ray reflection from a surface covered by a quasiliquid layer. The density variation across the interfaces is schematically indicated by the function $\rho(z)$. From ref. [31].

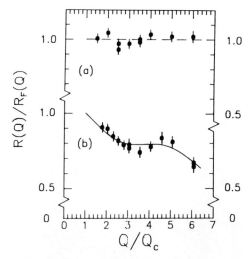

Fig. 9. Dependence of the normalized X-ray reflectivity from Pb{110} on the perpendicular momentum transfer Q. Measurements were taken in the scattering geometry of fig. 8. Curve (a) was measured at 300 K, curve (b) at 600.5 K (T_m = 600.7 K). From ref. [31].

Using this method Pluis et al. [31] have demonstrated that close to T_m a layer of lower density is present at the surface of a Pb{110} crystal. Figure 9 shows two curves of reflectivity versus momentum transfer, measured at 300 K and at $T_m - T = 0.2$ K. The data were taken at the synchrotron radiation source SRS (Daresbury, U.K.) using X-rays of 1.8 Å wavelength.

The reflectivities shown are normalized to the Fresnel reflectivity $R_F(Q)$, which is the reflectivity expected for a crystal abruptly terminated by a dry, flat, surface. The horizontal scale of fig. 9 shows the momentum transfer Q normalized to the value Q_c at the critical angle for total reflection. The horizontal curve $R(Q)/R_F(Q) \approx 1$ at 300 K shows that the surface at that temperature has no lowered density. But the curve obtained close to T_m clearly shows an oscillation, demonstrating the existence of two interfaces. The measured curve could be fitted to a calculation (solid curve) in which the quasiliquid layer is assumed to have 3.9 ± 1.2 % lower density and a thickness of 20 ± 3 Å. The derived thickness agrees with the MEIS data [27], but disagrees with the results from an X-ray diffraction study by Fuoss et al. [47]. The damping of the oscillation and the decay of the average signal with increasing Q/Q_c relate to the roughness of both interfaces (indicated by σ_{sl} and σ_{lv} in fig. 8).

3.4 Surface melting on small particles

In recent years several electron microscopy studies have been performed on micron-sized crystals The existence of a melt layer in these studies has revealed itself as a distinct change in the equilibrium crystal shape [32,33] or as a diffraction contrast in the skin of a particle [34, 52]. Conversely, surface melting may be completely absent if the crystal is bounded by non-wettable facets or capped with another material. Métois and Heyraud [36] prepared small facetted crystallites in the electron microscope by crystallizing molten droplets on a substrate and then depositing Pb. A small fraction of the microcrystals so produced were bounded by {111} facets only (fig. 10). These facets are known to be non-wettable, i.e. $\Delta\gamma^{\{111\}} < 0$ [15]. In the

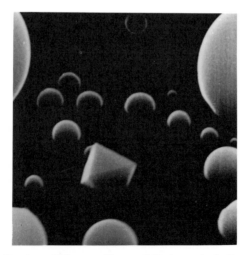

Fig. 10. A collection of Pb crystallites at 2 K above the bulk melting point. One sees coexistence of (overheated) sharp-edged polyhedra consisting of only {111} facets and molten spheres of Pb. From ref. [36].

absence of surface melting it proved possible to overheat these crystals by 2 to 4 K for several hours. All other crystals melted simultaneously at the bulk melting point T_m. Another successful attempt to overheat a crystal was reported by Daeges et al. [35]. Single-crystal Ag spheres coated with a thin continuous epitaxial layer of Au could be overheated by 25 K for one minute. Apparently, the absence of free surfaces on these crystals prevents surface melting to occur (analogously to Pb{110}, when capped with an epitaxial film of PbO [29]).

The results reported here are strongly suggestive of a one-to-one correspondence between the occurrence (absence) of superheating and the absence (occurrence) of surface melting.

4. OUTLOOK

The observations of surface melting reported here have substantially advanced our understanding of the phenomenon. The asymptotic temperature dependence of the melted layer thickness can generally be understood on the basis of phenomenological models of the Landau type. We may view surface melting as a wetting phenomenon. We have come across cases of complete wetting, non-wetting and incomplete wetting. The latter case presents a problem: the melting is usually blocked at very small thicknesses and the continuum model, so successful in accounting for complete wetting, will break down. Instead, one has to treat the disordering process on an atomic scale. Here lies the greatest challenge. Computer simulations of surface melting may give valuable insight in the atomic-scale mechanism of the initial stage of disordering. Molecular dynamics [53, 54] and Monte Carlo [55] simulations of surface premelting in Al{110} and Ni{110} have already indicated that the disordering is mediated by the creation of adatom-vacancy pairs. Interestingly, in the simulations by Stoltze {53} on Al{110} no evidence was found for the anisotropy of the type reported in the LEED experiment on Pb{110} [17] (see section 3.2). This, it seems, is a controversial issue.

The most attractive technique for the detection of surface melting appears to be X-ray scattering. But this is also the hardest experiment to do! Using X-ray scattering one can in principle accurately determine the exponents with which the various order parameters decay as $T \rightarrow T_m$ and one can, in the reflectivity mode, determine the density profile. With the advent of powerful synchrotron radiation sources (e.g. ESRF) these types of experiments will become less hard to do.

ACKNOWLEDGEMENTS

The author owes much gratitude to Joost Frenken, Arnoud Denier van der Gon and Bart Pluis for their important contributions to the content of this review. He enjoyed the stimulating discussions with Lev Mikheev. This work is part of the research program of the Stichting voor Fundamenteel Onderzoek der Materie (FOM) and is made possible by financial support from the Nederlandse Organisatie voor Wetenschappelijk Onderzoek (NWO)

REFERENCES

1. A.R. Ubbelohde, The Molten State of Matter (Wiley, New York, 1978), and references therein.
2. G. Borelius, in: *Solid State Physics, Advances in Research and Applications*, Vol. 6, Eds. F. Seitz and Turnbull (Academic Press, New York, 1958).
3. Sutherland, Phil. Mag., Vol.32 (Ser. 5) (1891) 31, 215, 524.
4. M. Born, J. Chem. Phys. 7 (1939) 591.
5. Y.P. Varshni, Phys. Rev. B2 (1970) 3952.
6. F.A. Lindemann, Z. Phys. 14 (1910) 609.
7. J. Frenkel, *Kinetic Theory of Liquids* (Dover, New York, 1955).
8. J. Lennard-Jones and A.F. Devonshire, Proc. Roy. Soc. A170 (1939) 464.

9. R.M.J. Cotterill, E.J. Jensen and W.D. Kristensen, in: *Anharmonic Lattices, StructuraL Transitions and Melting,* Ed. T. Riste (Noordhoff, Leiden, 1974), and references therein.

10. J.K. Kristensen and R.M.J. Cotterill, Phil. Mag. 36 (1977) 437.

11. D.P. Woodruff, *The Solid-Liquid Interface* (Cambridge Univ. Press, Londen, 1973).

12. Tammann, Z. Phys. Chem. 68 (1910) 205; Z. Phys. 11 (1910) 609.

13. J.W.M. Frenken and J.F. van der Veen, Phys. Rev. Letters 54 (1985) 134.

14. Da-Ming Zhu and J.G. Dash, Phys. Rev. Letters 57 (1986) 2959; Da-Ming Zhu and J.G. Dash, Phys. Rev. Letters 60 (1988) 432.

15. B. Pluis, A.W. Denier van der Gon, J.W.M. Frenken andJ.F. van der Veen, Phys. Rev. Letters 59 (1987) 2678.

16. J.F. van der Veen, B. Pluis and A.W. Denier van der Gon, in: *Chemistry and Physics of Solid Surfaces VII,* Springer Series in Surface Sciences, Vol. 10, Eds. R. Vanselow and R.F. Howe (Springer, Heidelberg,1988) p. 455.

17. U. Breuer, H.P. Bonzel, K.C. Prince and R. Lipowsky, Surface Sci. 223 (1989) 258.

18. U. Breuer, O. Knauff and H.P. Bonzel, to be published.

19. J.W.M. Frenken, B.J. Hinch, J.P. Toennies and Ch. Wöll, Phys. Rev. B41 (1990) 938.

20. R. Lipowsky, Ferroelectrics 73 (1987) 69, and references therein.

21. B. Pluis, D. Frenkel and J.F. van der Veen, to be published.

22. J.N.Israelachvili, *Intermolecular and Surface Forces* (Academic, San Diego, 1985).

23. K.D. Stock, Surface Sci. 91 (1980) 655.

24. J.W.M. Frenken, P.M.J. Marée and J.F. van der Veen, Phys. Rev. B34 (1986) 7506.

25. D. Beaglehole and D. Nason, Surface Sci. 96 (1980) 357.

26. Y. Furukawa, M. Yamato and T. Kuroda, J. Cryst. Growth 82 (1987) 665.

27. P. Pluis, T.N. Taylor, D. Frenkel and J.F. van der Veen, Phys. Rev. B40 (1989) 1353.

28. A.W. Denier van der Gon, R.J. Smith, J.-M. Gay, D.J. O'Connor and J.F. van der Veen, Surface Sci. 227 (1990) 143.

29. A.W. Denier van der Gon, B. Pluis, R.J. Smith and J.F. van der Veen, Surface Sci. 209 (1989) 431.

30. H.-N. Yang, T.-M. Lu and G.-C. Wang, Phys. Rev. Letters 63 (1989) 1621.

31. B. Pluis, J.-M. Gay, J.W.M. Frenken, S. Gierlotka, J.F. van der Veen, J.E. Macdonald, A.A. Williams, N. Piggins and J. Als-Nielsen, Surface Sci. 222 (1989) L845.

32. J.C. Heyraud, J.J. Métois and J.M. Bermond, J. Crystal Growth 98 (1989) 355.

33. A. Pavlovska, K. Faulian and E. Bauer, Surface Sci. 221 (1989) 233.

34. G. Devaud and R. H. Willens, Phys. Rev. Letters 57 (1986) 2683.

35. J. Daeges, H. Gleiter and J.H. Perepezko, Phys. Lett. A119 (1986) 79.

36. J. J. Métois and J.C. Heyraud, J. Phys. France 50 (1989) 3175.

37. J.G. Dash, Contemp. Phys. 30 (1989) 89, and references therein.

38. R. Chiarello and J. Krim, Langmuir 5 (1989) 567.

39. M. Bienfait and J.P. Palmari, in: *the Structure of Surfaces II*, Springer Series in Surface Sciences, Vol. 11, Eds. J.F. van der Veen and M. A. Van Hove (Springer, Heidelberg, 1988) p.559.

40. J.F. van der Veen, Surface Sci. Rept. 5 (1985) 199.

41. L. Mikheev and A. Trayanov, Phys. Rev. B, in press.

42. X.J. Chen, A.C. Levi and E. Tosatti, preprint.

43. E.G. McRae and R.A. Malic, Phys. Rev. B38 (1988) 13163.

44. A.W. Denier van der Gon, J.-M. Gay, J.W.M. Frenken and J.F. van der Veen, Surface Sci., in press.

45. F.F. Abraham and J.Q. Broughton, Phys. Rev. Letters 56 (1986) 734.

46. A.A. Chernov and L.V. Mikheev, Phys. Rev. Letters. 24 (1988) 2488; Physica A157 (1989) 1042.

47. O.H. Fuoss, L.J. Norton and S. Brennan, Phys. Rev. Letters 60 (1988) 2046.

48. R. Lipowsky, U. Breuer, K.C. Prince and H.P. Bonzel, Phys. Rev. Letters 62 (1989) 913.

49. H. Löwen, Phys. Rev. Letters, 64 (1990) 2104(C).

50. W. Dürr, D. Pescia, J.W. Krewer and W. Gudat, Solid State Comm.73 (1990) 119.

51. B. Pluis, A.W. Denier van der Gon, J.F. van der Veen and A.J. Riemersma, Surface Sci., in press
52. Y. Lereah, G. Deutscher, P. Cheyssac and R. Kofman, Europhys. Letters 12 (1990) 709.
53. P. Stoltze, J. Chem. Phys., in press.
54. E.T. Chen, R.N. Barnett and U. Landman, Phys. Rev. B41 (1990) 439.
55. A.W. Denier van der Gon, D.Frenkel, J.W.M. Frenken, R.J. Smith and P. Stoltze,to be published.

SURFACE MELTING AND DIFFUSION

Michel Bienfait and Jean-Marc Gay

CRMC2*-CNRS
Campus de Luminy, case 913
13288 Marseille cedex 09, France

Quasi-elastic scattering of neutrons or helium atoms are well-suited techniques to measure the mobility of molecules involved in surface melting processes. The recent results obtained by these techniques are critically reviewed in the case of thin films of methane or hydrogen adsorbed on graphite or MgO and for lead (110) surface. It is shown that the diffusion coefficient of the mobile layer, stable at the solid-vapor interface below the bulk melting point, is very large, and is in the 10^{-5} $cm^2 s^{-1}$ range. The corresponding thickness of this liquid-like film varies with temperature and is several molecular layer-thick about 1K below the melting point.

I - INTRODUCTION

Surface melting[1-3] has attracted much experimental and theoretical interest in the last few years. This phenomenon is characterized, at temperatures near but below the bulk melting point T_m, by the existence of a disordered and mobile film that wets the solid-vapor interface. The thickness of this film diverges as T_m is approached. Surface melting was observed on a variety of crystal faces except on the (111) surfaces of fcc metals. Several complementary techniques were used to analyse these surface instabilities below the melting point. Most of the techniques like ion scattering[4,5], heat capacity[2,6], neutron diffraction[7,8] and ellipsometry[9,10] were able to measure the temperature dependence of the "melted" thickness of the film and/or its long-range order. They showed that the disordered film thickness varied with T logarithmically or with a 1/3 power law depending on short-range or long-range surface interactions, respectively. However, the above techniques were unable to decide whether the observed disorder was static (microcrystalline or glassy) or dynamic (liquid-like). Accordingly, measurements of the molecular motion in the topmost layers of the crystal surfaces below T_m was required in order to determine if the

---

* Laboratoire associé aux Universités d'Aix-Marseille II et III

disordered surface layer could be correctly described as a "quasi-liquid".

Mobility in surface films can be studied by a variety of techniques[11-13]. Several of them have been used to analyze molecular motion in surface melting processes. The results obtained so far will be reviewed critically here.

The major part of the works performed on mobility measurements in surface melted layers were carried out by quasi-elastic neutron scattering (QENS)[14-19] and by quasi-elastic scattering of thermal-energy He atoms[20]. These studies mainly show that surface atoms or molecules attain liquid-like mobilities at temperatures as much as 0.8-0.9 T/T_m.

In addition, neutron scattering (QENS) is able to measure in the same experiment the T-dependence of both the thickness and the diffusivity of the "melted" layer. It also allows for the determination of the type of local motion. For instance, as far as the quasi-liquid thickness is smaller than two molecular layers the mobile layer can be regarded as a strongly correlated liquid or a lattice fluid. As the temperature is raised, the mobile layer depth increases and its diffusivity progressively tends towards that of a classical brownian motion. These points will be developed below.

II - TECHNIQUES AND MODELS

Quasi-elastic scattering of neutrons or He atoms are well-suited techniques for measuring the diffusivity of atoms or simple molecules on surfaces. The main difference between the two methods is their distinct penetration length. Neutrons can easily penetrate condensed matter and are almost insensitive to surfaces. However, the use of powders with large specific areas increases the surface-to-volume ratio and allows the experimentalist to record fairly intense surface QENS spectra. The experiments are carried out on multilayers of hydrogen-containing molecules condensed on highly divided substrates because hydrogen has a very strong incoherent cross-section for neutrons. The systems studied so far are methane[14-16,18] and hydrogen[17,19] thick films adsorbed on MgO(100) or graphite (0001) powders.

Unlike neutrons, thermal-energy He atoms do not penetrate bulk matter. They only probe the topmost atomic layer of a surface. Therefore they can yield information on the mobility of adatoms on the most external plane of single crystal surfaces whereas the measurements carried out by neutrons characterize the diffusivity averaged on the total melt depth. Hence the methods are complementary. Finally we must note that the actual energy resolution for He beams is poorer (80-170 µeV) than for thermal neutrons (20-50 µeV). Another severe limitation of He-atom scattering is its unability to collect data when the vapor pressure is larger than 10^{-5} torr, a pressure which is reached at melting for most of the materials. This limitation does not hold for neutron scattering which can work even if the vapor pressure surrounding the sample is a few atmospheres. The only mobility experiment using He-atom scattering was carried out on the (110) lead surface[20] whose vapor pressure at melting is in the 10^{-6} torr range.

When a beam of neutrons or He atoms is scattered by a fluid surface, some weak inelastic effects appear around the elastic peak. The broadening of the scattered energy distribution with respect to the incident energy is brought about by small energy transfers related to the diffusive motion of the surface atoms or molecules. The theory of quasi-elastic scattering is very similar for thermal-energy and He atom scattering, except for small differences explained below.

The recorded quasi-elastic neutron spectra are interpreted using the standard formalism of QENS[21] adapted to surface neutron scattering[13,22,23]. The assumption is made of a "two phase" system with a scattering law $S(\vec{Q},\omega)$ composed of two terms, the first one proportional to the solid fraction (1-x) of the film and the second one to the remaining part x of the liquid-like phase

$$S(\vec{Q},\omega) = (1-x)\, A(\vec{Q},\omega) + x A(\vec{Q},\omega) \otimes L(\vec{Q},\omega) \tag{1}$$

where A describes the isotropic rotational motion of the molecule. This function A is convoluted ($\otimes$) with a Lorentzian function $L(\vec{Q},\omega)$ describing the translational motion in the fluid phase

$$A(\vec{Q},\omega) = \left\{ j_0^2(Qr)\delta(\omega) + \pi^{-1} \sum_{\ell=1}^{\infty} (2\ell+1)\, j_\ell^2(Qr) \frac{D_r\, \ell(\ell+1)}{[D_r\, \ell(\ell+1)]^2 + \omega^2} \right\} e^{-Q^2\langle u^2\rangle} \tag{2}$$

j_ℓ is the spherical Bessel function of order ℓ ; $\vec{Q}$ the scattering vector ; $\hbar\omega = E - E_0$ is the gain or loss of energy E with respect to the incident energy E_0 ; r is the gyration radius of the molecule ; D_r is its rotational diffusion coefficient ; $\exp(-Q^2\langle u^2\rangle)$ is the Debye-Waller factor and $\langle u^2\rangle$ is the mean square displacement of the hydrogen protons.

For small Q values (Q < 1 Å^{-1}), the second term in eq(2) is negligible and $A(Q,\omega)$ results in a constant multiplying factor $j_0^2(Qr)\exp(-Q^2\langle u^2\rangle)$. On the other hand, the second term in eq(2) explains the small and broad contribution usually observed at large Q (Q $\geq$ 1 Å^{-1}) due to the rotational mobility of the molecules. This term has been used[18] to determine the rotational diffusion coefficient D_r in the solid phase at low temperature. But this determination is usually not the major purpose of the studies which are aimed to the measurement of the thickness of the liquid-like layer (fraction x) and of its translational diffusion coefficient.

The most important term in eq(1) is the Lorentzian function $L(\vec{Q},\omega)$ that represents the incoherent scattering by a mobile phase

$$L(\vec{Q},\omega) = \pi^{-1} f(\vec{Q}) / (f^2(\vec{Q}) + \omega^2) \tag{3}$$

with $f(\vec{Q})$ depending on the model : for a 2D liquid with brownian mobility

$$f(\vec{Q}) = DQ^2 \sin^2 \tau \tag{4}$$

where D is the translational diffusion coefficient of the fluid, τ is the angle between the normal of an individual graphite basal surface and the scattering vector $\vec{Q}$.

For a bulk liquid

$$f(\vec{Q}) = DQ^2 \tag{5}$$

Eqs.(4) and (5) are valid at small Q for all known diffusion models. To exploit the data in the whole Q-range, we can introduce more complicated expressions for $f(\vec{Q})$. These expressions have been established for hexagonal or square lattice models[24]. They are supposed to represent molecular motion of molecules in a strongly correlated fluid stabilized by the underlying crystal lattice.

Eq.(1) has to be averaged over the isotropic distribution of the crystallite planes and convoluted with the instrumental resolution and can then be tested against the experimental results. Although some of the equations written above may seem somewhat complicated, the final scattering function has only two adjustable parameters : x, the film liquid fraction, and D the translational diffusion coefficient of the mobile phase. The remaining parameters needed for eq.(1) are either given by the experimental conditions (Q and ω) or known from literature[16,18].

The above formalism valid for neutron scattering must be adapted to thermal-energy He atom scattering. The helium beam being reflected by the surface only probes the topmost fluid atomic layer. Hence x = 1 in eq. 1. Furthermore the only experiment reported so far was carried out on a monoatomic solid (Pb(110)). The rotational term (eq.2) is meaningless. Hence, $S(\vec{Q},\omega) = L(\vec{Q},\omega)$ with $f(\vec{Q})$ given by eq.(4) at small Q[20,25].

III - EXPERIMENTAL

III.1. Quasi-elastic neutron scattering experiments

In a QENS experiment the fluid part of a sample can be detected by a broad component in the scattering function (second term in eq.(1)). The remaining solid part of the sample yields a narrow resolution limited peak (first term in eq.(1)). Fig. 1 illustrates the above statements in the case of a ~10 layer thick film of methane condensed on graphite at three different temperatures. The wings in the scattering intensity distribution at ~1.5 and 0.5 K below the bulk melting temperature (90.66 K) are an unambiguous signature of the presence of a liquid-like phase in the film.

Several systems have been studied so far by QENS : $CH_4/MgO(100)$[14-16], CH_4/graphite(0001)[18] and HD/MgO(100)[17,19]. They have been selected for many practical reasons :
    - graphite and MgO powders can be prepared with a large specific

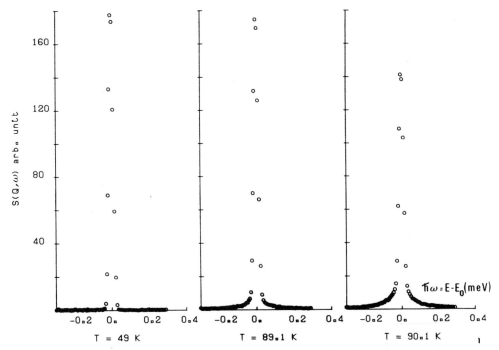

Fig. 1. Scattering functions of ~ 10 layers of CH_4/graphite at three different temperatures below T_m=90.66 K for Q=0.59 Å^{-1}. The graphite background has been subtracted. The film is fully solidified at 49 K. When the temperature is raised a liquid-like layer, responsible for the observed wings, coexists with an underlying solid phase. From Ref.18.

area (several m^2/cm^3). These substrates are very uniform and formed by graphite or MgO microcrystals with mostly (0001) or (100) facets exposed for adsorption respectively. The total adsorption area in a neutron cell is about 200 m^2.

- the adsorbates exhibit a layer-by-layer mode of growth up to at least 5 layers in the vicinity of the triple point. The information concerning the adsorption for the different systems can be found in the following references (CH_4/MgO[26,27], CH_4/graphite[28-30], H_2/MgO[17,31]). Some clustering can occur in particular conditions, especially at low temperatures. This problem will be addressed in paragraph V.

- diffraction measurements have shown that the topmost layer of the condensed film is either a (111) or a (100) plane in the case of CH_4 adsorption on graphite[32] or MgO[26] respectively. Hence, the use of the two different substrates permits the analysis of surface melting for two different crystallographic orientation of the CH_4 films. As for hydrogen condensed on MgO its interface between the film and the vapor is a (111) plane[33].

- the incoherent cross-section of hydrogen for neutrons is more than hundred times larger than for other atoms. Hence the QENS signal coming from the film is easily discriminated from that of the substrate. Besides the multilayer scattering spectra are obtained by subtracting the background due to the cell and the bare graphite or MgO substrates.

The experimental details of a QENS experiment have been reported several times [14,16,18,19,22,23]. The experiments were performed at the ILL (Grenoble) on the instrument IN5 or at the LLB (Saclay) on the instrument Mibemol using an incident wavelength of 8 Å (incident energy $E_0 = 1.278$ meV). Several banks of detectors were used at various scattering angles ranging from 0.3 to 1.4 $Å^{-1}$ on IN5 and from 0.3 to 0.9 $Å^{-1}$ on Mibemol. The instrumental resolution has a triangular shape with a FWHM of 27 µeV on IN5 and of 52 µeV on Mibemol.

III.2. Quasi-elastic He atom scattering experiment

The study of surface melting by He atom scattering experiment was performed on the (110) surface of a Pb single crystal. The experimental procedure for preparation, cleaning and measurement is described in Ref. 20. The incident energy was either 2.2 or 6.3 meV with corresponding resolutions of ~ 80 or 170 µeV respectively.

A selection of measured spectra is displayed in fig. 2 at temperatures of 446, 544 and 551 K. They can be compared to those reproduced in fig. 1 and obtained by QENS. Besides the fact that the He atom scattered spectra

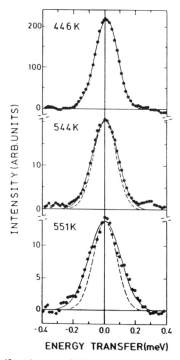

Fig. 2. Energy distributions of He atoms scattered from a Pb(110) surface, at three crystal temperatures, for Q = 0.64 $Å^{-1}$ along the [001] surface azimuth. The incident beam energy is 6.5 meV. The dashed curves show the experimental resolution of 163 µeV. The full curves are Gauss fits to the data ($T_m \simeq 600$ K). From Ref. 20.

are strongly resolution limited, they do not show the narrow central peak observed in fig. 1 and resulting from the scattering of the underlying solid ; He atoms only probe the very last atomic layer of the surface. Another difference which is , as far as we know, unexplained is the strong decrease of the He quasi-elastic signal at 544 and 551 K. This effect is not observed for neutrons in fig. 1. Still both quasi-elastic techniques clearly demonstrate an energy broadening increasing with temperature and consequently the existence of a mobile layer at the solid-vapor interface below melting temperature for CH_4 films and the Pb(110) surface.

IV - RESULTS

The most comprehensive results published so far on diffusivity measurements in surface melting processes have been obtained by QENS. The data reduction is made using eqs. (1)-(5) whose quantitative interpretation is very simple. The diffusivity D and the fraction x of the mobile phase are obtained from the width and from the relative intensity of the broad component respectively. One important point is checked before carrying out a detailed analysis of the recorded spectra. It is ensured that, at all temperatures, the reduction of the integrated intensity of the central peak is essentially equal to the integrated intensity of the wings. This means that part of the solid layer is transformed into a mobile phase when the temperature is raised. At low temperature, (49 K for CH_4), there is no wings (fig. 1) and the film is fully solidified. The two adjustable parameters are determined simultaneously at constant temperature, from the fit of the corresponding spectra. The results are reported below successively for the sake of simplicity.

IV.1. Translational mobility

IV.1.1.- Small Q analysis. The analysis of the spectra recorded at small $\vec{Q}$ yields the averaged translational diffusion coefficient of the mobile film below T_m and of the bulk liquid above. Some typical fits with a 3D brownian model (x = 0.65 ; D = 3.10^{-5} $cm^2 s^{-1}$) are represented in fig. 3 for a 6.5 layer thick film of HD condensed on MgO(100) at 16.49 K. The same detailed analysis has been carried out for all the recorded spectra. The obtained values of D are represented in an Arrhenius plot in figs. 4 and 5 for thin films of HD adsorbed on MgO and CH_4 adsorbed on MgO and graphite respectively. The left-hand part of the figure represents the T-dependence of the bulk liquid diffusion coefficient reported in the literature and obtained in the QENS experiments as well[16,18,19].The observed mobility of the surface film falls in the 10^{-5} $cm^2 s^{-1}$ range, which means that the measured translational coefficient is characteristic of a liquid-like phase. The same type of results, i.e. D ~ 10^{-5} $cm^2 s^{-1}$ have been obtained for the mobility measurements carried out by He atom scattering on the Pb(110) surface[20].

The analysis below T_m was performed using either a 2D or a 3D brownian model (eq. 4 or 5 respectively). The different fits cannot distinguish between the two types of motion, except at low T and above T_m for

CH_4 films[18]. The ratio between the obtained diffusion coefficients with the two models is close to 3/2 as expected from dimensionality arguments. Note that the measured diffusion coefficients do not depend on the initial thickness of the film within the experimental uncertainty.

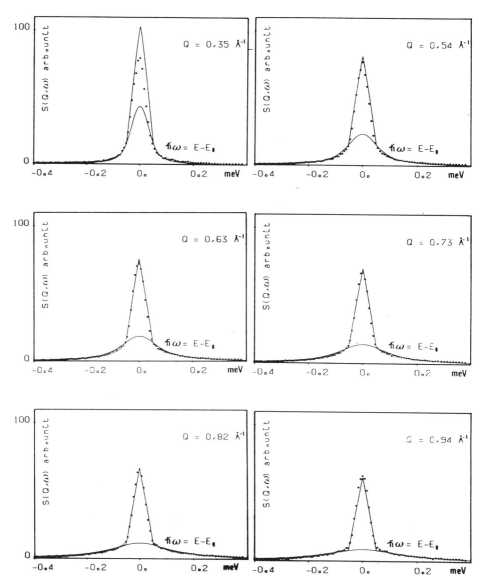

Fig. 3. Typical fits of the recorded QENS spectra for a 6.5 layer-thick film of HD condensed on MgO at 16.49 K with a 3D brownian model ("Liquid" thickness = 3.9 layers ; $D = 3.10^{-5}$ cm^2 s^{-1}). The bulk melting temperature for HD is T_m = 16.606 K. The Lorentzian component results from the presence of a fluid phase. The narrow resolution limited peak comes from the remaining part of HD solid on the MgO surface. From Ref. 19.

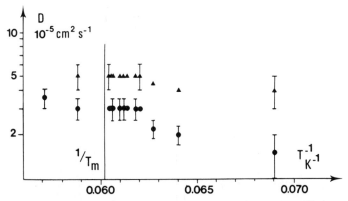

Fig. 4. Arrhenius plot of the translational diffusion coefficients of the liquid-like component of HD films on MgO. On the left-hand side of the figure, the bulk liquid diffusivity measured in this experiment ; on the right-hand side the mean diffusion coefficient of the HD mobile part of the film for a 2D (▲) and for a 3D (●) brownian model. From Ref.19.

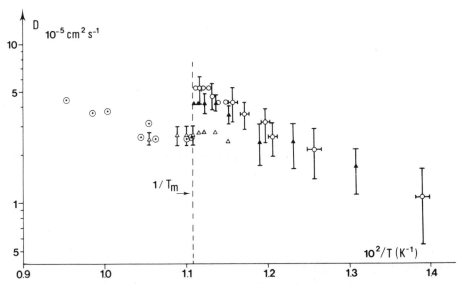

Fig. 5. Like fig. 4, for ~ 10 layer-thick films of CH_4 condensed on MgO and on graphite. On the left-hand side of the figure, the bulk liquid diffusivity is represented from literature (⊙) and from this study (△) ; on the right-hand side, the mean diffusivity of the mobile part of the CH_4 film adsorbed on graphite (▲) for a 2D and (△) for a 3D model compared to the values obtained for CH_4 adsorbed on MgO (○) are shown. From Refs. 16,18.

This ratio 3/2 between the 2D and 3D diffusion coefficients may be somewhat confusing. It brought about an apparent discontinuity in the temperature dependence of D at T_m due to the specific 2D model (see figs. 5 and 6). Considerable longer and more precise experiments giving information on the lateral and normal mobility as a function of the depth of the moving molecules with respect to the surface are needed to clarify this point.

Computer simulations of the surface mobility below T_m were carried out by several groups for solid rare gases[34,35] or methane thin films[36]. They generated interesting results:

- the diffusion coefficients in the surface mobile layer and in bulk liquid are similar, in agreement with QENS and He atom scattering results,
- the close packed (111) face for these fcc molecular solids exhibit surface melting as shown experimentally by QENS for the (111) vapor-solid interface of thin films of CH_4 condensed on graphite and of HD condensed on MgO. The surface melting of the (111) face of rare gases was also shown by other techniques[6,37]. These results contrast with the behavior of the (111) surfaces of fcc metals which do not undergo a melting transition[5,38,39]. This difference may be related to the type of interaction in the van der Waals solids and in metals.
- mobility calculations performed on the (111), (110) and (100) faces[34,36] yield similar averaged translational diffusion coefficient. The measurements obtained by QENS for the (111) and (100) surface of methane thin films (see fig. 5) agree with these results.

I V.1.2.- Large Q analysis. The above analysis has been performed in the small Q range where eq. 4 and 5 are valid. The data recorded at larger Q can yield information on local motion in the surface mobile film. In that case, several expressions of $f(\vec{Q})$ featuring different elementary jumps can be tested against the experimental data. For instance the expression of $f(\vec{Q})$ describing an hexagonal lattice fluid model is given below

$$f(\vec{Q}) = \{3 - 2 \,[\cos \vec{Q} \, \vec{a} + \cos \vec{Q} \, \vec{b} + \cos \vec{Q} \, (\vec{a} + \vec{b})] \,\} \,/\, 3t \qquad (6)$$

where $\vec{a}$ and $\vec{b}$ are the lattice vectors of the two-dimensional hexagonal lattice and t is the mean residence time on a site.

This model was selected here because, as stated above, in the case of adsorption of CH_4 on graphite, the solid surface has a (111) structure and the liquid-like layer covering the solid film is expected to have some crystalline order induced by the underlying lattice. As for the CH_4 films condensed on MgO, a square lattice fluid model was selected because the methane surface is a (100) plane in that case[16].

Notice the relation between t and D, for a two-dimensional fluid :

$$D = |a|^2 / 4t \qquad (7)$$

Only one parameter (t) is needed for the fitting. The remaining parameters are either given in section III (Q, ω, ..., D_r) or obtained from the

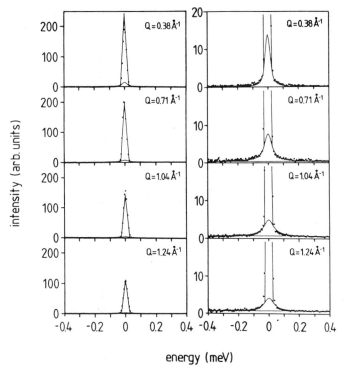

Fig. 6. Set of measured scattering functions of the CH_4 thin film condensed on graphite (~ 10 layers thick) for various scattering vectors Q at 84.1 K. The full lines correspond to the best fit with a hexagonal jump model (eqs. 1,2,3,6). The rotational diffusion contribution yields the horizontal line. The broad component results from the translational diffusion of the mobile part of the film. The narrow peak corresponds to the remaining instrumental broadened solid component ($x = 0.12$; $t = 2.10^{-11}$ s). From Ref. 18.

fit at small scattering vector. The agreement is very good for the thinnest liquid-like layers ($\leq$ 2 layer-thick).

Typical fits from eqs. (1)-(3),(6) are represented in fig. 6 for a few spectra chosen among the 20 used in the data reduction at 84.1 K for a ~ 10 layer-thick film of CH_4 condensed on graphite.

A convenient representation of the validity of the various models is to draw the width of the mobile component (FWHM) versus Q. According to eq. (6) the width for the hexagonal jump model exhibits a maximum at ~ $\pi/|\vec{a}|$. In our case, we have to average eq. (1,3,6) over the isotropic distribution of the graphite crystallites. The corresponding integration shifts the maximum to larger Q. The calculated FWHM for the jump model is represented in Fig. 7a and b for two temperatures 84.1 and 90.1 K with

317

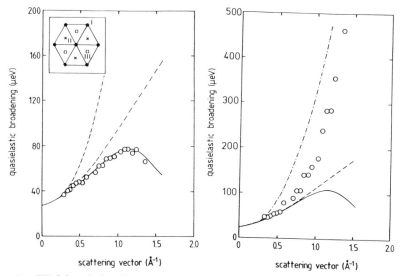

Fig. 7. Width of the Lorentzian component (mobile part) convoluted with the instrumental resolution versus scattering vector Q for a 2D hexagonal jump model (—), a 2D brownian model (---) and a 3D brownian model (-·-) respectively. Experimental results (o) at 84.1 K for ~ 1.2 melted layer (fig. 7a ; t = 2.10^{-11} s) and at 90.1 K for ~ 4 melted layers (fig. 7b ; t = $1.1 \ 10^{-11}$ s).

the widths corresponding to the two-dimensional (averaged eq.(4)) and three-dimensional (eq.(5)) brownian models as well. The experimental widths recorded for the mobile layer at these temperatures are reported on the same figure for CH_4 films adsorbed on graphite. The good agreement between our data at 84.1 K and the hexagonal jump model confirms the conclusions stated above about the existence of a hexagonal lattice fluid on the (111) surface, ~ 6 K below the melting point. At 90.1 K the experimental widths are intermediate between the one expected for the 2D and 3D models. This result shows that no simple model is appropriate to interpret correctly the QENS data at medium quasiliquid thickness. In fact, as the temperature increases, the mobility should be described as a continuous evolution from a two-dimensional jump diffusion to an isotropic 3D-diffusion.

Notice that eq. (6) has been established for a hexagonal lattice I with a = 4.25 Å corresponding to a close-packed $CH_4(111)$ plane (see box in fig. 7). Two triangular sublattices II and III are also present on the surface. Their jump distance is reduced by a factor $\sqrt{3}$, which means that the theoretical maximum of the quasi-elastic broadening would be shifted to larger Q by the same factor (expected maximum at ~ 2.1 Å^{-1} instead of ~ 1.2 Å^{-1}). The experimental width variation in fig. 7a is clearly inconsistent with such a model.

I V.2 Thickness of the liquid-like film

The relative intensity of the broad component recorded in a QENS experiment gives the fraction x of the mobile part of the film (eq.(1)). (Note that He atom scattering is unable to provide this information). Assuming a "liquid"-solid stratification, this fraction can be converted into the thickness L of the liquid-like layer stable at the film-vapor interface. The obtained values of L are represented in fig. 8 as a function of T for CH_4 films adsorbed on graphite and MgO. The thickness of the mobile layer on the (111) surface seems to be systematically slightly smaller than on the (100) surface although some of the error bars overlap. This trend may be related to the expectation that the densest plane (111) is more stable than the more open surface (100). It may also result from the residual field of the substrate which is known to be stronger on graphite than on MgO for CH_4 films.

One can also notice that the thickness of the liquid-like layer remains finite (5 to 6 molecular layers) even above the triple point. This observation is not surprising because the initial film thickness is limited to 10 layers and the substrate field tends to stabilize a few solid layers above the melting temperature, as already observed previously on several systems[6,16,40]. The stabilization of a few solid layers of an adsorbate by a crystal surface above T_m can be called presolidification or surface freezing.

The temperature dependence of the quasi-liquid thickness for HD films condensed on MgO can be found in ref. 19.

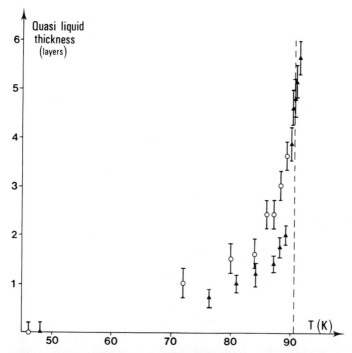

Fig. 8.   Quasi-liquid thickness versus T for ~ 10 layer-thick films of CH_4 adsorbed on graphite (▲) and on MgO (o). From Refs. 16,18.

As recalled in the introduction, the liquid thickness divergence depends on the type of interactions between surface atoms or molecules. In QENS experiments the limited number of experimental points and their corresponding error bars as well do not permit to draw definite conclusions on the value of the divergence exponent. Still the data obtained at low temperature (small liquid thickness) seem to favor a logarithmic divergence[18,19]. This behavior is probably due to the "proximity effect", the order of the solid propagating into the liquid at the interface[6,7].

V- DISCUSSION

QENS is a very powerful tool for measuring simultaneously the thickness and the mobility of the quasi-liquid layer. The large penetration depth of neutrons within condensed matter is used to probe the whole quasi-liquid phase, that cannot be done with He atoms which are sensitive to the only surface layer. QENS experiments are possible using powders with large specific areas. Secondary effects with respect to monocrystalline systems may occur and should be carefully analyzed. Up to now, the investigations have been carried out on powders on which a thin adsorbed film (about 10 layers) is deposited. With such systems, the experimentalist is faced to several questions:

a/ What is the growth mode of a thin film on a powder and the influence of capillary condensation?

b/ What may be the influence of size effects due to small crystallites or domains?

c/ Is surface melting observed on a thin adsorbed film, 10 layer thick, representative of surface melting on a semi-infinite crystal?

d/ The measured physical parameters are averaged over the isotropic powder distribution. Can we deduce unambiguously the effect at the solid-vapor interface of a single crystallite?

Other fundamental questions may be addressed. In particular, one may wonder whether the high translational mobility, similar to that of bulk liquid, is a strong enough argument to conclude the existence of surface melting. It is worth reminding the two phenomena predicted by the theoretical models: surface melting and surface roughening which may appear in close temperature ranges. Can we say that surface melting is actually observed in the QENS studies on CH_4/MgO, graphite and HD/MgO?

V.1. Characterization of the adsorbed film

For more than 10 years, numerous theoretical and experimental works[41-44] have been devoted to multilayer films adsorbed on different substrates, including graphite (0001) which has been largely used. Of interest is the problem of wetting: is it possible to condense a uniform film of infinite thickness (practically, of very large thickness) or is there a limited thickness of the adsorbate film that cannot be overrun. In the latter case, further deposition proceeds through crystallite or droplet condensation on top of the uniform film. Various experimental techniques can be employed for characterizing the wetting behavior: adsorption volumetry[42,45], calorimetry[6,46], ellipsometry[30], diffraction[8,27,32,47],... The adsorption isotherms are the more often used, but their sensitivity is limited to about 5 to 10 layers and they are usually unable to determine if the uniform film is thicker than this latter sensitivity threshold value. The other methods, sensitive to bulk signals, can say if bulk is present at higher coverages. The limitation comes usually from the similar properties of bulk and thick films that prevent any distinctions for thicknesses larger than about 10 layers. A minimum thickness of the uniform film is usually the only firm conclusion the experimentalist can draw, in case of complete wetting. For the systems studied using QENS and discussed in this paper, it is admitted a minimum thickness of 50 and 5 layers for methane on graphite[48] and MgO[8] respectively a few tens of degree below the melting point, and a minimum thickness of 7 layers of HD on MgO[17,19].

In reality, the wetting behavior can be perturbed by capillary condensation in powder systems[45,48]. This condensation cannot be avoided and appears generally simultaneously with the condensation of the uniform film beyond 4 or 5 layers. The capillarity effect can be seen on the adsorption isotherms. It explains the increase of the slope of the plateaus between successive steps. A measure of the capillary condensation can be deduced by this way. The isotherms recorded for CH_4/graphite and CH_4/MgO and reported on fig. 2 and fig. 1 in ref. 18 and 8 respectively, show that capillarity is negligeable for coverages lower than 5 layers. Beyond this statistical thickness, the presence of bulk crystallites is observed on neutron diffraction patterns measured on CD_4/MgO[8] and CD_4/graphite[47,49]. The neutron diffraction technique allows to determine precisely the amount of crystallites as a function of the total statistical thickness. Typically, 5 statistical layers of crystallites condense at 50 K for a nominal thickness of 10 statistical layers[8], for CD_4/MgO. Moreover, the mean size of the crystallites is estimated to be larger than 300Å (instrumental resolution) even when about 80% of the solid have disappeared some 1K below the bulk melting point. That means that these crystallites are certainly much larger than 300Å at lower temperatures. Their surface-to-volume ratio is small if compared to a uniform film. A straightforward calculation shows that the number of molecules involved in surface effects on the crystallites does not exceed 1/3 of the surface molecules on the uniform film, as long as the number of crystallites is not too large, i.e for crystallite statistical thicknesses smaller than 5 layers. The strong disappearance of solid crystallites when T is raised, as observed for CD_4/MgO[8] cannot be explained only by the surface melting phenomenon in the temperature range $0.80 < T/T_m < 0.99$. When the melting point is approached, some matter should be certainly transfered from the crystallites toward the uniform film through the liquid-like surface layer. In summary, bulk crystallites cannot be avoided in powder systems, but their surface effects remain less important than those of the uniform film for nominal thicknesses smaller than 10 layers.

V.2. Lowering of the melting temperature by size effects

Using very divided systems like powders obliges to consider size effects on the melting temperature[50]. This phenomenon is well known and can be easily estimated. Equation (8) expresses the melting temperature T for a crystallite with radius r :

$$T/T_m = \exp(-2\,\gamma_{sl}/\mu Lr)\qquad\qquad (8)$$

where T_m is the bulk melting temperature, γ_{sl} is the solid-liquid surface tension, μ is the volumic mass and L the melting latent heat.

For methane[8,49], the crystallites are always larger than 300Å and the melting temperature depletion is therefore only 1.5 K, at the very maximum (L = 220 cal/mol[51] and γ_{sl} is estimated to 20% of the liquid-vapor surface tension[48] $\gamma_{lg} = 13$ K/Å^2). The QENS and neutron diffraction measurements are consistent for revealing strongly mobile or disordered phases, respectively, from 10K below the bulk melting point and upward. Size effects may exist closer to T_m, but they cannot explain the clear-cut occurence of a liquid-like surface phase at lower temperatures. If the mobile phase was only made of droplets from the melting of small crystallites, its mobility should obey a 3D brownian model, that is not experimentally observed (fig.7).

V.3. Semi-infinite crystal surface and thin film surface

A uniform film of some 10 layers adsorbed on a substrate is generally considered as a model of semi-infinite crystal. This assumption is supported by the very small difference of chemical potential between the 10 layer thick film and bulk. The difference is, for instance, only 0.05% for methane/graphite and is certainly much smaller than the heterogeneities effects on the substrate surface[52]. If a precise balance of all the energetic terms involved in the equilibrium state of the adsorbed film is calculated, it shows that the substrate can strongly influence the surface melting phenomenon on thin films with respect to bulk. Different phase diagrams taking into account the substrate effect have been recently published[53]. The authors of this thermodynamic model argue for a specific behavior of methane/graphite. A first-order melting transition between homogeneous solid and liquid phases is expected, and the stratification of the film is not allowed. As a consequence of this prediction, substrate freezing should not occur that disagrees with the QENS[16] and neutron diffraction[49] measurements which show a solid phase at T>T_m. For methane/MgO and the other van der Waals systems studied up to now, the substrate should not perturb too much the surface melting effect and would act on the adsorbed film in stabilizing a solid layer at the substrate-adsorbate interface at temperatures higher than the bulk melting point (substrate freezing). Besides, a recent calorimetry study[54] shows that surface melting on thick films of Kr/graphite depends strongly on the total thickness of the film, in contrast with Ar and Ne/graphite. This observation is interpreted with a model involving strains within the adsorbed film due the substrate field. In conclusion, the substrate influence cannot be ignored and it may perturb surface melting in some cases. For the QENS studies we discuss here, the main effects observed are the limitation in the amount of the quasi-liquid layer thickness when the triple point is approached, simply due to the limited total thickness of the adsorbed film, and the existence of a frozen solid layer above the bulk melting point (substrate freezing).

V.4. The powder average

The physical parameters (thickness and mobility) deduced from the QENS studies are necessarily averaged over the isotropic powder distribution. Some information related to anisotropic effects are therefore lost. The power of the models used for small quasi-liquid thicknesses is noteworthy. The lattice fluid models, with a jumping distance defined by the underlying solid phase, account for the experimental measurements when the mobile phase is thinner than two layers. That result strongly favours the existence of a two-dimensional mobile layer coating the solid film, that is the unambiguous picture of a stratified film. The assumption of isotropic droplets is therefore ruled out. The peak shape of the neutron diffraction patterns[8] allows to determine the crystallinity profile at the solid film surface. This profile also implies the existence of a quasi-liquid layer on top of the solid phase.

One must admit that the QENS technique applied to isotropic powders cannot measure the component of the mobility parallel and perpendicular to the surface. It is also unable to distinguish the variation of mobility throughout the film thickness.

V.5. Surface melting and surface roughening

Surface melting and surface roughening are closely related phenomena which have been extensively described[1-3,55,56]. Surface melting is usually presented as the result of the balance of interfacial tensions which finally favors the formation of a liquid-like layer wetting the solid/vapor interface of a semi-infinite crystal at temperatures below the bulk melting point. The liquid-like character should be demonstrated by the vanishing of the shear modulus of the quasiliquid layer, by an enhancement of the diffusivity and a loss of long range order. A long standing confusion tends to equate surface melting and surface roughening. It is worth reminding that roughening is based on the vanishing of the pair vacancy-adatom energy on given crystallographic surfaces. Surface roughening preserves some long range order since the surface atoms remain located in lattice site positions even if a solid-vapor interface delocalization is expected. The surface disorder can be expressed in case of surface roughening by the divergence of the height-height correlation function at the surface. Unfortunately, no unique theory is able to describe both surface melting and roughening at the present time. Computer calculations[35] have dealt with these phenomena on Ar(110) facets. They reveal a strong decrease of the order parameter when the temperature is increased. Nevertheless, the order parameter retains a high nonzero value up to the melting point, indicating that the surface is disordered but still crystalline, i.e. rough but not melted. These simulations also show large diffusion coefficients similar to bulk liquid. Other experimental studies[57] have measured high diffusivity on crystalline surfaces below the bulk melting point which has been interpreted as a signature of roughening since the order parameter did not completely vanish. Some caution is therefore required for speaking of surface melting from the only QENS studies which are unsensitive to the order parameter. The lattice fluid models which describe well the thin mobile layer (about 1 layer), are probably nothing else than the pictures of a rough surface layer. When the bulk melting point is approached ($T/T_m > 0.94$), the thickness of the mobile layer agrees with that of the disordered layer, characterized by a short coherence length ($<30\,\text{Å}$), as deduced from the neutron diffraction study on CD_4/MgO. In this temperature range, liquid-like diffusivity and long range disorder appear to be closely related. This corresponds to surface melting rather fairly.

REFERENCES

1. J.F. van der Veen and J.W.M. Frenken, Surface Sci. 178:382 (1986)
2. J.G. Dash, Contemp. Phys. 30:89 (1989)
3. E. Tosatti, in : "The Structure of Surfaces II", vol. 11 of Springer Series in Surface Sciences, J.F. van der Veen and M.A. Van Hove eds. p. 535 Springer Verlag, Berlin, 1987
4. J.W.M. Frenken and J.F. van der Veen, Phys. Rev. Lett. 54:134 (1985) ; J.W.M. Frenken, P.M.J. Marée and J.F. Van der Veen, Phys. Rev. 34: 7506 (1986)
5. B. Pluis, A.W. Denier van der Gon, J.M.W. Frenken and J.F. van der Veen, Phys. Rev. Letters 59:2678 (1987)
6. D.M. Zhu and J.G. Dash, Phys. Rev. B38:11673 (1988)
7. J. Krim, J.P. Coulomb and J. Bouzidi, Phys. Rev. Lett. 58:583 (1987)
8. J.M. Gay, J. Suzanne, J.P. Coulomb, Phys. Rev. B41:11346 (1990)
9. Y. Furukawa, M. Yamamoto and T. Kuroda, J. Crystal Growth 82:665 (1987)
10. A.A. Chernov and V.A. Yakovlev, Langmuir 3:635 (1987)
11. "Surface Mobilities on Solid Materials", Vu Thien Binh (Plenum Pub. Corp. N.Y.) 1983
12. A.G. Naumovets and Yu.S. Vedula, Surface Sci. Reports 4:365 (1984)
13. M. Bienfait, in : "Dynamics of Molecular Crystals". J. Lascombe ed., Elsevier Sci. Pub. Amsterdam 353 (1987)
14. M. Bienfait, Europhys. Letters 4:79 (1987)
15. M. Bienfait and J.P. Palmari, in : "The Structure of Surface II", vol. 111 of Springer Series in Surface Sciences, J.F. Van der Veen and M.A. Van Hove Eds, p. 559 (Springer Verlag, Berlin, 1987)
16. M. Bienfait, J.M. Gay and H. Blank, Surface Sci. 204:331 (1988)
17. F.C. Liu, O.E. Vilches, M. Bienfait, P. Zeppenfeld, M. Maruyama and F. Rieutord, Bull. Am. Phys. Soc. 35:592 (1990)
18. M. Bienfait, P. Zeppenfeld, J.M. Gay and J.P. Palmari, Surface Sci. 226: 327 (1990)
19. P. Zeppenfeld, M. Bienfait, F.C. Liu, O.E. Vilches and G. Coddens, Journal de Physique, in press
20. J.W.H. Frenken, J.P. Toennies and C.H. Wöll, Phys. Rev. Lett. 60:1727 (1988)
21. M. Bée, Quasi-elastic Neutron Scattering (Hilger, Bristol, 1988)
22. J.P. Coulomb and M. Bienfait, J. Phys. (Paris) 47:89 (1986)
23. J.P. Coulomb, M. Bienfait and P. Thorel, Faraday Discuss. Chem. Soc. 80:81 (1985)
24. C.T. Chudley and R.J. Elliott, Proc. Phys. Soc. (London) 77:353 (1961)
25. A.C. Levi, R. Spadacini and G.E. Tommei, Surf. Sci. 121:504 (1982)
26. J.P. Coulomb, K. Madih, B. Croset and H.J. Lauter, Phys. Rev. Lett. 54:1536 (1985)

27. K. Madih, B. Croset, J.P. Coulomb and H.J. Lauter, Europhys. Lett. 8:459 (1989)
28. A. Thomy and X. Duval, J. Chim. Phys. 67:1101 (1970)
29. H.K. Kim, Q.M. Zhang and M.H. Chan, Phys. Rev. B34:4699 (1986)
30. H.S. Nham and G.B. Hess, Langmuir 5:575 (1989)
31. J. Ma, D.L. Kingsbury, F.C. Liu and O.E. Vilches, Phys. Rev. Lett. 61:2348 (1988)
32. J. Krim, J.M. Gay, J. Suzanne and E. Lerner, J. Physique (Paris) 47:1957 (1986)
33. D. Degenhardt, H.J. Lauter and R. Haensel, Japanese J. Appl. Phys., suppl. 26-3:341 (1987)
34. J.Q. Broughton and G.H. Gilmer, J. Chem. Phys. 79:5119 (1983)
35. V. Rosato, G. Ciccotti and V. Pontikis, Phys. Rev. B33:1860 (1986)
36. R.M. Lynden-Bell, Surface Sci. 230:311 (1990)
37. M. Maruyama, J. Cryst. Growth 89:415 (1988); 94:757 (1989)
38. K.D. Stock, Surface Sci. 91:655 (1980)
39. A.W. Denier van der Gon, R.J. Smith, J.M. Gay, D.J. O'Connor and J.F. van der Veen, Surface Sci.,227:143 (1990)
40. M. Bretz, J.G. Dash, D.C. Hickernell, E.O. McLean and O.E. Vilches, Phys. Rev. A8:1589 (1973);
    M. Bretz, in : Monolayer and Submonolayer Helium Films, J.G. Daunt and E. Lerner, Eds (Plenum Press, New York, 1973)
41. R. Pandit, M. Schick and M. Wortis, Phys. Rev. B26: 5112 (1982)
42. A. Thomy and X. Duval, J. Chim. Phys. 67:286 (1970)
43. J.A. Venables, J.L. Seguin, J. Suzanne and M. Bienfait, Surface Sci. 145:345 (1984)
44. J.A. Venables, G.D.T. Spiller and M. Hanbücken, Rept. Prog. Phys. 47:399 (1984)
45. F. Ser, Y. Larher and B. Gilquin, Molecular Phys. 67:1077 (1989)
46. M.S. Pettersen, M.J. Lysek and D.L. Goodstein, Surface Sci. 175:141 (1986)
47. J.Z. Larese, M. Harada, L. Passell, J. Krim and S. Satija, Phys. Rev. B 37:4735 (1988)
48. M.S. Pettersen and D.L. Goodstein, Surface Sci. 209:455 (1989)
49. J.M. Gay et al. , to be published
50. P. Buffat and J.P. Borel, Phys. Rev. A 13:2287 (1976)
51. J.H. Colwell, E.K. Gill and J.A. Morrison, J. Chem. Phys. 39:635 (1963)
52. J.G. Dash, J. Cryst. Growth 100:268 (1990)
53. M.S. Pettersen, M.J. Lysek and D.L. Goodstein, Phys. Rev. B40:4938 (1989)
54. D.B. Pengra, D.M. Zhu and J.G. Dash, preprint
55. J.G. Dash, in Proceed. Solvay Conf. on Surf. Sci., F.W. de Wette ed., Springer Verlag (1988)
56. D. Nenow and A. Troyanov, Surface Sci. 213:488 (1989)
57. H.P. Bonzel, Surface Sci. 21:45 (1970)

ON SURFACE MELTING

Andrea C.Levi

SISSA, Miramare
34014 Trieste
Italy

INTRODUCTION

You have already heard from Erio Tosatti the state of the art on surface melting. I wish to examine today some more special points and in particular:

1) How to describe surface melting thermodynamically?
2) How to compute the thickness of the molten layer?
This point should be divided into two parts:
    2a) Mean field
    2b) Fluctuations
3) What do molecular dynamics simulations teach us?
4) How does diffusion take place in the molten layer?
    4a) Experiments
    4b) Stochastic theories
    4c) Simulations

Many of the results discussed here refer to the Ph.D. work of Chen Xiaojie[1].

1   THERMODYNAMICS

Surface melting [2,3] takes place because, near the triple point, the chemical potential of the liquid, although higher, comes close to that of the solid and the vapour. Therefore the amount of free energy that is to be spent to build a molten layer may be compensated by the gain that is obtained by avoiding the drastic solid–vapour interface. It is immediate to see that if the interfaces had a finite thickness and if they did not interact with each other the molten layer might exist (provided $\gamma_{sv} > \gamma_{sl} + \gamma_{lv}$, where γ_{AB} is the interface free energy per unit area between phases A,B) but its thickness would remain finite when the triple point is approached (incomplete surface melting). The fact that the thickness increases and diverges at T_M, i.e. that the melting is in fact complete in most cases, implies that the thickness of an interface is infinite or that two interfaces interact or both.

We study this problem assuming the chemical potential μ to be a local quantity, in fact a function of thermodynamic variables. If the density is written as a periodic function

$$\rho = \sum_{\vec{g}} \rho_{\vec{g}} e^{i\vec{g}\cdot\vec{r}} \tag{1}$$

then the coefficients $\rho_{\vec{g}}$ are functions of the perpendicular coordinate z: deep in the solid they have their full value $\rho_{\vec{g}s}$ and they tend to zero as $z \to \infty$. The surface molten layer may be described as a range of z where $\rho_{\vec{g}}$ vanishes already for $\vec{g} \neq 0$

while ρ_0 remains fairly large. The natural (microscopic) thermodynamic variables are the Fourier coefficients $\rho_{\vec{g}}$, but in order to obtain a tractable problem only a small number of $\rho_{\vec{g}}$'s may be retained. The simplest theory would use only the mean density ρ_0. This is clearly insufficient (even the above definition of the molten layer would be inapplicable) but still teaches us something. The next simplest theory uses $\rho_0 = \rho$ and the set of $\rho_{\vec{g}}$'s with $|\vec{g}|$ minimal [4]. Assuming all these $\rho_{\vec{g}}$'s to be of equal magnitude c we are left with two variables: density ρ and crystal order c.
μ will be a function of ρ and c.

We may write the (Gibbs) free energy as a Ginzburg–Landau (or Cahn–Hilliard) functional

$$G = \int [J_\rho (\frac{d\rho}{dz})^2 + J_c (\frac{dc}{dz})^2 + \rho[\mu(\rho, c) - \mu_{eq}]dz = \int \mathcal{L} dz \qquad (2)$$

where the first two terms describe the cost of the surfaces in terms of Gibbs free energy and the third term the cost of interposing a non–equilibrium phase. Minimizing G is a variational problem yielding ρ and c as functions of z. The coefficients J_ρ and J_c are related to the γ's. E.g. if s_{ij} indicates the thickness of the interface between phases i and j

$$\gamma_{sv} = \frac{J_\rho \rho_s^2 + J_c c_s^2}{3 s_{sv}} \qquad \gamma_{\ell v} = \frac{J_\rho \rho_\ell^2}{3 s_{\ell v}} \qquad \gamma_{s\ell} = \frac{J_\rho (\rho_s - \rho_\ell)^2 + J_c c_s^2}{3 s_{s\ell}} \qquad (3)$$

where the auxiliary condition

$$\frac{\gamma_{sv} s_{sv} - \gamma_{s\ell} s_{s\ell}}{\gamma_{\ell v} s_{\ell v}} = \frac{2\rho_s}{\rho_\ell} - 1 \qquad (4)$$

is assumed to hold.
Let me first assume that the only relevant variable is ρ. Then the Euler equation is

$$\frac{d}{dz} \frac{\partial \mathcal{L}}{\partial \frac{d\rho}{dz}} = \frac{\partial \mathcal{L}}{\partial \rho} \qquad (5)$$

$$2 J_\rho \frac{d^2 \rho}{dz^2} = \frac{d}{d\rho} [\rho(\mu - \mu_{eq})] = -\frac{dV}{d\rho} \qquad (6)$$

This is the Newton equation for a point of mass $2 J_\rho$ in a potential V, z replacing time. V has the shape shown in Fig.1.

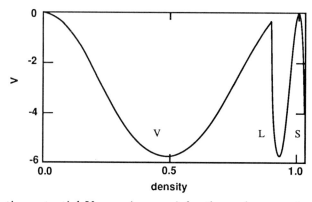

Fig.1 - The effective potential $V = \rho(\mu_{eq} - \mu)$ for three phases and one thermodynamic variable. V: vapour; L: liquid; S: solid.

$$V = \rho(\mu_{eq} - \mu) \qquad (7)$$

(in the vapour region $V = kT\rho - p - kT\rho \ln \frac{kT\rho}{p}$).

The important part of the profile is the liquid part, whose maximum level is $-\Delta$. The time (thickness) necessary to cross the liquid pass is logarithmic for a mass point that starting towards S from V with zero velocity, has energy zero. In the liquid region $V \approx -\Delta - A(\rho - \rho_{eq})^2$ and the liquid density range is $(\rho_{eq} - \delta, \rho_{eq} + \delta)$. The time spent on top of the liquid hill (i.e. the thickness of the molten layer) becomes very long near the triple point; it depends logarithmically on Δ:

$$\ell = \int \frac{d\rho}{v} = 2\sqrt{J} \int_{\rho_L \delta}^{\rho_L + \delta} \frac{d\rho}{\sqrt{E - V(\rho)}} = 2\sqrt{J} \int_{\rho_l - \delta}^{\rho_l + \delta} \frac{d\rho}{\sqrt{\Delta + A(\rho - \rho_L)^2}} = \qquad (8)$$

$$= 4\sqrt{\frac{J}{A}} \sinh^{-1}\left(\sqrt{\frac{A}{\Delta}}\delta\right) \approx \qquad (9)$$

$$\approx 2\sqrt{J/A}\,|\ln\Delta| + \text{const.} \qquad (10)$$

This logarithmic behaviour is quite general. More exactly $\ell = 2\sqrt{J_\rho/A}\,\ln(4A\delta^2/\Delta)$. It is also interesting to solve for Δ, obtaining

$$\frac{\alpha}{\sinh^2 \beta\ell} = \Delta \qquad (11)$$

$$\alpha = A\delta^2; \qquad\qquad \beta = \frac{1}{4}\sqrt{A/J} \qquad (12)$$

Let me now turn to a better theory, where two thermodynamic variables, density ρ and crystal order c, are involved [5]. The mechanical analogy, already introduced for one thermodynamic variable, becomes even more useful in this case.

MECHANICAL ANALOGY

mechanics	thermodynamics
time	level z
space coordinates	thermodynamic parameters ρ, c
mass	$2J_\rho, \quad 2J_c$
potential energy	$\rho(\mu_{eq} - \mu)$
action	free energy functional G
Newton Equations	Euler equations

The Euler–Newton equations are

$$2J_\rho \frac{d^2\rho}{dz^2} = \frac{\partial[\rho(\mu - \mu_{eq})]}{\partial \rho} \qquad 2J_c \frac{d^2 c}{dz^2} = \frac{\partial[\rho(\mu - \mu_{eq})]}{\partial c}. \qquad (13)$$

The problem is then to study the trajectories from the vapour V to the solid S (trajectories that don't reach S and get lost have no physical significance). Only a finite number of physical trajectories exists (usually 1 or 3).

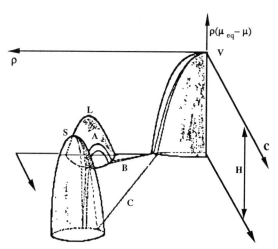

Fig.2 - $\rho(\mu_{eq} - \mu)$ as a function of two parameters: ρ (density) and c (crystal order). Three trajectories, two of the surface melting type (A,B) and one of the sublimation type (C) run from V (vapour) to S (solid).

A typical situation is as follows (see Fig.2): One of the trajectories passes far from the liquid point L (sublimation trajectory). At low temperature only the sublimation trajectory exist. At higher temperatures there appear also trajectories passing near L and, in fact, approaching L more and more closely as the temperature is increased towards the triple point. These are surface melting trajectories. In this case the stable thermodynamic state corresponds to the trajectory having the lowest Gibbs free energy (usually the sublimation trajectory); the other trajectories represent metastable states: surface melting may be metastable.

Another possibility, however, is that when the triple point is approached, the sublimation trajectory ceases to exist: then either there is only one trajectory which is of the surface melting type, or all trajectories are of such type. This is the true surface melting case. Such a scenario is expected when the non–equilibrium chemical potential is very high far from the equilibrium phases, i.e. when the energy depends strongly on the exact mutual positions of atoms, e.g. for a two–body potential system.

The thickness ℓ of the molten layer is given by the time the trajectory spends near the liquid point L. The displacements near L depend exponentially on time; hence the latter depends logarithmically on Δ.

More precisely

$$\ell = 2\sqrt{J_{c/a}} \ln\left(\omega^{-1} + q\omega^{2\sqrt{b_L}-2}\right) \approx \tag{14}$$

$$\approx 2\sqrt{J_{c/a}}[\ln\omega^{-1} + q\omega^{2\sqrt{b_L}-1}] \tag{15}$$

where $\frac{\partial^2(\rho\mu)}{\partial\rho^2} = a$ in proximity of the equilibrium phases

$$b_p = \frac{J_\rho}{aJ_c}\frac{\partial^2(\rho\mu)}{\partial c^2}\bigg|_p \tag{16}$$

for $p = S, L, V$

$$q = \frac{4b_s b_L}{b_s + \sqrt{b_s b_\ell} + b_\ell} \tag{17}$$

$$\omega = \frac{\Delta}{ax_\ell(x_s - x_\ell)}. \quad (x_p = \sqrt{\frac{J_\rho}{J_c}}\rho_p \quad \text{for} \quad p = S, L, V) \tag{18}$$

When long–range Van der Waals forces are included, the dependence of ℓ on Δ change from logarithmic to power–law.

The thermodynamic cost (per particle) of creating a molten layer is reduced by Van der Waals forces from Δ to $\Delta - (2H/\ell^3)$ (where H is *Hamaker's constant*). For example, in the naive approximation where only one thermodynamic parameter is used, the equation $\alpha/(\sinh^2 \beta\ell) = \Delta$ becomes

$$\frac{\alpha}{\sinh^2 \beta\ell} + \frac{2H}{\ell^3} = \Delta. \tag{19}$$

If $H = 0$, ℓ depends logarithmically on Δ.

If $H > 0$, the second term asymptotically wins and $\ell \sim (2H/\Delta)^{1/3}$.

If $H < 0$ i.e. if the liquid is stronger than the solid (e.g. for Bi, Ge: for germanium see below, however) ℓ is bounded, $\ell < \ell_{max}$ (blocked melting). For $|H|$ small ℓ_{max} is of the order of $\ln |H|/(2\beta)$.

If, however, H becomes even more negative, less than a critical value $H_c \approx -\alpha/2\beta^3$, no solution exists and ℓ vanishes up to the triple point (non–melting).

A useful concept is that of the effective potential $V(\ell)$. The Gibbs free energy can be viewed as a function of ℓ; the system tries to minimize $V(\ell) = V_0(\ell)$ (the equilibrium part) $+\ell\Delta$ (the chemical potential part). Thus the equation $\alpha/(\sinh^2 \beta\ell) + 2H/\ell^3 = \Delta$ corresponds to minimization of $V = V_0 + \ell\Delta$ where the equilibrium part is given by

$$V_0(\ell) = \frac{2\alpha}{\beta} \frac{1}{e^{2\beta\ell} - 1} + \frac{H}{\ell^2}. \tag{20}$$

Minimizing the effective potential is the content of the mean–field approximation. However, the effective potentials presented here are oversimplified. Probably the most important effect that has been overlooked is layering: the effective potential contains an oscillatory contribution. When $\Delta \to 0$ it may happen that only a finite number of minima become negative. Then the melting is blocked for quite different reasons from above, where the blocking was caused by Van der Waals forces. The blocking due to layering is only provisional if the Hamaker constant is positive and closer to the triple point first–order surface melting is expected. If H is negative its effect reinforces that of layering and surface melting is blocked altogether.

In germanium melting was observed by LEED and by Rutherford back scattering to be blocked at about two molten layers. This is probably too thin to be due directly to the negative Hamaker constant and may be attributed to layering; however, due to the negative Hamaker constant, the blocking is expected to persist up to the triple point.

2 HAMAKER CONSTANT CALCULATIONS

For molecular crystals the Hamaker constant may be obtained, at least in first approximation, by summing the long–range attractive potential of all the molecules. But for more general solid systems this procedure becomes of course meaningless and the evaluation of the Hamaker constant requires more careful consideration.

The problem was treated in full generality by Lifshitz and coworkers in terms of electromagnetic fluctuations in a sandwich where a thin medium, 3, is located between two half-spaces respectively filled with media 1 and 2 [6]. Assuming 1 to be vapour, 2 solid and 3 liquid the result is

$$
\begin{aligned}
H &= \frac{\hbar}{32\pi^2} \int_0^\infty d\xi\, a(\xi) \int_0^\infty dx \frac{x^2 e^{-x}}{1 + a(\xi)e^{-x}} \\
&= \frac{\hbar}{16\pi^2} \int_0^\infty d\xi \sum_{n=1}^\infty (-1)^{n+1} n^{-3} a^n(\xi)
\end{aligned} \tag{21}
$$

where

$$a(\xi) = \frac{\epsilon_s(\xi) - \epsilon_l(\xi)}{\epsilon_s(\xi) + \epsilon_l(\xi)} \cdot \frac{\epsilon_l(\xi) - 1}{\epsilon_l(\xi) + 1}. \tag{22}$$

Here ϵ_A is the complex dielectric function for phase A, evaluated at the imaginary frequency $\omega = i\xi$ (the analytic continuation is unquivocal, because ϵ_A is just a sum of simple poles).

The Hamaker constant H is positive, corresponding to interface repulsion, if the liquid is less conducting than the solid ($\epsilon_l(\xi) < \epsilon_s(\xi)$ for most ξ). This is the usual situation; in some cases, however, e.g. for Ge which turns metallic at melting, the liquid is more conducting than the solid and H is negative, giving rise to an effective interface attraction and ultimately to blocked melting.

It must be stressed that considerably extensive data have to be used (up to the far ultraviolet range) because the value of the Hamaker constant is controlled by the difference between the solid and the liquid dielectric functions. Indeed if the Drude approximation (valid at low frequencies) was used, a result would be obtained in excess by a factor of 3 over the true value.

When the correct results (which include very small corrections for the T-dependence of the electromagnetic properties and for the contribution of core levels) are used to evaluate the thickness of the molten layer, the agreement with the data of Van der Veen and coworkers is excellent [7] (see Fig.3). This also indicates that mean-field is a good approximation near the triple point (the upper critical dimension is less than 3). On the other hand it becomes a bad approximation for larger Δ, especially for molecular systems, as the MD simulations indicate (see below).

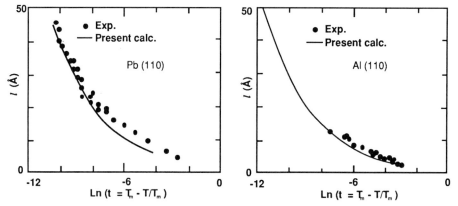

Fig.3 - The thickness of the molten layer computed for lead and aluminum using the calculated Hamaker constant.

3   MOLECULAR DYNAMICS SIMULATIONS

Surface melting is a difficult problem–too difficult for analytic theory, at least at the present stage. A reasonably convincing model was presented by Trayanov and Tosatti, but it could only be treated mean field [8]. The fluctuation corrections are in a worse shape still. We will come back to those problems later.

It is possible, however, to explore the problem of surface melting by simulation. This was first done by Broughton and Gilmer, whose thorough study was presented in a series of papers [9]. Some discussion, however, remained inconclusive, because the small thickness which the molten layer could achieve made many of its properties (included the thickness itself) subject to large relative fluctuations. For this reason we decided to extend such molecular dynamics studies in Trieste: this work was performed by Chen Xiaojie.

The systems considered were of two kinds. To study the dependence of the thickness ℓ on Δ, and similar properties that Broughton and Gilmer had left indecided, an artificial "supermelting" potential, described below, was introduced. This potential has the property that very thick molten layers are generated, so that the numerical fluctuations are relatively largely suppressed. On the other hand dynamical properties such as diffusion and, more generally, the quasi–elastic structure factor for neutron or atom scattering, were studied on the unmodified (apart from cutting the attractive tail smoothly at a conveniently chosen distance) Lennard–Jones potential.

Supermelting

For a molecular system where the attractive intermolecular potential behaves asymptotically as $-C_s/r^s$, the "repulsive" Hamaker term causing the increase of the molten layer thickness is given by H/ℓ^{s-4} if $s > 4$ or by $-H \ln \ell$ if $s = 4$. $H = p_s C_s \rho_{liq} (\rho_{sol} - \rho_{liq})$ where $p_s = 2\pi/[(s-2)(s-3)(s-4)]$ if $s > 4$ and $p_4 = \pi$.

The case $s = 3$ is pathological, because in the present picture it would imply a macroscopic thickness for the molten layer at a temperature $T < T_M$, which is contradictory. Thus 4 is the lowest possible integer.

The thickness of the molten layer is obtained asymptotically from

$$\frac{(s-4)H}{\ell^{s-3}} = \Delta \quad \ell = \left[\frac{(s-4)H}{\Delta} \right]^{\frac{1}{s-3}} = \left[\frac{2\pi C - s\rho_{liq}(\rho_{sol} - \rho_{liq})}{(s-2)(s-3)\Delta} \right]^{\frac{1}{s-3}} \quad (\text{for} \quad s > 4) \tag{23}$$

$$\frac{H}{\ell} = \Delta \quad \ell = \frac{H}{\Delta} = \frac{\pi C_4 \rho_{liq}(\rho_{sol} - \rho_{liq})}{\Delta} \quad (\text{for} \quad s = 4), \tag{24}$$

which is the value the previous expression takes in the limit $s \to 4$. Thus in all cases

$$\ell = \left[\frac{2\pi}{(s-2)(s-3)} \frac{C_s \rho_{liq}(\rho_{sol} - \rho_{liq})}{\Delta} \right]^{\frac{1}{s-3}}. \tag{25}$$

Thus for $s = 4$ the thickness increases rapidly as $(T_m - T)^{-1}$. This fast thickness increase is called **supermelting** [11].

The simulations have the following characteristics:

1) The constancy of temperature can be achieved with good precision.

2) The parameter O_4 (another measure of crystallinity) is 1 at low z, small at intermediate z showing the passage from the crystal to the molten layer.

3) The density as a function of z shows layering. The last–but–one peak corresponds to the outer layer: diffusion and other phenomena are studied there. The last peak corresponds to adatoms flying over the surface.

4) The thickness of molten layer is shown in Fig.4: its increase is first logarithmic, then power–law as expected, but the absolute values are considerably larger than mean–field predictions. This may be due to two reasons: a) either the elementary mean–field is incorrect b) or fluctuations are important; or more probably both!

a) A true, microscopic mean–field theory was presented by Trayanov and Tosatti [8]. The results, given in Fig.4, show an improvement over naive mean field theory, but not sufficient. b) Fluctuations should be considered.

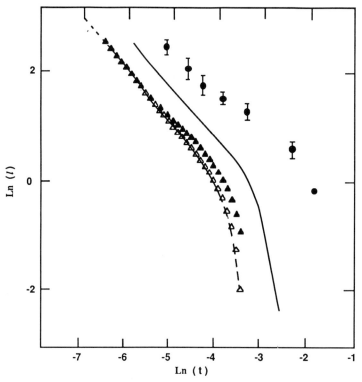

Fig.4 - Molten layer thickness. Dashed curve and open triangles: naive mean-field theory; Solid curve: Trayanov and Tosatti; Filled triangles: fluctuation theory described in the text. Data points from simulation.

4 THE ROLE OF FLUCTUATIONS

Considering fluctuations is in many ways similar to passing from Newton's mechanics to quantum mechanics. Instead of choosing the configuration that minimizes the free energy, we accept all configurations and sum over them.

Thus a functional integral partition function is generated, similar to Feynman's path integral:

$$Z = \int D\{\rho, c\} \exp(-\beta G\{\rho, c\}\xi^2) \tag{26}$$

when ρ, c depend on xyz and where the characteristic length ξ is introduced because G is Gibbs' free energy per unit area.

In the approximation (considered by Lipowsky) where the only fluctuating quantity is the thickness $\ell(x, y)$ [12], Z reduces to

$$Z = \int D\ell(x, y) \exp(-\beta \xi^2 G\{\ell(x, y)\}) \tag{27}$$

(G is now a functional of $\ell(x, y)$). The surface is not flat (especially on the solid side, see Fig.5): there are fluctuations both in the centre of mass of the molten layer (generalized capillary waves) and in the thickness.

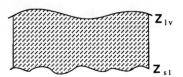

$\mathbf{z}_{1v}$

$\mathbf{z}_{sl}$

Fig.5 - A fluctuating quasi-liquid layer.

Moreover, ρ and c may fluctuate locally. Thus considering only ℓ–fluctuations is a rather drastic approximation. The surface average of the thickness $< \ell >$ is another functional of $\ell(x,y)$. The mean thickness $<< \ell >>$ is given by

$$<< \ell >> = \frac{1}{Z} \int D\ell \exp(-\beta \xi^2 G) < \ell > . \tag{28}$$

The mean–field approximation ℓ_0 is obtained by taking the minimum of G. Following a suggestion by Baskaran, expand ℓ in plane waves (in the plane)

$$\ell = \ell_0 + \frac{1}{\sqrt{S}} \sum_{\vec{K}} (A_{\vec{K}} \ell^{i\vec{K}\vec{R}} + \delta_{\vec{K}}) \tag{29}$$

$\delta_{\vec{K}}$ is the displacements of the centre of oscillation.

For each $\vec{K}$ the contribution is optimized by minimizing

$$< V(\ell_0 + \frac{1}{\sqrt{S}} A_{\vec{K}} \cos\varphi + \delta_{\vec{K}}) > \tag{30}$$

with the result (for small amplitudes)

$$\delta_{\vec{K}} = -\frac{1}{4\sqrt{S}} \frac{V'''(\ell_0)}{V''(\ell_0)} |A_{\vec{K}}|^2. \tag{31}$$

Then

$$G = \sum_{\vec{K}} [\frac{1}{2}\sigma K^2 + V''(\ell_0)] |A_{\vec{K}}|^2 \tag{32}$$

where by equipartition we find $|A_{\vec{K}}|^2$ and $\delta_{\vec{K}}$:

$$< |A_{\vec{K}}|^2 > = \frac{kT}{\frac{1}{2}\sigma K^2 + V''(\ell_0)} \tag{33}$$

$$\delta_{\vec{K}} = -\frac{kT}{\sqrt{S}} \frac{V'''(\ell_0)}{V''(\ell_0)[\frac{1}{2}\sigma K^2 + V''(\ell_0)]}. \tag{34}$$

Then

$$<< \ell >> = \ell_0 + \frac{1}{\sqrt{S}} \sum_{\vec{K}} < \delta_{\vec{K}} > = \tag{35}$$

$$= \ell_0 + \frac{\sqrt{S}}{4\pi^2} \int\int < \delta_{\vec{K}} > d^2 K = \tag{36}$$

$$= \ell_0 - \frac{kT}{8\pi\sigma} \frac{V'''(\ell_0)}{V''(\ell_0)} \ln[1 + \frac{\sigma K_{max}^2}{2V''(\ell_0)}] \tag{37}$$

(the correction is positive because $V'''(\ell_0)$ is negative).

The effective surface tension σ is to be found separately (this can be done exactly by treating the fluctuations systematically and stopping at first order, where everything is translationally invbariant in the z–direction, i.e. the only fluctuations are capillary waves and the molten layer thickness is constant over the surface).

This is not really necessary, however. $V''(\ell_0)$ is an increasing function of Δ. Near the triple point the fluctuation is negligible anyway (we are above the upper critical dimension for the problem). Far from the triple point the logarithm may be expanded and σ disappears:

$$<< \ell >>= \ell_0 - \frac{kT}{16\pi} K^2 V'''(\ell_0) \tag{38}$$

and for an attractive potential in r^{-s}

$$<< \ell >> \ell_0 + \frac{s-2}{s-3} \frac{kT K_{max}^2}{16\pi\Delta}. \tag{39}$$

For supermelting $(s = 4)$ $kT K_{max}^2/8\pi\Delta = H'/\Delta$ is of the same order as $\frac{H}{\Delta}$ (mean field) and the Hamaker constant is effectively increased from H to $H + H'$.

It may be seen from Fig.4, however, that the fluctuation correction, as calculated here starting from the naive mean-field theory, is too small (a better mean-field starting point, of the quality of the Trayanov-Tosatti theory, would be required anyway).

5   SURFACE DIFFUSION AND THERMODYNAMIC STRUCTURE FACTOR

Diffusion in the molten layer is an important problem for several reasons. First of all it is only by a diffusion experiment that a distinction can be made between a true molten layer and a simple disordered layer. Secondly, there is a wealth of experiments in this field: macroscopic deformation experiments [13], scattering experiments by Frenken et al. with atoms [14] and by Bienfait et al. with neutrons [15].

I want, in particular, to discuss the scattering experiments and the related theoretical and simulation work.

The structure factor $S(\vec{Q},\omega)$ comprises a self and a distinct part

$$S(\vec{Q},\omega) = S_s(\vec{Q},\omega) + S_d(\vec{Q},\omega) \tag{40}$$

(approximately corresponding to diffuse and coherent scattering respectively). Let the width of the self part $S_s(\vec{Q},\omega)$ be $\omega_{1/2}(\vec{Q})$. Let us choose the x direction and study $\omega_{1/2}(Q_x)$.

For free diffusion $\omega_{1/2}(Q_x) = DQ_x^2$.

For jump diffusion $\omega_{1/2}(Q_x) = 2D1 - \cos a_x Q_x/a_x^2$.

Reality must be in between. Let us consider the experimental data (by Frenken et al. [14] and Bienfait et al. [15]), the solution of the Smoluchowski equation in a periodic potential (Ferrando) [16] and the simulation data (Chen Xiaojie).

Diffusion in a periodic potential was studied extensively (but mostly in the small $|\vec{Q}|$ limit).

A general theory should include both position and velocity variables. The resulting Fokker–Planck equation applies to probability density in phase space.

But averaging over velocities (physically: assuming the velocities to be in equilibrium: this is accurate for smooth potentials, long times), the Fokker–Planck equation is replaced by the much simpler *Smoluchowski equation* in the <u>spatial</u> density only:

$$\frac{\partial n}{\partial t} = D_0 \nabla^2 n - \beta D_0 \nabla \cdot (n\vec{F}) \tag{41}$$

$$\vec{F} = -\nabla V, \quad V \quad \text{periodic.} \tag{42}$$

The equation is solved expanding in eigenvectors of the operator at the r.h.s.; the eigenvalues give the decay rates.

For small Q

$$\omega_{1/2}(\vec{Q}) = DQ^2 \quad \text{where} \quad D \text{ is a renormalized diffusion coefficient.} \quad (43)$$

In 1–dimension Festa and Galleani [17] find:

$$D = \frac{D_0}{< \exp(\beta V) >< \exp(-\beta V) >} < D_0. \quad (44)$$

$$\text{averages over the unit cell} \quad (45)$$

For large Q $\omega_{1/2}(\vec{Q})$ lies between free diffusion and jump diffusion. The envelope of $\omega_{1/2}(\vec{Q})$ is similar to the free diffusion parabola DQ^2, but $\omega_{1/2}(\vec{Q})$ vanishes at the reciprocal lattice vectors as in jump diffusion. Physical meaning: when $\vec{Q}$ equals a reciprocal lattice vector $\vec{G}$ all atoms in the same positions relative to their unit cells scatter in phase, so that the neutron (or the atom) cannot know whether they have diffused or not, so the correlation times become extremely long, so the frequency width $\omega_{1/2}$ becomes extremely narrow (see Fig.6).

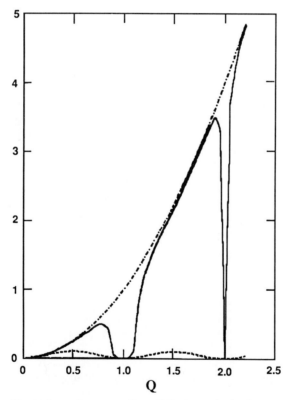

Fig.6 - $\omega_{1/2}(\vec{Q})$. Dash-dotted curve: free diffusion; dashed curve: jump diffusion; solid curve: Smoluchowski equation.

On the (110) surface (Frenken's experiments) there are channels along the $[\bar{1}10]$ direction one should distinguish diffusion across the channels ($[001]$ direction) and along the channels.

For the former a cosine is found, both experimentally (Frenken et al.) and by simulation (Chen Xiaojie). Physical interpretation: the motion across the channels is difficult, the barriers are high, and diffusion is simply solid–like jump diffusion. The same is found with the Smoluchowski equation.

Along the channels the situation is more interesting. Both simulations and the Smoluchowski equation show a much larger width $\omega_{1/2}$ in the second Brillouin zone than in the first. Moreover the experiments show a minimum at zero boundary, interpreted as double jumps. The same effect is found by Chen Xiaojie at the boundary between the first and the second Brillouin zones, and again between the second and the third. This feature cannot be reproduced by the Smoluchowski equation, because the latter, assuming instantaneous equilibration of velocities, cannot describe double jumps. Double jumps imply persistence of velocities, and a flight over the first well to reach the next [18].

ACKNOWLEDGMENTS

It is a pleasure to express my gratitude to Erio Tosatti and to Chen Xiaojie. The present lecture is basically a short account of part of their work.

REFERENCES

1 X.J.Chen,Ph.D.Thesis,SISSA,Trieste,in preparation
2 J.W.M.Frenken and J.F.van der Veen,Phys.Rev.Letters **54** (1985)134
3 Da-Ming Zhu and J.G.Dash,Phys.Rev.Letters **57**(1986)2959
4 D.W.Oxtoby and A.D.J.Haymet,J.Chem.Phys. **76**(1982)6262
5 A.C.Levi and E.Tosatti, Surf.Sci. **189/190**(1987)641
6 I.E.Dzyaloshinsky, E.M.Lifshitz and L.P.Pitaevsky, Soviet Phys.JETP **10**(1961)161
7 X.J.Chen,A.C.Levi and E.Tosatti, to be published
8 A.Trayanov and E.Tosatti, Phys.Rev.Letters **59**(1987)2207
9 J.Q.Broughton and G.H.Gilmer, J.Chem.Phys. **79** (1983)5095,5105,5119
10 X.J.Chen, F.Ercolessi and E.Tosatti, Helv.Phys.Acta **62** (1989)824
11 Chen Xiaojie, A.Trayanov and E.Tosatti, to be published
12 R.Lipowsky, Z.Phys. **B55**(1984)345
13 See e.g. Vu Thien Binh and P.Melinon, Surf.Sci. **161** (1985)234
14 J.W.H.Frenken,J.P.Toennies and C.H.Wöll, Phys.Rev. **B41**(1990)938
15 M.Bienfait,P.Zeppenfeld,J.M.Gay and J.P.Palmari, Surf.Sci. **226**(1990)327
16 R.Ferrando,*Tesi di laurea*,University of Genova,1988; R.Ferrando, R.Spadacini, G.E.Tommei and A.C.Levi, Physica A, to be published
17 R.Festa and E.Galleani d'Agliano, Physica **90A**(1978)229
18 G.E.Tommei, private communication.

FROST HEAVE AND THE SURFACE MELTING OF ICE

J.G.Dash

Department of Physics, University of Washington
Seattle, Washington 98195

I. INTRODUCTION.

Ice is the most common substance exhibiting premelting phenomena. Observations by a variety of experimental techniques have shown that liquid water persists in porous media, to temperatures as low as $-40°$ C. The melting of ice at surfaces has extremely important consequences in the environment, including the low frictional resistance of ice, the sintering of snow, and the creep of glaciers. There are serious non equilibrium effects associated with the persistence of liquid water at subzero temperatures; migration of this water under temperature gradients accounts for frost heave, which damages man made structures and acts as an agent for extensive geological change in temperate and subpolar climates. The causes of the persistence of unfrozen water and its migration in temperature gradients have been actively debated for many years, but a considerable advance in their explanation can be made in terms of recent developments in the fundamental theory of surface melting. This paper reviews the environmental effects and laboratory studies, and discusses the phenomena in the context of the physics of surface melting.

II. FROST HEAVE and ICE SEGREGATION

It has long been known that much of the geomorphic development of arctic and alpine areas is dictated by freezing and thawing in soils and rocks. A variety of dynamical effects includes frost heave and ice segregation, patterned ground formation, size sorting in soils, upfreezing of stones, and frost weathering of rock. (Anderson 1973, Anderson 1988, Taber 1930, Miller 1978, Gilpin 1980, O'Neil 1985, Hallet 1988, Van Vliet -Lanoe 1988, Walder, 1986). These dynamical effects commonly go under the heading of 'frost heave'. A description of frost heave and its serious consequences are given in a National Research Council report (Polar Research Board 1984):

"Nearly everyone living in the northern and southern temperate zones has experienced the effects of ice segregation and frost heaving through the destruction of roads and highways, the displacement of foundations, the jamming of doors, the misalignment of gates, and the cracking of masonry. Many people often have simply and mistakenly assumed that these effects result solely from the expansion of pore water on freezing.

*When confined, water can rupture pipes, break bottles, and crack rocks as it freezes.
However, most of the destructive effects of frost heaving are caused by "ice segregation",
a complex process that results from the peculiar behavior of water and other liquids as
they freeze within porous materials. In particular, water is drawn to the freezing site
from elsewhere by the freezing process itself. When this water accumulates as ice,
it forces the soil apart, producing expansion of the external soil boundaries, as well
as internal consolidation. The dynamic process of ice segregation and the expansion
resulting from freezing of the in situ pore water, together, cause frost heaving".*

Typical frost heave conditions in the natural environment are shown schematically,
in Fig.1. The diagram is of a section of ground overlying an aquifer at $T > T_o$, with
temperature decreasing at higher ground, to subfreezing at ground level. Above the
aquifer is a region of wet unfrozen ground, followed by a layer of partly frozen soil
(the 'frozen fringe'), and then a stratum of solid ice (an 'ice lens'). Under steady
state conditions water flows upward from the aquifer, to crystallize as ice on the lower
surface of the lens. The transport is driven through the frozen fringe by the 'frost
heave pressure', and through the unfrozen soil by capillary forces. The frost heave
pressure is opposed by the weight of the overburden. As long as the heaving pressure
dominates and the temperature gradient and supply of liquid are maintained, the flow
continues and the ground surface steadily rises.

The seriousness and persistence of the engineering problems are described in the
Council report:

*"Major engineering projects have focused attention on problems created by the freezing
of water in earth materials. These problems received detailed attention during the
design of the Trans-Alaska oil pipeline, and additional aspects have been encountered
during the design of the Alaska Natural Gas Transportation System.... Efforts to design
roadways, airfields, buildings, powerlines, and other structures to prevent damage from
frost heaving date from the early 1900's..."*

A review published some years ago (Anderson 1973) is a useful introduction to the
subject, with a description of field observations, laboratory studies, and an appraisal of
theories current at that time. The physical basis for the persistence of unfrozen water
in soils below the normal melting point has been attributed to various mechanisms,
including electrostatic forces, density variations, impurities, and pressure melting. In
1984 there was no consensus on the fundamental causes: according to the 1984 Council
report, *"....no model enjoys universal or general acceptance...".*

Active controversy continues to this day.

There is an extensive literature of frost heave, especially in journals of soil science,
glaciology, and hydrology. In the following sections we present an explanation of frost
heave in terms of the theory of surface melting, a review of the experimental evidence
for premelting of ice at interfaces with solid walls, and a description of quantitative
studies of transport and frost heave pressures in porous media.

III. THEORY

The fundamental principles of surface melting are well established, and several
reviews are available (Broughton 1983; Nenow 1986; van der Veen 1988; Schick 1988;
Dash 1988, 1989a). A number of discussions of frost heave have been presented, but

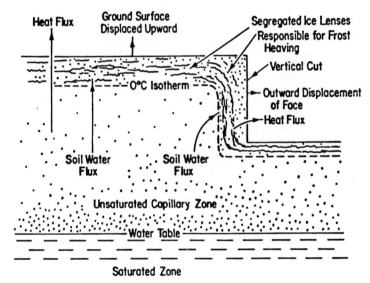

Fig. 1. Typical conditions of frost heave and ice segregation in the natural environment (Polar Research Board 1984).

in journals less familiar to the surface physics community (Derjaguin 1986; Forland 1988; Gilpin 1980b; Kuroda 1985). The frost heave papers generally focus on the ice-water system to the exclusion of others, whereas the theory shows that analogous phenomena should occur in other systems. Therefore, in the following we cast the thermodynamics of surface melting in terms that are convenient for bringing out the motivation for pressure and migration effects (Dash 1989b).

Consider a solid in equilibrium with vapor or a wall at temperature T and pressure P (usually, one considers an interface with a low density vapor phase, but the thermodynamics holds equally well for solid wall interfaces). If the interface is wetted by a macroscopic layer of the melt (m) liquid the free energy of the layer is composed of bulk and surface terms:

$$G_m(T, P, d) = [\rho_\ell \mu_\ell(T, P)]d + \Delta\gamma f(d) \cdot \gamma_s \tag{1}$$

where d is the thickness, ρ_ℓ and μ_ℓ are the density and chemical potential of the bulk liquid (ℓ). γ_s is the interfacial coefficient of the unwetted solid (s), $\Delta\gamma$ is the difference between the interfacial coefficients with and without the melted layer, and $f(d)$ is the thickness dependence of the coefficients. In general, $f(d)$ is a positive monotonic function tending to unity at infinite thickness. For example, in simple van der Waals materials with unretarded potentials, $f(d) = (1 - \sigma^2/d^2)$, where σ is a constant on the order of a molecular diameter.

In thermodynamic equilibrium, the chemical potential of the melted layer is equal to that of the solid. Differentiating Eq.1,

$$\mu_m(T, P, d) = \frac{1}{\rho_\ell}\left(\frac{\partial G_m}{\partial d}\right)_{T,P} = \mu_\ell(T, P) + \left(\frac{\Delta\gamma}{\rho_\ell}\right)\frac{\partial f}{\partial d} = \mu_s(T, P) . \tag{2}$$

The interfacial energy term in μ_m introduces an effective *thermomolecular pressure* δP_m acting on the melted layer. This can be seen from the general thermodynamic effect on the chemical potential of a system by a change in its external environment. If the unperturbed chemical potential of a system is $\mu(T, P)$ and the field energy per molecule is u, the chemical potential is changed to

$$\mu'(T, P, u) = \mu(T, P) + u = \mu(T, P) + \delta P/\rho . \tag{3}$$

Hence the interfacial energy term introduces an added pressure,

$$\delta P_m = \Delta\gamma(\partial f/\partial d) = \rho_\ell[(\mu_s)_{T,P} - (\mu_\ell)_{T,P}] . \tag{4}$$

At the transition (T_o, P_o) of the bulk material, $\Delta\mu \equiv (\mu_s - \mu_\ell) = 0$, hence δP_m vanishes, but $\Delta\mu \neq 0$ off bulk coexistence. To evaluate $\Delta\mu$ at other temperatures and pressures we expand in T, P about (T_o, P_o). To first order,

$$\Delta\mu(T, P) = \left[\frac{\partial\Delta\mu}{\partial T}\right]_o (T - T_o) + \left[\frac{\partial\Delta\mu}{\partial P}\right]_o (P - P_o) = -q_m t + \left[(1/\rho_s - 1/\rho_\ell)\right](P - P_o) , \tag{5}$$

where q_m is the latent heat of melting per molecule and $t \sim (T_o - T)/T_o$. In our original derivation we neglected the pressure shift. This term is generally much smaller than the temperature shift, but its form and importance depend on whether the melting occurs at an interface with vapor or a solid wall. I am grateful to H. Wagner and to S. Dietrich and M. Napiorkowski for pointing out the omission and for calculating the pressure correction for vapor interfaces. In the following we include the pressure effect for both vapor and solid wall interfaces.

At vapor interfaces the system exists along the solid-vapor equilibrium line, hence the pressure term can be expressed in terms of the temperature shift and the slope $(dP/dT)_{sv}$ of the boundary. Then combining Eqs.(4) and (5), the thermomolecular pressure is

$$\delta P_m = \rho_\ell t[-q_m + (1/\rho_s - 1/\rho_\ell)T_o(dP/dT)_{sv}] . \tag{6}$$

Making use of the Clausius-Clapeyron relation for the bulk melting transition, Eq.(6) yields the thermomolecular pressure for vapor interfaces:

$$(\delta P_m)_v = -\kappa_v \rho_\ell q_m t , \tag{7a}$$

where the pressure factor is

$$\kappa_v = \left[1 - \frac{(dP/dT)_{sv}}{(dP/dT)_{s\ell}} \right] . \tag{7b}$$

The magnitude of κ_v is typically a very minor correction; in H_2O, for example, it differs from unity by about 3 ppm.

At interfaces with solid walls the pressure shift depends on the thermomolecular pressure. At local equilibrium ('maximum heave pressure'), $(P - P_o) = -\delta P_m$. With this subsitution, Eqs.(4) and (5) yield the thermomolecular pressure for interfaces with solid walls:

$$(\delta P_m)_w = -\rho_s q_m t . \tag{8}$$

Thus the correction factor turns out to be the ratio (ρ_s/ρ_ℓ), which typically causes a change of 5 to 10 percent. In most substances the correction increases the coefficient, but in water it is a decrease.

The thickness of the liquid film depends on the type of interface and the nature of the intermolecular forces. For example, in the case of van der Waals forces, substituting the specific form of f(d) given earlier in Eq.(4), and combining with Eq.(5), we obtain for vapor interfaces

$$d_{sv} = \left(-2\sigma^2 \frac{\Delta\gamma}{\kappa_v \rho_\ell q_m} \right)^{1/3} t^{-1/3} . \tag{9a}$$

For surface melting at a wall,

$$d_w = \left(-2\sigma^2 \frac{\Delta\gamma}{\rho_s q_m} \right)^{1/3} t^{-1/3} . \tag{9b}$$

In the case of short range forces, $\partial f/\partial d \propto \exp(-cd)$, where c is a constant, the result is a logarithmic temperature dependence, $d \propto |\ell n t|$. Other types of interactions, such as dipolar forces, can produce other, similarly specific temperature dependences.

We see that in contrast to the specificity of $d(T)$, the thermomolecular pressure is independent of the nature of the intermolecular forces. The negative sign reflects the fact that the interfacial free energy is lowered by surface melting, so that $\Delta\gamma < 0$. It may appear that the interfaces are mechanically unstable under the influence of a negative pressure acting on the liquid, tending to pull the interfaces apart. However, such thickening of the liquid layer would require the conversion of a quantity of solid to liquid, which would raise the free energy of the system above its equilibrium value. Conversely, the stability of a liquid layer at temperatures below the normal melting point necessitates a negative thermomolecular pressure, i.e. a suction.

The thermomolecular pressure can be detected if a temperature gradient is imposed along the surface. The resulting pressure gradient will cause mass transport in the liquid layer, toward lower temperature (see Fig.2). If the interface is a single, low area surface the mass transport may be very slow and difficult to detect, but it will be much more apparent in a porous medium with large interfacial area, such as in fine soil or clay.

The theory predicts that thermomolecular effects can occur in other substances. Table 1 lists the thermomolecular pressure coefficients for several materials, according to Eq.(8). It should not be inferred from its inclusion in the list that a particular substance does surface melt, but if any facets do premelt, the pressure coefficient will be the same value for all. A thermomolecular effect can occur even in the absence of surface melting, if liquid and solid can be made to coexist under conditions displaced from the bulk phase equilibrium melting line. In the case of surface melting, the displacement is caused by interfacial free energies. Another possible cause is supercooling, which can be more than $30°$ in water. Supercooling has been employed in this way in some studies of frost heave, as described in Sec. V. In this connection, it is interesting that a 'frost heave' effect of He^4 in Vycor glass has been observed recently (Hiroi 1989), although helium does not have a solid-liquid-vapor triple point. The effect was made possible by the displacement of the melting pressure of helium confined to micropores. The measured coefficient was found to be in good agreement with the theoretical value over a limited temperature range.

We note that since the density and latent heat are functions of P and T the local pressure coefficient will vary along an extended region subjected to a finite temperature difference. Also, it is important to point out that although Eqs. 7 and 8 are completely independent of the nature of the interactions, their universality rest on the assumption that the melt liquid is identical with the bulk. This is an approximation, since a liquid is modified in the proximity of its boundaries, where it is more ordered (Broughton 1981). The proximity effect is believed to be a common phenomenon in liquids adjacent to

Table 1. Thermomolecular coefficients of various substances at solid wall interfaces.

Substance	T_m (K)	dP_m/dT (bar/K)
Argon	83.	5.7
Mercury	234.	7.1
Water	273.	11.2
Gallium	292.	2.6
Lead	600.	4.5

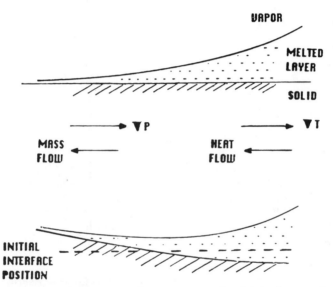

Fig. 2. Mass and thermal transport along an unconfined interface undergoing surface melting.

solid boundaries. It has been studied extensively in theory and computer simulations. Few measurements have been made, although the effect should be important in the dynamics of thin films, such as in the transport of liquid during frost heave. A recent study has shown that the range of the effect in liquid rare gases is on the order of a few molecular layers (Zhu 1988). Since the range of the proximity effect depends on the nature of the liquid and the wall, bulk properties are appropriate only in the limit $T \rightarrow T_m$, where d diverges. As T decreases the layer thins and the proximity effect becomes more important, causing the pressure to fall below the asymptotic relation, in a non universal manner.

In the next section we review the experimental evidence for surface melting of ice in contact with solid walls.

IV. UNFROZEN WATER at SUBZERO TEMPERATURES.

Many studies have detected the presence of unfrozen water persisting to temperatures well below $0°C$. Much of the evidence has been obtained from saturated soils and other porous media, with additional results from flat or nearly planar low area substrates. These results have been interpreted by some observers as surface melting, but others claim that they indicate static disorder. Even where the surface layers are admitted to be liquid, there is still no general acceptance that they are due to surface melting of pure H_2O) at equilibrium (i.e. the interfacial energy-driven mechanism discussed in Sec.III). Other causes, which have been proposed as dominant or strongly contributory, include pressure melting, curvature and solute effects, and charge separation at active substrate interfaces. Although each of these mechanisms may be present along with surface melting in soils, the evidence strongly points to surface melting as the dominant process over most of the temperature range of frost heave phenomena. In addition a number of recent experiments with more completely characterized substrates present an increasingly convincing case for the effect.

Several techniques have been used for studies of the unfrozen water content in various soils, including dilatometry and calorimetry (Anderson 1973), nuclear magnetic resonance (Tice 1982,1984,1990; Black 1989), acoustic transmission (Deschatres 1989), and em pulse transmission (van Loon 1990). The results generally show that the liquid fraction is proportional to the specific surface area. The liquid fraction increases with temperature, and in fine soils reaches 100 percent a few tenths of a Kelvin below the bulk melting point. Fig. 3 shows measurements between $-10°$ and $0°$ in an assortment of soils with different average particle sizes and size distributions . The temperature dependence of the liquid fraction has been empirically fitted by logarithmic or power law functions of the difference $(T_l - T)$ (Anderson 1973).

In nmr studies, the persistence of unfrozen water is detected by the persistence of a narrow spectral component down to temperatures well below $0°$. The sharp component is attributed to motional narrowing due to a combination of rotational and translational diffusion. Tice, et al (Tice 1982,1984) found that the signal strength of the narrow component in a clay soil had a temperature dependence qualitatively similar to the unfrozen water content determined by other methods. Two additional important results were obtained. The unfrozen water content remained constant as the total amount of water was reduced, as long as some ice was present, i.e. the reduction was at the expense of the ice phase. This is consistent with what is expected from surface melting, since the liquid layer thickness is determined by surface coefficients, and the

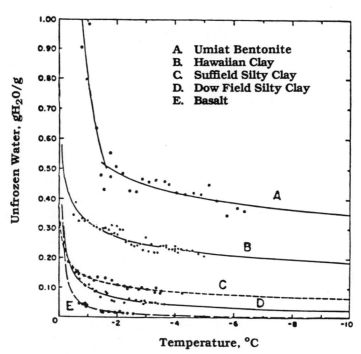

Fig. 3. The unfrozen water content in various soils, determined by nuclear magnetic resonance (Anderson, 1973).

quantity of the underlying solid, if more than a minimal thickness, is unimportant. In the later report, Tice, et al. measured the unfrozen water content in several frozen silts containing varying compositions of natural solutes. Their results showed that although the dissolved salts caused appreciable freezing point depression, an important appreciable thickness of mobile component remained when the salts were leached out of the system.

Nuclear resonance measurements of pure H_2O in finely divided silica indicate that liquid fraction rises smoothly with T; the estimated thickness of liquid increases from about 20 $\mathring{A}$ at $-6°$, to 50 $\mathring{A}$ at $-2°$ (Barer 1977). These results were later used in the analysis of macroscopic motion, in which the freezing characteristics at quartz interfaces was calculated from the pressure-driven sliding of ice columns in fine capillaries (Barer 1980). The movement was interpreted in terms of a lubricating layer of unfrozen water at the ice-quartz boundary; the thickness and viscosity was consistent with values determined in the nmr work.

Careful quantitative measurements of the thickness of surface melted layers have been obtained by "wire regelation" a classic experiment that is sometimes used to demonstrate *pressure melting* to undergraduates. The experiment shows that a weighted wire will sink slowly through a block of ice, leaving no groove after its passage. The traditional interpretation of the motion has been that local melting occurs under the wire due to the local pressure, and refreezing occurs above, the rate being controlled by heat conduction across the wire (Zemansky 1944). If T and P are high enough, pressure melting certainly contributes to the motion. However, in a careful quantitative study (Gilpin 1980a) of several wire diameters and materials, the applied pressures were kept too low to cause pressure melting except at temperatures very close to 0°. Nevertheless, there was appreciable drift of the wire at temperatures as low as -35°C. Gilpin analyzed the motion as due to viscous flow of premelted liquid around the wire, dependent on the thickness and viscosity of the fluid. He concluded that the thickness could be described by a power law in $(T_o - T)$, with an empirical exponent 1/2.4 for all wire diameters and materials over the entire temperature range (Fig.4). A reexamination of Gilpin's data (Dash 1989a) shows that they are also consistent with an exponent 1/3, i.e. the theoretical value for surface melting controlled by van der Waals forces (see Sec.III) over a narrower temperature range, where the calculated thickness is at least 20 Å, and a more rapid decrease of thickness below this temperature. Rather than a more rapid drop in thickness, the decreased flow rate can be due to an effective increase of viscosity in the very thin liquid, because of the proximity effect.

A current study applied quasi elastic neutron scattering to the problem. In this method very low energy neutrons are analyzed for their energy shifts after scattering from a target. The method is especially useful for exploring the very low energy states which are characteristic of liquids, where the scattered intensity has a Lorentzian line shape with a width proportional to the self diffusion constant (Page 1972). Fig.5 shows Intensities were recorded and analyzed for 30 discrete scattering angles between 0 and 2 radians, but are here shown as the summed intensities for all scattering angles. the intensity distributions of neutrons scattered from a target of water in graphitized carbon black powder (Maruyama 1990), illustrating the evolution of an increasingly broad component as T rises. At the highest temperatures the diffusion constant agrees closely with the value for bulk water, but as T falls D drops well below the bulk value

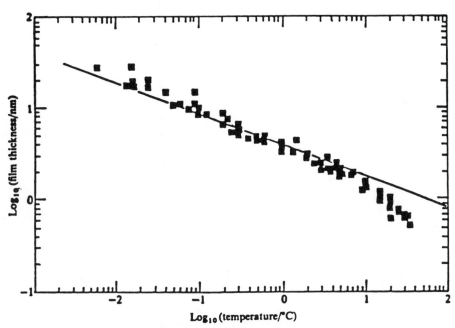

Fig. 4. Thickness of the liquid layer around wires imbedded in ice, calculated from measurements of 'wire regelation' (Gilpin 1980). The line is a power law, with exponent 1/3, corresponding to the theoretical T dependence for surface melting governed by dispersion forces. The more rapid decrease of apparent thickness at low temperature is due to the decreased fluidity in very thin films (Dash 1989a).

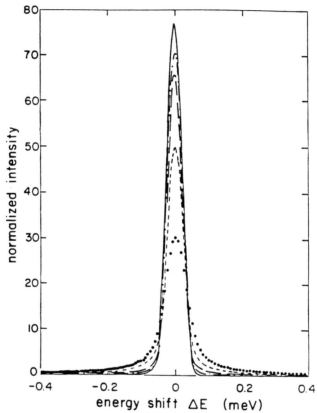

Fig.5. Intensity profiles of very low energy neutrons scattered from a mixture of H_2O and graphitized carbon black. (Maruyama 1990). Temperatures in °C of the individual curves are: ————— , –50; —— · —— · ——, –2.5 —— —— ——, –1.4; – – – – – – – , –0.8; · · · · · · +2.3

for the same T, due to the thinning of the layer. The choice of powder material was made for several reasons. The adsorption and surface properties of graphite are well known; it is chemically inert, its surfaces are microscopically flat, being dominantly basal plane facets, and these facets have a crystal structure fine grained relative to that of ice. Thus, the graphite presents well characterized microscopically uniform boundaries, with no specific chemical interactions at the interfaces.

Recent nmr studies of the unfrozen water content in various soil/water mixtures have yielded empirical power law exponents approximately equal to the theoretical value $(-1/3)$ for surface melting governed by van der Waals forces (Black 1989). In Fig.6 we show the results of another series of such measurements by the same group, plotted so as to demonstrate that the theoretical power law is closely followed over a span of more than three decades of temperature.

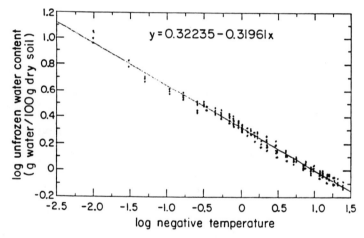

Fig. 6. The unfrozen water in a fine soil, determined by nuclear magnetic resonance (Tice 1990). The soil is nearly pure carbonates, primarily $CaCO_3$; average particle size is about 6 micrometers. Measurements were on several samples of the soil, taken from the same source.

Several studies of fine ice particles have been made by nmr techniques, showing diffusion narrowed components at various starting temperatures as low as $-100°$, and increasing at higher T (Kvilividze 1974, Ocampo 1983, Mizuno 1987). The relative strength of the narrow component was greater in more finely dispersed samples, indicating that it was associated with the interfacial area. However, although a portion of the narrowing may be due to surface melting (a part is attributed to rotational diffusion), it cannot be identified uniquely with the ice-vapor, ice-substrate, or ice grain boundary interfaces.

V. PRESSURE and FLOW at SUBZERO TEMPERATURE.

There have been many measurements of transport rates and heaving pressures in soils of varying composition and fineness, and of water containing varying amounts of chemical impurities. In general, the results show that the flow rate and maximum heaving pressure are proportional to the temperature difference across the sample, but analysis is often complicated by uncertainties due to chemical impurities and particle size distribution. Some studies have been carried out under simpler conditions, where liquid has been maintained at lower than normal temperatures by supercooling (see Sec.III.). Such measurements have yielded pressure differences agreeing with the thermomolecular coefficient for water, as shown in Fig.7, while flow rates through capillaries between chambers of ice and supercooled water are shown in Fig.8.

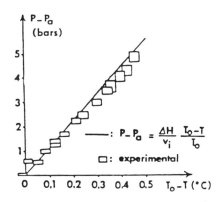

Fig. 7. Measured pressure difference between reservoirs of ice and supercooled water through a microporous membrane (Buil 1981).

Frost heave pressures in a fine clay soil, illustrated in Fig. 9, show a linear trend with temperature near 0°, with a slope in reasonable agreement with the theory. Other studies show that the maximum pressure near 0° is independent of the nature of the chemical composition and particle size distribution in the soil, although the rate of pressure buildup is slower in coarser soils. In Fig. 9 the data fall below the theory as the temperature decreases below about $-10°$, which may be due to a combination of proximity effect and incomplete equilibrium.

Applications of the theory for calculating flow rates in real systems, particularly in natural geological deposits, require extensive characterization of details such as impurity levels, particle size and connectivity, and go beyond the scope of our discussion. The several journal articles and symposium proceedings which are listed in this review can provide an introduction to the practical applications of the basic principles.

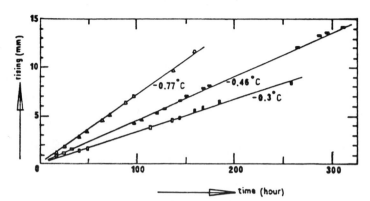

Fig. 8. Mass transport through a narrow channel between reservoirs of ice and super-cooled water (Vignes 1974).

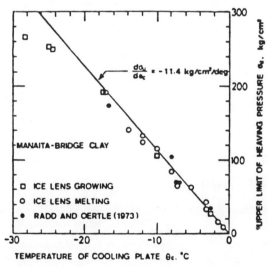

Fig. 9. Maximum ice heave pressure versus temperature difference in fine silt (Takashi 1981).

ACKNOWLEDGMENTS

I am grateful to several colleagues and students for their important contributions to this review. Marcia Baker introduced me to the general subject of ice in the environment, and motivated me to explore the theoretical connection between surface melting and thermomolecular pressure. Bernard Hallet introduced me to frost heave, and is a valued colleague in these studies. Michael Schick has been a colleague and friendly critic in this work as in our many years of common interest in surface physics. Michael Elbaum and Hai Ying Fu, have stimulated me by their questions, criticisms, and research results. I give warm thanks to H. Wagner and to S. Dietrich and M. Napiorkowski, who pointed out an omission in an earlier version of the theory. The support of The National Science Foundation, through Grants DMR-89133454 and INT-8612187, is gratefully acknowledged.

REFERENCES

Anderson,D.M. and Morgenstern, N.R., 1973, in *Permafrost, Proc. 2nd Int. Conf.*, Natl. Acad. Sci., Wash. D.C., pp257-288.

Anderson,S.P., 1988, Bull. Geol. Soc. Am.**100**, 609; Proc. 5th Int. Conf. Permafrost, Trondheim, 666.

Barer, S.S., Kvilividze, V.I., Kurzaev, A.B., Sobolev, V.D., and Churaev, N.V., 1977, Dokl. AN SSSR **235**, 601.

Barer, S.S., Churaev ,N.V., Derjaguin, B.V., Kisileva, O.A., and Sobolev, V.D., 1980, J. Colloid Interface Sci. **74**, 173.

Black, P.B. and Tice, A.R. 1989, Wat. Resources. Res. **25**, 2205.

Broughton, J.Q., Bonissent, A.and Abraham, F.F., 1981,J. Chem. Phys.**74**, 4029; Oxtoby, D. and Haymet, A.D.J., 1982, J. Chem. Phys. **76**, 6262; Israelachvili, J.N. and McGuiggan, P.M., 1988, *Science* **241**, 795 (1988); Tarazona,P., and Vicente, L., 1985, Mol. Phys. **56**, 557.

Broughton, J.Q. and Gilmer, G.H. 1983, J. Chem. Phys. **79**, 5119(1983).

Buil,M., Soisson, A. and Aguirre-Puente,J., 1981, C.R.Acad.Sci.(Paris) **293**, ser.II, 653.

Cary, J.W. 1987, Water Resources Res.**23**, 1620.

Colbeck,S.C., 1979, J.Colloid Interface Sci.**72**, 371.

Dash, J.G., 1988, in *Proc. Solvay Conference on Surface Science*, F.W. de Wette, ed., Springer-Verlag; 1989a, Contemp.Phys. **30**, 89; 1989b, Science **246**, 1591.

Derjaguin B.V.and Churaev,N.V., 1986, Cold Regions Science and Technology **12**, 57.

Deschatres, M.H., Cohen-Tenoudji, F., and Aguirre-Puente, J., 1989, 5th Int. Symp. on Ground Freezing, Nottingham, p.551.

Dietrich, S., (Academic, London 1988) in *Phase Transitions and Critical Phenomena*, ed. by C. Domb and J. Lebowitz, Vol. 12.

Dietrich, S. and Napiorkowski, M. 1990, private communication.

Forland,K.S. ,Forland,T., and Ratkje, S.K.,1988, *Irreversible Thermodynamics*, J. Wiley and Sons, pp.235-244.

Gilpin, R.R., 1980a, J. Colloid Interface Sci. **77**, 435; 1980b, Wat. Resources Res. **16**, 918.

Hallet, B., 1976, J. Glaciology **17**, 209; Hallet, B., Anderson, S.P., Stubbs, C.W., and Gregory, E.C., 1988, Proc. 5th Intl. Conf. on Permafrost, Trondheim, 770.

Hiroi,M. ,Mizusaki,T., Tsuneto,T., Hirai,A.,and Eguchi,K., 1989, Phys. Rev.B**40**, 6581

Kisileva,O.A. and Sobolev,V.D., 1980, J. Coll. Interf. Sci. 1980, **74**, 173.

W. R. Koopmans and R. D. Miller, Proc. Soil Sci. Soc. Am. **30**, 680 (1966).

Kuroda,T. and Lacmann,R.,1982, J.Cryst. Growth **56**, 189.

Kuroda,T.,1985, 4th Int. Symp. Ground Freezing, Sapporo; 1987, J.de Physique **48**, C1-487.

Kvilividze, V.I., Kisilev, V.F., Kurzaev, A.B., and Ushakova, L.A., 1974, Surf. Sci. **44**, 60.

Lipowsky, R., 1982, Phys. Rev. Lett. **49**, 1575.

Low, P.F. 1976,1979,1980, Soil Sci. Soc. Am. J. **40**, 500; **43**, 651; **44**, 667.

Maruyama,M., Bienfait, M., Coddens, G., and Dash, J.G., 1990, private communication.

Miller,R.D., 1978, Proc. 3rd Int. Conf. on Permafrost, National Research Council, Ottawa.

Mizuno, Y. and Hanafusa, N., 1987, J. de Physique **C1-48**, c1-511.

Mulla,D.J., 1983a, J. Coll. Interf. Sci. **100**, 576; Mulla, D.J. and Low, P.F. 1983b, J. Coll. Interf. Sci. **95**, 51.

Nenow, D. and Trayanov,A., 1986, J. Cryst. Growth **79**, 801.

Ocampo,J. and Klinger,J., 1983, J.Phys.Chem. **87**, 4325.

Oliphant, J.L. and Low, P.F. 1982, 1983, J. Coll. Interf. Sci. **89**, 366; **95**, 45.

O'Neill, K. and Miller, R.D., 1985, Water Resources Res. **21**, 281.

Page, D.L. 1972, in *Water, a Comprehensive Treatise, v.1*, (F. Franks, ed.), Plenum Press.

Piper, D., Holden, J.T., and Jones, R.H. 1988, 5th Int. Conf. on Permafrost, Norway, p.370.

Polar Research Board 1984. *Ice Segregation and Frost Heaving*, Nat.Acad.Press, Wash.D.C.

Radd, F.J. and Oertle, D.H., 1973, 2nd Int. Symp. Permafrost, Yakutsk, p.377.

Riecke, R.D., Vison, T.S., Mageau, D.W., 1983, Proc. 4th Intl. Conf. on Permafrost, Fairbanks, 1066. 2734.

Schick, M., 1988, in *Liquids at Interfaces*, Les Houches Summer Scool Lectures, Session XLVIII, to be published.

Shreve,R.L. 1984, J.Glaciology **30**, 341.

Smith, M.W. and Tice, A.R. 1988, 5th Int. Conf. on Permafrost, Trondheim

Taber, S., 1929, J. Geol. **37**, 428; 1930, J. Geol. **38**, 303.

Takashi, T., Ohrai, T., Yamamoto, H., and Okamoto, J., 1981, Eng. Geol.**18**, 245.

Tice, A.R., Oliphant, J .L., Nakano, Y., and Jenkins, T.F. 1982, Cold Regions Research and Engineering Laboratory Report 82-15, Office of the Chief of the US Army Corps of Engineers.

Tice, A.R., Zhu Yuanlin, and Oliphant, J.L., 1984, CRREL Report 84-16.

Tice, A.R. and Hallet, B., 1990, private communication.

van der Veen, J.F., Pluis, B., and Denier van der Gon, A.W., 1988, in *Chemistry and Physics of Solid Surfaces, Vol. VII*, Springer-Verlag.

van Loon, W.K.P., Perfect, E., Groenevelt, P.H., and Kay, B.D. 1990, Proc. Int. Symp. on Frozen Soil Impacts on Agricultural, Range, and Forest Lands, Spokane, Washington, (K.R.Cooley, ed.), CRREL Special Rpt. 90-1, p.186.

Van Vliet-Lanoe, B., 1988, Proc. 5th Intl. Conf. on Permafrost, Trondheim, 1088.

Vignes,M. and Dijkema,K.M.1974,J. Coll. Interface Sci. **49**,165

Wagner, H. 1990, private communication.

Walder, J., and Hallet, B., 1985, Geol. Soc. of America, Bulletin **96**, 336; 1986, Arctic and Alpine Res., **18**, 27.

Weber, T.A. and Stillinger, F.H., 1983, J.Phys.Chem. **87**, 4277.

Van Vliet-Lanoe, B., 1988, Proc. 5th Intl. Conf. on Permafrost, Trondheim, 1088.

Vignes,M. and Dijkema,K.M.1974,J. Coll. Interface Sci. **49**,165

Zemansky, M.W. 1944, private communication. One of the author's most vivid memories as an undergraduate at The City College of New York (now CUNY) is of Prof. Mark Zemansky demonstrating wire regelation to the Physics Club. After explaining the physical principles according to the pressure melting mechanism, he admitted that the actual process "seems to be more complicated than that".

D. M. Zhu and J. G. Dash, Phys. Rev. Rev. Lett. **60**, 432 (1988).

MULTILAYER PHYSISORBED FILMS ON GRAPHITE

George B. Hess

University of Virginia
Charlottesville, VA 22901

ABSTRACT

This paper reviews experimental work in the multilayer range on physisorption of classical rare gases and simple molecular gases on graphite. The first part outlines ideas about various phenomena which occur or may occur in these systems, such as wetting and limitations to complete wetting, surface roughening and surface melting as they are manifest in films of finite thickness, and substrate-induced effects such as interface-induced freezing and layering in liquid films. Next a number of specific systems are examined from an experimental point of view for evidence of these phenomena. For example, layer critical points in solid rare gas and methane films on graphite have been associated with surface roughening of (111) facet of the corresponding bulk crystal. Surface melting of certain facets has been identified with behavior seen in argon and other films on graphite as well as on MgO slightly below the bulk melting temperature of the adsorbate. On the other hand, freezing of several layers of thick neon and argon films above the bulk melting point is attributed to interaction with the graphite substrate. Certain liquid films, such as those of ethane, ethylene, and oxygen on graphite, exhibit a strong tendency for layering in the proximity of the substrate. Complex phase diagrams may occur in layered films due to interaction of layering with melting and possibly also orientational ordering; examples include oxygen and tetrafluoromethane on graphite. Finally, certain experimental difficulties of a general nature are discussed.

INTRODUCTION

The introduction of exfoliated graphite a little over two decades ago provided a large-area adsorption substrate which was chemically inert and much more homogeneous than those available previously, such as graphitized carbon black. This was particularly evident in the sharpness of layer condensation steps in adsorption isotherms, and the resolution of finer features such as layer melting [1-3]. Compressed or confined forms of exfoliated graphite, such as Grafoil, Papyex, ZYX, and graphite foam*, provided thermal conductivity adequate for heat capacity measurements. This initiated a period of expanded interest and rapid progress in physisorption studies, employing both classical methods and techniques new to this area, and with strong interaction with theory. Some of this work is reviewed in [4-11]. At the same time innovative but less extensive work was done on other substrates, including metal single crystals and films, alkali halides, powders of lamellar halides, and MgO; this work will not be reviewed here.

---

*Grafoil and ZYX are products of Union Carbide Corp.; Papyex is a product of Le Carbone Lorraine.

Although adsorption isotherm studies often extended to about five layers, a large fraction of work in the 1970's and even up to the present has concentrated on monolayer films. Apart from the limitations of some techniques, this is natural in view of the rich phenomena found in monolayers and important theoretical input for quasi-two-dimensional systems. Work on multilayer films has been motivated in part by systematic progression to higher coverage and in part by interest in phenomena relating to the transition from two to three dimensions, such as wetting, surface roughening, and surface melting.

Techniques

It is not possible to describe here the many techniques which have been applied to the study of physisorbed films. However it may be useful to mention some characteristics and limitations of a few widely employed techniques which effect their ranges of applicability, especially for the multilayer regime.

Among thermodynamic techniques, vapor pressure isotherm and heat capacity measurements will be mentioned. A family of vapor pressure isotherms can be thought of as determining the two-dimensional equation of state; that is, the adsorbate coverage n (molecules per unit area) or surface excess density, as a function of temperature T and chemical potential μ. In equilibrium the chemical potential is equal for the adsorbate in the film and that in the coexisting three-dimensional vapor, which can usually be treated as an ideal gas. In that case

$$\mu = k_B T \ln p - k_B T \ln (k_B T/\lambda^3) \qquad (1)$$

where $\lambda = h(2\pi m k_B T)^{-1/2}$ is the thermal de Broglie wavelength. For convenience we will henceforth omit Boltzmann's constant k_B and measure chemical potential in temperature units. In the usual experiment the temperature is regulated, the vapor pressure p is measured, and the coverage is either inferred from knowledge of the amount of gas admitted to the cell (the volumetric technique) or measured directly, for instance by ellipsometry (e.g., [12]) or by the mass loading of a mechanical oscillator of which the adsorbing substrate is part (e.g., [13,14]). It becomes difficult to make precise vapor pressure measurements below roughly 10^{-3} to 10^{-5} Torr. Thermomolecular corrections also become important in this range and somewhat higher, if the pressure gauge is not at the temperature of the adsorption cell, and cause additional uncertainty in pressures and pressure ratios. Finally, at pressures somewhat below this range, equilibration via vapor transport may become prohibitively slow, particularly for a large-area powder sample such as exfoliated graphite. Thus vapor-pressure-isotherm measurements are limited to sufficiently high vapor pressures and hence, for a given adsorbate, sufficiently high temperature. On the other hand, heat capacity measurements can be extended to the lowest temperatures, provided that care is taken in the cool-down procedure to ensure that the adsorbate is uniformly distributed over the adsorbent. There may be limitations on the high-vapor-pressure side if the ratio of adsorbent surface area to cell volume is not sufficiently large. This is because an increasing portion of the adsorbate charge is required to occupy the vapor volume of the cell, leading to a large desorption contribution to the measured heat capacity. This does not prohibit conventional adiabatic heat-pulse calorimetry up to vapor pressures of order one atmosphere (e.g., [15,16]), but may be a limitation when other considerations impose a less favorable surface-to-volume ratio.

A diffraction technique is required to obtain structural information. Diffraction probes may be classified as those which interact strongly with the adsorbate and those which interact weakly. The former include beams of atoms, which see strictly the top atomic layer of the diffracting sample, and low energy electrons (LEED), which typically sense to a depth of two or three layers. The vapor pressure must be sufficiently low to allow transmission of the beam, not more than about 10^{-4} Torr for LEED and in practice much less for atom scattering. Consequently these probes have been applied predominantly to monolayer films. To avoid severe electron-induced desorption in LEED, the primary beam current density must be very low (~ 1 $\mu a/cm^2$), which requires an area detector of high sensitivity. Reflection high energy electron diffraction (RHEED) has been used to study the wetting behavior of thick films, although only at very low vapor pressures. The grazing-incidence diffraction patterns distinguish whether the adsorbate-vapor interface presents a flat surface parallel to the substrate (giving modulated streaks) or the adsorbate is clustered into compact crystallites (giving a pattern of spots). Some results will be cited later. Transmission high energy electron diffraction (THEED), using an electron microscope with a special sample stage, has been

358

applied to the growth of noble gas films on graphite and other substrates, also at low pressures [17, 18].

X-ray and neutron diffraction are weakly interacting probes, which consequently can be applied in a high vapor pressure environment and thus are broadly applicable to multilayer films. In general, large-specific-area powder substrates are required for sufficient diffracted intensity, so that information on bond orientation with respect to the substrate is lost. However the brightness of synchrotron x-ray sources is sufficient that diffraction from films on a single crystal surface has been studied (e.g., [19, 20]). High-Z adsorbates on a low-Z substrate such as graphite are favorable for x-ray diffraction, but there has been extensive work on adsorbates as light as O_2 and C_2H_4. For neutron diffraction a sufficiently large coherent scattering cross section is required; favorable adsorbates include [36]Ar, nitrogen, chlorine, and deuterium or deuterated hydrocarbons (e.g., [21]).

Inelastic scattering can provide information on the dynamics of films. Neutron and atom beams are used at thermal energies, and energy transfer measurements, by time-of-flight or momentum dispersion of the scattered beam, can provide phonon spectra of adsorbate layers as well as bare surfaces. Quasielastic neutron scattering (QENS) can measure translational or rotational mobilities of adsorbate molecules, and can, for instance, provide a direct indication of surface melting [22, 23]. Nuclear magnetic resonance relaxation times also provide information on translational and rotational mobility (e.g., [24]).

Wetting

A liquid or solid film of adsorbate can form at a substrate-vapor interface if the adsorbate molecule-substrate attractive potential $u \cdot f(z)$ is sufficiently strong compared to the adsorbate-adsorbate molecular interaction $v \cdot F(r)$. Here u and v give the energy scales of the respective interactions, while functions f and F which specify the distance dependence. For sufficiently large u/v the equilibrium film might be expected to grow uniformly to macroscopic thickness as the vapor pressure is increased to saturation. This is termed "complete wetting" or simply "wetting." On the other hand if u/v is small the adsorbate may form only a limited number of layers at vapor pressures up to saturation, at which point any additional adsorbate will condense into three-dimensional droplets or crystallites [25, 26]. This is termed "incomplete wetting" (or "nonwetting" if the number of layers is zero). As the actual criterion involves comparison of free energies, a system which exhibits incomplete wetting at low temperatures might wet completely above some wetting temperature T_w. This was studied systematically by Pandit, Schick, and Wortis [27]. It is an interesting experimental question whether there exist in physisorption systems any such wetting transitions which are not tied to phase transitions of the bulk adsorbate. (See, for instance, the review by Dietrich, [28]).

The preceding discussion should apply to liquid films. In the case of a solid film there may be an additional obstacle to wetting in that the structure favored in the first monolayer of adsorbate may not be compatible with any plane of the bulk crystal [29]. In this case, for continued growth of the film, the bottom layer must restructure when more layers are added, or else there must be a structural discontinuity between lower and higher layers. Either will introduce an extra interfacial energy which is likely to prevent wetting. This presumably accounts, for example, for the failure of the low temperature α-solid phases of oxygen and nitrogen to wet graphite beyond two or so layers, even though u/v is relatively large for these systems. Even if the lattice structure and molecular orientations in the monolayer and bulk solid are compatible, a strongly attractive substrate interaction will compress the monolayer to a lattice constant smaller than that of the bulk adsorbate crystal. As the film grows, this surface-induced stress will be shared by all of the layers, so long as they remain mutually commensurate, and the strain will decrease as n^{-1}, where n is the number of layers. However the resulting strain energy decreases more slowly ($\sim n^{-1}$) than the integrated attractive van der Waals energy ($\sim n^{-2}$) and should eventually prevent wetting [30-32]. However this may occur at several tens of layers and be indistinguishable in practice from complete wetting [31].

Table 1 lists various adsorbates on graphite arranged according to the value of u/v and classified by reported wetting behavior. Here u is represented by the isosteric heat of adsorption near zero coverage, q_{st}^0, and v is represented by the low-temperature heat of sublimation, h_0, of the bulk solid adsorbate. Other summaries of wetting behavior are given in Table 5.1 of [28] and in [9]. The experimental results are in general agreement with the

Table 1. Wetting properties of various absorbates on graphite. The relative strength (u/v) adsorbate-subtrate and adsorbate-adsorbate interactions is measured by the ratio of the isosteric heat of adsorption at zero coverage, q_{st}^0, to the bulk cohesive energy, h_0 (taken from Krim [13]). For present purposes the "wetting mode" is "1" if thick films, of order six or more layers of the indicated phase, have been reported (see text); "2" means only thinner films have been reported, up to saturation. Thus "1" includes the candidates for complete wetting. Parentheses indicate uncertain interpretations or extrapolations.

Absorbate	q_{st}^0/h_0 =u/v	Wetting mode		
		Low–T solid	Higher–T solid	Liquid
H_2O	0.31	–	–	2
CO_2	0.46	2	–	–
C_2H_4	0.42-0.79	2	–	1
CF_4	0.73	2	(1)	1
O_2	0.93	2	1	1
C_2H_6	0.99	2	–	1
Xe	1.04	1	–	1
Kr	1.17	1	–	1
Ar	1.23	1	–	1
N_2	1.32	2	1	1
Ne	1.39	2	1	(1)
CH_4	1.51	1	–	1
CO	1.55	2	1	1
H_2	4.6	1	–	(1)

suggestion of Bienfait et al. [29] that complete (or nearly complete) wetting by solid adsorbates will occur only for an intermediate range of u/v, although the behavior of some systems remains in dispute. On the weak-substrate side (low u/v), CO_2 is found not to wet graphite beyond one layer [33] and H_2O less than a monolayer [34,35]. Ethylene and ethane form a maximum of one and two layers respectively at low temperatures, although complete wetting is observed above the bulk melting point. At the other end, the extreme cases of high u/v are ^{3}He and ^{4}He on graphite, for which the degree of wetting by the liquid has been controversial; this is reviewed by Dietrich [28]. More recently Zimmerli and Chan [36] reported complete wetting by both isotopes. In passing we call attention to studies of wetting by various solid helium phases in a pressurized liquid ambient, also reviewed by Dietrich [28]. The next system included in Table 1 on the high u/v side is neon on graphite. RHEED experiments [29,37] found that neon does not wet graphite beyond two layers at 8 K. However volumetric adsorption measurements [38] were interpreted as showing thick films at 13.5 K and above.

The systems at intermediate values of u/v, of order 0.8 to 1.3 on the scale of Table 1, remain as candidates for complete wetting on the sublimation curve. Later in this chapter we will discuss in detail some of these systems. Argon, krypton, xenon, methane, nitrogen, carbon monoxide, and oxygen on graphite all have been reported to form thick, if not wetting, solid films over at least some temperature interval.

Layering

Adsorption isotherms may exhibit either continuous or stepwise increase in coverage with vapor pressure. Figures 1 and 2 show examples of stepwise isotherms obtained by ellipsometry and volumetry, respectively. The height of each step corresponds to approximately one molecular layer and the step represents crossing a region of coexistence of two phases, an n-layer film and an (n+1)-layer film. In the simplest cases this can be thought of as a vapor-liquid (or vapor-solid) transition in the (n+1)st layer, on top of n Minert layers. However the transition may involve significant changes in more than just the top layer; examples will be cited in CF_4 and oxygen films where addition of a layer is accompanied by freezing of one or more underlayers.

With increasing temperature, such a coexistence region will terminate at a critical point, or, if the coexisting phases have different symmetry, at a tricritical or multicritical point. In the case of no symmetry change, the layering transition is expected to belong to the universality class of the 2D Ising model and the slope of the isotherm at half-layer coverage above T_c should vary as $(T-T_c)^{-\gamma}$ with $\gamma = 7/4$. The critical temperature can be estimated from a family of adsorption isotherms on the high temperature side by extrapolating back to zero the inverse slope (e.g., [39]). Alternatively, heat capacity scans will have peaks at crossing of the coexistence boundary, so that closely spaced scans can map out the high-temperature termination of the coexistence region (e.g., [40]).

Even in a first-order condensation in region, isotherm steps will not be perfectly vertical due to heterogeneity of the substrate and also to finite lateral size of substrate grains. Heterogeneity effects have been analyzed quantitatively by Ecke et al. [41]. Experimentally, the narrowest second-layer step widths on cleaved or exfoliated graphite are of order $\delta\mu = T\delta p/p \simeq 0.2$ K (e.g. [42-44]). Slow relaxation effects in a solid film may contribute to the apparent width if insufficient time is allowed for equilibration, even on a single surface at relatively high vapor pressure (e.g., [45]).

In the simplest picture, the plateaus between steps might be expected to be horizontal. This is not necessarily true for several reasons. First, a 2D gas phase on top of the

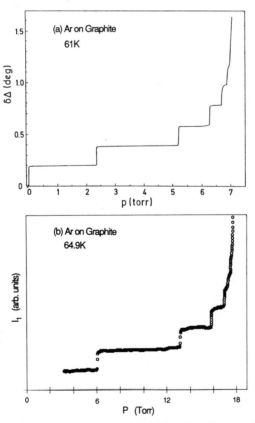

Fig. 1. Adsorption isotherms for argon on graphite obtained by ellipsometry:
    (a) Isotherm at 61K, showing the ellipsometric measure of coverage, $\delta\Delta$, as a function of pressure. (From Faul, Volkmann, and Knorr [119].)
    (b) Isotherm at 64.9K, showing an ellipsometric measure of coverage designated I_1 as a function of pressure. (From Youn [121].)

highest condensed layer may attain significant density before condensation of the next layer. Second, this may be accompanied by significant compression of the underlayer. In more extreme cases this whole picture of condensation transitions in the top molecular layer may be an oversimplification: Examples will be cited later in which first-order condensation appears to occur rather deep under an amorphous layer, which itself grows continuously with vapor pressure. In addition there may be an extrinsic contribution to the slope of plateaus; for instance, in volumetric measurements on a powder substrate there may be a contribution from capillary condensation.

The ratio of the critical temperature of the 2D square Ising model to that the 3D cubic Ising model is $T_c(2D)/T_c(3D) = 0.50$ [46]. It has been pointed out that, for a wide variety of simple adsorbates on graphite, the ratio of monolayer critical point to 3D critical point is very close to 0.4 [3,47]. Monolayer critical points are somewhat affected by lateral interactions with the substrate [48]. Critical points of the second, third, or fourth layer tend to be a few degrees higher than that of the monolayer and fall near 0.45 $T_c(3D)$. Some examples are displayed in Table 2. The ratio of 3D melting point, $T_m(3D)$ to $T_c(3D)$ is much more variable, so that the layer critical points may fall either above or below the bulk melting point. For Ar, Kr, and Xe, $T_{c,4}/T_m(3D) \simeq 0.82$; for CH_4, N_2, and CO, $T_{c,4}/T_m(3D) \simeq 0.95$; for C_2H_4, O_2, and CF_4, $T_{c,2}/T_m(3D) \simeq 1.11$; and for C_2H_6 $T_{c,2}/T_m(3D) = 1.35$. This is a major determinant of the character of multilayer films: In the first two groups there is a range of

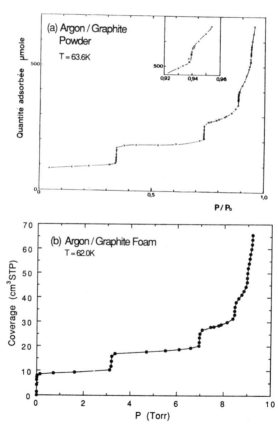

Fig. 2.   Adsorption isotherms for argon on graphite obtained by volumetric measurements:

(a) Adsorption of argon on exfoliated graphite powder at 63.6K. (From Ser, Larher, and Gilquin [118]; Gilquin [117]).

(b) Adsorption of argon an exfoliated graphite foam at 62.0K. (From Larese et al. [127].)

Note, in comparison to Fig. 1, the rises on the higher plateaus, attributed to a contribution to the coverage due to capillary condensation.

temperatures below the melting point where the growth of solid films is continuous and almost featureless, whereas for the last groups liquid films over a range above the bulk melting temperature are layered, as indicated by stepwise growth, to a surprising thickness, of order 7 layers.

Surface Roughening

In a wetting solid film, the layer critical points are related asymptotically to a property of the corresponding face of the bulk crystal, surface roughening [49-51]. This is unlocking of the crystal surface from a particular lattice plane through proliferation of steps, the free energy of which goes to zero at the roughening temperature T_R. This results in the loss of faceting in thermal equilibrium. The transition, which is of inverted Kosterlitz – Thouless type, is preceded on the low temperature side by the appearance and proliferation of vacancies in and adatoms on the initially smooth surface plane. As the temperature increases towards T_R, the characteristic size of clusters of adatoms and vacancies diverges, and the clusters gain their own (divergent) adatom and vacancy clusters. Roughening is thus an essentially many-layer phenomenon, but in a sense is triggered by sufficient disordering on a scale of one layer. Monte Carlo simulations [50,52], reproduced in Fig. 3, reveal no abrupt change at T_R in the local surface structure, in keeping with the long lateral range associated with the divergence. The surface of a finite film differs from that of the free crystal in that the roughening transition is necessarily truncated, and also the layering is stiffened by the gradient of the substrate potential. A succession of 2-D Ising critical points $T_{c,n}$ is expected in films of different coverages near (n - 1/2) layers. Using different methods, Huse [53] and Nightingale et al. [54] have shown that the sequence $T_{c,n}$ should converge to the roughening temperature, with $T_R - T_{c,n} \sim (\ln n)^{-2}$. Experimental results and computer simulations for rare gases on graphite are discussed below.

The surface of a liquid should always be rough due to excitation of capillary waves. Therefore, in the case of a liquid film showing multiple layering critical points, the sequence $T_{c,n}$ is expected to decrease towards the bulk melting temperature or lower.

Table 2. Bulk melting temperature T_m(3D) of various adsorbate compared with the critical temperature $T_{c,n}$ of the nth layer on graphite. Values of the ratio of $T_{c,n}$ to T_c(3d), the bulk critical temperature, span a relatively narrow range [47]. For Ar, Kr, and Xe the lower layer critical points are tabulated. If instead the critical points terminating the reentrant first-order regions were used, then $T_{c,n}/T_m$(3D) would be 0.92 in each case. The values used for N_2 and possibly CO correspond to upper critical points.

	T_m (3D)	$T_{c,n}$	$\dfrac{T_{c,n}}{T_c(3D)}$	$\dfrac{T_{c,n}}{T_m(3D)}$		
	(K)	(K)			n	Ref.
Ne	24.56	18.0	0.40	0.73	3	[103]
Ar	83.81	67.3	0.45	0.80	3,4,5	[117,120]
Kr	115.76	95	0.45	0.82	3,4	[145,146]
Xe	161.39	133	0.46	0.82	4,5	[154]
N_2	63.15	60	0.48	0.95	4,5	[173]
CO	68.2	65	0.49	0.95	4,5	[145]
CH_4	90.68	85	0.45	0.94	4,5,6	[145,158]
O_2	54.36	60	0.39	1.10	3	[117]
CF_4	89.53	101	0.44	1.13	2	[93,196]
C_2H_4	103.97	113	0.40	1.09	2	[76,77]
C_2H_6	90.35	122	0.40	1.35	2	[42,93]

A crystal face is said to exhibit surface melting if a "quasi-liquid" layer exists on the equilibrium surface at a temperature below the bulk melting point T_m, and its thickness increases with temperature, diverging as T approaches T_m [55-57]. Thus the phenomenon may be viewed as wetting of the crystal by its melt, driven by a lower surface free energy of the composite interface compared to a simple solid-vapor interface, which balances a higher bulk free energy contribution from the undercooled liquid in the layer. The quasi-liquid layer must be translationally disordered, but will have a periodic density modulation induced by the underlying crystal; it is characterized by substantial density between lattice sites and by mobility

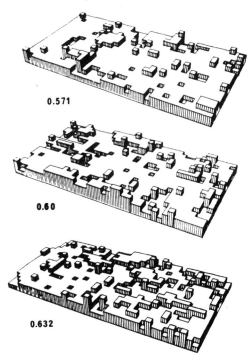

Fig. 3.   Computer drawings of the (001) surface of the cubic SOS model, generated by Monte Carlo calculations.   The numbers give the temperature, scaled by the interaction strength; the roughening temperature is T_R = 0.60.   (From Weeks and Gilmer [52].)

comparable to that of a liquid. Although first-order melting of individual layers is possible in principle, most simulations suggest continuous melting, and there need be no thermodynamic phase transition other than that at T_m. Surface melting is expected in thick solid films if it occurs on the corresponding face of the bulk adsorbate crystal, and has been suggested to occur in several systems including methane on graphite [58] and MgO [22], and neon and argon on graphite [16,59]. In the latter experiments the interpretation is uncertain because there is evidence that capillary condensed adsorbate is melting. In all of these cases melting of the top layer occurs several degrees below T_m.

A complementary effect is the substrate-induced freezing above T_m of one or more layers of a liquid film at the substrate interface. Zhu and Dash [60] report evidence of at least three solid layers above T_m in argon and neon films on graphite, and neutron diffraction shows similar behavior for methane on MgO [61]. For systems in which both liquid and solid wet, Pettersen et al. [62] constructed a slab model in which surface melting and substrate-induced freezing are treated on a par, so that for suitable parameter values one might have diverging solid layer thickness on approaching T_m from above.

Substrate Potential

The simplest model for a multilayer film is a slab of the appropriate bulk phase of the adsorbate, placed in the attractive potential $V(z)$ of the substrate. This is essentially the Frankel-Halsey-Hill model (see [5], p. 238ff). The adsorbate chemical potential μ is the energy required to add one adsorbate molecule at the free surface, which is at a distance z above the substrate surface. If at least an outer portion of the adsorbate slab is identical to the corresponding surface region of the bulk adsorbate, as hypothesized, then $\mu(z) = \mu_0 + V(z)$, where $\mu_0(p, T)$ is the chemical potential of the bulk phase. Assuming $V(z)$ is negative and is asymptotically monotone increasing, this model gives complete wetting and $\mu(z) - \mu_0$ is independent of temperature; that is, lines of constant coverage, including layer condensation lines, are parallel to the bulk vapor coexistence line in the μ-T phase diagram. This is equivalent to saying that the partial entropy of the film, given by $(\partial S/\partial N)_T = -(\partial \mu/\partial T)_N$, is equal to the adsorbate bulk phase entropy per molecule.

The practical question is whether one can be justified, in cases where the experimental layer lines are parallel to the bulk, in identifying $\mu_n - \mu_0$ with V_n, the potential at the distance of the nth layer. Clearly it is not true in general. For instance, the substrate-induced strains which may prevent wetting do so by giving additional contributions to the chemical potential of the film, and such a contribution may be present even if it does not prevent wetting at moderate thickness. Detailed calculations, where they exist (e.g., [63]), indicate the importance of other contributions to the chemical potential. Nevertheless it may be useful in these cases to compare $\mu_n - \mu_0$ with models for V_n, if only as a test for completeness of wetting. Such a comparison was made by Thomy and Duval [2], using $V_n = -Cn^{-x}$ for $2 \leq x \leq 4$. The van der Waals interaction, integrated over the half-space occupied by the substrate, gives this form with $x = 3$. In the case of graphite a better approximation may be to integrate over layers and sum the layer contributions. This gives [64]

$$V_n = -3C_3 d_0 \sum_{m=0}^{\infty} (md_0 + nd_1)^{-4} \tag{2}$$

where d_1 is the adsorbate layer spacing and $d_0 = 3.348$ Å is the graphite layer spacing. As this potential is to represent the difference between graphite and continued adsorbate, C_3 is an effective van der Waals coefficient.

EXAMPLES OF MULTILAYER FILMS ON GRAPHITE

The following sections are brief reviews of the state of experimental knowledge in the multilayer range for several of the more widely studied adsorbates on graphite. Monolayers are discussed only as required to provide a connection to the multilayer phases. These systems differ in u/v, the relative substrate attraction; in $T_m/T_c(3D)$, which determines the relation of layering to melting; in the importance of non-spherical shape or interactions; and in molecular diameter relative to the graphite lattice constant, which is known to be very important for the monolayer phase diagram. Helium and hydrogen are omitted here, but are discussed by Lauter et al. [65].

Ethylene/Graphite

This system has became the prototype for triple point wetting. We will, however, approach the multilayer phase diagram from the thin film side. In monolayer ethylene on graphite, neutron diffraction [66] and heat capacity experiments [67] support the identification of three solid phases, which differ in molecular orientation and structure. The high density (HD) phase, with the ethylene C=C axes perpendicular to the substrate, extends to coexistence with bulk solid up to about 74 K. Thus wetting is incomplete at low temperatures. The monolayer melts in a continuous transition, which proceeds from the disordered low density phase (DLD or simply LD) in which the molecules are rotating with the C=C axes parallel to the substrate [23, 24, 67-69]. The melting temperature is near 67.5 K in the submonolayer and increases continuously to a maximum of about 90 K in the full monolayer. The existence of an intermediate density phase (ID) which is distinct from LD is suggested by heat capacity peaks but has been questioned on the basis of an x-ray diffraction study [70, 71].

Volumetric isotherms by Menaucourt et al. [72] revealed the appearance of a second layer at a temperature estimated as 79.9 K, and a third layer at 98.3 K. Although no further isotherm steps were resolved, the coverage approaching saturation continued to increase with temperature such as to suggest a wetting transition at or very close to the bulk melting point at 103.97 K. Subsequent volumetric measurements have resolved a fourth layering step above 101 K [73,74]. This sequence of layer-appearance transitions was confirmed by x-ray diffraction measurements: Appearance of the second layer at 75 K and the third at 98 K was reflected in step decreases of the diffraction intensity from the remaining bulk crystallites [70,71]. A scattering profile characteristic of bilayer solid (2S) and coexisting bulk crystals was seen between 80 and 93 K, but 2S was replaced by a 2D-liquid profile at 97 K. The liquid profile is quite narrow (FWHM ~ 0.25 Å^{-1}), indicating fairly extended short-range order. The primary diffraction peak of the 2S structure is at a smaller wave vector (Q_0 = 1.58 Å^{-1}) than even the ID (or compressed LD) phase (Q_0 = 1.63 Å^{-1}), which suggests that the molecules have considerable freedom to rotate [71]. Heat capacity scans for C_2D_4 at sufficiently large coverage show peaks at 77, 79.7, 96, and 99 K [67]. These are identified as, respectively, the monolayer HD-to-ID transition, second layer appearance, bilayer melting, and third layer appearance. Possible further layering peaks are not resolved from bulk melting. A proposed phase diagram is reproduced in Fig. 4. One should note the considerable variation in estimates of the second-layer appearance temperature, and of the phase boundaries in this neighborhood.

Drir et al. [75] reported ellipsometric coverage isotherms for a temperature range around the melting point. Distinct layer condensation steps were observed up to the seventh or eighth layer at temperatures up to 107.8 K. The layer lines in the μ-T plane are parallel to the bulk liquid line within experimental error and, for layers above the second, are terminated at the low temperature end by bulk solid without evidence for film freezing. At the high temperature end, heat capacity measurements indicate termination of the layer-condensation lines at critical points, located for the second through fifth layers of C_2D_4 at $T_{c,2}$ = 115.3 ± 1.0 K, $T_{c,3}$ = 111 ± 1 K, $T_{c,4}$ = 108.4 ± 1.0 K, and $T_{c,5}$ ≃ 105 K [76]. It was suggested that this sequence is approaching T_m. New ellipsometric measurements in our lab [77] extend the results of Drir to both lower and higher temperatures. The location of layer steps are shown in Fig. 5. The second layer appears at 74 K and appears to have decreases in slope near 80 K and 95 K, which would suggest phase transitions in the bilayer. The upper transition agrees with previous

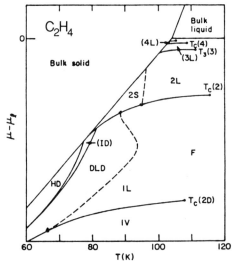

Fig. 4. A proposed phase diagram for ethylene on graphite in the chemical potential-temperature plane. μ_ℓ is the chemical potential of bulk liquid-vapor coexistence, which has been extrapolated into the solid region. HD, ID, and DLD are monolayer solid phases, which are discussed in the text. (From Zhang, Feng, Kim, and Chan [76].)

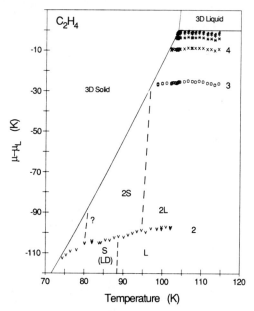

Fig. 5.   Phase diagram for ethylene on graphite in the multilayer region, showing data of Abbott and Hess [77] for chemical potentials at condensation of the second through sixth layers.   Dashed lines are suggested phase boundaries.

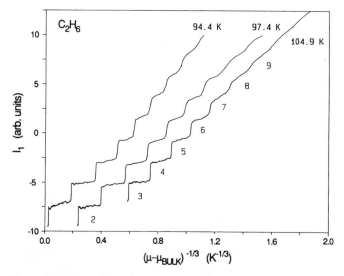

Fig. 6.   Three isotherms for ethane on graphite above the bulk melting temperature, T_m = 90.35 K. I_1 is the ellipsometric coverage and μ_{Bulk} (T) is the chemical potential at bulk liquid-vapor coexistence. The form of the abscissa was chosen to give approximately equal intervals for each layer. (From Nham [93].)

identifications of bilayer melting. It is possible that the two lower heat capacity peaks of Kim et al. [67] should be reinterpreted as second layer appearance followed by a solid-solid transition in the bilayer at 80 K. Even below 80 K the slope of the second layer line is much less than that of the bulk sublimation line, presumably reflecting a large entropy of molecular rotation in the bilayer. At higher temperatures the layer critical points can be estimated from data on the increase in the slopes of layer-condensation steps. We find $T_{c,2} = 113.1 \pm 0.5$ K, $T_{c,3} = 111 \pm 1$ K, $T_{c,4} = 110.5 \pm 1$ K, $T_{c,5} = 108 \pm 1$ K, $T_{c,6} = 107 \pm 1$ K, in reasonable agreement with the heat capacity results. Some difference between C_2H_4 and C_2D_4 should be expected, as $T_{c,1}$ is higher by 2.3 K for the former in similar heat capacity scans [67].

Ethane/Graphite

Bulk solid ethane on the sublimation curve has a monoclinic structure up to 89.68 K, a second ordered structure over a very narrow range, and then a plastic phase with cubic structure between 89.78 K and the melting point, $T_m = 90.32$ K [78]. No effects of the high temperature solid phases on film growth have been reported.

Monolayer ethane on graphite has been studied by neutron diffraction [79], LEED [80-84], inelastic incoherent neutron scattering [85,86], QENS [87], and heat capacity [88,89]. Regnier et al. [90] obtained volumetric isotherms above 87 K. This work has revealed a rich structure in the monolayer, including three solid phases (S_1, S_2, S_3) and two partially ordered liquid phases (I_1, I_2). Some of this is reviewed by Gay et al. [83].

The multilayer regime bears a resemblance to that of ethylene, in that a maximum of two solid layers form at saturation, presumably due to structural mismatch with the monoclinic solid [91], while thick, layered liquid films are observed at and well above the bulk melting point. Ethane, in fact, provides an extreme example of a layered liquid, with $T_{c,2}$ about 30 K above T_m [42,90]. This can be attributed to the large value of the ratio T_c (3D)/$T_m = 3.38$. The critical points of successively higher layers decrease: $T_{c,2} = 121.7 \pm 0.3$ K, $T_{c,3} = 117.0 \pm 0.6$ K, $T_{c,4} = 114 \pm 1$ K, $T_{c,5} = 108 \pm 1$ K, $T_{c,6} = 100 \pm 2$ K, $T_{c,7} = 95 \pm 2$ K, and the extrapolated value for $T_{c,8}$ is below T_m [42]. This progression appears to be

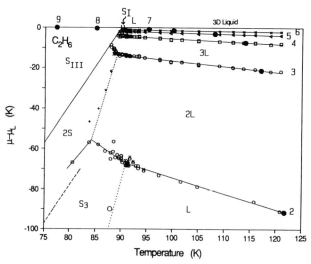

Fig. 7.   Phase diagram for ethane on graphite in the chemical potential-temperature plane. Symbols fit by solid lines are locations of isotherm steps corresponding to condensation of the second through the sixth layers. Dotted lines are possible melting transitions inferred from other experiments (between S_3 and L) or conjectured (between 2S and 2L). Plusses and the large circle near these lines represent small features in isotherms. The dashed line is an extrapolation of the second layer condensation line reported at lower temperatures from LEED attenuation. (From Nham [93].)

consistent with the theoretical model of Chernov and Mikeev [92]. However the impression of layering is enhanced by another effect: The rate at which isotherm steps broaden above their critical points decreases dramatically with increasing layer number [42]. Examples of isotherms above T_m are shown in Fig. 6.

The multilayer phase diagram of ethane on graphite is shown in Fig. 7, with data from ellipsometric isotherms [93] as well as phase transition lines from other experiments. The phase labeled S_3 is the most dense monolayer solid, in which the molecules are $\sqrt{3} \times \sqrt{3}$ commensurate with the substrate, standing on end, and rotating about their vertical C-C axes. S_3 melts to a correlated liquid (I_2) in which the molecules retain their orientation perpendicular to the substrate [81]. This transition is first order at low temperatures and is believed to be continuous at higher temperatures [83]. The slopes of the liquid layer-condensation lines on the right side of Fig. 7 can be attributed tentatively to the gradual loss of this orientational order in (the bottom layer of) thicker liquid films.

The condensation of the second layer, followed by bulk solid, was observed in the temperature range 64 to 75.4 K by LEED isotherms in which the attenuation of the graphite (01) spot intensity was monitored [83]. The second layer condenses at a nearly constant chemical potential offset of $\delta\mu \simeq 42$ K below bulk solid. An extrapolation of this line (dashed line in Fig. 7) joins reasonably well with the bend at 84 K in the ellipsometric second layer line. This implies that the bilayer solid has little or no rotational entropy, and, at coexistence with S_3, orientational disordering and bilayer melting occur as a single or as closely spaced transitions. Heat capacity measurements by Zhang and Migone [94] show a strong heat capacity peak at 82.5 K and a weaker peak at 85.5 K which appear to be bilayer transitions at coexistence with S_3. At higher coverages the lower peak moves up to 83.5 K, while the upper peak is more difficult to trace but might go to a peak at 88 K. An upward bend of the third layer condensation line at 89 K appears to indicate the upper termination of bilayer melting. Nham [93] has attempted to analyze the partial entropies of these transitions. From the third layer up, this system appears to provide an example of simple triple point wetting, in which thick layered liquid films continue below T_m until they meet the bulk sublimation curve.

Rare Gases on Graphite

The classical rare gases should, to a first approximation, all have the same bulk phase diagram except for different scales of length and energy, according to the law of corresponding states. Their respective films on graphite, however, may differ because they have different values of relative substrate attraction u/v, which is inversely proportional (again in the simplest model) to the atomic polarizability. In addition the atomic diameters match differently to the period of the graphite lateral potential (its strength also differs), resulting in major differences in the monolayer films: The monolayer region of krypton on graphite is dominated by the $\sqrt{3} \times \sqrt{3}$ R30° commensurate phase, while neon, argon, and xenon have liquid-vapor coexistence regions and freeze to an incommensurate solid. Submonolayer freezing is strongly first order in xenon and neon, but continuous or very weakly first order in argon [95-97]. The extent to which the multilayer phase diagrams differ due to u/v or the lateral potential is examined below.

Neon/Graphite

This system has been studied in the monolayer range by heat capacity measurement [98,99], LEED [100,101], and neutron diffraction [102]. The monolayer critical point is at 15.8 K, with triple point melting at 13.5 K. Volumetric adsorption isotherms covering the range 13.5 to 24 K by Lerner and coworkers [38,103] found thick films, with at least four discrete steps at the lower temperatures. Layer critical points were estimated as $T_{c,1} = 16 \pm 1$ K, $T_{c,2} = 19 \pm 1$ K, and $T_{c,3} = 18 \pm 1$ K. On the other hand RHEED experiments at 8 K [37] found only two uniform layers before condensation of bulk crystallites. If both experiments are correctly interpreted, there may be a wetting transition or a thin-thick film transition between 8 and 13.5 K.

Zhu and Dash [59,60,104] made a series of heat capacity scans covering approximately 15 to 27.5 K and 1 to 12.5 statistical layers. Figure 8 shows scans for several coverages between 5.0 and 6.5 layers. The weak peaks near 19 K are identified with layer critical points (more precisely, the upper terminations of layer coexistence regions), consistent with isotherm results at two and three layers. These critical points are expected to accumulate at the roughening transition of the neon (111) face at high coverage. The numbered peaks are attributed to melting of discrete layers near the graphite substrate; under a sufficiently thick film, the melting temperatures of the bottom three layers are shifted above the bulk melting point, $T_m = 24.56$ K. The most prominent feature above 3.3 layers is the remaining sharp peak (M), which grows and shifts upwards towards T_m with increasing coverage. This was attributed to the manifestation in the film of surface melting of the neon (111) face, of which a detailed analysis is given by Zhu and Dash [60]. Critique of this interpretation will be deferred until discussion of the similar heat capacity peaks seen in argon.

Argon/Graphite

The monolayer has been studied extensively by volumetric isotherms [96, 105, 106], heat capacity [97, 107], neutron diffraction [108, 109], x-ray diffraction [95, 110, 111], and LEED [112-114]. Primary emphasis is on the nature of melting. Among the highest resolution contributions to this is the heat capacity study of Migone et al. [97], which found submonolayer melting via a weak narrow peak at 47.2 K followed by a larger broad peak centered at 49.5 K. An adsorption isotherm study by Zhang and Larese [115] gives similar results at monolayer completion. The monolayer liquid-vapor critical point is $T_{c,1} = 55$ K.

Thick films have been seen over the whole temperature range studied: RHEED experiments at 10 to 20 K indicated uniform growth to coverages of at least 10 layers on a graphite single crystal [37, 116]. A volumetric isotherm study with exfoliated graphite powder found at least five layers at 63.6 K [117, 118]. An ellipsometric isotherm on HOPG at 59 K showed at least six discrete steps and coverage up to 12 layers before light scattering indicated bulk crystallization [119]. Similar measurements [120, 121] found 11 ± 1 layers between 64 and 83 K. On the other hand a study of mass loading of a graphite fiber found only four layers at 61 K, increasing to complete wetting at and above T_m [122]. Two additional layers were found at 61.8 K with a more uniform fiber [123]. It is possible that defects of the fiber surface may account for less complete wetting than on the cleaved or exfoliated surfaces [124].

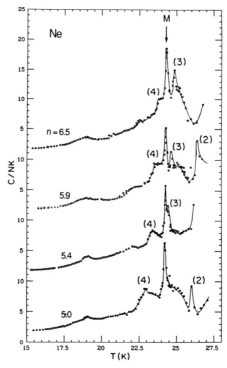

Fig. 8.   Heat capacity of neon films of four different coverages, given in nominal layers by the numbers at the left side of each trace. Certain peaks are labelled by integers for discussion in the text. (From Zhu [104].)

Stepped isotherms indicate layer-by-layer condensation up to about 68 K: Volumetric experiments [117] locate $T_{c,2} = 70.0 \pm 0.5$ K and $T_{c,3} = 68.0 \pm 0.5$ K. Ellipsometric step widths, fit assuming that the compressibility exponent is that of the 2D Ising model, $\gamma = 7/4$, yield layer critical temperatures $T_{c,2} = 69.4 \pm 0.2$ K and $T_{c,3} = 67.6 \pm 0.2$ K, in agreement with [117]; and $T_{c,4} = 67.0 \pm 0.2$ K and $T_{c,5} = 67.4 \pm 0.4$ K [120]. This is consistent with weak heat capacity peaks seen by Zhu and Dash [16,59,60] at 67.5 to 68.5 K for nearly all coverages. It therefore seems clear that the surface layer disorders at this temperature, which according to Huse [53] and Nightingale et al. [54] should indicate a roughening transition of the Ar (111) face near 68 K.

Neutron diffraction data for various coverages of ^{36}Ar on exfoliated graphite powder at T = 10 K confirm the layer coexistence picture and give some structural details [125]. The diffraction profiles can be fit by a succession of monolayer and mutually commensurate bilayer, trilayer, and four-layer structures. In three and four layers films, there is some admixture of ABA with the expected ABC stocking, and a bulk contribution appears at four layer films which is attributed to capillary condensation. Ser et al. [118] identify the small rise on their isotherm plateau preceding the third-layer step and the larger rises preceding the fourth and higher steps as a contribution from capillary condensation. This is supported by comparison with the flatter plateaus in ellipsometric isotherms [119,120], which employ a single surface and so are not subject to this problem.

The multilayer phase diagram derived from the heat capacity data of Zhu and Dash [16,59,60], reproduced in Fig. 9, is nearly identical to that for neon apart from the temperature scale. The numbered transition lines are again identified with melting of layers near the substrate. This interpretation has been tested at low coverages by neutron diffraction results of [126], which give both the structure of the solid portion of the film and, from the intensity in a window on the low-q side of the first solid diffraction peak, a measure of liquid (or structurally disordered) coverage. This is shown in Fig. 10. For a total coverage of three (incommensurate solid) layers, the midpoint of the third-layer melting is at 68 K, second-layer melting at 83 K, and first-layer at 94 K. For coverage of two layers, the second layer melts around 69 K and the first at 92 K. These temperatures agree well with Zhu and Dash. The picture slightly above T_m, then, is a liquid film in which the bottom few layers are solidified by interaction with the graphite substrate when the film is not too thin, designated

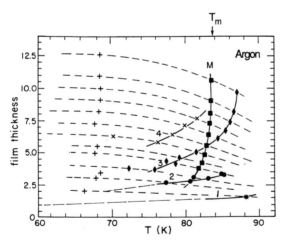

Fig. 9.   Diagram for argon on graphite in the coverage-temperature plane, showing the locations of heat capacity anomalies. Dashed lines are trajectories of heat capacity scans. The plusses are peaks identified with termination of layer coexistence regions. Lines labelled 2, 3, and 4 and the associated symbols are attributed to melting of the corresponding layers, counted up from the substrate. The line of squares marked M locate sharp peaks identified with melting of the rest of the film. (From Zhu and Dash [60], Zhu [104].)

substrate(-induced) freezing by Pettersen et al. [62]. The required thickness near 83 K ranges from a fraction of a liquid layer for one solid layer to a few liquid layers for three solid layers. These transitions appear to be continuous across the neighborhood of the bulk melting temperature. More specifically, the third-layer melting line is continuous across the line of sharp heat capacity peaks (M) which descends from the bulk melting point, and this interpretation of the data is strengthened by similar results for neon. Thus the numbers in Fig. 9 also designate the (discrete) number of solid layers present in each region of the phase diagram. The amount of liquid, on the other hand, must be able to vary continuously unless layering coexistence regions appear, a point to which we will return later.

Then what do the sharp transitions M represent? Figure 11a shows these peaks for the higher coverages. They originally were analyzed on the basis of a phenomenological theory of surface melting. Specifically, the heat capacity on the low temperature side of the peak was related to the temperature-dependent thickness of the quasi-liquid layer, and the truncation on the high temperature side, hence the peak temperature, was related to the approach of this thickness to the total film thickness [16,60]. Additional information was extracted from the integrated area under the peak, [59,60].

The number of crystalline solid layers does not change across M at least as high in coverage as the melting lines are well defined, which is true up to about 8 statistical layers total coverage. For lower coverage several interpretations are possible. One is that these peaks are not related to the uniform film, but represent melting of bulk solid which is capillary condensed at contacts between graphite flakes, with melting temperature depressed by concave surface curvature. This peak appears abruptly at a desorption-corrected coverage of 3.7 layers, which is close to the appearance of a bulk contribution to neutron diffraction on the same substrate (graphite foam), reported by Larese et al. [125] at 4 layers at T = 10 K. As noted earlier, Ser et al. [118] find evidence in volumetric isotherms of substantial capillary condensation below the fourth layer on a graphite powder substrate, which could have somewhat different pore geometry (Fig. 2a). A similar isotherm, obtained by Larese et al. [127] on graphite foam, is shown in Fig. 2b. A second possibility is that the non-crystalline layers, while structurally disordered, are not liquid and undergo a further transition. This is not supported by computer simulations, discussed below, which indicate that liquid-like mobility appears with structural disordering. Finally, there could be a structural transition within the ordered layers, although its occurrence at the melting temperature would be coincidental. Dash [128] now believes that these peaks in argon and neon are related to capillary-condensed adsorbate. Consequently, line M and the other features in Fig. 9 arise from different parts of

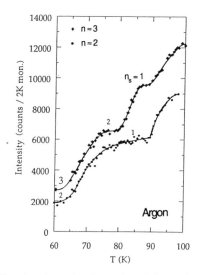

Fig. 10. Neutron diffraction intensity from 3 and 2 nominal layers of argon on graphite in a fixed Q-window which includes scattering only from liquid. The numbers 1, 2, 3, on the traces indicate the number of solid layers remaining at the corresponding plateaus. (From Larese and Zhang [126].)

the sample, and the coverage scale above three or four layers is too high as applied to the uniform part of the film. At greater coverages one would anticipate a contribution to the peaks near T_m from capillary condensation, but in addition there are higher layers of film which must melt at or near this temperature.

The ellipsometric measurements of Youn [120, 121] revealed a reappearance of first-order layering transitions in the temperature range 71 or 73 K to 77 K for coverages above three layers. A few isotherms are shown in Fig. 12, illustrating the evolution of the higher steps with temperature. The step labeled 4′ appears at a lower chemical potential and fractionally lower coverage than step 4, and similarly for higher layers, but there is a well defined though broad third step up to at least 73 K. The widths of the fourth, fifth, and sixth layer condensation steps are shown in Fig. 13. The chemical potential relative to bulk solid of the steps in the reentrant region are independent of temperature, which means the layer being added has essentially the entropy of bulk solid.

Figure 14 is the coverage-temperature diagram of Zhu and Dash, modified to include these coexistence regions. A few of the heat capacity features below 77 K attributed to

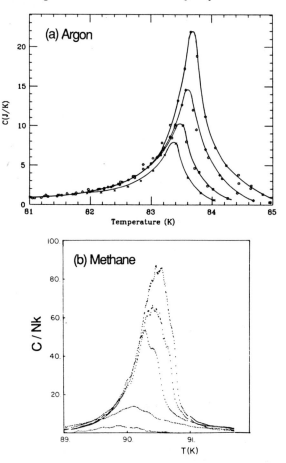

Fig. 11. (a) Heat capacity of argon on graphite in the neighborhood of the melting peak (M) for four coverages: n = 10.5, 8.8, 7.9, and 7.1 nominal layers. A correction for desorption has been subtracted. (From Zhu and Dash [16], Zhu [104].)

     (b) Heat capacity of methane an graphite in the neighborhood of the bulk melting temperature, obtained from continuous heating scans at nominal coverages of 18.3, 14.1, 9.3, 6.5, and 5.0 layers. A correction for desorption has been subtracted. (From Pettersen, Lysek, and Goodstein [164].)

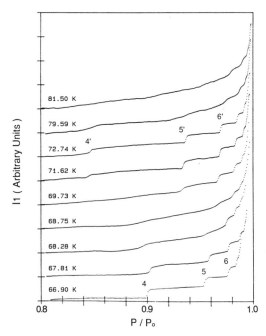

Fig. 12. Vapor pressure isotherms for argon on graphite, starting from three-layer films on the left. I_1 is the ellipsometric coverage and P/P_0 is the pressure normalized to saturated vapor pressure. (From Youn [121].)

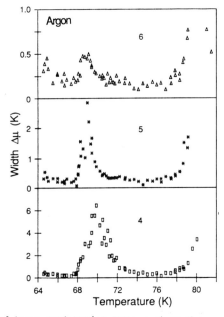

Fig. 13. Widths of layer condensation steps, such as those seen in Fig. 12, as functions of temperature for the fourth, fifth and sixth layers. (From Youn and Hess [120].)

layer melting now appear to be associated with crossing boundaries of coexistence regions. The third-layer melting line now meets the upper-right corner of a coexistence region (4′) whose phases we identify as two and three solid layers under approximately one disordered layer. This resembles the generic multilayer phase diagram, but with the critical point moved up to 77 K. The relation of the melting line to the termination of the coexistence region could resemble that in the monolayer of either argon, krypton, or xenon; or melting could be entirely continuous, as there is no change in symmetry. The fourth-layer melting line in Fig. 9 is one layer too high at 77 K; it should meet the top of the 3-to-4 solid layer coexistence region as sketched in Fig. 14. That is, the total coverage there must be about five instead of six layers. It seems plausible that the difference could be in capillary condensation.

The picture which emerges is that a disordered surface layer begins to form near 68 K and grows continuously over a range of a few degrees to somewhat less than one molecular layer. For some reason this disordered profile does not automatically allow continuous growth of the film, and in the range 71-77 K film growth is predominantly by first order transitions adding successive solid layers under the disordered surface. This might be an example of preroughening, proposed by den Nijs [129]. Approximately another molecular layer is added to the disordered surface over a few degrees above 77 K. Assuming the disordered layer is in fact quasi-liquid, this constitutes the start of surface melting.

Computer simulations should be of great value in interpreting and interpolating experimental observations, particularly where relatively simple interactions give rise to complex phenomena. There have been a number of molecular dynamics investigations of Lennard-Jones (111) and other surfaces [130-133] with fairly consistent results. On increasing temperature the first few layers at the surface disorder successively. In each layer the process starts with atom promotion to produce vacancies, followed by structural disordering within the layer, accompanied by a large increase in mobility within the layer. Similar behavior is found in simulations of argon films of various numbers of layers on graphite [134-137]. These studies also reproduce such effects as substrate-induced freezing, and should eventually allow detailed correlation with experimental phase diagrams.

The question of where is the roughening transition is not entirely clear. Ordinarily disordering on the scale of one layer is sufficient to unlock the interface from the crystal planes. Thus the roughening transition coincides with the critical temperature of the top layer of the film in the limit of large thickness [53, 54]. However in this case the presence of reentrant layering transitions indicates that this unlocking does not occur until 77 K. The

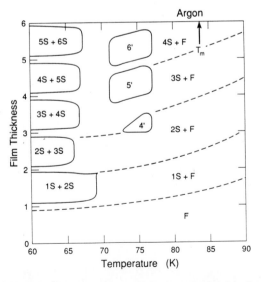

Fig. 14. Schematic phase diagram for argon on graphite in the coverage-temperature plane. Coexistence regions (enclosed by solid lines) are inferred primarily from isotherm data and melting lines (dashed lines) are primarily from heat capacity data. The coexistence regions designated 4′, 5′, 6′, correspond to similarly labeled steps in Fig. 12.

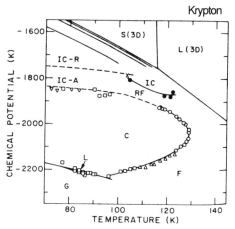

Fig. 15. Phase diagram for krypton on graphite in the chemical potential-temperature plane. G = gas, F = fluid, C = commensurate monolayer solid, IC = incommensurate monolayer solid, RF = reentrant fluid. (From Specht et al. [20].)

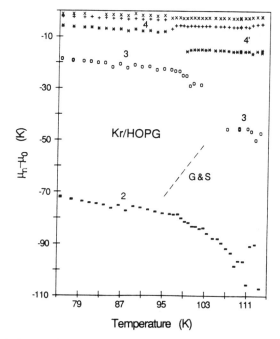

Fig. 16. Chemical potential-temperature diagram for krypton on graphite in the multilayer regime, showing layer condensation lines from ellipsometric isotherms. The dashed line is the weak feature seen by Gangwar and Suter. (From Youn, Li, and Hess, [146].)

interruption of layering at 68 K becomes decreasingly conspicuous in higher layers (Fig. 13). On the other hand, macroscopic observations of the loss of faceting in small krypton and xenon crystals seem to give roughening temperatures near the lower critical point for films of those adsorbates [138].

376

Krypton/Graphite

Figure 15 shows the μ-T phase diagram for krypton, with phase transition curves in the extended monolayer regime as established by heat capacity [139], vapor pressure isotherm [1-3, 43, 140, 141], LEED [142], and x-ray diffraction studies [20, 143]. The multilayer regime of present interest is in the upper left, adjacent to 3D solid, and has been subject of a number of vapor pressure isotherm studies and a heat capacity study.

Thomy and Duval [1, 2, 7] observed five sharp layer steps in a volumetric isotherm at 77.3 K. Thick uniform films were found by transmission electron microscopy at 40 K [18] and uniform films up to at least 10 layers at 15 to 50 K by RHEED [29, 116]. Ellipsometric isotherms between 57 and 77 K resolved 5 or 6 layers and, at the higher temperatures, found evidence of bulk crystallites coexisting with a uniform coverage equivalent to 9 ± 1 layers [45]. The chemical potential of the fifth layer condensation was essentially temperature-independent relative to bulk solid. Volkmann and Knorr [43], in ellipsometric measurements between 55 and 95 K, resolved up to 7 steps and saw evidence of light scattering due to bulk crystallites at about 16 layers. At 95 K the third and fourth steps have broadened considerably. On the other hand, Zimmerli and Chan [144] found that a maximum of four layers absorb on a high quality graphite fiber at 78 K. Gangwar and Suter [44] have made a high-precision volumetric study of the region of the second and third layers. They find $T_{c,2} = 98.2 \pm 0.3$ K and $T_{c,3} = 96.2 \pm 0.3$ K . A weak feature between the second and third layer condensation steps is identified as second-layer or bilayer melting. Larher and Angerand [145] reported layer critical points at $T_{c,2} = 97.5$ K, $T_{c,3} = 95$ K, and $T_{c,4} = 95$ K without giving details of their volumetric data.

New ellipsometric measurements [146] are consistent with the studies just described, but in addition find reentrant first-order layering in the fourth and higher layers over the range 100 to 107 K, quite similar to the argon case. Figure 16 gives the μ-T layering phase diagram. The dashed line is the weak transition of Gangwar and Suter [44], which we do not resolve. Figure 17 gives the coverage at the midpoints of the layer steps, relative to the midpoint of the second layer. It appears that the layer designated 4' intrudes under the old fourth layer, after some disordered surface layer growth in the range 93 to 99 K.

Pengra, Zhu, and Dash [147] have made heat capacity scans at high coverages in the neighborhood of T_m and see melting peaks which differ from the argon case, as shown in Fig. 18. With decreasing coverage the peaks shift more rapidly to lower temperature, and do not follow a common curve on the low temperature edge. This they attribute to strain in the krypton film, which would imply that this adsorbate remains predominently layered.

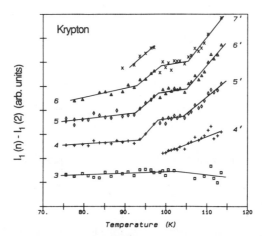

Fig. 17. Coverage-temperature diagram for krypton on graphite in the multilayer regime. Symbols indicate the ellipsometric coverage at the center of the indicated step relative to the center of the second-layer condensation step. (From Youn, Li, and Hess, [146].)

377

Xenon/Graphite

Issues in the extended monolayer regime include the low temperature structure as a function of coverage (e.g., [148]) and the order of the melting transition of the compressed bottom layer (e.g., [149-151]).

The question of wetting of graphite by xenon has been controversial. Early volumetric adsorption isotherms showed at least four distinct layers at 109 K [2,7]. Transmission electron diffraction found uniform film growth to macroscopic thickness [17,18], but not necessarily below saturation. RHEED experiments also showed uniform film growth at 10 to 60 K [37,116]. Volumetric measurements by Inaba et al., [152] reported a maximum of four layers at 108 K, but subsequent measurements by Ser et al. [118] found thicker films, consistent with wetting. In an x-ray diffraction study of xenon on single-crystal graphite, Hong and Birgeneau [153] found very thick films near 100 K, but a limit of 13 layers at 86 K, 7 layers at 69 K, and 6 layers below 55 K. Below 55 K it was found that the bottom layer of the trilayer film was commensurate with the graphite, while the higher layers were incommensurate.

An ellipsometric isotherm study by Youn and Hess [154] finds top layer disordering at about 133 K. Above three layers there is a reappearance of first order layering transitions between about 138 and 150 K, so that the phase diagram is similar to those of argon and krypton. Data on the loci of layer condensation steps are shown in Fig. 19.

Methane/Graphite

Information on the monolayer phase diagram of methane on graphite is summarized in recent reports of calorimetric [155] and LEED studies [156]. A submonolayer commensurate solid phase transforms continuously near 48 K to expanded incommensurate solid, which melts

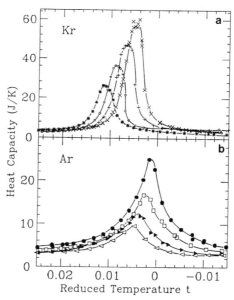

Fig. 18. (a) Heat capacity of krypton on graphite in the neighborhood of the bulk melting temperature, T_m, plotted against reduced temperature, $t = (T_m-T)/T_m$. The coverages are 11.8, 10.3, 8.7, and 7.2 layers.
(b) Heat capacity of argon an graphite, from Fig. 11, plotted the same way.
(From Pengra, Zhu, and Dash [147].)

at a triple point at 56 K. The liquid-vapor critical point is near 69 K [155]. At higher coverages there is a dense incommensurate solid. It has been suggested that the first layer melting line may bend back to join the second layer condensation line, [157] but confirmation is lacking.

Volumetric isotherms covering the range 77 to 96 K [2, 15, 58, 145] find films of at least four layers and layer critical points $T_{c,2}$ = 82.5 K, $T_{c,3}$ = 82.5 K, $T_{c,4}$ = 86.7 K, $T_{c,5}$ = 87.5 K [145]. Heat capacity peaks attributed to the termination of layer coexistence are observed in films from two to at least four layers at significantly lower temperatures, near 78 K [15]. QENS from a 10-layer film shows that one layer is disordered and mobile down to at least 76 K [58]. X-ray diffraction measurements indicate the appearance of one disordered layer in the bilayer film between 68 and 78 K [127]. Thus it appears that top-layer melting (or disordering), at least in few-layer films, precedes the layer critical points and thus occurs several degrees below the roughening temperature of methane (111). In contrast to the rare gases, there is no indication of broadening of layer condensation steps at the lower temperature and no offset of the layer chemical potentials; instead $\mu_n - \mu_0$ has negative slope above 78 K, especially for n = 2 and n = 3, indicating a partial entropy greater than the entropy of bulk solid [15]. This has been confirmed in our lab [158]. Neutron diffraction study of films in the bilayer and trilayer range by Larese et al. [159], mostly at 25 K, showed that the bilayer is compressed relative to bulk methane and not perfectly ordered between layers, whereas the trilayer is slightly expanded. These features may couple to the top-layer melting behavior.

Kim, Zhang, and Chan [155] made a calorimetric study of the molecular orientational ordering transitions in multilayer films of CH_4 and CD_4 near 20 to 27 K, which are believed to be related to the ordering transitions of the bulk solids. Chan [160] now believes that capillary condensation makes an important contribution to these peaks. Inaba and Morrison [161, 162] reported volumetric measurements showing incomplete wetting below 75 K and only three layers at 65 K, but it appears that they failed to reach saturation. Ellipsometric measurements show a discrete fifth layer down to 47 K and a limiting thickness of 10 ± 2 layers over 61 to 76 K [45]. RHEED measurements covering 14 to 40 K showed uniform films

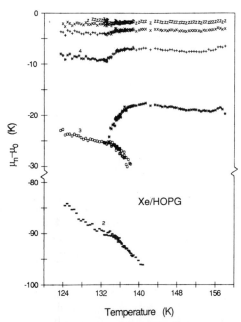

Fig. 19. Loci of layer condensation steps in the chemical potential - temperature plane for xenon on graphite. The chemical potential on the bulk sublimation line is μ_0. (From Youn and Hess [154].)

to at least 10 layers, but evidence of bulk crystallites above 15 statistical layers [163]. Zimmerli and Chan [144] find incomplete wetting of a graphite fiber below the triple point, with about five layers at 78 K, similar to their results for the rare gases.

Pettersen, Lysek, and Goodstein [62, 164] measured the heat capacity of thick methane films in the neighborhood of the bulk melting temperature. Some of their data are shown in Fig. 11b. These peaks differ in detail from the argon heat capacity peaks of Zhu and Dash (See Fig. 11a). An analysis based on a continuum slab model with interfacial energies led to the interpretation that wetting solid films melt to incompletely wetting liquid slightly below T_m, and the liquid films then wet at T_m and above. It was argued on the basis of heat capacity and NMR data that capillary condensation, although it would be present in equilibrium, did not occur. The issue is discussed by Goodstein, La Madrid, and Lysek in this volume.

Nitrogen/Graphite

Submonolayer nitrogen melts from a commensurate, orientationally disordered phase at an "incipient triple point" near 49 K [165, 166]. Orientational ordering to a herringbone structure occurs below 28 K [167, 168]. At higher coverages there are uniaxial and triangular incommensurate phases.

Bulk nitrogen has a low-temperature cubic α-solid phase, with four molecules per unit cell in different orientations, and an orientationally disordered hexagonal β-solid phase. Below the bulk α-β transition at 35.6 K the film thickness is limited to a few layers. RHEED and LEED experiments indicate a maximum of about two layers below 20 K [116]. Heat capacity measurements were interpreted as evidence for three-layer films coexisting with bulk near 35 K [169]. A neutron diffraction study, [170] suggested a film of one solid and three fluid layers in this region. LEED attenuation isotherms near 33 K indicated at least four layers below saturation [171].

At higher temperatures much thicker films are seen. Ellipsometric isotherms near 50 K show at least five steps and coverage equivalent to at least ten layers below saturation [119]. An earlier calorimetric study [172] found beyond three-layer coverage a peak at 53 K, attributed to second-layer melting, and a growing peak near 61.5 K, attributed to bulk crystallites. Preliminary ellipsometric measurements in our lab [173] find five layer-condensation steps near 35 K, and near 48 K a limiting thickness of 10 or 11 layers. The third and fourth layer-condensation steps broaden near 54 K and the fifth and sixth near 60 K. The fourth layer appears to have a reentrant region of sharp transitions around 58 to 60 K.

Carbon Monoxide/Graphite

Bulk CO has α- and β-solid structures similar to nitrogen, and there are general similarities in the film phases as well. Monolayer CO exhibits "incipient triple point" melting near 49 K from a $\sqrt{3} \times \sqrt{3}$ orientationally disordered commensurate phase, with a reentrant fluid region separating the commensurate and incommensurate solids, as in krypton [174, 175]. Orientationally ordered phases exist below 25 K [176, 177]. The heat capacity measurements of Feng and Chan [174] also provide evidence for second-layer melting at a triple point at 53 K, with a second-layer critical point at 60 K.

Larher et al. [178] have measured a series of volumetric adsorption isotherms between 51 and 63 K, spanning the bulk α-β transition at 61.6 K. Above $T_{\alpha\beta}$ the chemical potential of the fifth layer step is found to be parallel to that of bulk β-solid at $\mu_5 - \mu_\beta = -5$ K, suggesting the possibility of wetting by this phase. Incomplete wetting is found in the α-solid region, with fourth-layer appearance at 55.6 K. Layer critical points are located at $T_{c,2} = 59.7$ K, $T_{c,3} = 58.7$ K, and $T_{c,4} = T_{c,5} = 65$ K [145]. Zimmerli and Chan [144] find incomplete wetting of a graphite fiber even in the β-solid region.

Oxygen/Graphite

Bulk oxygen has three solid phases: Monoclinic antiferromagnetic α-oxygen is stable below 23.9 K. β-oxygen, stable between 23.9 and 43.8 K, is rhombohedral with all molecules preferentially aligned along and precessing about the three-fold c-axis. γ-oxygen, stable between 43.8 and 54.36 K, is cubic, with eight molecules per unit cell which have four different rotational constraints [179].

A large number of monolayer phases of oxygen on graphite have been identified (see, for instance, [180]), of which only one, the ζ_2 solid, will be mentioned here, because its relation to bilayer phases has been the subject of considerable work. This dense, high-temperature monolayer phase has a triangular lattice with the oxygen molecules preferentially oriented perpendicular to the substrate. This structure is nearly identical to the close-packed plane of bulk β-oxygen, which suggests a possibility of complete wetting [181, 182].

Oxygen films have been studied in the multilayer regime by calorimetry [183-185], vibrating fiber mass loading [186], ellipsometry [187, 188], x-ray diffraction [189-192], neutron diffraction [182], RHEED [29, 116], and LEED [193]. The ellipsometric study of Youn and Hess [188] maps the phase diagram down to 37 K and is useful in interpreting some earlier diffraction and thermodynamic measurements. Thick, layered liquid films are observed above T_m [187], consistent with the large value of $T_c(3D)/T_m$. Layer critical temperatures were not examined closely, but sharp second through seventh layer condensation steps were seen at least up to 58 K. Gilquin [117] found critical temperatures for layers two, three, and four near 60 K.

Below T_m down to 46 K the film remains liquid, as indicated by values of the partial entropy, but higher layers are successively preempted by bulk γ-solid: The fifth layer appears at 51 K, the sixth at 52.5 K, etc. Heat capacity measurements of Stoltenberg and Vilches [183] show peaks in this temperature region which, with increasing coverage, increase in size and shift upward towards T_m, as would be expected from conversion of γ-solid to liquid film at unresolved layer appearance transitions. Melting of a capillary-condensed portion

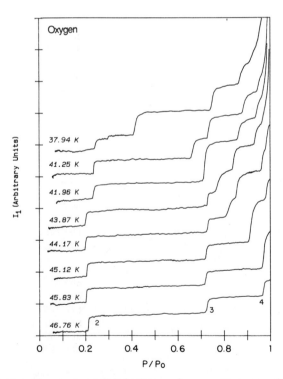

Fig. 20. Ellipsometric coverage (I_1) isotherms for oxygen on graphite, plotted against reduced pressure p/p_0 where p_0 is the saturated vapor pressure. There is one layer of oxygen present at the left end of each trace. Layer condensation steps are labeled for the bottom isotherm. (From Youn and Hess [188].)

might also contribute. Rather similar heat capacity peaks were found just below $T_{\alpha\beta}$, and many signal a triple point wetting transition there as well. However, RHEED observations suggest incomplete wetting on both sides of $T_{\alpha\beta}$ [29].

Figure 20 shows a series of ellipsometric isotherms in the temperature range of film melting. The smaller steps, such as those in the 46.76 K isotherm, correspond to addition of one liquid layer. There are also larger steps spanning an unusual range of heights. This is a consequence of the preferential alignment of molecules perpendicular to the substrate in solid layers. On one hand this allows closer packing and a relatively large increase in areal density from liquid to solid, and on the other hand it gives enhanced ellipsometric sensitivity to the solid due to the highly anisotropic molecular polarizability of O_2. The medium-size steps towards the upper right correspond to addition of solid layers. Larger steps result from concurrent layer addition and multiple layer freezing.

The loci of steps in the μ-T plane are shown in Fig. 21. Identification of the phases in terms of the number of liquid and solid layers is based on the step heights and the slopes of the coexistence lines. Solid layer condensation lines in the upper left are parallel to the bulk β-solid sublimation line and within the resolution of the experiment the film thickness appears to diverge approaching saturation. The top layer in films thicker than two layers appears to melt continuously. Otherwise the entire film melts at a first-order transition line extending through the points labeled A through F. For instance, the four-layer film transforms at a short segment above layer triple point A from L/3S to 4L. Heat capacity scans in this region should tend to follow layer coexistence lines and have sharp peaks at the lettered layer triple points. This is qualitatively what was seen by Stoltenberg and Vilches [183] but they have not analyzed these data in detail. Chiarello et al. [182] have analyzed neutron diffraction data to extract

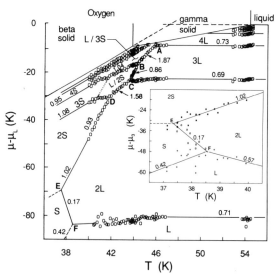

Fig. 21. Chemical potential - temperature phase diagram for oxygen films on graphite. Bulk condensed phases are located at the top. Circles are locations of steps in isotherms, such as those in Fig. 20. The numbers associated with the phase boundary lines designate the average step height for that transition, normalized to unity for addition of a nominal solid layer. The inset shows data for the low temperature region, plotted relative to the chemical potential of third-layer condensation. Phases are designated by the number of liquid and solid layers; these identifications are based on an interpretation of the step heights and the coexistence line slopes, which are related to coverage and entropy differences. (From Youn and Hess [188].)

the number of solid and number of liquid layers at six points in the phase diagram between 34.5 and 46.4 K. These data are consistent with Fig. 21, and in particular one point supports the identification of 2S/L.

The phase S at the lower left corner of Fig. 21 is identified with ζ_2. There have been extensive x-ray studies of this phase and its neighbors. Figure 22 shows x-ray diffraction profiles of Mochrie et al. [191] for five temperatures at a nominal coverage slightly larger than bilayer solid, 2S. There is evidence that the nominal coverage scale in these experiments is too high [192,193], and the true coverage appears to be between that of S and 2L. The system then follows a path from S-2S coexistence through E and F to L-2L coexistence. The "η-phase" suggested by Stephens et al. [189] is the coexistence region of S and 2L between triple points E and F. In this region a single sharp diffraction peak is seen at 2.19 Å^{-1}, corresponding to a triangular structure for S (i.e., ζ_2). At lower temperatures, along S-2S coexistence, there are two additional peaks near 2.15 and 2.25 Å^{-1} [190,191] which have been interpreted alternatively as mutually incommensurate layers or a distortion from triangular structure in 2S.

Heat capacity peaks and an increase in magnetic susceptibility [183,184] indicate a phase transition near 46 K, probably in the 2L phase, which is not shown in Fig. 21. The lower-temperature phase is called "Fluid II." Morishige et al. [192] report additional broad x-ray structure associated with this phase. This transition was not resolved in the ellipsometric study.

Tetrafluoromethane/Graphite

Bulk CF$_4$ at sublimation pressure exists as an orientationally ordered monoclinic solid (α) below 76.23 K and a plastic solid (β) up to the melting point T_m = 89.56 K (see [194] for references). Neutron and x-ray diffraction and heat capacity studies of the monolayer on graphite, summarized by Zhang et al. [195], found at least four solid phases. The highest-coverage monolayer is hexagonal incommensurate, and this phase has shown no evidence of an orientational disordering transition between 18 and 97 K, where the first step of a two-stage

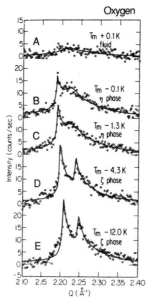

Fig. 22. High-resolution x-ray diffraction profiles for two nominal layers of oxygen an graphite. The reference temperature T_m is identified with the temperature of triple point F in Fig. 21, approximately 39 K. According to the present interpretation, these profiles correspond to the regions of phase coexistence of L+2L (A), S+2L (B and C), and S+2S (D and E). (From Mochrie et al. [191].)

melting process is identified [195]. In the bilayer range, this heat capacity study finds peaks near 57 and 61 K and five peaks between 86 and 101 K. Since an ellipsometric study showed that the second-layer critical point is associated with the highest temperature peak [196], there is a generous supply of peaks which can be attributed to orientational disordering and melting in the bilayer.

RHEED and LEED observations [197, 198] were interpreted as indicating a wetting transition at 37 K. Further study suggested that the observations might be due to a change in growth mode of 3D crystallites [91]. Volumetric measurements found only four layers at 77 K [199], in agreement with later ellipsometric isotherms, which also found a maximum of two layers between 52 and 72 K [200].

Layering transitions found in the ellipsometric study are shown in Fig. 23. The critical points of layers two through five are in the neighborhood of 101 K, in agreement with the data of Dolle et al. [199]; they obtained somewhat higher numbers due to different assumptions used in the analysis. Thick, apparently wetting, films are seen in the liquid region, while below T_m the maximum number of layers decreases. In contrast to the oxygen case, melting of CF_4 films should produce only a weak ellipsometric signature due to density difference, and this was not resolved. Nevertheless, changes in slopes of layer condensation lines, notably for the fourth and fifth layers, give evidence of a complex pattern of layer-by-layer melting. A unique feature of this system is the appearance of a double-layer step, from 2S to 4S, between 72 and 75 K. The slope of this condensation line indicates that the second bilayer has greater entropy than even β-solid, hence is orientationally disordered from its first appearance. Above 75 K the third and fourth layers split, with the third layer apparently condensing as liquid. There is as yet no structural information in this regime.

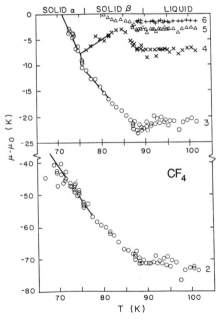

Fig. 23. Chemical potential-temperature diagram for CF_4 on graphite. Symbols locate layer condensation steps for the layers designated on the right. The chemical potentials are referenced to the saturated bulk phase occurring at the particular temperature. Between 71 and 75K the third and fourth layers condense together in a direct bilayer to four-layer transition. (From Nham, Drir, and Hess [200].)

DISCUSSION

Several parameters have been suggested as important in determining the general features of multilayer phase diagrams. The ratio (u/v) of adsorbate-substrate to adsorbate-absorbate interaction must be sufficiently strong in order to produce wetting [26]. On the other hand, too large a value of u/v should produce strain in a solid film, which will again prevent complete wetting [29]. In practice only a few simple molecular adsorbates are known to fail to wet graphite in their liquid phase (e.g., water) and the same is true for gold (111) [13]. Non-wetting by solid adsorbates is more commonly attributable to structural compatibility of the completed monolayer or bilayer film with the bulk solid than to the value of u/v, as has been emphasized by Larher and coworkers [73]. However neon seems to be an exception, and the strain effect may well impose a limiting thickness in the range above 10 layers for a number of other solid adsorbates, as suggested by calculations [31]. Larher et al. [178] suggested that plastic solid phases are likely candidates for wetting, e.g., β-CO. However, the situation is reversed in the case of oxygen, and a more direct compatibility criterion seems necessary.

Dimensional incompatibility between adsorbate and substrate lateral structure is important for monolayer films on graphite, as it determines the possibility of a commensurate phase and hence the melting behavior. For substrates with stronger lateral potential this parameter may be important in determining the maximum film thickness [201]. However it does not seem to play a significant role for multilayer films on graphite.

The ratio of melting temperature to critical temperature of the bulk adsorbate seems to be a useful predictor of the form of the multilayer phase diagram, due to its connection with the ratio of roughening to melting temperatures. This depends on the empirical observation that the higher-layer critical points are generally close to $0.45\ T_c(3D)$. On this basis the adsorbates which have been reviewed fall into groups (listed in the introduction) with similar layering characteristics.

Considerable controversy has existed regarding the importance of capillary condensation in various experiments. For a simple geometry and a liquid adsorbate of known surface tension and wetting properties, the amount of capillary condensed material can be calculated as a function of reduced vapor pressure [202]. However the appropriate description of the contacts between flakes in powdered graphite materials is uncertain. For solid adsorbates the surface tensions are not known. Empirically there is evidence, even in as open a substrate as graphite foam, for appreciable capillary condensation before the fourth layer, and a substantial contribution beyond four layers. This comes from comparison of adsorption on powder and single-surface substrates, as in Figs. 1 and 2, and in certain cases from the appearance of bulk solid diffraction peaks. In some cases hysteresis in coverage is seen [161] but more often it is not. The possibility of capillary condensation complicates the interpretation of multilayer melting behavior, as we have seen in the cases of the rare gases and methane, and introduces uncertainties in the coverage scales in other experiments, in addition to confusing the determination of complete wetting. This has been discussed recently by Ser et al. [118].

Even on a single-surface substrate, determination of the completeness of wetting, or conversely the limiting film thickness at saturation, becomes a very subtle matter beyond 10 or 15 layers. The excess substrate potential at the film surface is very small: Typically $V_{10} \simeq 0.4$ K and $V_{15} \simeq 0.1$ K. This implies a very small depression of the coexisting vapor pressure from saturated vapor pressure, $(p_0-p_n)/p_0 = V_n/T$. The Clausius-Clapyron equation gives $\delta p_0/p_0 = (L/k_B T)\ (\delta T/T)$, where L is the heat of vaporization per molecule. For argon just below its melting point, for example, $L/k_B T = 11.2$. This puts very stringent requirements on temperature uniformity and stability for any type of equilibrium experiment, in addition to the manometry requirements in the case of adsorption isotherms. Certain techniques, such as RHEED, THEED, or ellipsometry, may reveal uniform film growth at temperatures so low that p_0-p_n is not measurable, but they then do not distinguish wetting film growth from epitaxial adsorbate crystal growth beyond saturation.

ACKNOWLEDGMENTS

The author is grateful for suggestions and contributions of many colleagues, including M. Chan, D. Goodstein, O. Vilches, Y. Larher, J.M. Phillips, J. Krim, M. Bienfait, M. La Madrid,

L. Bruch, and M. den Nijs. I particularly thank J. Larese, J.G Dash, R. Suter, U. Volkmann, A. Migone, and G. Torzo, who, in addition, allowed use of experimental results from their labs prior to publication. To those whose work is not mentioned due to arbitrariness in selection or to inadvertence, I apologize. For unpublished work at Virginia and for assistance in preparation of this paper I am indebted to D. Abbott, H.S. Nham, G. Reynolds, D.M. Li, E. Updike, and especially to H.S. Youn. This work was supported by the Low Temperature Physics Program of the National Science Foundation under grants DMR-8617760 and DMR-9004108.

REFERENCES

1. A. Thomy and X. Duval, J. Chim. Phys. 66, 1966 (1969).
2. A. Thomy and X. Duval, J. Chim. Phys. 67, 286 (1970).
3. A. Thomy and X. Duval, J. Chim. Phys. 67, 1101 (1970).
4. J.G. Dash, Films on Solid Surfaces (Academic, New York 1975).
5. W.A. Steele, The Interaction of Gases with Solid Surfaces (Pergamon, Oxford, 1974).
6. O.E. Vilches, Ann. Rev. Phys. Chem. 31, 463 (1980).
7. A. Thomy, X. Duval, and J. Regnier, Surf. Sci. Repts. 1, 1 (1981).
8. M. Bienfait, in Phase Transitions in Surface Films, edited by J. G. Dash and J. Ruvalds (Plenum, New York, 1981) p. 29.
9. M. Bienfait, Surf. Sci. 162, 411 (1985).
10. R. Marx, Phys. Repts. 125, 1 (1985).
11. K. Strandburg, Rev. Mod. Phys. 60, 161 (1988).
12. G. Quentel, J.M. Rickard, and R. Kern, Surf. Sci. 50, 343 (1975).
13. J. Krim, J.G. Dash, and J. Suzanne, Phys. Rev. Lett. 52, 635 (1984); J. Krim, Ph.D. thesis, Univ. of Washington (1984), unpublished.
14. P. Taborek and L. Senator, Phys. Rev. Lett. 57, 218 (1986).
15. J.J. Hamilton and D.L. Goodstein, Phys. Rev. B 28, 3838 (1983).
16. D.M. Zhu and J.G. Dash, Phys. Rev. Lett. 57, 2959 (1986).
17. G.L. Price and J.A. Venables, Surf. Sci. 49, 264 (1975).
18. H.M. Kramer, J. Crystal Growth 33, 65 (1976) .
19. E.D. Specht et al., J. Phys. (Paris) 46, L561 (1985).
20. E.D. Specht et al., Z. Phys. B 69, 347 (1987).
21. J.K. Kjems et al., Phys. Rev. B 13, 1446 (1976).
22. M. Bienfait, Europhys. Lett. 4, 79 (1987).
23. J.Z. Larese et al., Phys. Rev. Lett. 61, 432 (1988).
24. J.Z. Larese and R.J. Rollefson, Surf. Sci. 127, L172 (1983); Phys. Rev. B 31, 3048 (1985).
25. J.G. Dash, Phys. Rev. B 15, 3136 (1977).
26. D.E. Sullivan, Phys. Rev. B 20, 3991 (1979).
27. R. Pandit, M. Schick, and M. Wortis, Phys. Rev. B 26, 5112 (1982).
28. S. Dietrich, in Phase Transitions and Critical Phenomena, vol. 12. edited by C. Domb and J.L. Lebowitz (Academic Press, London, 1988), p. 1.
29. M. Bienfait et al., Phys. Rev. B 29, 983 (1984).
30. R.J. Muirhead, J.G. Dash, and J. Krim, Phys. Rev. B 29, 5074 (1984).
31. F.T. Gittes and M. Schick, Phys. Rev. B 30, 209 (1984).
32. D.A. Huse, Phys. Rev. B 29, 6985 (1984).
33. A. Terlain and Y. Larher, Surf. Sci. 125, 304 (1983).
34. C. Pierce, R.N. Nelson, J.W. Wiley, and H. Cordes, J. Am. Chem. Soc. 73, 4551 (1951).
35. A.V. Kiselev and N.V. Kovaleva, Bull. Acad. Sci. USSR, Div. of Chem. Sci., No. 2, 955 (1959).
36. G. Zimmerli and M.H.W. Chan, Phys. Rev. B 38, 8760 (1988).
37. J.L. Seguin et al., Phys. Rev. Lett. 51, 122 (1983).
38. E. Lerner, F. Hanono, and C.E.N. Gatts, Surf. Sci. 160, L524 (1985).
39. Y. Nardon and Y. Larher, Surf. Sci. 42, 299 (1974).
40. H.K. Kim and M.H.W. Chan, Phys. Rev. Lett. 53, 170 (1984).
41. R.E. Ecke, J. Ma, A.D. Migone, and T.S. Sullivan, Phys. Rev. B 33, 1746 (1986).
42. H.S. Nham and G.B. Hess, Phys. Rev. B 38, 5166 (1988).
43. U.G. Volkmann and K. Knorr, Surf. Sci. 221, 379 (1989).
44. R. Gangwar and R.A. Suter (1990), preprint.

45. H.S. Nham and G.B. Hess, Langmuir $\underline{5}$, 575 (1989).
46. J.D. Weeks, G.H. Gilmer, and H.J. Leamy, Phys. Rev. Lett. $\underline{31}$, 549 (1973).
47. Y. Larher, Phys. Rev. A $\underline{20}$, 1599 (1979).
48. F. Millot, Y. Larher, and C. Tessier, J. Chem. Phys. $\underline{76}$, 3327 (1982).
49. W.K. Burton, N. Cabrera, and F.C. Frank, Phil. Trans. Roy. Soc. London $\underline{243A}$, 299 (1951).
50. J.D. Weeks in Ordering in Strongly Fluctuating Condensed Matter Systems, edited by T. Riste (Plenum, New York, 1979), p. 293.
51. H. van Beijeren and I. Nolden, in Structure and Dynamics of Surfaces II, edited by W. Schommers and P. von Blanckenhagen (Springer-Verlag, Berlin, 1987), p. 259.
52. J.D. Weeks and G.H. Gilmer, Adv. Chem. Phys. $\underline{40}$, 157 (1979).
53. D.A. Huse, Phys. Rev. B $\underline{30}$, 1371 (1984).
54. M.P. Nightingale, W.F. Saam, and M. Schick, Phys. Rev. B $\underline{30}$, 3830 (1984).
55. J.F. van der Veen, B. Pluis, and A.W. Denier van der Gon, in Chemistry and Physics of Solid Surfaces VII, edited by R. Vanselow and R. Howe (Springer-Verlag, Berlin, 1988), p. 455.
56. A. Trayanov and E. Tosatti, Phys. Rev. B $\underline{38}$, 6961 (1988).
57. J.G. Dash, Contemp. Phys. $\underline{30}$, 89 (1989).
58. M. Bienfait, P. Zeppenfeld, J.M. Gay, and J.P. Palmari, Surf. Sci. $\underline{226}$, 327 (1990).
59. D.M. Zhu and J.G. Dash, Phys. Rev. Lett. $\underline{60}$, 432 (1988).
60. D.M. Zhu and J.G. Dash, Phys. Rev. B $\underline{38}$, 11673 (1988).
61. J.M. Gay, J. Suzanne, and J.P. Coulomb, Phys. Rev. B $\underline{41}$, 11346 (1990).
62. M.S. Pettersen, M.J. Lysek, and D.L. Goodstein, Phys. Rev. B $\underline{40}$, 4938 (1989).
63. L.W. Bruch, J. Chem. Phys. $\underline{87}$, 5518 (1987).
64. S. Chung, N. Holter, and M.W. Cole, Phys. Rev. B $\underline{31}$, 6660 (1985).
65. H.J. Lauter, V.L.P. Frank, H. Godfrin, and P.Leiderer (1990), this volume.
66. S.K. Satjia et al., Phys. Rev. Lett. $\underline{51}$, 411 (1983).
67. H.K. Kim, Y.P. Feng, Q.M. Zhang, and H.M.W. Chan, Phys. Rev. B $\underline{37}$, 3511 (1988).
68. H.K. Kim, Q.M. Zhang, and M.H.W. Chan, Phys. Rev. Lett. $\underline{56}$, 1579 (1986).
69. J.Z. Larese, L. Passell, and Ravel, Can. J. Chem. $\underline{66}$, 633 (1988).
70. M. Sutton, S.G.J. Mochrie, and R.J. Birgeneau, Phys. Rev. Lett. $\underline{51}$, 407 (1983).
71. S.G.J. Mochrie et al., Phys. Rev. B $\underline{30}$, 263 (1984).
72. J. Menaucourt, A. Thomy, and X. Duval, J. Phys. (Paris) $\underline{38}$, C4-195 (1977).
73. I. Bassignana and Y. Larher, Surf. Sci. $\underline{147}$, 48 (1984).
74. I. Arakawa, Y. Koga, and J.A. Morrison, in Adv. in Phase Transitions, ed. by J.D. Embury and G.R. Purdy (Pergamon, Oxford, 1988), p. 145.
75. M. Drir, H.S. Nham, and G.B. Hess, Phys. Rev. B $\underline{33}$, 5145 (1986).
76. Q.M. Zhang, Y.P. Feng, H.K. Kim, and M.H.W. Chan, Phys. Rev. Lett. $\underline{57}$, 1456 (1986).
77. D.S. Abbott and G.B. Hess (1990), unpublished .
78. M.H.M. Schutte, K.O. Prins, and N.J. Trappeniers, Physica $\underline{144B}$, 357 (1987).
79. J.P. Coulomb et al., Phys. Rev. Lett. $\underline{43}$, 1878 (1979).
80. J. Suzanne, J.L. Seguin, H. Taub, and J.P. Biberian, Surf. Sci. $\underline{125}$, 153 (1983).
81. J. Suzanne, J.M. Gay, and R. Wang, Surf. Sci. $\underline{162}$, 439 (1985).
82. J.M. Gay, J. Suzanne, and R. Wang, J. Phys. (Paris) $\underline{46}$, L-425 (1985).
83. J.M. Gay, J. Suzanne, and R. Wang, J. Chem. Soc., Faraday Trans. II $\underline{82}$, 1669 (1986).
84. J.W. Osen and S.C. Fain, Jr., Phys. Rev. B $\underline{36}$, 4074 (1987).
85. F.Y. Hansen et al., Phys. Rev. Lett. $\underline{53}$, 572 (1984).
86. F.Y. Hansen and H. Taub, J. Chem. Phys. $\underline{87}$, 3232 (1986).
87. J.P. Coulomb and M. Bienfait, J. Phys. (Paris) $\underline{47}$, 89 (1986).
88. S. Zhang and A.D. Migone, Phys. Rev. B $\underline{38}$, 12039 (1988).
89. S. Zhang and A.D. Migone, Surf. Sci. $\underline{222}$, 31 (1989).
90. J. Regnier, J. Menaucourt, A. Thomy, and X. Duval, J. Chim. Phys. $\underline{78}$, 629 (1981).
91. J.M. Gay, J. Suzanne, J.G. Pepe, and T. Meichel, Surf. Sci. $\underline{204}$, 69 (1988).
92. A.A Chernov and L.V. Mikeev, Phys. Rev. Lett. $\underline{60}$, 2488 (1988).
93. H.S. Nham, Ph.D. thesis, Univ. of Virginia (1989), unpublished.
94. S. Zhang and A.D. Migone (1990), preprint.
95. J.P. McTague, J. Als-Nielsen, J. Bohr, and M. Nielsen, Phys. Rev. B $\underline{25}$, 7765 (1982).
96. Y. Larher, Surf. Sci. $\underline{134}$, 469 (1983).
97. A.D. Migone, Z.R. Li and M.H.W. Chan, Phys. Rev. Lett. $\underline{53}$, 810 (1984).
98. G.B. Huff and J.G. Dash, J. Low Temp. Phys. $\underline{24}$, 155 (1976).
99. R.E. Rapp, E.P. de Souza,, and E. Lerner, Phys. Rev. B $\underline{24}$, 2196 (1981).

100. S. Calisti and J. Suzanne, Surf. Sci. 105, L255 (1981).
101. S. Calisti, J. Suzanne, and J.A. Venables, Surf. Sci. 115, 455 (1982).
102. C. Tiby, H. Wiechert, and H.J. Lauter, Surf. Sci. 119, 21 (1982).
103. F. Hanono, C.E.N. Gatts, and E. Lerner, J. Low Temp. Phys. 60, 73 (1985).
104. D.M. Zhu, Ph.D. thesis, Univ. of Washington (1988), unpublished.
105. Y. Larher and B. Gilquin, Phys. Rev. A 20, 1599 (1979).
106. J.L.M. Demetrio de Souza and E. Lerner, J. Low. Temp. Phys. 66, 367 (1987).
107. T.T. Chung, Surf. Sci. 87, 438 (1979).
108. H. Taub et al., Phys. Rev. B 16, 4511 (1977).
109. C. Tiby and H.J. Lauter, Surf. Sci. 117, 277 (1982).
110. M. Nielsen et al., Phys. Rev. B 35, 1419 (1987).
111. K.L. D'Amico, J. Bohr, D.E. Moncton, and D. Gibbs, Phys. Rev. B 41, 4368 (1990).
112. C.G. Shaw, S.C. Fain, Jr., and M.D. Chinn, Phys. Rev. Lett. 41, 955 (1978).
113. C.G. Shaw and S.C. Fain, Jr., Surf. Sci. 83, 1 (1979).
114. C.G. Shaw and S.C. Fain, Jr., Surf. Sci. 91, L1 (1980).
115. Q.M. Zhang and J.Z. Larese (1990), preprint.
116. J.A. Venables et al., Surf. Sci. 145, 345 (1984).
117. B. Gilquin D. Sc. thesis, Nancy; CEA Note 2091 (1979).
118. F. Ser, Y. Larher, and B. Gilquin, Molecular Phys. 67, 1077 (1989).
119. J.W.O. Faul, U.G. Volkmann, and K. Knorr, Surf. Sci. 227, 390 (1990).
120. H.S. Youn and G.B. Hess, Phys. Rev. Lett. 64, 918 (1990).
121. H.S. Youn, Ph.D. thesis, Univ. of Virginia (1989), unpublished.
122. L. Bruschi, G. Torzo, and M.H.W. Chan, Europhys. Lett. 6, 541 (1988).
123. L. Bruschi and G. Torzo (1990), this volume.
124. J.G. Dash, J. Crystal Growth 100, 268 (1990).
125. J.Z. Larese et al., Phys. Rev. B 40, 4271 (1989).
126. J.Z. Larese and Q.M. Zhang, Phys. Rev. Lett. 64, 922 (1990).
127. J.Z. Larese et al. (1990), private communication.
128. J.G. Dash (1990), private communication.
129. M. den Nijs, Phys. Rev. Lett. 64, 435 (1990).
130. J.Q. Broughton and G.H. Gilmer, J. Chem. Phys. 79, 5105; 5119 (1983).
131. J.Q. Broughton and G.H. Gilmer, J. Chem. Phys. 84, 5741 (1986).
132. V. Rosato, G. Ciccotti, and V. Pontikis, Phys. Rev. B 33, 1860 (1986).
133. S. Valkealahti and R.M. Nieminen, Phys. Scr. 36, 646 (1987).
134. Y.J. Nikas and C. Ebner, J. Phys. C : Cond. Matter 1, 2709 (1989).
135. C.D. Hruska and J.M. Phillips, Phys. Rev. B 37, 3801 (1988).
136. J.M. Phillips, Langmuir 5, 571 (1989).
137. A.L. Cheng and W.A. Steele, Langmuir 5, 600 (1989).
138. M. Maruyama J. Crystal, Growth 89, 415 (1988).
139. D.M. Butler, J.A. Litzinger, and G.A. Stewart, Phys. Rev. Lett. 44, 466 (1980).
140. Y. Larher and A. Terlain, J. Chem. Phys. 72, 652 (1980).
141. R.M. Suter, N.J. Colella, and R. Gangwar, Phys. Rev. B 31, 627 (1985).
142. S.C. Fain, Jr., M.D. Chinn, and R.D. Diehl, Phys. Rev. B 21, 4170 (1980).
143. P.W. Stephens, P. Heiney, R.J. Birgeneau, and P.M. Horn, Phys. Rev. Lett. 43, 47 (1979).
144. G. Zimmerli and M.H.W. Chan (1990), preprint.
145. Y. Larher and Angerand, Europhys. Lett. 7, 447 (1988).
146. H.S. Youn, D.M. Li, and G.B. Hess (1990), unpublished.
147. D. Pengra, D.M. Zhu, and J.G. Dash (1990), submitted for publication and private communication.
148. H. Hong et al., Phys. Rev. B 40, 4797 (1989).
149. R.J. Birgeneau and P.M. Horn, Science 232, 329 (1985).
150. R. Gangwar, N.J. Colella, and R.M. Suter, Phys. Rev. B 39, 2459 (1989).
151. A.J. Jin, M.R. Bjurstrom, and M.H.W. Chan, Phys. Rev. Lett. 62, 1372 (1989).
152. A. Inaba, J.A. Morrison, and J.M. Telfer, Molecular Phys. 62, 961 (1987).
153. H. Hong and R.J. Birgeneau, Z. Phys. B 77, 413 (1989).
154. H.S. Youn and G.B. Hess (1990), unpublished.
155. H.K. Kim, Q.M. Zhang, and M.H.W. Chan, J. Chem. Soc., Faraday Trans. II 82, 1647 (1986).
156. J.M. Gay, A. Dutheil, J. Krim, and J. Suzanne, Surf. Sci. 177, 25 (1986).
157. D.L. Goodstein, J.J. Hamilton, M.J. Lysek, and G. Vidali, Surf. Sci. 148, 187 (1984).
158. H.S. Youn, G.G.Reynolds, E. Updike, and G.B. Hess (1990), unpublished.

159. J.Z. Larese et al., Phys. Rev. B 37, 4735 (1988).
160. M.W.H. Chan (1990), private communication.
161. A. Inaba and J.A. Morrison, Chem. Phys. Lett. 124, 361 (1986).
162. A. Inaba, Y. Koga, and J.A. Morrison, J. Chem. Soc., Faraday Trans. II. 82, 1635 (1986).
163. J. Krim, J.M. Gay, J. Suzanne, and E. Lerner, J. Phys. (Paris) 47, 1757 (1986).
164. M.S. Pettersen, M.J. Lysek, and D.L. Goodstein, Surf. Sci. 175, 141 (1986).
165. A.D. Migone, M.H.W. Chan, K.J. Niskanen, and R.B. Griffith, J. Phys. C : Solid State Phys. 16, L1115 (1983).
166. K.D. Miner, M.H.W. Chan, and A.D. Migone, Phys. Rev. Lett. 51, 1465 (1983).
167. R.D. Diehl and S.C. Fain, Jr., Surf. Sci. 125, 116 (1983).
168. Q.M. Zhang, H.K. Kim, and M.H.W. Chan, Phys. Rev. B 32, 1820 (1985).
169. Q.M. Zhang, H.K. Kim, and M.H.W. Chan, Phys. Rev. B 33, 413 (1986).
170. S.K. Wang et al., Phys. Rev. B 39, 10331 (1989).
171. R.D. Diehl and S.C. Fain, Jr., J. Chem. Phys. 77, 5065 (1982).
172. T.T. Chung and J.G. Dash, J. Chem. Phys. 64, 1855 (1976).
173. D.S. Abbott, H.S. Youn, and G.B Hess (1990), unpublished.
174. Y.P. Feng and M.H.W. Chan, Phys. Rev. 64, 2148 (1990).
175. A. Terlain and Y. Larher, Surf. Sci. 93, 64 (1980).
176. H. You and S.C. Fain, Jr., Surf. Sci. 151, 361 (1985).
177. K. Morishige, C. Mowforth, and R.K. Thomas, Surf. Sci. 151, 289 (1985).
178. Y. Larher, F. Angerand, and Y. Maurice, J. Chem. Soc., Faraday Trans. I 83, 3355 (1987).
179. G.C. DeFotis, Phys. Rev. B 23, 4714 (1981).
180. M. Toney and S.C. Fain, Jr., Phys. Rev. B 36, 1248 (1987).
181. M. Nielsen and J.P. McTague, Phys. Rev. B 19, 3096 (1979).
182. R. Chiarello, J.P. Coulomb, J. Krim, and C.L. Wang, Phys. Rev. B 38, 8967 (1988).
183. J. Stoltenberg and O.E. Vilches, Phys. Rev. B 22, 2920 (1980).
184. D.D. Awschalom, G.N. Lewis, and S. Gregory, Phys. Rev. Lett. 51, 586 (1983).
185. U. Köbler and R. Marx, Phys. Rev. B 35, 9809 (1987).
186. C.E. Bartosch and S. Gregory, Phys. Rev. Lett. 54, 2513 (1985).
187. M. Drir and G.B. Hess, Phys. Rev. B 33, 4758 (1986).
188. H.S. Youn and G.B. Hess, Phys. Rev. Lett. 64, 443 (1990).
189. P.W. Stephens et al., Phys. Rev. Lett. 45, 1959 (1980).
190. P.A. Heiney et al., Surf. Sci. 125, 539 (1983).
191. S.G.J. Mochrie et al., Surf. Sci. 138, 599 (1984).
192. K. Morishige, K. Mimata, and S. Kittaka, Surf. Sci. 192, 197 (1987).
193. H. You and S.C. Fain, Jr., Phys. Rev. B 33, 5886 (1986).
194. D. van der Putten, K.O. Prins, P.J. Kortbeek, and N.J. Trappeniers, J. Phys. C: Solid State Phys. 20, 3161 (1987).
195. Q.M. Zhang, H.K. Kim, and M.H.W. Chan, Phys. Rev. B 34, 8050 (1986).
196. H.S. Nham and G.B. Hess, Phys. Rev. B 37, 9802 (1988).
197. J. Suzanne, J.L. Sequin, M. Bienfait, and E. Lerner, Phys. Rev. Lett. 52, 637 (1984).
198. J.M. Gay, M. Bienfait, and J. Suzanne, J. Phys. (Paris) 45, 1497 (1984).
199. P. Dolle, M. Matecki, and A. Thomy, Surf. Sci. 91, 271 (1980).
200. H.S. Nham, M. Drir, and G.B. Hess, Phys. Rev. B 35, 3675 (1987).
201. Y. Larher and Millot, J. Phys. (Paris) Colloque 38, C4-189 (1977).
202. E. Cheng and M.W. Cole, Phys. Rev. B 41, 9650 (1990).

FLUID INTERFACES: WETTING, CRITICAL ADSORPTION, VAN DER WAALS TAILS, AND THE CONCEPT OF THE EFFECTIVE INTERFACE POTENTIAL

S. Dietrich

Fachbereich Physik
Bergische Universität Wuppertal
Postfach 100127
D-5600 Wuppertal 1
Federal Republic of Germany

ABSTRACT

A transparent description of wetting phenomena can be obtained from the so-called effective interface potential $\Omega_s(l)$ which is the surface free energy of a wetting film with a prescribed thickness l. From $\Omega_s(l)$ one obtains directly the various types of wetting transitions as well as their associated thermal singularities. Furthermore $\Omega_s(l)$ allows one to trace back the character of these interfacial phase transitions to its dependence on atomic interactions. To a large extent this approach has been carried out analytically for wetting of a wall by a one-component fluid and for interfacial wetting of the vapor phase in binary liquid mixtures. Particular attention is paid to the presence of van der Waals tails as well as to the competition between wetting phenomena and critical adsorption.

I. INTRODUCTION

Consider the α - γ interface between two phases α and γ of condensed matter. If a third phase β becomes thermodynamically stable a β-like film of thickness l_0 may form at the α - γ interface. If upon a change of temperature or pressure l_0 attains a macroscopic value, it is said that β wets the α - γ interface. In that case the α - γ interface splits into an α - β and a β - γ interface.

Fig. 1 describes this situation in the case that α is an inert wall and β and γ are the liquid and gas phase, respectively, of a simple one-component fluid. (If not stated otherwise in the following I consider such a system.) Fig. 2 explains the notions of critical, complete, first-

Phase Transitions in Surface Films 2
Edited by H. Taub *et al.*, Plenum Press, New York, 1991

order, and prewetting transitions which may occur at the α - γ interface. (For a more complete account of wetting phenomena the reader is referred to a number of review articles about this subject which have appeared recently [1-4].) These wetting phenomena are phase transitions within the wall-gas interface structure. They can be described transparently by introducing the effective interface potential $\Omega_s(l)$. It is defined as the substrate-gas surface tension under the restriction that the liquidlike wetting layer has a prescribed thickness l. The

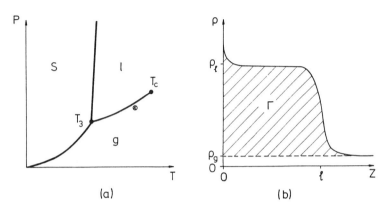

(a)   (b)

FIG.1: (a) The generic bulk phase diagram in the pressure-temperature variables for a simple one-component system with the solid, liquid, and gas phase. At the triple point T_3 the melting- and the sublimation-curve meet the gas-liquid coexistence line, which ends at the critical point T_c. ⊗ denotes a typical equilibrium situation leading to a number density profile $\rho(z)$ as in (b). (b) The density profile $\rho(z;p,T)$ of a gas close to a substrate. Γ denotes the coverage of the substrate. For reasons of simplicity density oscillations close to the substrate as they occur due to packing effects are suppressed. The system is assumed to be homogeneous parallel to the surface so that ρ depends only on z, which measures the distance from the mean position of the nuclei in the top layer of the substrate s. Thus $\rho(z)$ vanishes smoothly for $z \to 0$, which is not shown in (b). ρ_l is the number density of the liquid at coexistence and ρ_g is the actual gas density in the bulk. $l = \Gamma/(\rho_l-\rho_g)$ is one of the possible definitions for the thickness of the wetting film.

actual substrate-gas surface tension is the minimum of $\Omega_s(l)$: $\sigma_{sg} = \min_l \Omega_s(l) = \Omega_s(l_o)$. For $\sigma_{sg} < \sigma_{sl} + \sigma_{lg}$ the substrate is nonwet, for $\sigma_{sg} = \sigma_{sl} + \sigma_{lg}$ it is wet. σ_{sl} and σ_{lg} are the substrate-liquid surface tension at coexistence and the liquid-gas surface tension, respectively. The qualitative behavior of $\Omega_s(l)$ for critical, first-order, and complete wetting is shown in Fig.3. There it becomes apparent that, apart from the exceptional case in Fig.3(h), both the occurence and the singularities of continuous wetting transitions follow from the behavior of $\Omega_s(l)$ at large l.

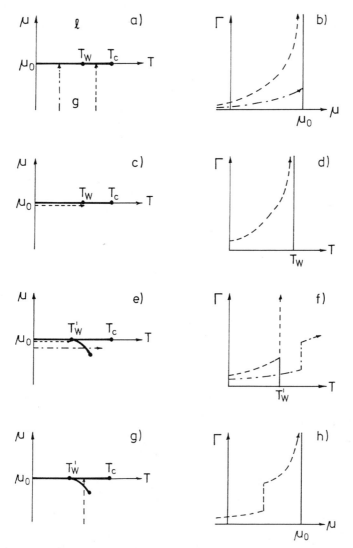

FIG.2: The coverage Γ along several typical paths in the bulk phase diagram in terms of chemical potential μ and temperature T. The liquid-gas coexistence line ($\mu=\mu_0$, $T_3 \leq T \leq T_c$) separates the liquid ($\mu>\mu_0$) from the gas phase ($\mu<\mu_0$). For reasons of simplicity the influence of the solid phase is omitted and the coexistence line is straightened out. Below the wetting transition temperature T_w there is incomplete wetting, for $T \geq T_w$ there is complete wetting: (a) and (b). (c) and (d) correspond to critical wetting, whereas (e) and (f) describe a first-order wetting transition. (e) - (h) give a characterization of the prewetting line which is attached tangentially at T_w' of a first-order wetting transition. (a) - (d) and (e) - (h) belong to different substrate potentials but to the same fluid. Therefore T_c and μ_0 are the same for (a) - (h), whereas T_w in (a) - (d) is different from T_w' in (e) - (h).

In a simple fluid the particles have spherically symmetric pair interactions $w(|\mathbf{r}-\mathbf{r}'|)$ with $w(r\to\infty)\sim r^{-6}$. The substrate potential they are exposed to behaves like $V(z\to\infty)\sim z^{-3}$. In such a system the asymptotic behavior of $\Omega_s(l) = \sigma_{sl}+\sigma_{lg}+\omega(l)$ takes on the following form for large l [1]$(\Delta\rho=\rho_l-\rho_g)$:

$$\omega(l) = (\mu_o - \mu)\Delta\rho\cdot l + \frac{a_2}{l^2} + \frac{a_3}{l^3} + \frac{a_4}{l^4} +\ldots . \tag{1}$$

According to the discussion given above the necessary conditions for critical wetting at $T = T_w$ are: $a_2(T=T_t)<0$, $a_2(T=T_c)\geq0$, and $a_3(T=T_w)>0$ where T_w is defined implicitly by $a_2(T=T_w)=0$ (see Fig.3 and Ref.1).(Tricritical wetting requires $a_2(T=T_w) = a_3(T=T_w)=0$.) Eq.(1) renders directly the thermal singularities at critical wetting $(\mu=\mu_o,T\to T_w)$:

$$\Gamma/\Delta\rho = \frac{3}{2}\frac{a_3}{|a_2|}(1+ \frac{8}{9}\frac{a_4}{a_3^2}|a_2|+\ldots)$$
$$= \hat{l}_o\tau^{-1}(1+ \tau/\hat{\tau}_c +\ldots) . \tag{2}$$

Here $\tau \equiv (T_w-T)/T_w\to 0$ with $\hat{l}_o = 3a_{3,0}/|2a_{2,1}|$ and $\hat{\tau}_c = |a_{3,1}/a_{3,0} - a_{2,2}/a_{2,1}+ 8a_{4,0}a_{2,1}/(9a_{3,0}^2)|^{-1}$ where the expansion $a_k = \sum_{i\geq0} a_{k,i}\tau^i$ and $a_{2,0} = 0$ are used for k = 2,3,4....

As one can see in Fig.3(b) and (g) the minimum of $\Omega_s(l)$ becomes shallower for $l_o\to\infty$. This signals the build up of long range parallel correlations within the depinning interface, because the parallel correlation length is given within mean-field theory by $\xi_\parallel=\sigma_{gl}^{1/2}/[\partial^2\Omega_s(l)/\partial l^2|_{l=l_o}]^{1/2}$. Thus one finds for critical wetting [1]

$$\xi_\parallel\sim\tau^{-\nu_\parallel}, \nu_\parallel = \frac{5}{2} , \tag{3}$$

and

$$\sigma_{sg} - \sigma_{sl} - \sigma_{lg}\sim -\tau^{2-\alpha_s},\alpha_s =- 1 . \tag{4}$$

The analogous analysis for complete wetting yields [1] $(T > T_w, \Delta\mu = \mu_o-\mu \to 0)$:

$$\rho_s\sim(\Delta\mu)^{\beta_s}, \beta_s =- \frac{1}{3},$$
$$\xi_\parallel\sim(\Delta\mu)^{-\nu_\parallel}, \nu_\parallel = \frac{2}{3}, \tag{5}$$
$$\sigma_{sg} - \sigma_{sl} - \sigma_{lg}\sim(\Delta\mu)^{2-\alpha_s}, \alpha_s = \frac{4}{3} .$$

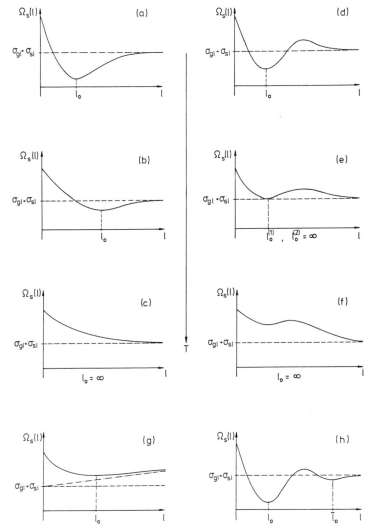

FIG.3: Effective interface potential $\Omega_s(l)$ as function of the thickness l of the liquidlike wetting film. $\sigma_{sg} = \Omega(l_0)$ where l_0 is the position of the minimum of $\Omega_s(l)$. (a) - (c) correspond to a system which undergoes a critical wetting transition upon a rise in temperature T: l_0 diverges continuously for $T \rightarrow T_w$ and $l_0 = \infty$ for $T > T_w$ as shown in (c). (d) - (f) describe a different system exhibiting a first-order wetting transition. As shown in (e), $\Omega_s(l, T=T_w)$ has two global minima, $l_0^{(1)}$ and $l_0^{(2)} = \infty$. Again, (f) is a situation above T_w. In all figures except (g) the systems are at gas-liquid coexistence. (g) shows a situation above T_w but off coexistence, so that $\Omega_S(l \rightarrow \infty) \sim l$. (h) corresponds to a third system for which $\Omega_s(l)$ has a local minimum at $\tilde{l}_0$ and the global one at l_0. Even if $\tilde{l}_0 \rightarrow \infty$ upon an increase of T, there might be no critical wetting because the global minimum can remain at $l = l_0$.

Fig.3 and the considerations leading to Eqs.(2-5) show that the effective interface potential makes transparent the nature of wetting transitions (i.e. the order of the transition) and yields the singularities associated with them. Therefore the following sections are devoted to the derivation of Eq.(1) for various physical systems and to the discussion of the coefficients a_k, which are functions of T and μ and functionals of $\{w(r)\}$ and $\{V(z)\}$.

II. WETTING OF A WALL BY A ONE-COMPONENT SIMPLE FLUID

The following density functional for inhomogeneous fluids serves as an adequate starting point for the problem under consideration:

$$
\Omega[\{\rho(\mathbf{r})\};T,\mu;\{w(|\mathbf{r}-\mathbf{r}'|)\},\{V(\mathbf{r})\}] = \int d^3r\, f_h(\rho(\mathbf{r}),T) +
$$
$$
+ \frac{1}{2}\int d^3r\int d^3r'\ \tilde{w}(|\mathbf{r}-\mathbf{r}'|)\rho(\mathbf{r})\rho(\mathbf{r}') + \int d^3r(\rho_w V(\mathbf{r}) - \mu)\rho(\mathbf{r}). \tag{6}
$$

$\Omega[\{\rho(\mathbf{r})\};T,\mu]$ is the grand canonical potential for a given number density $\rho(\mathbf{r})$ of the fluid, which is confined to the halfspace $V_+ = \{\mathbf{r} = (\mathbf{r}_{||},z) \in \mathbb{R}^3 \,|z \ge 0\}$. (For a derivation see, e.g., Ref. 5) The fluid particles interact via spherically symmetric pair potentials $w(r)$, which are attractive at large distances r where they decay $\sim r^{-6}$. (Here I do not discuss retardation effects due to which $w(r) \sim r^{-7}$ for distances $r \gtrsim 100$Å.) Their divergent repulsive part at small distances is treated by the introduction of a reference system of hard spheres with diameter r_0. Thus $f_h(\rho,T)$ is the bulk Helmholtz free energy density of a homogeneous reference system with number density ρ. According to the Weeks-Chandler-Andersen theory [6] one has $\tilde{w}(r \ge \bar{r}_0) = w(r)$ where $\bar{r}_0$ is the position of the minimum of $w(r)$; $\tilde{w}(r \le \bar{r}_0) = w(\bar{r}_0) = \tilde{w}(\bar{r}_0)$. The optimum choice for the hard sphere diameter r_0 is determined by a certain correlation function, which depends on $\{w(r \le r_0)\}$, the mean density ρ, and temperature T [6]. Thus Eq. (7) is uniquely determined by the interaction $w(r)$ between the fluid particles and the substrate potential $\rho_w V(\mathbf{r})$. For convenience in the latter the mean number density ρ_w of the wall has been introduced. Since the corrugation effects in the substrate potential decay exponentially [7] and since one is interested in the power law behavior of $\Omega_s(l) - \Omega_s(\infty)$ for large l the dependence of V on $\mathbf{r}_{||}$ can be ignored so that $V(\mathbf{r}) \equiv V(z)$. Consequently the actual density profile $\rho_0[z;T,\mu;\{w(r)\},\{V(z)\}]$, which minimizes $\Omega[\{\rho(\mathbf{r})\};T,\mu]$ for a given boundary condition at $z = +\infty$, depends on the spatial variables only through the distance z from the wall. The wall occupies the half space $V_- = \{\mathbf{r} \in \mathbb{R}^3 \,|z \le 0\}$ so that $V(z) \sim z^{-3}$ for large z. $z = 0$ is given by the mean position of the nuclei of the top substrate layer. The actual grand canonical potential $\Omega[T,\mu;\{w(r)\},\{V(z)\}]$ is the minimum value of Eq. (6) at $\rho(\mathbf{r}) = \rho_0[z;T,\mu;\{w(r)\}, \{V(z)\}]$.

Although in principle the density functional theory is an exact formalism for describing fluids [8], its actual implementation, like the version used in Eq.(6), corresponds to a certain mean-field approximation. Consequently both those fluctuations, which cause bulk properties to deviate from their classical behavior near critical points, and capillary waves of the emerging liquid–vapor interface are discarded. Whereas in the present context the former fluctuations are relevant only in the case of a close vicinity of T_w and T_c or, in the case of complete wetting, close to T_c, capillary waves are excited even at low temperatures. However, in the presence of long-range forces – as they govern the fluid systems considered here – capillary waves are known to be irrelevant for the thermal singularities of continuous wetting transitions [9]. Furthermore in real systems gravity suppresses the build up of capillary waves with arbitrary long wavelength, which are not captured by a mean-field description. Correspondingly in the presence of gravity any actual density profile is given by a convolution of the so-called intrinsic density profile with a Gaussian distribution of the mean interface position (see the discussion of Eq.(2.39) in Ref.10). In this sense here and in the following I analyze the mean-field approximation of the *intrinsic density profile*.

Finally I want to state that Eq.(6) does not allow for the formation of a solid phase. In the derivation of Eq.(6) the two-particle density of the reference system is replaced by its asymptotic form $\rho(\mathbf{r}) \cdot \rho(\mathbf{r}')$ for $|\mathbf{r}-\mathbf{r}'| \to \infty$ and the free energy of the inhomogeneous reference system is evaluated in the local density approximation by using the free energy density for a homogeneous reference fluid. As an approximate analytic expression for $f_h(\rho,T)$ one may adopt the Percus-Yevick [6] or the Carnahan-Starling [6] formula, which, however, are not capable of describing the freezing transition of the hard-sphere reference system. (Up to now no analytic expression for this freezing transition is known.) Here, however, I focus only on fluid phases, i.e., $T > T_3$. Thus in the present context the main deficiency of Eq.(6) consists in the absence of density oscillations near the wall caused by packing effects (see Sect.III.B in Ref.1 for a list of the relevant references). Here I ignore them because I am interested in the algebraic decay of $\Omega_s(l)$ - $\Omega_s(\infty)$ for large l (see Eq.(1)) and because these density oscillations decay exponentially even in the presence of long-range van der Waals interactions [11]. This is a reasonable approximation for continuous wetting transitions as considered here. However, in the case of first-order wetting transitions, for which the behavior of $\Omega_s(l)$ for small values of l may be important, it is indispensable to take these density oscillations into account [12,13]. The application of the weighted density functional theory [14] to inhomogeneous systems may prove to be a promising way to solve such problems.

Keeping all these caveats in mind at present Eq.(6) represents the most adequate and managable approach for the problem under consideration. In order to obtain the effective interface potential, Eq.(6) is evaluated for suitable trial functions $\rho(z,l;T,\mu)$, which correspond at least approximately to the formation of a liquidlike layer with the prescribed thickness l. In systems with long-ranged forces this approach has been carried out systematically by applying the sharp-kink approximation [15] and the soft-kink approximation [10] (see Fig.4).

Since within these approximations the effective interface potential could be determined *analytically*, a much broader insight into the mechanisms, which drive wetting transitions, was made possible compared with full numerical solutions of Eq.(6), which can be obtained only for a vanishing small set of parameters within the relevant *function space* $\{w(\mathbf{r}), V(\mathbf{r})\}$. Ultimately such analytic solutions are the only justification for introducing the effective interface potential in the first place, because for a given system $\{w_0(\mathbf{r}), V_0(\mathbf{r})\}$ Eq. (6) can be solved numerically according to standard procedures. However, this function space cannot be explored numerically. This exploration becomes feasible as soon as $\Omega_s(l)$ is − at least within certain approximations − analytically known. Even if one was interested in only one specific physical system, such an exploration would be indispensable because the interaction potentials $w(\mathbf{r})$ and $V(z)$ are only crudely known so that one must understand what will happen if they are varied.

Despite the success [1,4,10,15] of the sharp- and soft-kink approximations for the relevant trial functions, they display many shortcomings (see Fig.4): there are discontinuities, van der Waals tails (see below) are missing; there are dependences on − in principle − unspecified input parameters like d_w and χ (see Fig. 4); the structure of the fluid very close to the wall is not taken properly into account; and the actual trial functions deviate for finite values of l from a superpostion of the wall-liquid and the free liquid-gas density profiles.

In a recent effort these shortcomings have been cured [16,17] by using the following trial function at coexistence $\mu = \mu_0^-$:

$$
\rho(z,l;T) = \begin{cases} \rho_{wl}(z) & , & z \leq \kappa(l) - \lambda(l) \\ G(\kappa(l) - z, l), & \kappa(l) - \lambda(l) \leq z \leq \kappa(l) + \lambda(l) \\ \rho_{lg}(z - l) & , & z \geq \kappa(l) + \lambda(l). \end{cases} \tag{7}
$$

This trial function consists of the free liquid-gas interface

$$
\rho_{lg}(z) = \begin{cases} \rho_1 - \sum_{k \geq 3} A_k^{(1)} |z|^{-k}, & z \leq -\xi \\ F(z) & , |z| \leq \xi \\ \rho_{lg} + \sum_{k \geq 3} A_k^{(g)} z^{-k}, & z \geq \xi, \end{cases} \tag{8}
$$

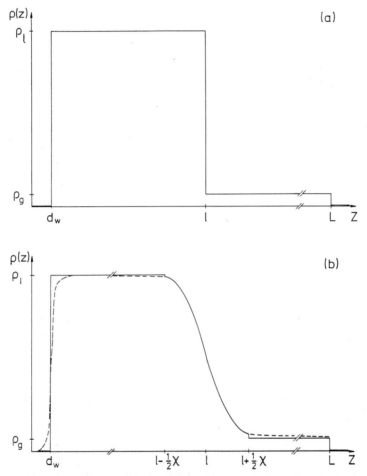

FIG.4: Sharp kink (a) and soft kink (b) approximation for the density profile. ρ_l and ρ_g are the bulk densities of the coexisting liquid and gas phases, respectively. $\rho(z<d_w) = 0$ due to the repulsive part of the substate potential. In (b) and for $|z-l| < \chi/2$ $\rho(z)$ coincides with the density profile of the free liquid-gas interface. The broken line in (b) indicates a more realistic density profile; $\rho(z>L) = 0$ for computational convenience.

whose center z* is positioned at z=l, and the wall-liquid interface

$$\rho_{wl}(z) = \begin{cases} q(z) & ,z \le \xi \\ \rho_1 + \sum\limits_{k=3}^{\infty} Q_k^{(1)} z^{-k} & ,z \ge \xi \end{cases} \tag{9}$$

near the wall (see Figs.5 and 6). They match smoothly in a transition region of width $2\lambda(l)$, which is described by an unspecified function $G(z,l)$. Since for large l the wall-liquid interface must emerge, the postion of the transition region has to move to infinity, too. Therefore one has $\kappa(l \to \infty) \sim \kappa_0 l$ with $0 < \kappa_0 < 1$. For similar reasons one can assume that the width of the transition region vanishes as $\lambda(l \to \infty) \sim l^{-(1+\varepsilon)}$, $\varepsilon > 0$ [17]. In Eqs.(8) and (9) both $F(z)$ and $q(z)$ are also unspecified functions. ξ is the bulk correlation length and the terms $\sim A_k$ and $\sim Q_k$ represent the so-called *van der Waals tails* whose algebraic decay is characteristic for systems governed by long-ranged van der Waals forces. The coefficients A_3 and Q_3 are known analytically [17]. In view of my focus on the algebraic decay of $\omega(l)$ in Eq.(1) the additional exponential tails in $\rho_{lg}(z)$ and $\rho_{wl}(z)$ are omitted. In the final results (c.f. Eqs.(14-21)) they can easily be incorporated.

As indicated in Fig. 6 only those wetting films are considered whose thickness l is large compared with the bulk correlation length:

$$l > \xi. \tag{10}$$

According to Eqs.(7-9) $\xi = \xi(T)$ is the correlation length *at* coexistence.

The trial function in Eq.(7) is taken at coexistence. In order to describe complete wetting one has to generalize it for $\mu < \mu_0(T)$. The simplest form is

$$\rho(z,l;T,\mu) = \rho(z,l;T) + \Theta(z - l)[\rho_g(T,\mu) - \rho_g(T,\mu_0(T))]. \tag{11}$$

Eq.(11) guarantees that $\rho(z \to \infty, l; T, \mu)$ approaches the correct bulk value $\rho_g(T,\mu)$. However this simple choice exhibits a discontinuity at $z = l$ which is proportional to the undersaturation. Therefore the results obtained from Eq.(11) are reliable only for small undersaturations $\Delta\mu = \mu_0 - \mu$.

By using these trial functions one obtains after very complicated manipulations the following expression for the effective interface potential $\Omega_s(l) = \sigma_{sl} + \sigma_{lg} + \omega(l)$ at large values of l [17]:

$$\omega(l) = \Delta\mu(\Delta\rho \cdot (l + d_{lg}^{(1)}) - \rho_1 d_{wl}^{(1)}) + \sum\limits_{k=2}^{4} a_k l^{-k} + O(l^{-5} \ln l). \tag{12}$$

400

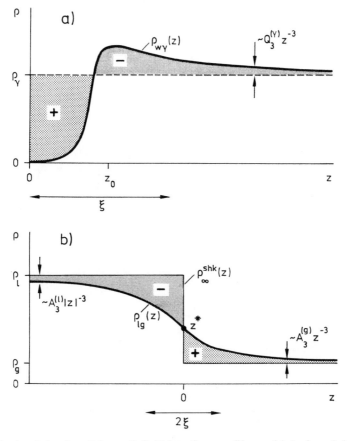

FIG.5: Qualitative behavior of the wall-fluid interface profile $\rho_{w\gamma}(z)$ (γ=l or g) (a) and of the free liquid-gas interface profile $\rho_{lg}(z)$ (b). In (a) $\rho_{w\gamma}(z)$ vanishes for $z \to 0$ with an essential singularity, reaches a maximum at $z = \bar{z}_0$, which may coincide with the minimum of $V(z)$, and decays $\sim Q_3^{(\gamma)} \cdot z^{-3}$ towards the bulk value ρ_γ at $z = +\infty$. The asymptotic behavior of $\rho_{w\gamma}(z)$ starts for $z \geq \xi$. $Q_3^{(\gamma)}$ can also be negative so that $\rho_{w\gamma}(z)$ approaches ρ_γ from below. In (b) $\rho_{lg}(z)$ reaches ρ_l as $A_3^{(l)} \cdot |z|^{-3}$ for $z \to -\infty$ and ρ_g as $A_3^{(g)} \cdot z^{-3}$ for $z \to +\infty$. The main variation of $\rho_{lg}(z)$ occurs in a slab of thickness 2ξ around its center z^* located at $z = 0$. $\rho_\infty^{shk}(z) = \rho_l\theta(-z) + \rho_g\theta(z)$ is the sharp-kink approximation for $\rho_{lg}(z)$. For $T \to T_c$ ξ diverges both in (a) and (b), $\rho_\gamma \to \rho_c$ and $\rho_l - \rho_g \to 0$. The coefficients $A_3^{(\gamma)}$ are always positive. In (a) and for $\gamma = l$ the difference between the positively and the negatively marked area divided by ρ_l is a microscopic length $d_{wl}^{(1)}$ which enters the coefficients of the effective interface potential. (c.f. Eq. (20)). In (b) the difference between the positively and the negatively marked area divided by $\rho_l - \rho_g$ is a microscopic length $d_{lg}^{(1)}$ corresponding to the free liquid gas interface (c.f. Eq. (21)). The value of $d_{wl}^{(1)}$ is uniquely fixed whereas $d_{lg}^{(1)}$ depends on the choice of z^*.

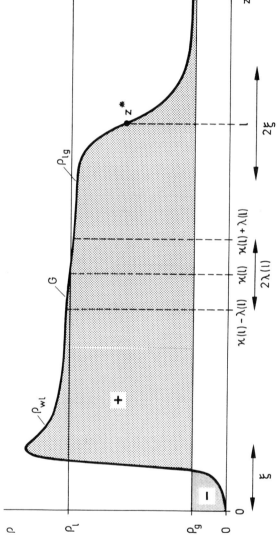

FIG.6: Schematic drawing of the trial density profile $\rho(z,l)$ from Eq.(7) for approximating the effective interface potential $\Omega_s(l)$. The wall-liquid interface profile $\rho_{wl}(z)$ is followed by a transition region described by $G(z)$ which is centered around $z = \kappa(l)$ and has a width $\lambda(l)$. For $z \geq \kappa(l) + \lambda(l)$ the liquid-gas interface profile starts. The center z^* of the latter is positioned at $z = l$. ρ_l and ρ_g are the liquid and gas density, respectively, at coexistence. ξ is the bulk correlation length (see Fig.5). The l-dependences of $\kappa(l)$ and $\lambda(l)$ are discussed in the main text. The coverage Γ is the difference between the positively and the negatively marked area.

(Note the differences to Eq.(1), which follow from the cruder sharp-kink approximation.)
The coefficients take on the following form:

$$a_2 = \frac{1}{2}\Delta\rho(\rho_w u_3 - \rho_1 t_3),$$

(13)

$$a_3 = a_3^{(0)} - 2a_2 d_{lg}^{(1)},$$

(14)

and

$$a_4 = a_4^{(0)} - 3a_3 d_{lg}^{(1)} + 3a_2(d_{lg}^{(2)} - 2(d_{lg}^{(1)})^2)$$

(15)

with

$$a_3^{(0)} = \frac{1}{3}\Delta\rho(\rho_w u_4 - \rho_1(t_4 + 3t_3 d_{wl}^{(1)}))$$

(16)

and

$$a_4^{(0)} = \frac{1}{4}\Delta\rho(\rho_w u_5 - \rho_1(t_5 + 4t_4 d_{wl}^{(1)} + 6t_3 d_{wl}^{(2)})).$$

(17)

Thus the coefficients a_k depend on those determining the substrate potential at large z,

$$V(z) = -\sum_{k\geq 3} u_k z^{-k},$$

(18)

and the fluid-fluid interaction at large distances,

$$t(z) = \int_z^\infty dz' \int d^2 r_{\parallel} \, \tilde{w}(\sqrt{r_{\parallel}^2 + z'^2})$$

$$= -\sum_{k\geq 3} t_k z^{-k},$$

(19)

as well as on the mean number density ρ_w of the wall, the liquid density $\rho_l(T,\mu=\mu_o)$ at coexistence, the gas density $\rho_g(T,\mu\leq\mu_o)$, and on moments of the wall-liquid and the liquid-gas interface profiles, respectively, both at coexistence (i=1,2):

$$d^{(i)}_{wl} = i\int_o^\infty dz\, z^{i-1}(1 - \rho_{wl}(z)/\rho_l) \tag{20}$$

and

$$d^{(i)}_{lg} = \frac{i}{\Delta\rho}\int_{-\infty}^\infty dz\, z^{i-1}\,(\rho_{lg}(z) - \rho^{shk}_\infty(z)), \tag{21}$$

where $\rho^{shk}_\infty(z) = \Theta(z)\rho_g + \Theta(-z)\rho_l$ is the sharp-kink approximation for the free interface. A geometrical interpretation of the lengths $d^{(1)}_{wl}$ and $d^{(1)}_{lg}$ is given in the caption to Fig.5. Owing to the van der Waals tails in $\rho_{wl}(z)$ and in $\rho_{lg}(z)$ both $d^{(i)}_{wl}$ and $d^{(i)}_{lg}$ do not exist for $i \geq 3$. This is the reason behind the fact that in Eq.(12) the quoted correction terms are proportional to $l^{-5}\ln l$ instead of l^{-5}.

As discussed in detail in Ref.17 Eqs.(12-21) correspond to a whole family of different effective interface potentials which are parameterized by $d^{(1)}_{lg}$. The value of $d^{(1)}_{lg}$ depends on the choice of z* as the center of the free liquid-gas interface profile. According to our construction scheme (see Eg.(7) and Figs.5 and 6) this choice corresponds to choosing one of the various possible definitions for the thickness of a given wetting film. All physical properties of the system (the value of T_w given by $a_2(T_w)$, the separatrix between first- and second-order wetting given by $a_2(T_w) = a_3(T_w) = 0$, the wall-gas surface tension given by $\rho_{wg} = \min_l \Omega_s(l)$, the coverage $\Gamma = \int_o^\infty dz(\rho(z) - \rho_g)$ etc.) must be independent of this choice. As has been shown in Ref.17, the effective interface potential given in Eq.(12) fulfills all these invariance requirements.

The explicit results in Eqs.(12-21) allow one to draw several interesting conclusions. First, for a continuous wetting transition the transition temperature is given exactly by the simple implicit equation:

$$\rho_l(T = T_w)/\rho_w = u_3/t_3. \tag{22}$$

Surprisingly this temperature, as defined by Eq.(22), marks not only a continuous phase transition of the wall-gas interface profile but also a qualitative change of the wall-liquid transition (see Fig.3). If, however, $\rho_c/\rho_w < u_3/t_3 < \rho_{13}/\rho_w$ a_2 is negative at T_3 and positive close to T_c. Thus there is a temperature T_w with $T_t < T_w < T_c$ such that $a_2(T{=}T_w) = 0$. If in addition $a_3(T{=}T_w) > 0$ T_w is indeed the transition temperature for a continuous wetting transition [1].

Eq.(24) therefore states that for a continuous wetting transition to occur the wall-liquid density profile must approach ρ_l from below at the triple point and from above close to T_c. Thus one obtains the surprising result that critical wetting on the *gas side* of $\mu_o(T)$ goes along with a change of sign for the asymptotic behavior of the wall-liquid density profile on the *liquid side* of $\mu_o(T)$. This change of sign occurs at the wetting transition temperature T_w. Above T_w the wall prefers the liquid phase and $Q_3^{(l)} > 0$ whereas $Q_3^{(l)} < 0$ below T_w. interface profile ($-$ provided that Eq.(22) has a solution between T_3 and T_c). This can be seen from the expression for the van der Waals tail of the wall-liquid interface [17] (see Eq.(9)):

$$Q_3^{(l)} = \rho_1^2(\rho_w u_3 - \rho_1 t_3)\kappa_T^{(l)}. \tag{23}$$

Here $\kappa_T^{(l)}$ is the isothermal compressibility $VN^{-2}(\partial N/\partial\mu)_{T,V} = -V^{-1}(\partial V/\partial p)_{T,N} = \rho_1^{-2}[\dfrac{\partial^2 f_h}{\partial\rho^2}(\rho_1,T) + w_o]^{-1}$ in the liquid phase; p is the pressure, N is the mean number of particles and $w_o = \int d^3r\, \tilde{w}(r)$. The combination of Eqs.(13) and (23) shows that $Q_3^{(l)}$ and a_2 have the same sign and that they both change sign at $T = T_w$:

$$Q_3^{(l)} = 2(\rho_1 - \rho_g)^{-1}\rho_1^2 \kappa_T^{(l)} \cdot a_2 . \tag{24}$$

Thus for $u_3/t_3 < \rho_c/\rho_w$, with $\rho_c = \rho_1(T{=}T_c)$, a_2 and $Q_3^{(l)}$ are negative for all temperatures $T_3 \leq T \leq T_c$. In this case $a_2 < 0$ and $l = \infty$ is always a maximum of $\Omega_s(l)$ and wetting on the gas side can never occur. Instead a drying transition on the liquid side of $\mu_o(T)$ is possible. This checks naturally with the fact $Q_3^{(l)} < 0$ because in this case $\rho_{wl}(z)$ approaches ρ_1 from below corresponding to the preference of the wall for the gas phase. If $u_3/t_3 > \rho_{13}/\rho_w$,with $\rho_{13} = \rho_1(T{=}T_3)$, $Q_3^{(l)}$ is positive for all temperatures which reflects the fact that a strong substrate potential favors the liquid phase. Since in this case $a_2 > 0$ for all T one can conclude that if the wall-liquid density profile approaches ρ_1 from above, the system is either wet on the gas side of $\mu_o(T)$ for all temperatures or it undergoes a first-order wetting

The same kind of arguments leading to Eq.(23) show that the van der Waals tail of the wall-gas interface, $\rho_{wg}(z\to\infty) = \rho_g + Q_3^{(g)}z^{-3}+...$, has the following form

$$Q_3^{(g)} = \rho_g^2(\rho_w u_3 - \rho_g t_3)\kappa_T^{(g)},\tag{25}$$

where $\kappa_T^{(g)}$ is the isothermal compressibility of the gas phase. Eq.(25) holds quite generally even in the presence of a wetting film with arbitrary but finite thickness l. Eqs.(24) and (25) show that if the wall-liquid interface profile approaches its bulk value ρ_l from above the same is true for the wall-gas interface profile. This holds for all temperatures.

After this discussion of the connection between the asymptotic behavior of the effective interface potential and of the wall-liquid and wall-gas interface profiles, I want to address the question of wetting close to T_c as far as it can be inferred from the above results for the effective interface potential. In the first instance in accordance with the present approach (see the discussion following Eq.(6)), these conclusions are valid only within mean-field theory. However, there is reason to expect that some of these conclusions remain valid beyond mean-field theory (see below).

In the following I consider critical wetting ($\Delta\mu=0$, $\tau\to 0$) with a transition temperature T_w close to T_c, i.e. $\hat{t} = (T_c-T_w)/T_c \ll 1$, and complete wetting ($\Delta\mu\to 0$, $T>T_w$) close to T_c, i.e. $t = (T_c-T)/T_c \ll 1$. Close to T_c the bulk quantities entering the coefficients a_k vary as $\Delta\rho \sim t^\beta$, $\rho - \rho_l \sim t^\beta$, $\beta \cong 0.33$ ($= 0.5$ within mean-field theory), and $\xi \sim t^{-\nu}$, $\nu \cong 0.64$ ($=0.5$ within MFT). At low temperatures the quantities $d_{wl}^{(i)}$ are determined by atomic lengths. However, close to T_c they diverge owing to the phenomenon of critical adsorption (see Refs. 18 and 19 as well as Sect. IX.B in Ref.1) according to [17]:

$$|d_{wl}^{(i)}(T \to T_c)| \sim t^{-(i\nu-\beta)}, i = 1,2.\tag{26}$$

Combining these results one finds (see Eq.(2))[17]

$$\hat{l}_0(T = T_w \to T_c) \sim \hat{t}^{1-\nu} \to 0\tag{27}$$

and

$$\hat{\tau}_c \sim \hat{t} \to 0.\tag{28}$$

Thus the strength $\hat{l}_0$ of the τ^{-1}-divergence for critical wetting vanishes close to T_c. Simultaneously the critical region $\tau < \tau_c$ below T_w, within which the asymptotic behavior $\sim \tau^{-1}$ of the coverage dominates, shrinks $\sim \hat{t}$ due to the vicinity of the critical point (see Fig. 7). The combination of these formulae yields the restriction $l_0 > \hat{l}_0 \hat{\tau}_c^{-1} \sim \hat{t}^{-\nu} \sim \xi$ in accordance with Eq.(10). Therefore the presence of a critical point impedes the identification of a continuous wetting transition if it happens to occur close to T_c. This prediction is in accordance with recent experimental results for the wetting transition in the cyclohexane-acetonitrile mixture for which $\hat{t} \cong 9 \cdot 10^{-3}$ [20]. These measurements are concerned with the contact angle Θ, which vanishes in the case of critical wetting according to $\Theta = \Theta_0 \cdot \tau^{3/2}$ with $\Theta_0 = [\frac{8}{27} |a_{2,1}|^3 a_{3,0}^{-2} (\sigma_{gl}(T_w))^{-1}]^{1/2}$. With the above results and the fact that $\sigma_{gl} \sim t^{2v}$ one has, for $\hat{t} << 1$, $\Theta_0(T_w) \sim t^{-\frac{3}{2}+\beta}$. With the condition $\tau \lesssim \hat{t}$ (see above) one is lead to the requirement $\Theta < \Theta_{max} \sim \hat{t}^\beta \to 0$ for the range of contact angles within which the asymptotic singularity of critical wetting dominates.

The application of Eq.(10) to the case of complete wetting leads, together with Eq.(5), to the requirement $\Delta\mu < \Delta\mu_c \sim t^{3v} \to 0$. Therefore upon approaching T_c the $(\Delta\mu)^{-1/3}$–law for complete wetting becomes confined to a rapidly decreasing region $\Delta\mu < t^{1.92}$ (see Fig.7). Outside this region one finds critical adsorption with $l_0 \sim \xi \sim (\Delta\mu)^{-v/\Delta} = (\Delta\mu)^{-0.402}$. (Note that $\xi(t,\Delta\mu) \sim t^{-v} \cdot f(\Delta\mu \cdot t^{-\Delta})$ with $f(0)=$ const. and $f(x\to 0) \sim x^{-v/\Delta}$, $\Delta \cong 1.57$.) These considerations show *explicitly* how the tails of the van der Waals interactions, which lead to the $(\Delta\mu)^{-1/3}$–law, become irrelevant upon approaching T_c compared with the critical phenomena, which lead to the more singular $(\Delta\mu)^{-0.402}$–law.

Thus the competition between critical adsorption and complete wetting can be described as follows (see Fig. 8). In the case of critical adsorption at a wall the density profile approaches its bulk value as $z^{-\beta/v}$ up to $z \cong \xi$; for larger z it behaves $\sim \exp(-z/\xi)$. Thus the coverage diverges as $\xi^{-\beta/v+1}$. This means that along *any* path towards $(T_c, \mu_0(T_c))$ outside the curves $|\Delta\mu| = const \cdot |t|^\Delta \sim |t|^{1.57}$ the coverage diverges as $|\Delta\mu|^{(\beta-v)/\Delta} = |\Delta\mu|^{-0.19}$. For any path towards the critical point from *above* T_c and within the wedge $|\Delta\mu| \leq const.(-t)^\Delta \sim (-t)^{1.57}$ the coverage diverges as $(-t)^{\beta-v} = (-t)^{-0.30}$. *Below* T_c one has to distinguish the wet and non-wet side of the coexistence curve. (Without loss of generality I assume that the wall is wetted upon approaching coexistence from the gas side of coexistence, i.e. for $\Delta\mu \to 0^+$.) If T is approached from below on the liquid side of coexistence within the wedge formed by $\mu=\mu_0(T)$ and $\Delta\mu = const \cdot t^\Delta \sim t^{1.57}$ the coverage diverges $\sim t^{\beta-v} = t^{-0.30}$ as from above T_c. However, on the gas side of coexistence below T_c this latter power law $t^{-0.30}$ holds only within the narrow wedge formed by the curves $\Delta\mu = const \cdot t^{3v} \sim t^{1.92}$ and $\Delta\mu = const \cdot t^\Delta \sim t^{1.57}$. Outside this curved wedge towards the gas side the coverage diverges as $(\Delta\mu)^{-0.19}$ (see above).

However, within the wedge on the gas side formed by the coexistence curve $\mu = \mu_0(T)$ and the curve $\Delta\mu = const \cdot t^{3v} \sim t^{1.92}$ the coverage is governed by wetting phenomena. For $\mu = \mu_0^-$ the wall is wet and the coverage is infinite (provided $T_w < T \leq T_c$). For any isothermal path $T = const < T_c$ within this wedge $\rho_s \sim \Delta\rho \cdot l_0 \cong \rho_{s,0} \cdot (\Delta\mu)^{-1/3}$ with $\rho_{s,0} \sim (u_3\rho_w - t_3\rho_l)^{1/3} t^\beta$. If one considers now a path γ within this wedge towards T_c, e.g. $\Delta\mu = const \cdot t^x$ with $x > 3v$ in order to stay in this wedge, the coverage along this path will diverge $\sim t^{\beta-x/3}$. This divergence is stronger than for any path within one of the other wedges and regions mentioned above. Furthermore, within those wedges and regions the divergences do

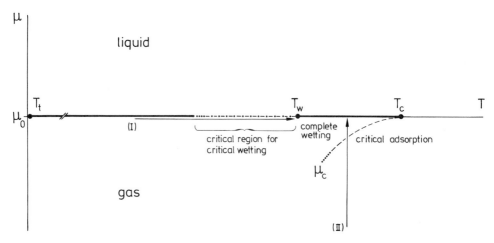

FIG.7: Schematic drawing of the liquid-gas coexistence curve $\mu_0(T)$ in the bulk phase diagram. For reasons of simplicity it is taken to be a straight line between the triple point T_t and the critical point T_c. T_w is the transition temperature for a critical wetting transition. Along path (I) at μ_0^- the thickness l of the wetting film diverges upon approaching T_w. Within the critical region for critical wetting this divergence is dominated by the asymptotic power law $l \sim (T_w-T)^{-1}$. The width of this critical region is proportional to $T_c - T_w$. Whatever the order of the wetting transition at T_w, l diverges also along the path (II). Within the indicated wedgelike region l diverges as $(\Delta\mu)^{-1/3}$ according to complete wetting; $\Delta\mu = \mu_0 - \mu$. Below that region and sufficiently close to T_c, l grows as $(\Delta\mu)^{-v/\Delta}$ with $v/\Delta \cong 0.402$ according to critical adsorption. The crossover between these two power laws along path (II) occurs at $\mu_c(T) \cong \mu_0 -$ const. $\cdot (T_c-T)^{3v}$ with $3v \cong 1.92$. Thus for $T \to T_c$ the $(\Delta\mu)^{-1/3}$-law for complete wetting is confined to a rapidly decreasing region. In fact the region denoted as critical adsorption is further subdivided by another crossover line (see Fig. 8), $\mu \cong \mu_0 -$ const. $\cdot t^\Delta$ with $\Delta \cong 1.57$, below which l grows indeed as $(\Delta\mu)^{-v/\Delta}$ for $\Delta\mu \to 0$. However, between $\tilde{\mu}_c$ and μ_c, l levels off as function of $\Delta\mu$ before it starts to grow as $(\Delta\mu)^{-1/3}$ between μ_0 and μ_c. The crossover lines $(-\Delta\mu) \sim t^\Delta$ below T_c on the liquid side and $|\Delta\mu| \sim (-t)^\Delta$ above T_c are shown in Fig. 8.

not depend on the choice of a particular path towards the critical. In this wetting dominated wedge, however, the divergence does depend on the choice of γ (see above).

The boundaries between the various wedges and regions mentioned above give rise to obvious crossover phenomena for the behavior of the coverage $\Gamma(\Delta\mu,t)$, which can easily be inferred from the above discussion (see Fig. 8). As a particular example I consider a path parallel to the coexistence curve on the gas side with a small undersaturation $\Delta\mu = const > 0$ and with $T \to T_c$. For $T < T_w$, $\Gamma(T)$ will be small, perhaps including layering transitions or even a thin-thick transition. In the case that T_w corresponds to a continuous wetting transition, Γ will rise smoothly around T_w towards a finite value according to the scaling laws with respect to $\Delta\mu$ and τ which apply close to a critical wetting transition (see Ref.1). In the case that T_w corresponds to a first-order wetting transition, Γ will not change much at $T = T_w$ but there will be a thin-thick transition upon crossing the prewetting line attached to T_w at $T > T_w$, provided $\Delta\mu$ is not too large to miss it. Note that the possible thin-thick transition below T_w as mentioned above may stem from the rest of the prewetting line, whose anchor point slides below T_w if the interaction potentials in the system are changed such that the first-order wetting transition is transformed into a continuous wetting transition without a prewetting line at T_w. Upon a further increase of temperature at fixed $\Delta\mu > 0$ the thickness of the complete wetting film increases only slightly according to the factor $(u_3\rho_w - t_3\rho_l)^{1/3}$ (see above), so that due to the factor $\Delta\rho$ the coverage decreases $\sim t^\beta$ until one reaches the first crossover boundary $\Delta\mu \sim t^{3\nu}$ entering the critical adsorption regime where Γ now starts to diverge $\sim t^{\beta-\nu}$ until one reaches the next crossover boundary $\Delta\mu \sim t^\Delta$ beyond which $\Gamma(\Delta\mu,t)$ levels off as function of temperature at a large value $\sim (\Delta\mu)^{(\beta-\nu)/\Delta}$. (Note that within MFT $3\nu = \frac{3}{2} = \Delta$, so that there these two boundaries coincide up to a possibly different prefactor.) Finally, above T_c one encounters the last crossover boundary beyond which Γ decreases again $\sim (-t)^{\beta-\nu}$ upon a further increase in temperature. Thus this particular path shows the rich variety of growth modes for a wetting film competing with critical adsorption.

The algebraic singularities for critical and complete wetting are governed by the power law decay of the van der Waals interactions $\sim r^{-6}$ between the particles (see Eqs.(2-5)). In the sense of renormalization group theory these power law tails of the van der Waals interactions become irrelevant close to T_c [21] so that critical exponents become universal numbers as it is apparent for the phenomenon of critical adsorption. Thus the competition, as discussed above, between wetting and critical adsorption can serve as a paradigm how - also at interfaces - universality is restored explicitly close to T_c. The van der Waals tails in the interface profiles (see Eqs.(8) and (9)) provide another example thereof. The amplitude $Q_3^{(1)}$ (see Eq.(23)) diverges, within MFT, $\sim t^{-1}$ for $t \to 0$. However, these van der Waals tails are present only for $z > \xi$ so that $Q_3^{(1)} \cdot z^{-3} < Q_3^{(1)} \cdot \xi^{-3} \sim t^{-1}t^{+3/2} = t^{1/2} \to 0$. (Similar

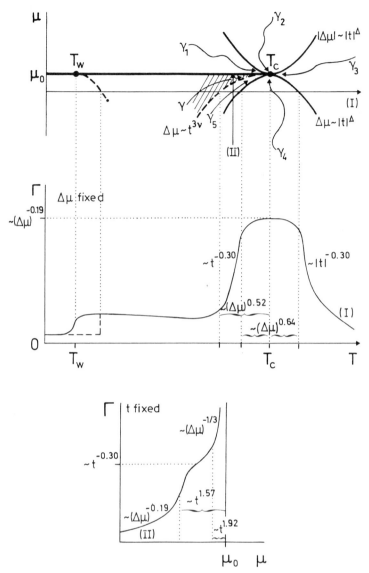

FIG.8: The competition between complete wetting and critical adsorption as well as the role of the various crossover lines discussed in the main text are displayed by the behavior of the coverage Γ along a path of constant undersaturation (I) and along an isotherm (II). The hatched area is the complete wetting regime. There the strength of the divergence of Γ along a certain path γ towards T_c depends on the particular choice of this path. In contrast to that along any path γ_1, γ_3, and γ_5 the coverage Γ diverges as $|t|^{-(\nu-\beta)} = |t|^{-0.30}$ and along any path γ_2 and γ_4 Γ diverges as $|\Delta\mu|^{-(\nu-\beta)/\Delta} = |\Delta\mu|^{-0.19}$. The various asymptotic power laws are indicated. Note that $\Delta \approx 1.57$, $3\nu \approx 1.92$, $\nu-\beta \approx 0.30$, $(\nu-\beta)/\Delta \approx 0.19$, $1/\Delta \approx 0.64$, and $1/(3\nu) \approx 0.52$. If T_w is a first-order wetting transition there is a prewetting line (dashed line), otherwise it is absent.

arguments hold for the van der Waals tails in ρ_{gl} [17].) This checks with the fact that at T_c the density profile decays $\sim z^{-\beta/\nu}$, $\beta/\nu \cong 0.51$, towards ρ_c which dominates any z^{-3}–law due to the van der Waals interactions. The possibilities to infer this power law decay $\sim z^{-\beta/\nu}$, together with the prefactor, uniquely from reflectivity measurements has been discussed in Refs.22-24, see also Refs.1 and 25.

I close this section with a discussion of the limiting behavior of the actual wall-gas interface profile $\rho_{wg}(z;l)$, which may include a wetting film of thickness l. It exhibits the following properties:

(i) For $l \rightarrow \infty$ and z fixed, $\rho_{wg}(z;l)$ reduces to the wall-liquid interface profile: $\lim_{l \rightarrow \infty} \rho_{wg}(z;l)$
= $\rho_{wl}(z)$.

(ii) For $l \rightarrow \infty$ and $z \rightarrow \infty$, $\rho_{wg}(z;l)$ reduces to the emerging free liquid-gas interface profile: $\lim_{l \rightarrow \infty} \rho_{wg}(z=l+y;l) = \rho_{lg}(y)$.

(iii) For l fixed and $z \rightarrow \infty$, $\rho_{wg}(z;l)$ behaves like the wall-gas interface without a wetting film: $\lim_{l \rightarrow \infty} [z^3(\rho_{wg}(z;l)-\rho_g)] = Q_3^{(g)}$, independent of l.

The properties (i) and (ii) follow from Antonov´s rule and (iii) is the content of Eq.(25) [17]. The trial function with which Eq.(6) is evaluated in order to obtain the effective interface potential should display these properties. As can be checked immediately the trial function $\rho(z,l;T)$ used in Eq.(7) does fulfill the requirements (i) and (ii) but not (iii). Instead, one has $\lim_{z \rightarrow \infty} [z^3(\rho(z,l;T)-\rho_g] = A_3^{(g)}$, which is independent of the substrate potential whereas $Q_3^{(g)}$ does depend on it (see Eq.(25)). Therefore, for any finite value of l, the actual wall-gas density profile cannot be represented only as an appropriate superposition of the wall-liquid and the liquid-gas density profile as it is the case for the trial function $\rho(z,l;T)$ in Eq.(7). In order to cure this defect one can consider a new and improved trial function $\rho^*(z,l;T)$:

$$\rho^*(z,l \; ;T) = \rho(z,l;T) + \Theta(z - (\kappa(l) - \lambda(l))) \sum_{k \geq 3} D_k z^{-k} , \qquad (29)$$

with

$$D_3 = Q_3^{(g)} - A_3^{(g)} = \rho_g^2(\rho_w u_3 - \rho_l t_3)\kappa_T^{(g)} , \qquad (30)$$

and $\rho(z,l;T)$ given by Eq.(7); the coefficients D_k are taken to be independent of l but otherwise unspecified for $k \geq 4$. It is straightforward to check that $\rho^*(z,l;T)$ fulfills all three requirements (i) - (iii).

The surprising fact is that the asymptotic behavior of the effective interface potential following from the trial function ρ^* is identical to that following from trial function ρ, i.e., the coefficients a_2, a_3, and a_4 remain unchanged (see Eqs.(12-15)) [17]. Therefore one can conclude that the leading behavior of the effective interface potential for large l up to and including terms $\sim l^{-4}$ depends only on the following quantities: the bulk densities, the asymptotic behavior of the interaction potentials, the wall-liquid and liquid-gas surface tensions, and the zeroth and first moments of the wall-liquid and liquid-gas interface profile, respectively. This corresponds to the picture that continuous wetting transitions lead to a smooth evolution of the wall-liquid and liquid-gas interfaces out of the wall-gas interface structure. At coexistence $\Omega_s(l)$ approaches the constant value $\rho_{sl} + \rho_{lg}$ (see Eq.(12)) which is determined equally by the substrate-liquid and the liquid-gas surface tension. At first glance also the approach towards this constant seems to depend equally on both interface structures via $d_{wl}^{(i)}$ and $d_{lg}^{(i)}$. However, as can be seen from Eqs.(13-17), at a continuous wetting transition with $a_2(T=T_w) = 0$ a_3 is independent from $d_{lg}^{(i)}$ and at a tricritical wetting transition with $a_2(T=T_w) = a_3(T=T_w) = 0$ even a_4 is independent of $d_{lg}^{(i)}$. This implicates that both the separatrix between first and second order wetting transitions, given by $a_2 = a_3 = 0$, and the separatrix between second and third order wetting, given by $a_2 = a_3 = a_4 = 0$, do not depend on the structure of the emerging free liquid-gas interface. Thus the order of wetting transitions of *wall-gas* interfaces depends, in addition to the bulk densities and interaction potentials, only on the structure of the corresponding *wall-liquid* interfaces.

As it should, the effective interface potential is free from – in principal – arbitrary input parameters like d_w and χ in the sharp- and soft-kink approximations (see Fig.4); κ, λ, and ξ drop out and d_w is replaced by a well-defined feature of the wall-liquid interface, which exhibits even a singular temperature dependence (see above). Surprisingly the van der Waals tails of the wall-liquid and liquid-gas interfaces do not show up explicitly in the expression for the effective interface potential, but only implicitly in the quantities $d_{lg}^{(i)}$ and $d_{wl}^{(i)}$. Thus in view of the above discussions one is lead to the suspicion that Eqs.(12-21) capture the asymptotic behavior of the effective interface potential exactly.

III. INTERFACIAL WETTING IN BINARY LIQUID MIXTURES

There is a conceptual advantage to studying wetting phenomena at a wall because in this case the position of the interface one is interested in is prescribed geometrically by a given substrate potential. However, the comparison between theory and experiments is particularly difficult for solid-fluid interfaces. A hard wall induces strong density oscillations close to it, which resemble solidlike adsorbate layers, which depend delicately on the corrugation of the substrate and the associated strain effects. The substrate itself depends both on temperature and pressure. The adsorption process modifies the surface structure of the substrate compared with the situation where it is clean. There is interdiffusion between wall and adsorbate particles. Defects and surface roughness lead to randomness in the substrate potential. These difficulties are not addressed properly either by the model used in Sect.II or by more sophisticated density functionals.

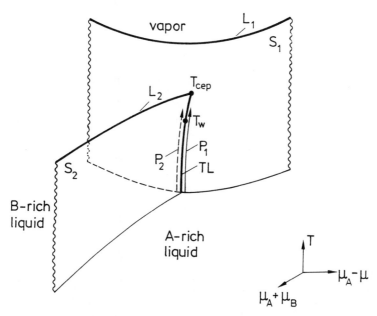

FIG.9: Bulk phase diagram of a simple binary liquid mixture of type II (see Ref.5) in the space of temperature T and the chemical potentials μ_A and μ_B of the A and B particles, respectively. The sheet S_1 is the locus of first-order phase transitions separating the liquid phases from the vapor whereas S_2 separates the A-rich liquid phase from the B-rich liquid phase. S_1 and S_2 are bounded by lines L_1 and L_2 of critical points. The critical end point T_{cep} is the intersection between L_2 and S_1. The sheets S_1 and S_2 meet at the triple line TL which ends at T_{cep}. Interfacial wetting transitions occur at T_w either along path p_1 or p_2 which for reasons of clarity are taken slightly off the triple line but both are lying on S_1. Along p_1 at T_w the B-rich liquid phase wets the A-rich liquid–vapor interface whereas along p_2 the A-rich liquid wets the B-rich liquid–vapor interface. If the wetting transition is first-order, a prewetting line is attached to T_w which lies on S_1 and which joins TL tangentially.

All these difficulties arise because in reality the wall is not inert but tends towards thermal equilibrium with the adsorbed fluid. Therefore either one carefully chooses a wall material, which minimizes these disturbing effects and which remains in a restricted thermal equilibrium during the relevant observation time, or one strives for a full study of the coupled wall-fluid system. Up to now the second option is out of reach. If, however, the confining material is a fluid itself, this option can be fully implemented by considering the interface between, say, the A-rich liquid phase and the vapor phase of a binary liquid mixture consisting of A and B particles along the triple line (see Fig.9). In this case one can study the wetting of the intrinsic A-rich liquid–vapor interface by the B-rich liquid phase. (The experimental and theoretical status of these *interfacial wetting phenomena* together with the corresponding literature is discussed in Sects.V.D and IV.C, respectively, in Ref.1; see also Ref.5 and the references therein.)

In the analysis of these interfacial wetting phenomena at the liquid-vapor interface of binary liquid mixtures all the aforementioned difficulties are no longer present. Instead, new ones arise.The parameter space is increased because there are three relevant interaction potentials: w_{AA}, w_{BB}, and w_{AB}. The translational invariance is no longer broken explicitly by a wall but more indirectly by imposing boundary conditions: for $z \to -\infty$ there is, e.g., the A-rich liquid phase, denoted as the α phase, and for $z \to +\infty$ there is the vapor phase denoted as the γ phase; the wetting phase is denoted by β. Now there are six relevant bulk number densities ($\rho_{A,\alpha}$, $\rho_{A,\beta}$, $\rho_{A,\gamma}$, $\rho_{B,\alpha}$, $\rho_{B,\beta}$, and $\rho_{B,\gamma}$) and two relevant density profiles $\rho_A(z)$ and $\rho_B(z)$. Interfacial wetting is tied to the triple line (the vapor phase is *not* air); this requires the fixing of two of the three relevant thermodynamic variables T, μ_A, and μ_B compared with one out of T and μ in Sect.II. The wetting of the $\alpha - \gamma$ interface leads to the formation of two fluid interfaces, $\alpha - \beta$ and $\beta - \gamma$, which *both* exhibit capillary fluctuations compared with one in the case of wetting of a wall. As in the latter case it turns out that these additional interface fluctuations are also irrelevant for the thermal singularities of continuous wetting transitions (see Ref.9 and Sect.IV.C in Ref.1). Therfore one can again apply a mean-field type density functional theory, c.f. Eq.(31), which generalizes Eq.(6) to the case of a two-component system. One should note that owing to the absence of the aforementioned density oscillation at the emerging $\alpha - \beta$ and $\beta - \gamma$ interfaces, Eq.(31) is even more appropriate for studying the wetting phenomena under consideration than Eq.(6) for the case of wetting a wall. Finally, as discussed above, the analysis of wetting a wall is impeded by the approach of the wall towards thermal equilibrium. For interfacial wetting phenomena, however, one faces the opposite problem. Since the theoretical predictions apply only for complete thermal equilibrium, a comparison between theory and experiment requires a careful equilibration of the experimental system. This can be accomplished by avoiding the formation of thick metastable wetting films (see Ref.26 and Sect.XII in Ref.1).

Thus one can conclude that with respect to a number of aspects the study of interfacial wetting offers several advantages compared with wetting of a wall. The binary liquid mixtures are more complex than a one-component fluid. However, as discussed above, this additional complexity can be brought under control both in theory and experiment.

As stated above, the following density functional is a suitable starting point for studying inhomogeneous binary liquid mixtures (i=A,B; j=A,B):

$$\Omega[\{\rho_i(\mathbf{r})\};T,\mu_i;\{w_{ij}(|\mathbf{r}-\mathbf{r}'|)\}] = \int d^3r \, f_h(\rho_i(\mathbf{r}),T) +$$

$$\frac{1}{2}\sum_{i,j}\int d^3r \int d^3r' \, \tilde{w}_{ij}(|\mathbf{r}-\mathbf{r}'|)\rho_i(\mathbf{r})\rho_j(\mathbf{r}') - \sum_i \mu_i \int d^3r \, \rho_i(\mathbf{r}) \quad . \tag{31}$$

Eq.(31) is the generalization of Eq.(6). Its derivation, the definition of f_h, and the connection between the functions $\{\tilde{w}_{ij}(r)\}$ and the spherically symmetric pair potentials $w_{AA}(r)$, $w_{BB}(r)$, and $w_{AB}(r)$ can be found in Ref.5. In analogy to Eq.(19) one has

$$t_{ij}(z) = \int_z^\infty dz' \int d^2r_{||} \, \tilde{w}_{ij}(\sqrt{r_{||}^2 + z'^2})$$

$$= -\sum_{k\geq 3} t_{ij,k} z^{-k}, \quad z \gg \max(\sigma_A, \sigma_B), \tag{32}$$

so that $\hat{w}_{ij} := -\int d^3r \, \tilde{w}_{ij}(r) = 2t_{ij}(0)$. σ_i is the hard-core diameter of the particles of type i, which is determined by the repulsive part of the interaction potential $w_{ij}(r)$ (see Ref.5).

In order to obtain the effective interface potential $\Omega_s^{(\alpha,\gamma)}(l)$ for wetting of the $\alpha-\gamma$ interface by the β phase Eq.(31) has to be minimized under the restriction that at the intrinsic $\alpha-\gamma$ interface a β-like film of thickness l is formed. In practice the asymptotic behavior of $\Omega_s^{(\alpha,\gamma)}(l)$ for large l is obtained by evaluating Eq.(31) for a suitably chosen pair of trial functions $\rho_i(z,l;T,\mu_{0,i}^{(\alpha,\gamma)})$; for $\mu_i = \mu_{0,i}^{(\alpha,\gamma)}$ the system is at $\alpha-\gamma$ coexistence, i.e., on the sheet S_1 in Fig.9. For $\mu_i = \tilde{\mu}_i(T)$ the system is on the triple line. In analogy to Eq.(11) the simplest ansatz is

$$\rho_i(z,l \; ;T,\mu_{0,i}^{(\alpha,\gamma)}) = \rho_i(z,l \; ;T) + \Theta(z-l)[\rho_{i,\gamma}(T,\mu_{0,i}^{(\alpha,\gamma)}) - \rho_{i,\gamma}(T,\tilde{\mu}_i(T))] +$$

$$+ \Theta(-z)[\rho_{i,\alpha}(T,\mu_{0,i}^{(\alpha,\gamma)}) - \rho_{i,\alpha}(T,\tilde{\mu}_i(T))] \quad . \tag{33}$$

Here $\rho_{i,\alpha}$ and $\rho_{i,\gamma}$ are the appropriate bulk number densities of the particles of type i in the phases α and γ, respectively. $\rho_i(z,l;T)$ corresponds to a pair of density profiles at the triple line which are a suitable superposition of the $\alpha-\beta$ and the $\beta-\gamma$ interface profiles *at the* triple line:

$$\rho_i(z, l; T) = \begin{cases} \rho_{i;\alpha,\beta}(z) \quad, & z \leq \kappa(l) - \lambda(l) \\ G_i(\kappa(l) - z, l), & \kappa(l) - \lambda(l) \leq z \leq \kappa(l) + \lambda(l) \\ \rho_{i;\beta,\gamma}(z - l) \quad, & z \geq \kappa(l) + \lambda(l). \end{cases} \tag{34}$$

Thus the pair of trial functions consists of the density profiles of the free $\alpha-\beta$ interface centered around $z = 0$ and those of the free $\beta-\gamma$ interface centered around $z = l$, $(\kappa,\nu) = (\alpha,\beta),(\beta,\gamma)$, ξ_i are two correlation lengths,

$$\rho_{i;\kappa,\nu}(z) = \begin{cases} \rho_{i,\kappa} - \sum_{k\geq 3} A_{i,k}^{(\kappa;\kappa,\nu)} |z|^{-k} & , z \leq -\xi_i \\ F_i^{(\kappa,\nu)}(z) & , |z| \leq \xi_i \\ \rho_{i,\nu} + \sum_{k\geq 3} A_{i,k}^{(\nu;\kappa,\nu)} |z|^{-\nu} & , z \geq \xi_i \end{cases} \qquad (35)$$

Eqs.(34) and (35) have the same structure as Eqs.(7) and (8) for the case of wetting of a wall. In both cases the functions G_i, F_i, G, and F describe the transition region between the emerging interfaces and they remain unspecified. The coefficients $A_{i,3}$ of the van der Waals tails in Eq.(35) are known analytically [17]. For the functions $\lambda(l)$ and $\kappa(l)$ the same restrictions are valid as before.

In a large-scale analytic effort [17] it is possible to determine the asymptotic behavior of the effective interface potential $\omega_{\alpha\gamma}(l) = \Omega_s^{(\alpha,\gamma)}(l) - \sigma_{\alpha\beta} - \sigma_{\beta\gamma}$ based on Eqs.(33-35):

$$\omega_{\alpha\gamma}(l) = \sum_i (\tilde{\mu}_i - \mu_{o,i}^{(\alpha,\gamma)})\{(\rho_{i,\beta} - \rho_{i,\gamma})(l + d_{i,\beta\gamma}^{(1)} - d_{i,\alpha\beta}^{(1)}) +$$
$$(\rho_{i,\alpha} - \rho_{i,\gamma})d_{i,\alpha\beta}^{(1)}\} + \sum_{k=2}^{4} a_k l^{-k} + O(l^{-5}\ln l) \qquad (36)$$

The first sum over $i = A,B$ arises in the case of complete interfacial wetting. The coefficients a_k are given by [17]:

$$a_2 = a_2^{(o)}, \qquad (37)$$

$$a_3 = a_3^{(o)} + \sum_{i,j} T_{ij,3}(d_{i,\alpha\beta}^{(1)} - d_{j,\beta\gamma}^{(1)}), \qquad (38)$$

and

$$a_4 = a_4^{(o)} + \sum_{i,j}[\frac{3}{2}T_{ij,3}(d_{i,\alpha\beta}^{(2)} + d_{j,\beta\gamma}^{(2)} - 2d_{i,\alpha\beta}^{(1)}d_{j,\beta\gamma}^{(1)}) + T_{ij,4}(d_{i,\alpha\beta}^{(1)} - d_{j,\beta\gamma}^{(1)})], \qquad (39)$$

with

$$a_k^{(o)} = \frac{1}{k}\sum_{i,j} T_{ij,k+1} \qquad (40)$$

where

$$T_{ij,k} = (\rho_{i,\alpha} - \rho_{i,\beta})(\rho_{j,\beta} - \rho_{i,\gamma})t_{ij,k} \qquad (41)$$

416

and

$$d_{i,\kappa v}^{(n)} = \frac{n}{\rho_{i,\kappa} - \rho_{i,v}} \int_{-\infty}^{\infty} dz\, z^{n-1} [\rho_{i;\kappa,v}(z) - \rho_{i;\kappa v}^{shk}(z)] \; ; \tag{42}$$

$\rho_{i;\kappa,v}^{shk}(z) = \Theta(z)\rho_{i,v} + \Theta(-z)\rho_{i,\kappa}$ is the sharp-kink approximation for the number density profile of the particles of type i for the free $\kappa - v$ interface. In Eq.(41) $\rho_{i,\beta}$ takes on its actual value for $\mu_i = \mu_{o,i}^{(\alpha,\gamma)}$; all other quantities in Eqs.(36-42) are evaluated at the triple line $\mu_i = \tilde{\mu}_i(T)$.

The structure of this effective interface potential is discussed in detail in Ref.17. There are many – in principle an infinite number of – possible definitions for assigning a thickness l to the $\alpha - \gamma$ interface compound; each definition gives rise to a different effective interface potential. According to Eq.(34) in the present approach each such definition fixes the center of the $\alpha - \beta$ and of the $\beta - \gamma$ interface and therewith the values of $d_{i,\kappa v}^{(n)}$. However, it can be shown explicitly [17] that all these possible and different effective interface potentials lead to the *same* predictions for all physical property like the value of T_w, separatrices etc.. Surprisingly it turns out that the fact of the presence of two distinct number density profiles for interfacial wetting effectively gives rise to a similar excluded volume effect as d_w does in the case of wetting of a wall (see Fig.4 and Eq.(16) with $d_{wl}^{(1)} \cong d_w$ at low temperatures; here $d_{B,\kappa v}^{(1)} - d_{A,\kappa v}^{(1)}$, which is an intrinsic property of the $\kappa - v$ interface, plays a similar role.) In sum it turns out that both the *transition temperature* for critical interfacial wetting and the *separatrix* between first- and second-order wetting are determined completely by the *bulk* densities along the triple line, by the *leading* term of each interaction potential, and by certain zeroth moments of the emerging *free* interfaces.

IV. DISCUSSION

The comparison between Sect.II and Sect.III reveals a conceptual difference between wetting of a wall and interfacial wetting. In both cases the grandcanonical free energy splits into a volume contribution ~ V and into a surface contribution ~ A. In the case of a fluid near a wall one has

$$\Omega[\mu,T;\{w(r)\},\{V(z)\}] = V \cdot \Omega_b[\mu,T;\{w(r)\}] + A \cdot \Omega_s[\mu,T;\{w(r)\},\{V(z)\}] \tag{43}$$

whereas for a binary liquid mixture in contact with its own vapor one has

$$\Omega[\mu_i, T; \{w_{ij}(r)\}] = V \cdot \Omega_b[\mu_i, T; \{w_{ij}(r)\}] + A \cdot \Omega_s[\mu, T; \{w_{ij}(r)\}] . \qquad (44)$$

Eq.(43) shows that for a one-component fluid near a wall the surface contribution is a functional of both the interaction potential between the adparticles and the substrate potential whereas the bulk contribution depends only on $\{w(r)\}$. On the other hand for fluid interfaces in binary liquid mixtures Eq.(44) shows that both the bulk and the surface contribution depend on the *same* set of parameters. Although these latter dependences are different one can raise the question to which extent interfacial wetting is determined by *bulk* properties. In the case of wetting a wall this question cannot be posed due to the additional dependence of Ω_s, which determines the wetting behavior, on the substrate potential.

This particular issue of interfacial wetting has been explored in detail in Ref.5. Within mean-field theory (see Eq.(31)) all bulk properties depend parametricly on the integrated strengths of the interaction potentials ($\hat{w}_{AA}$, $\hat{w}_{BB}$, $\hat{w}_{AB}$)and on the hard-core diameters of the particles (σ_A, σ_B). If the energies (Ω, μ_i, k_BT) are measured in units of $\hat{w}_{AB}$ and the number densities in units of σ_A^{-3}, all bulk properties are fixed by *three* parameters:
$$\bar{w}_{AA} = \hat{w}_{AA} / \hat{w}_{AB}, \quad \bar{w}_{BB} = \hat{w}_{BB} / \hat{w}_{AB} \quad \text{and}$$

$$r = \sigma_{BB}/\sigma_{AA}. \qquad (45)$$

This means that in this three-dimensional parameter space every point corresponds to a distinct physical system and it specifies uniquely its bulk phase diagram. It turns out [5] that all physical systems, which exhibit a bulk phase diagram of the type shown in Fig.8, lie in a subspace which forms a tube with a finite cross section for $r = $ const. (see Figs.15-18 in Ref.5).

As becomes apparent from Eqs.(36-42) the wetting behavior depends on many more details of the interaction potentials than the bulk properties do. The leading term $\sim a_2$ in the effective interface potential depends on the six bulk number densities along the triple line $\bar{\mu}_i(T)$ and on $t_{ij,3}$ (see Eqs.(37,40,41)). Thus, as far as the bulk quantities are concerned, a_2 depends on $\sigma_A^{-3}, \bar{w}_{AA}, \bar{w}_{BB}, r$, and $k_BT/\hat{w}_{AB}$.

Although interfacial wetting is determined by the whole effective interface potential $\Omega_s^{(\alpha,\gamma)}(l)$, the knowledge of the coefficient a_2 alone allows one already to deduce the following features of the wetting phenomena under consideration (see Fig.3):

(i) $a_2(T) < 0$ for all T: This means that $l = \infty$ is always at least a local maximum of $\Omega_s^{(\alpha,\gamma)}(l)$. Thus, it never can be its global minimum. Therefore, in this case, there is no wetting transition at all.

(ii) $a_2(T) \geq 0$ for all T: This means that $l = \infty$ is always at least a local minimum of $\Omega_s^{(\alpha,\gamma)}(l)$. If it happens to be its global one for all T, the $\alpha - \gamma$ interface is wet for all T. Otherwise, due to higher-order terms in $\Omega_s^{(\alpha,\gamma)}(l)$, there will be a first-order wetting transition at $T_w^* < T_{cep}$.

(iii) $a_2(T) < 0$ for low temperatures and $a_2(T_{cep}) \geq 0$: In this case, the $\alpha - \gamma$ interface is not wet at low temperatures. Furthermore, there is a temperature $T_w < T_{cep}$ such that $a_2 (T>T_w) > 0$, i.e., that $l = \infty$ is at least a local minimum of $\Omega_s^{(\alpha,\gamma)}(l)$ for $T > T_w$. This fulfills the necessary condition for critical wetting. (If $a_3(T_w) > 0$, the sufficient condition for critical wetting is also fulfilled, provided it is not spoiled by terms in $\Omega_s^{(\alpha,\gamma)}(l)$ of still higher order.) If critical wetting does occur, $a_2(T_w) = 0$ is an exact, implicit equation for the wetting transition temperature.

The aim now is to classify binary liquid mixtures with respect to these catagories (i) - (iii). For that purpose one has to study the sign of $a_2(T)$ for low T and for $T \to T_{cep}$ both along path p_1 and p_2 (see Fig.9). If the freezing temperature T_4 of the binary liquid mixture is sufficiently low, the bulk densities at that temperature take on the following approximate values:

$$\left. \begin{array}{ll} \rho_{A,\alpha} \cong \rho_A^{(o)} \quad, & \rho_{B,\alpha} \cong 0 \\[2mm] \rho_{A,\beta} \cong 0, \quad, & \rho_{B,\beta} \cong \rho_B^{(o)} \end{array} \right\} \text{ at } T_4 \text{ on } p_1, \tag{46}$$

$$\left. \begin{array}{ll} \rho_{A,\alpha} \cong 0 \quad, & \rho_{B,\alpha} \cong \rho_B^{(o)} \\[2mm] \rho_{A,\beta} \cong \rho_A^{(o)} \quad, & \rho_{B,\beta} \cong 0 \end{array} \right\} \text{ at } T_4 \text{ on } p_2, \tag{47}$$

and

$$\rho_{A,\gamma} \cong 0 \quad, \quad \rho_{B,\gamma} \cong 0. \quad \text{at } T_4 . \tag{48}$$

$\rho_A^{(o)}$ and $\rho_B^{(o)}$ represent the densities of the one-component A fluid and of the one-component B fluid at their corresponding triple points. In principle $\rho_A^{(o)}$ and $\rho_B^{(o)}$ depend in a complicated way on $\hat{w}_{AA}$ and $\hat{w}_{BB}$, respectively, and on the sizes of the particles. However, in a dense fluid its density is mainly determined by the repulsive part of the interaction potential. Thus, in practice the ratio $\rho_A^{(o)}/\rho_B^{(o)}$ depends only weakly on $\hat{w}_{AA}$ and $\hat{w}_{BB}$, so that

$$\rho_A^{(o)}/\rho_B^{(o)} \cong (\sigma_{BB}/\sigma_{AA})^3 = r^3. \tag{49}$$

With Eqs.(46) and (47) one has

$$a(T_4) \cong -\frac{1}{2}\rho_B^{(o)}(\rho_B^{(o)}t_{3,BB} - \rho_A^{(o)}t_{3,AB}) \quad \text{on } p_1 \tag{50}$$

and

$$a(T_4) \cong -\frac{1}{2}\rho_A^{(o)}(\rho_A^{(o)}t_{3,AA} - \rho_B^{(o)}t_{3,AB}) \quad \text{on } p_2 . \tag{51}$$

Since $t_{3,ij} > 0$ the condition $a(T_4) < 0$ for the $\alpha - \beta$ interface to be nonwet at low temperatures takes on the following form:

$$\bar{t}_{3,BB} > r^3 \quad \text{along } p_1 \tag{52}$$

and

$$\bar{t}_{3,AA} > r^{-3} \quad \text{along } p_2 , \tag{53}$$

where $\bar{t}_{3,ij} = t_{3,ij} / t_{3,AB}$.

On the other hand close to T_{cep} one has due to Eqs.(37,40,41) and according to what was said in the second paragraph after Eq.(44):

$$\lim_{T \to T_{cep}} \text{sign}(a_2(T)) = \lambda(\bar{w}_{AA}, \bar{w}_{BB}, r, \bar{t}_{3,AA}, \bar{t}_{3,BB}) . \tag{54}$$

Eqs.(52-54) show that the various wetting properties (i) - (iii) are determined by a five-dimensional parameter space spanned by the arguments of λ in Eq.(54). Strictly speaking, these quantities are independent from each other. In practice, however, they are correlated with each other. A typical form for $\tilde{w}_{ij}(r)$ is [5, 6]

$$\tilde{w}_{ij}(r) = \begin{cases} 4\varepsilon_{ij}[(\sigma_{ij}/r)^{12} - (\sigma_{ij}/r)^6] & \text{for } r/\sigma_{ij} \geq 2^{1/6} \\ -\varepsilon_{ij} & \text{for } r/\sigma_{ij} \leq 2^{1/6} \end{cases} \tag{55}$$

with $\sigma_{AB} = (\sigma_{AA} + \sigma_{BB})/2$. With this form one finds [15]

$$\bar{t}_{3,AA} = \frac{8}{(1+r)^3}\bar{w}_{AA} \tag{56}$$

420

and

$$\bar{t}_{3,BB} = \frac{8r^3}{(1+r)^3} \bar{w}_{BB} \cdot \qquad (57)$$

With these two relations the dimension of the relevant parameter space is reduced from five to three: $\bar{w}_{AA}$, $\bar{w}_{BB}$, and r. According to the second paragraph after Eq.(44) this is identical to the parameter space which determines the bulk properties. Therefore the knowledge of the bulk phase diagram of a binary liquid mixture allows one to predict which of three wetting classes (i) - (iii) this particular system will exhibit.

This classification subdivides the aforementioned tube in the parameter space ($\bar{w}_{AA}$, $\bar{w}_{BB}$, r), within which all systems lie having a bulk phase diagram of the type shown in Fig.9, into three sections [5]. Under the assumption that all possible binary liquid mixtures are distributed homogeneously within this tube, a thorough analysis allows one to draw, inter alia, the following conclusions [5]:

(1) There are systems whose wetting transition temperatures coincide with T_{cep}; in general, these systems are not those whose number densities of the two liquid phases are equal.
(2) For equal sizes of the two particles critical wetting is found for those systems whose interactions between like particles are similar. For different sizes of the particles the interaction between the smaller particles must decrease and between the larger particles it must increase in order to obtain critical wetting.
(3) It is possible that the liquid phase with the higher number density wets the liquid-vapor interface. This requires the wetting phase to be rich in small particles. In this case and for $r < 0.75$ wetting occurs only by the liquid phase with the higher number density.
(4) If the wetting phase is rich in large particles it always has the lower number density.
(5) For $r \leq 0.7$, and if the wetting layer is rich in small particles, there is an increasing possibilitiy for a dewetting transition.
(6) For $r \leq 0.55$ critical wetting is no longer possible by a liquid phase which is rich in small particles.
(7) If the sizes of the particles are very different only first-order wetting transitions can occur.
(8) The overall chances of finding critical wetting are best for $r = 0.93$. It is more likely to find it at the interface between the liquid rich in small particles and the vapor; there the chances are best for $r = 0.80$. For small values of r, critical wetting becomes very unlikely.

The purpose of these considerations is to provide a guideline for a *systematic* experimental exploration of interfacial wetting in binary liquid mixtures and to respresent a starting point for more detailed numerical studies of a few carefully selected model systems.

For example, this guideline is designed to spur the experimental search for a critical wetting transition, which has not yet been successful for the few systems studied so far [1]. Even a clear-cut first-order wetting transition is not yet known [1]. Once such systems are found they can be scrutinized theoretically in more detail in order to check theory against the experimental evidence. In view of the wealth of theoretical predictions there is, apart from a few promising examples, a painful dearth of precise and systematic experimental studies. It is hoped that the above analysis, as well as the previous sections, will encourage more experiments in order to achieve a deeper understanding of these interesting phenomena at interfaces.

ACKNOWLEDGMENTS

In this subject I have enjoyed fruitful and very pleasant collaborations with Arnulf Latz, Marek Napiórkowski, Peter Nightingale, Rüdiger Schack, and Michael Schick. I have benefitted from many stimulating discussions with Wilhelm Fenzl and Herbert Wagner.

REFERENCES

1. S. Dietrich, in *Phase Transitions and Critical Phenomena*, edited by C. Domb and J.L. Lebowitz (Academic, London, 1988) and references therein.
2. D.E. Sullivan and M.M. Telo da Gama, in *Fluid Interfacial Phenomena*, edited by C.A.Croxton (Wiley, New York, 1986), p. 45 and references therein.
3. P.G. de Gennes, Rev. Mod. Phys. **57**, 827 (1985).
4. M. Schick, in *Liquids at Interfaces*, Les Houches Summer School Lectures, Session XLVIII, edited by J. Charvolin, J.F. Joanny, and J. Zinn-Justin (Elsevier, Amsterdam, 1989).
5. S. Dietrich and A. Latz, Phys. Rev. B**40**, 9204 (1989).
6. J. Barker and D. Henderson, Rev. Mod. Phys. **48**, 587 (1976).
7. W.E. Carlos and M.W. Cole, Surf. Sci. **91**, 339 (1980) and references therein.
8. R. Evans, Adv. Phys. **28**, 143 (1979).
9. S. Dietrich, M.P. Nightingale, and M. Schick, Phys. Rev. B**32**, 3182 (1985).
10. M. Napiórkowski and S. Dietrich, Phys. Rev. B**34**, 6469 (1986).
11. A. Trayanov and E. Tosatti, Phys. Rev. Lett. **59**, 2207 (1987).
12. B.C. Freasier and S. Nordholm, Mol. Phys. **54**, 33 (1985).
13. T.F. Meister and D.M. Kroll, Phys. Rev. A**31**, 4055 (1985).
14. W.A. Curtin and N.W. Ashcroft, Phys. Rev. Lett. **56**, 2775 (1986).
15. S. Dietrich and M. Schick, Phys. Rev. B**33**, 4952 (1986).
16. M. Napiórkowski and S. Dietrich, Europhys. Lett. **9**, 361 (1989).

17. S. Dietrich and M. Napiórkowski, preprint (1990).

18. H.W. Diehl in *Phase Transitions and Critical Phenomena*, edited by C. Domb and J.L. Lebowitz (Academic, London, 1986), Vol. 10, p. 75 and references therein.

19. A. Ciach and H.W. Diehl, Europhys. Lett. (1990), in press.

20. L.M. Trejo, J. Gracia, C. Varea, and A. Robledo, Europhys. Lett. **7**, 537 (1988).

21. P. Pfeuty and G. Toulouse, *Introduction to the Renormalization Group and Critical Phenomena* (Wiley, Chichester, 1977), Sects. 2.3 and 2.4.

22. S. Dietrich and R. Schack, Phys. Rev. Letter **58**, 140 (1987).

23. S. Dietrich, Colloque de Physique **C7**, 233 (1989).

24. S. Dietrich, Physica A (1990), in press.

25. A.J. Liu and M.E. Fisher, Phys. Rev. A**40**, 7202 (1989).

26. R.F. Kayser, M.R. Moldover, and J.W. Schmidt, J. Chem. Soc., Faraday Trans. II **82**, 1701 (1986).

ADSORPTION STUDIES WITH A GRAPHITE FIBER MICROBALANCE

L. Bruschi and G. Torzo

Dipartimento di Fisica - Università di Padova
and
Istituto Nazionale Fisica della Materia
Padova - Italy

INTRODUCTION

The present work describes the graphite fiber microbalance, a sophisticated and powerful technique that we used to investigate the growth behavior of film adsorbed onto graphite. Our first result obtained with this method was the detection of a triple point wetting transition for argon [1]: here we report several improvements we introduced, during the last three years, both in the experimental set-up and in the data reduction procedure. This technique is in fact a rather recent acquisition in the field of experimental adsorption studies: a simpler and less accurate version was first used by C. Bartosh and S. Gregory in 1985 to investigate the oxygen adsorption close to the bulk triple temperature[2], and later by P. Taborek and L. Senator[3], and by G. Zimmerli and M.W. Chan[4] to study the wetting transition in liquid helium. The improved version here described includes: 1) a detailed analysis of the effects due to the interaction of the oscillating probe with the vapor phase, 2) a different driving mechanism, 3) a sophisticated frequency-lock apparatus, and 4) a cell geometry optimized for reduction of the thermal gradients.

DESCRIPTION OF THE METHOD

The main features of this technique are common to the older and well consolidated technique of the vibrating wire viscometer[5]. In a schematic distinction one may say that, when using the vibrating wire as a viscometer, one is focused to measure the *resonance width*, that is affected by the fluid viscous drag, while in a microbalance experiment one is mainly interested in the *resonant frequency shift*, that measures the changes of the wire effective mass (due both to the adsorbed layer and to the hydrodynamic mass).

The common features of both devices are the following: a conducting wire is stretched inside a transverse static magnetic field, and it is driven into resonant vibration by an appropriate exciting force.

Phase Transitions in Surface Films 2
Edited by H. Taub *et al.*, Plenum Press, New York, 1991

The magnetic field induces at the wire ends a voltage V_o proportional to the time derivative of the displacement: thus the vibration frequency can be measured, and V_o can be used to lock the frequency at resonance by a feedback loop. The resonant frequency v depends on the tension τ applied to the wire, on its effective mass M_e, and its length l: $v=\sqrt{\tau/4lM_e}$. Therefore, at constant τ and l, when a film is adsorbed onto the wire, the mass loading produces a frequency shift. Even in the absence of film adsorption, the density and the viscosity of the fluid affects the motion of the wire, producing both a resonance broadening and a frequency shift due to the hydrodynamic mass (the mass of the fluid displaced by the wire).

As a consequence, in order to achieve accurate measurements of the adsorbed mass with a vibrating wire microbalance, a detailed analysis of hydrodynamic effects must be performed. Most of the mathematics required by this analysis has been worked out by G.Stokes[6], and it was fruitfully used for viscosity measurements in liquid helium at the lambda transition and in methane at high density[7].

The correction for hydrodynamic effects, however, in a graphite fiber microbalance experiment requires the knowledge of several parameters: the radius r and density ρ_w of the fiber, and the density ρ and viscosity η of the fluid. The fluid density is generally well known for rare gases in a wide range of temperatures and pressures. On the contrary the viscosity data in the literature cover only a restricted thermodynamic range. Nevertheless it is possible, using this technique, to obtain an experimental value for η from frequency measurements.

Both the fiber radius and density depend on the type of fiber: r ranges between 5 and 7 μm, and ρ_w between 1.9 and 2.2 g cm^{-3}, for P100, P120 and PX7 mesophase-pitch-derived fibers[8]. Scanning electron micrographs show that the radius of a single fiber is practically constant for its whole length, but there is a considerable spread in the radius values, within the same fiber batch. It is therefore reasonable to assume, for the density of the particular fiber used in each microbalance experiment, a deviation from the (average) nominal value of the batch. The actual values of r and ρ_w, however, can also be experimentally obtained by a procedure that is outlined hereafter.

The Stokes analysis, that is valid in perfectly hydrodynamic regime (i.e. when the molecule mean free path λ is much shorter than the fiber radius), gives for the resonant frequency v and for the resonance width Δv the following expressions[9]:

$$v = v_o\sqrt{\frac{M_w}{M_w + KM + \Delta M}} = v_o\sqrt{\frac{M_w}{M_e}} \tag{1}$$

$$\Delta v = \frac{Mv}{M_w + KM + \Delta M}\left(K' + \frac{M_w \Delta v_o}{Mv}\right) \tag{2}$$

where v_o and Δv_o are the resonant frequency and the resonance width in vacuum; M_w, M and ΔM are the fiber mass, the mass of fluid displaced by the fiber, and the mass adsorbed onto the fiber, respectively; $K(m)$ and $K'(m)$ are known functions that can be numerically calculated from the parameter m given by:

$$m = \frac{r}{2\delta} = \frac{r}{2}\sqrt{\frac{2\pi\nu\rho}{\eta}} \qquad (3)$$

where $\delta = \sqrt{,\eta/(2\pi\nu\rho)}$ is the viscous penetration depth.

THE CALIBRATION PROCEDURE

As both r and ρ_w are only approximately known, a calibration must be performed with gases with known density and viscosity, with $\Delta M=0$, i.e. at temperature $T \gg T_c$ where there is no adsorption. Letting $\Delta M=0$ in relation (1) and (2) we obtain for K and K' the relations:

$$K = \frac{\rho_w}{\rho}\left[\left(\frac{v_o}{v}\right)^2 - 1\right] \qquad (4)$$

$$K' = \frac{\rho_w}{\rho}\left[\left(\frac{v_o}{v}\right)^2 \frac{\Delta v}{v} - \frac{\Delta v_o}{v}\right] \qquad (5)$$

The ratio K'/K does not depend on ρ_w, and therefore it can be calculated from the measured v, v_o, Δv and Δv_o. It is also a known function of the parameter m : $K'/K = G(m)$. Inverting this function we obtain $m = G^{-1}(K'/K) = M(K'/K)$. This calculated value of m_c, through relation (3), gives the fiber radius r. Using $K(m_c)$ or $K'(m_c)$, from (4) or (5) yields the fiber density ρ_w.

The adsorbed mass is obtained by solving the relation (1) with respect to ΔM:

$$\frac{\Delta M}{M_w} = \left[\left(\frac{v_o}{v}\right)^2 - 1\right] - \frac{\rho}{\rho_w}K(m) \qquad (6)$$

Relation (6) gives the adsorbed layer thickness (as relative change of the fiber mass $\Delta M/M_w$) in terms of the measured v and v_o, and of the hydrodynamic term $K\rho/\rho_w$, once the fiber radius r, the viscosity η and the density ρ_w are known.

The above relations have been derived assuming a *perfect hydrodynamic regime* for the fluid surrounding the oscillating fiber. However, at low gas densities, where the molecule mean free path λ becomes comparable with the fiber radius, the system enters a *semi ballistic regime,* and the previous equations must be properly modified. Using an empirical approach[10] we simply replaced relations (1) and (2) by formally identical relations, where the Stokes functions K and K' are substituted by new functions F and F', that must satisfy the conditions:

$$\lim_{\lambda/r \to 0} F = K \quad \text{and} \quad \lim_{\lambda/r \to 0} F' = K'$$

Again letting $\Delta M=0$, e.g. for temperatures much higher than the critical temperature, and for gases with known density and viscosity, we calculate F from relation (4), and K from the m values given by relation (3).

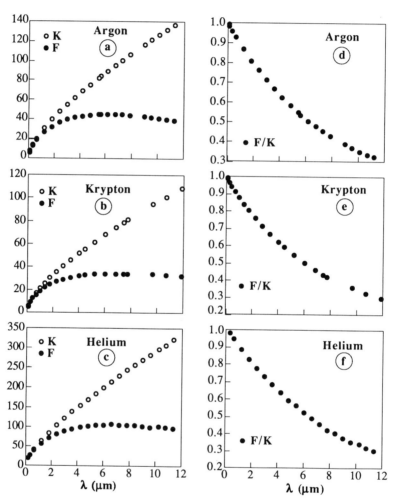

Figure 1. (a,b,c) The Stokes function K and the function F vs the mean free
path λ, calculated at room temperature for argon, krypton and helium; (d,e,f) The
correction factor F/K calculated for argon, krypton and helium.

The results for Argon, Krypton, and Helium, at t= 30 °C are reported in figure 1(a,b, c) respectively. As expected, $K_{(\lambda \to 0)} = F_{(\lambda \to 0)}$. At small densities (i.e. large λ) the departure of F from the Stokes function K is well described by a simple exponential $F = Ke^{-\lambda/(\alpha 2r)}$, as shown in figure 1(d, e, f).

A unique value of the fitting parameter for the different gases (α=0.72 with the PX7 fiber and α=0.74 with the P120 fiber) gives, in the whole density range (up to $\lambda/r \approx 2$), a good agreement between the measured resonant frequencies and the frequencies calculated from (1) with the function K replaced by F (figure 2). In other words, for ΔM=0, the behavior of the resonant frequency as a function of density is well described, in the whole density range, by the relation:

$$v^2 = v_o^2 \frac{\rho_w}{\rho_w + Ke^{-\lambda/(\alpha 2r)}\rho} \tag{7}$$

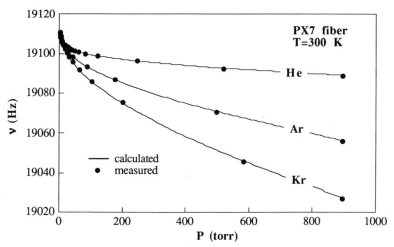

Figure 2 . The measured values of the resonant frequencies (dots) compared to the values (lines) calculated from relation [7].

For ΔM=0 the hydrodynamic correction may be measured as $(v_o/v)^2-1$, or calculated as explained above : in figure 3 we report the measured values, compared with the calculated ones (the data are the same as those in figure 2). The hydrodynamic correction is plotted here in units of equivalent number of layers, to give an idea of the importance of this term in the adsorption isotherm measurements.

Once this *calibration* has been performed, we are able to evaluate the adsorbed mass by means of the relation:

$$\frac{\Delta M}{M_w} = \left[\left(\frac{v_o}{v}\right)^2 - 1\right] - \frac{\rho}{\rho_w} Ke^{-\lambda/(\alpha 2r)} \tag{8}$$

where the exponential correction to the Stokes hydrodynamic shift becomes important in the low density range.

The mean free path is calculated in c.g.s. units as $\lambda = \eta/(0.499\rho\,\bar{v})$, where $\bar{v} = \sqrt{8RT/\pi M_a}$ is the molecular mean square velocity, R is the molar gas constant, and M_a is the molecular mass[11]. When the viscosity value at the working temperature T is unknown, it may be calculated from the resonance widths Δv and Δv_0 measured at the highest density and under vacuum, respectively. (Using relation (6), and the actual value of ρ_w one gets a value for K' and the corresponding value of m. Relation (3) then yields η)

THE EXPERIMENTAL APPARATUS

In the fiber microbalance used by other authors [2,3,4] the exciting force is produced by the interaction between the magnetic field and the a.c. current forced through the fiber. A

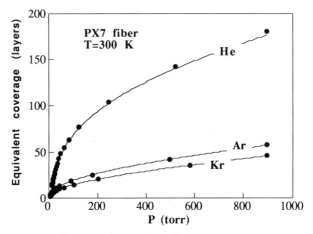

Figure 3 . The hydrodynamic correction evaluated at room temperature where $\Delta M = 0$: the dots are the measured values of $(v_0/v)^2 - 1$, the lines are the calculated values of the function $F = K\exp(-\lambda/2\alpha r)$ for the three gases .

drawback of this technique is that the resulting (large) synchronous signal, due to the fiber electric resistance, must be compensated by a bridge circuit. The balance is usually temperature dependent. Moreover the joule heating produced by the exciting current may give an offset of the fiber temperature with respect to the measured cell temperature, expecially at the lowest densities. Both these effects are avoided in our apparatus where the fiber is mechanically excited through a piezo transducer (figure 4).

The fiber is stretched by a soft spring made of a Be-Cu thin strip to which the fiber is glued by conductive epoxy. This set-up provides a strong reduction of the temperature dependence of the applied tension τ induced by unmatched expansion coefficients: the resulting temperature coefficient of the resonant frequency in vacuum ranges from 10 to 100 p.p.m.K^{-1}. The straight-vertical position of the strip avoids the dependence of τ on the vapor density due to changes of the hydrostatic pressure.

A second improvement has been introduced in the frequency-lock technique[12], to achieve the necessary high accuracy in the measurement of the resonant frequency. In the standard phase-locked-loop the working frequency F is locked *very close to* but not *exactly at* the resonant frequency v. The residual error F-v =$f(Q, \phi, V_p)$ is a function of the quality factor Q, of the residual phase-error ϕ (between the reference signal and the signal measured

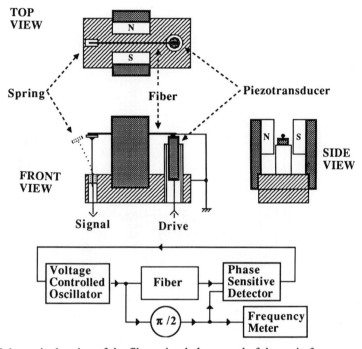

Figure 4 . Schematic drawing of the fiber microbalance and of the main frequency-lock-loop.

by the lock-in amplifier), and of the in-phase pick-up signal V_p due to imperfect shielding between the driving and the output signals. By driving the fiber oscillation through a piezotransducer the in-phase pick-up is made negligible, and the phase error, that becomes particularly important when Q is low and/or it changes during the measurements (e.g. with the gas density), is cancelled by modulating the reference signal at low frequency and by using a second phase-loop.

The accuracy in the resonant frequency measurement is better than 1 p.p.m., corresponding to a mass resolution of 3×10^{-12} g, or to a coverage of 0.02 layer in argon.

The experimental cell is designed to minimise thermal gradients. An inner thick-walled copper cell is soldered inside an outer thick-walled copper cell. The last is thermally coupled to the vacuum can wall by radiation, the space between the two cells being evacuated, and it is thermoregulated at the working temperature within 0.1 mK. This set-up, with the measuring cell anchored by a single thermal path to the thermoregulated outer cell (figure 5), provides an excellent temperature uniformity: the maximum ΔT, estimated in the worst case between any two points in the inner cell, is less than 1 μK. Several checks performed by varying the thermoregulating power from 0.01 W up to 0.5 W leaves unchanged the measured saturated vapor pressure. The temperature of the stainless steel filling capillary is

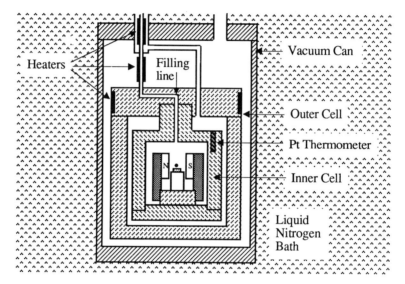

Figure 5 . Schematic drawing of the experimental cell.

kept slightly higher than the working temperature to avoid possible gas condensation onto a cold spot during the measurements close to the saturated vapor pressure Po.

EXPERIMENTAL RESULTS

Two isothermal adsorption measurements taken in argon below the triple bulk temperature ($T_c = 83.4$ K) with different fibers are shown in figure 6. At T=61.19 K, with a P120 fiber, only the first three steps are clearly observed, and one may estimate a maximum of 5 layers at the saturated vapor pressure P_0. At T=61.84 K, with a PX7 fiber, also the 4th and 5th steps show up, and the curve slope between the first steps is smaller, thus indicating a weaker capillarity effect. At this temperature a maximum coverage of 6 layers is observed at P_0. The thickness of the adsorbed film at P_0, unlikely in volumetric measurements, may be here measured by waiting for equilibrium conditions : once P_0 is reached (i.e. further gas admitted into the cell does not change the pressure), also the resonant frequency stabilizes at a constant value (usually after a transient upward jump probably due to bulk solid growth at the fiber ends). An evaluation of the substrate uniformity may be obtained, following the standard proposed by J.G. Dash[13], from the slope sharpness of the second adsorption step: we estimate the average size of the uniform surface domain for the PX7 fiber to be larger than 300 Å, corresponding to a change of the chemical potential $\delta\mu \approx RT\delta P/P \approx 13$ J/mole. The small dots in figure 6 represent the measured quantity $[(v_0/v)^2-1] \approx 2(v_0-v)/v$, and the circles represent the fractional mass change of the fiber, calculated from relation (8).

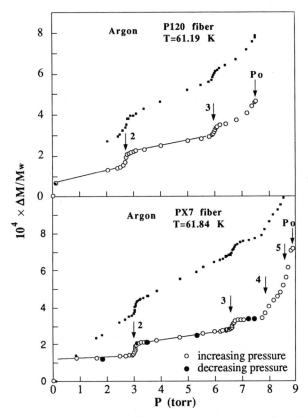

Figure 6 . The fractional change of the fiber mass measured as a function of pressure in Argon at two temperatures below the bulk triple temperature. The small dots are values calculated without the hydrodynamic shift correction. The circles are values calculated from relation [8].

The PX7 run also shows a very good agreement between data taken while increasing pressure and data taken while decreasing pressure. This feature characterizes the fiber microbalance technique with respect to the volumetric adsorption technique, where the measurement may be easily performed only by increasing the coverage, because the data taken during *desorption* exhibit usually considerable hysteresis[14].

The investigation of the film growth behavior in the very low coverage region becomes difficult when using the volumetric adsorption method, because extremely long time is required to reach termodynamic equilibrium in the whole exposed surface. This region, however, may be easily investigated by the oscillating fiber microbalance technique, where a much smaller adsorbing surface is involved.

A set of data taken in the millitorr range is shown in figure 7: here the pressure scale has been calibrated versus the amplitude V of the signal induced at resonance. For a given driving power, V is proportional to the quality factor, and therefore it is approximately inversely proportional to the gas pressure: this provides us with a *local* pressure measurement. The calibration is performed by measuring V as a function of the pressure P, corrected for the thermomolecular effect[15], at four pressure values up to 25 millitorr. The adsortpion curve is then obtained by measuring the frequency v_0 as a function of V. The mass loading is simply calculated as $\Delta M/M_w = -2(v_0-v)/v$, because the hydrodynamic frequency shift is negligible in this pressure range. The pressure value corresponding to the first layer completion is in good agreement with the results obtained by volumetric adsorption isotherms on grafoil[16]. The same agreement[17] is found for the pressure values corresponding to the 2nd and 3rd layer completion shown in figure 6.

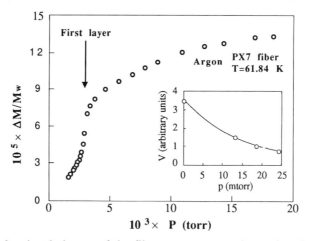

Figure 7 . The fractional change of the fiber mass measured as a function of pressure in Argon in the low coverage region. The pressure scale is calibrated versus the signal amplitude V (insert), and the hydrodynamic shift is neglected.

As a conclusion we resume the main advantage offered by the fiber microbalance technique with respect to volumetric adsorption measurements: 1) the substrate may be easily cleaned by joule heating with a small current passed through the fiber, 2) a fast settling time at very low pressure may be achieved by sealing off the cell and using the signal amplitude as a *local* pressure sensor, 3) the film thickness may be monitored both while increasing and decreasing the pressure without hysteresis, 4) the coverage may be measured also at constant pressure as a function of temperature, 5) future developments seem to be promising (e.g. using different substrates as metal-plated quartz whiskers or metal wires). In this case the *fiber* microbalance might result a useful complementary technique to the *quartz-crystal* microbalance, extensively used for adsorption studies by J.G. Dash[18] and J.Krim[19].

REFERENCES

[1] L.Bruschi, G.Torzo and M.W.Chan,Wetting Behavior of Argon on Graphite, Eur.Phys. Lett. 6: 541 (1988)

[2] C.Bartosh and S.Gregory, Mechanical Properties of Oxygen Multilayers on Graphite, Phys. Rev.Lett. 54: 2513 (1985)

[3] P.Taborek and L.Senator, Wetting Transition in Liquid Helium, Phys. Rev. 57: 218 (1986)

[4] G.Zimmerli and M.W.Chan, Complete Wetting of Helium on Graphite, Phys. Rev.B 38: 8760 (1988)

[5] L.Bruschi and M.Santini, Vibrating Wire Viscometer, Rev. Sci. Instrum. 46: 1560 (1975); J.T.Tough, W.D.McCormick and J.G.Dash, Viscosity of Liquid Helium II, Phys. Rev. 132: 2373 (1963)

[6] G.Stokes, *Mathematical and Physical Papers* , vol III ,Cambridge University Press, London, (1922)

[7] L.Bruschi, G.Mazzi, M.Santini and G.Torzo, The Behavior of the He^4 Viscosity near the Superfluid Transition, J. Low Temp. Phys. 29: 63 (1977); P.S. Van der Gulik, R. Mostert and H.R. van der Berg, The Viscosity of Methane at 25 C up to 10 kbar, Physica A 151: 153 (1988)

[8] Fibers provided by R.Bacon of Amoco Performance Products Inc., Parma Technical Center, Ohio

[9] L.Bruschi, Behaviour of Shear Viscosity near the Critical Point of Carbon Dioxide, Il Nuovo Cimento, 1: 361 (1982)

[10] For a different approach to this problem see for example A.M.Guenault, V.Keith, C.J.Kennedy, S.G.Musset and G.R.Pickett, The Mechanical Behavior of a Vibrating Wire in a Superfluid ^{3}He-B in the Ballistic Limit, J. Low Temp. Phys 60: 511 (1986), and references therein.

[11] J.Kestin and W.Leidenfrost, An Absolute Determination of the Viscosity of Eleven Gases over a Range of Pressures, Physica 25: 1033 (1959)

[12] L.Bruschi and G.Torzo, Method for Accurate Resonant Frequency Measurements with a Phase-Modulated Feedback Loop, Rev. Sci. Instrum. 58: 2181 (1987); L.Bruschi, R.Storti and G.Torzo, A Simple Design for a Voltage Controlled Frequency Independent Phase Shifter, Rev. Sci. Instrum. 58: 1960 (1987).

[13] J.G. Dash, On a Standarized Measure of Substrate Uniformity, in *Phase Transition in Surface Films*, J.G. Dash and J. Ruvalds Ed., Plenum Press, New York (1979).

[14] A. Inaba and J.A. Morrison, The Wetting Transition and Adsorption/Desorption Hysteresis for the Methane/Graphite System, Chem.Phys.Lett. 124: 361 (1986)

[15] T. Takaishi and Y. Sensui, Thermal Transpiration Effect of Hydrogen, Rare Gases and Methane, Trans.Far. Soc. 59: 2503 (1963)

[16] F. Millot, Adsorption of the First Layer of Argon on Graphite, J.Physique 40,L9 (1979)

[17] B. Gilquin, Etude Thermodynamique de l'Adsorption Physique de l'Oxygene sur la Face de Clivage de Cristaux Lamellaires, Note CEA-CEN 2091 (1979)

[18] A.D. Migone,J.G. Dash, M.Schick and O.E. Vilches : Triple Point Wetting of Neon Films, Phys. Rev B, 34, 6322 (1986)

[19] J. Krim and A. Widom, Damping of a Crystal Oscillator by an Adsorbed Monolayer and its Relalation to Interfacial Viscosity, Phys. Rev B, 38, 12184 (1988)

THE METHANE/GRAPHITE PHASE DIAGRAM

D. L. Goodstein, M. A. La Madrid and M. J. Lysek

Department of Physics
California Institute of Technology
Pasadena, CA 91125

INTRODUCTION

Methane adsorbed on graphite has been one of the most intensely studied of all multilayer film systems. A rather detailed phase diagram for the system was proposed by Goodstein *et al.* in 1984,[1] elaborated and embellished by Wortis in 1984,[2] and modified by Pettersen *et al.* in 1986.[3] The purpose of proposing the phase diagram was to stimulate further research by presenting hypotheses to be tested and drawing attention to open questions. The authors of the phase diagram had to confront some of the most interesting and vexing issues in the field of adsorbed films today: Does the adsorbed solid wet the substrate, and if so, why? Is there evidence of roughening, capillary condensation, or surface melting? Most intriguing of all, can one observe a dimensional crossover, from 2D to 3D behavior in the melting transition? The authors proposed answers to all of those questions. The purpose of this paper is to reexamine those proposed answers in the light of subsequent experimental and theoretical work on methane/graphite and other related systems. As we shall see, although much has been learned, many of the questions remain open. This is very much a work in progress.

Figure 1 shows the phase diagram offered by Wortis,[2] as modified by Pettersen *et al.*,[3] in the μ (chemical potential) versus T (temperature) plane. Starting at the lower left of the figure, the first layer is known to have a triple point at 56K, and a critical point, $T_c(1)$, at 70K.[4] In the figure, solid curves are based on at least some evidence, while dashed curves are pure conjecture guided by plausibility and thermodynamics. Thus, the first layer melting curve becomes dashed above about 105K, the highest temperature at which it has been observed. Above that temperature, it is conjectured to become reentrant as shown, turning around to meet the bilayer and bulk melting curves.

The diagram postulates a wetting transition, T_W, at about 10K. Below that temperature 3 layers form as μ is increased from below, then the bulk coexistence curve is encountered, meaning that bulk crystals start to grow in equilibrium with the 3-layer film (the exact number of layers is not important here, so long as the number is small). Above 10K, an infinite, or at least, very large number of layers is formed as bulk coexistence is approached from below at constant T. These layers form in distinct layering transitions, the solid curves meaning that a given layer goes from essentially empty to essentially full at constant μ. On an adsorption isotherm (μ or vapor pressure P versus amount adsorbed, N, at constant T) this kind of layering transition would show up as a vertical step, N increasing by one layer at fixed P or μ. These layering transitions end in the $\mu - T$ plane in critical points, $T_c(n)$ for the nth layer. As n approaches infinity, $T_c(n)$ approaches T_R, the bulk roughening temperature, $T_R \approx 78K$. Above T_R, the equilibrium shape of the bulk solid is spherical like a liquid droplet, rather than faceted like a crystal at low temperature.[5]

The melting transition is observed (by means of heat capacity peaks) from very thick films down to a few layers,[3] where the latent heat of melting becomes equal to zero, at what

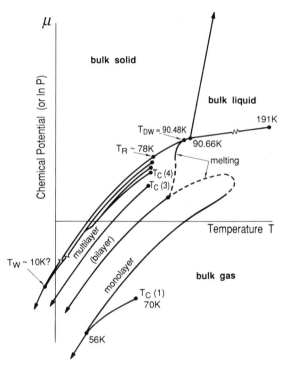

Figure 1. Sketch of methane/graphite phase diagram, from Refs. [1],[2],[3]. Solid curves represent experimental data while dashed curves are conjecture.

is assumed to be a tricritical point.[6] Below that it is conjectured to continue on (the dashed part of the curve) as a higher order phase transition, to meet up with the reentrant part of first layer melting and with bilayer formation. At the thick film limit it was initially imagined to approach the bulk triple point, but theoretical considerations indicating that it should not do so[7,8] were supported by new experiments.[19] The 1986 modification to the phase diagram[3] was that the multilayer melting transition approaches the bulk sublimation curve at T_{DW}, a point slightly below the triple point. This phenomenon is known as triple point dewetting. Thus, the multilayer melting curve, from the dewetting transition down to the tricritical point, would be what is known as a "prewetting" transition, perhaps the only one observed so far.

In the next sections, we review all the experimental data known to us concerning more than one layer of methane adsorbed on graphite. As we shall see, there are a number of key issues on which various authors disagree. We shall take up in turn the issues of Roughening, Wetting, Melting and Capillary Condensation.

ROUGHENING

Roughening is a thermodynamic phase transition that occurs when it becomes energetically favorable for the face of a crystal to form steps. Layering transitions in adsorbed films

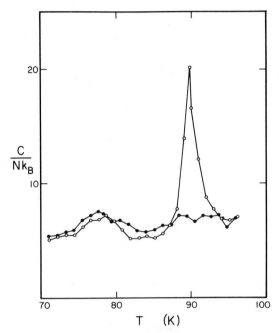

Figure 2. Specific heat of methane/graphite for 2.5 layers (lower curve) and 5 layers (upper curve). (Ref. [9].)

are expected to have critical points that approach the roughening temperature in the limit of infinite thickness.

The phase diagram, Fig. 1, is based largely on a thermodynamic survey by Hamilton and Goodstein (HG).[9] They reported heat capacities and vapor pressures for methane on uncompressed graphite foam in the range of about 1 to 6 nominal layers and at temperatures from about 70K to 95K. Figure 2 shows typical heat capacity data, thermodynamically corrected to isostericity at roughly 2.5 and 5 layers. Both isosteres exhibit bumps of about the same size at $T \approx 78$K. Quite similar bumps are evident in the data for all isosteres at coverages above about 2 layers. The bumps were interpreted by HG to signal the emergence of the system from the two phase region of the layering transition shown in Fig. 1. If that interpretation is correct, the bumps must occur at temperatures $T \leq T_c(n)$ within each layer. Since all the bumps were observed to occur in the range 76-78K, the guess was offered that the methane roughening temperature would be $T_R \approx 78$K. Zhu and Dash (ZD)[10] observed strikingly similar behavior in argon and neon adsorbed on graphite and proposed the same explanation with $T_R \approx 69$K for argon, and 19K for neon. Lahrer and Angerand[11] have reported evidence of layering transitions in both methane/graphite and argon/CdCl$_2$ from vapor pressure isotherms, obtaining $T_R \approx 69$K in agreement with ZD for argon, but $T_R \approx 87$K, somewhat higher than proposed by HG, for methane. Pettersen et al.,[6] reporting measurements of the nuclear magnetic resonance (NMR) transverse relaxation time, T_2, of methane adsorbed on Grafoil have interpreted their data to indicate a two-fold increase in the mobility of the uppermost few layers, consistent with a roughened surface, in the temperature range from about 75K-90K, as one would expect from HG. Bienfait et al.,[12] reporting measurements of quasi-elastic neutron scattering (QENS) interpret their data to indicate a disordered upper layer, which they model as a "lattice fluid", at temperatures above 76K. This observation too appears consistent with the interpretation of a roughening transition in this temperature range. Thus, all authors appear to agree on the basic phenomenon of a roughening transition, but there is some disagreement about the value of T_R.[37] ZD[10] have pointed out that $T_R/T_t \approx 0.8$ for a number of solids including methane, argon, and neon. On the other hand Lahrer and Angerand's value of T_R for methane [11] leads to a substantially higher value of $T_R/T_t \approx .96$. In addition, there is also disagreement on the mobility[6] of this disordered layer, the NMR measurement reporting a solid like mobility while the QENS measurement reports a liquidlike mobility.[12]

WETTING

There seems to be general agreement that liquid methane wets graphite at all temperatures above T_t, the bulk triple point. Much more interesting, however, is the question of whether, to what extent, and at what temperature solid methane wets graphite. The phase diagram, Fig. 1, proposes that solid methane wets graphite at all temperatures above $T_W \approx$ 10K. We now examine the evidence.

Krim et al.[13] have reported indications from low energy electron diffraction (LEED) data that, below $T \approx 14$K, no more than one molecular layer forms while a few layers (3 or 4) form above 14K. On the other hand, Kim et al.[14] interpret heat capacity data to mean that up to 6 layers grow below 11K and as many as 14 layers at $20K < T < 40K$. Krim et al.,[13] on the basis of reflection high energy electron diffraction data (RHEED), estimate that a maximum of 15 layers grow in the range $14K < T < 40K$.

The low temperature situation thus seems a bit confused. There appears to be a thin-thick transition, or perhaps even a complete wetting transition as postulated in Figure 1 at $T \approx 11$-14K. Above $T \approx 14$-20K and up to 40K it seems clear that fairly thick films have been observed. An ellipsometry study by Nham and Hess[15] in the range $42K < T < 76K$ yields incomplete wetting with films of 8-11 layers growing. Inaba and Morrison[16] have reported evidence of a wetting transition at 75.5K, but they also report hysteresis in their data, and they may not have baked their graphite at a high enough temperature. It might help to sort out some of the confusion in these results to note that intrinsically low surface area techniques (LEED, RHEED, ellipsometry) invariably yield lower maximum film thicknesses than do high area techniques. This may be because low area measurements become extremely sensitive to small temperature fluctuations when thick films are attempted. On the other hand, high surface area substrates may be susceptible to capillary condensation.

Starting at temperatures around 70K, there are some indications that wetting might be complete. Adsorption isotherms by Thomy and Duval,[17] by Bienfait *et al.*,[12] and by Lahrer and Angerand[11] all show clear layer formation up to at least 5 layers at P and μ below the bulk coexistence values. The survey data of HG also give this result, and do so over a wide range of temperatures. Adsorption isotherms from HG, taken at 1K intervals from 80K to 95K are shown in Fig. 3, where N is plotted versus $\mu - \mu_o$. Here $\mu_o(T)$ is the chemical potential of the bulk phase at coexistence, solid for $T \leq 90$K, liquid for $T \geq 91$K. Note that $\mu - \mu_o$ is always negative, signifying that film growth is not complete. Even more interesting,

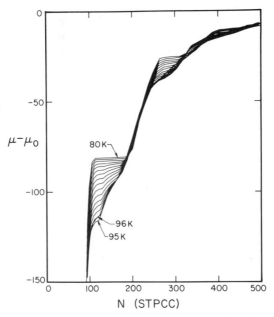

Figure 3. Adsorption isotherms of methane/graphite. (Ref. [9]).

the fact that this plot reduces all of the data to an essentially universal curve, especially at large N, means there is no difference in wetting behavior below and above $T_t = 90.66$K. This point is especially important because in many other systems a phenomenon called triple point wetting occurs,[18] the maximum permitted film thickness growing to infinity as T approaches T_t from below. In fact, Zimmerli and Chan[35] have reported evidence of triple point wetting of methane on a vibrating graphite fiber. However, Fig. 3 indicates that triple point wetting does not occur in methane/graphite, except perhaps at film thicknesses considerably greater than the 6 nominal layers shown in the data.

For thicker films, adsorption isotherms become an ineffective test of wetting because $\mu - \mu_o$ becomes too small to measure accurately. A better test is the isosteric heat capacity, C_N, which, for the case of triple point wetting, is expected to rise to a maximum at a temperature $T_p(n)$ such that $T_t - T_p(n) \propto n^{-3}$. Lysek *et al.*[19] have reported measurements of C_N in methane films up to 18 nominal layers, observing peaks like that shown in the 5-layer data of HG in Figure 2, but measured with much better temperature resolution. Their results are plotted in the form $T_p(n)$ versus n^{-3} in Figure 4. We use n to denote the amount adsorbed, measured in layers. Rather than a straight line extrapolating to T_t as n^{-3} goes to zero, as expected for triple point wetting, the data are not linear and extrapolate to a temperature $T_{DW} = 90.48K$. This result is not only inconsistent with triple-point wetting, it

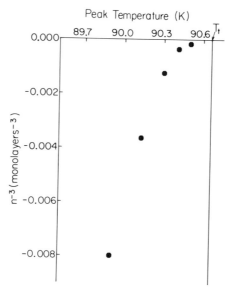

Figure 4. Peak temperature vs. coverage for specific heat data in Figure 5. (Refs. [3],[19].)

is also inconsistent with any picture of non-wetting by the solid at any thickness less than 18 layers. If there were crystals of bulk solid present, the heat capacity would have a delta-function peak at T_t. No such peak was observed.

One experiment yields evidence for very much thicker solid films. Pettersen and Good-stein (PG)[20] have reported measurements of the NMR spin-lattice relaxation time, T_1, in multilayer methane on graphite. It is known that T_1 is up to two orders of magnitude shorter in monolayer methane on graphite than in bulk methane.[21] PG find that T_1 varies smoothly as a function of nominal film thickness Z according to $T_1^{-1} = (T_1^b)^{-1} + aZ^{-1}$ where T_1^b is the bulk value of T_1 at the same temperature, and a is a constant, independent of T and Z. This simple formula fits hundreds of data points for film thicknesses from 1 to 51 layers and temperatures from 70K to 105K. The formula is derived, and the value of a correctly estimated, from a simple model in which film growth is uniform, and spins are relaxed by localized moments known to exist in the graphite substrate. The raw data for the decay of spin echos give no indication of more than one relaxation time, as one would expect if the system were partly film and partly bulk. Thus, PG conclude, uniform films grow up to at least 50 layers for $70K < T < 105K$.

All of these results, indicating that solid methane wets graphite, either completely or at least to very thick films, are perplexing because quite general theoretical arguments have been advanced that indicate that solids should not wet solid substrates.[22] The basic argument is that because the intermolecular spacing in the first adsorbed layer is governed by the strength of the adsorbate-substrate interaction, it cannot except by accident be equal to the intermolecular spacing of the unstrained bulk solid. But subsequent layers are locked into the spacing of the first layer. Therefore all solids grown on a solid substrate must be strained, and will at some film thickness give way to growth of bulk crystals rather than get any thicker.

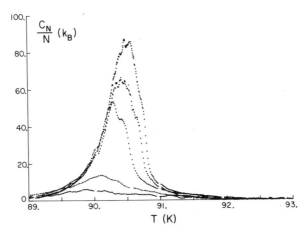

Figure 5. Specific heat data of
methane/graphite for 5-18 layers.
(Ref. [3]).

To overcome this dilemma and explain how solid methane might wet graphite, the authors of the phase diagram, Fig. 1, proposed that at second layer completion a phase transition occurs into a mutually commensurate bilayer, incommensurate with the substrate, and having the correct intermolecular spacing to permit unstrained bulk solid methane to grow. Some evidence concerning this idea may be found in neutron diffraction measurements by Larese et al.,[23] in the range $20K < T < 50K$. The positions of the diffraction peaks indicated that at first layer completion, as expected, the intermolecular spacing was smaller than that of the bulk solid. The second layer data were complicated, appearing to have the character of a mixture of two phases. In addition, computer simulations by Phillips and Hruska[24] show that the first and second methane layers on graphite are incommensurate with the substrate and with each other. However, at completion of the third layer, Larese et al. observed an A-B-C stacking sequence (corresponding to the fcc structure of bulk methane) at the correct bulk intermolecular spacing. If these results are correct, there is no apparent reason why solid methane should not grow very thick, if not infinitely thick, films.

To summarize what we have said in this section:

There is some evidence for a wetting transition, or perhaps a thin-thick transition at $T \approx$ 11-14K. At higher temperatures, say, above 20K, most authors observe fairly thick films of 10 layers or more. There is much evidence of quite thick films above 70K. For a solid to grow very thick films, the first few layers must provide a template at the correct intermolecular spacing. Figure 1 proposes that the system organizes itself into this condition at completion of the second layer at least near 70K. The data (at lower temperature) indicate this may actually occur at third layer completion. In any case, very thick solid films do seem to grow at $T < T_t$. Except for the vibrating fiber experiment, the evidence does not seem to favor triple-point wetting in this system.

MELTING

In the 1984 version of Fig. 1, the multilayer melting curve was thought to be an extension of the bulk melting curve, encountering bulk coexistence at the triple point, T_2. However, on general thermodynamic grounds, Pandit and Fisher[7] argued that film melting should not be an extension through the triple point of bulk melting, and Ebner[8] reached a similar conclusion based on a 6-state Potts lattice gas model calculation. To check these theoretical ideas, Lysek et al.[19] performed a series of heat capacity measurements of very thick films (5-18 layers) with very high temperature resolution (points spaced at about .01K, compared to the 1K spacing of the HG survey). They also took the important additional step of calibrating their thermometers independently at the triple point of bulk methane. These data are shown in Fig. 5. The peak temperatures for each curve of constant n, $T_p(n)$ was shown plotted versus n^{-3} in Fig. (4). As we discussed earlier, if these peaks were due to triple point wetting, or as we shall see below, if they were due to surface melting or stratified melting, Fig. (4) should result in a straight line with intercept at $T_t = 90.66$K. As the figure shows, it ends instead at a point $T_{DW} = 90.48 \pm .02$K. This point is known as the dewetting transition. At temperatures between T_{DW} and T_t, the bulk phase is solid and the film phase is liquid; the film cannot become bulk by growing to infinite thickness. Instead, bulk crystals coexist with liquid film along the bulk sublimation curve. Thus, the system does not wet in this region, even though the solid wets at $T < T_{DW}$ and the liquid wets at $T > T_t$. That is why the phenomenon is called dewetting.

In order to understand the specific heat data, a thermodynamic model correct in the limit of thick films was constructed by Petterson et al. [3]. The film is pictured as an incompressible continuum bulk phase formed in the external potential of the substrate. The film thickness is assumed to be given by the Frenkel-Halsey-Hill equation $\mu(T, Z) = \mu_i(T) - \Delta C_3^{iw}/Z_i^3$ where μ_i is the chemical potential of the bulk phase at coexistence with its vapor and ΔC_3^{iw} is a constant characteristic of the methane/graphite system.

The surface contribution to the Landau potential, $\Omega = F - \mu N$, is given by

$$\Omega_s(T, \mu) = -\int_{-\infty}^{\mu} N_s d\mu = -A \int_0^{n_i} n \frac{d(\mu - \mu_i)}{dn} dn$$

where the coverage or surface excess density is $n = N/A = (\rho_i - \rho_g)Z_i$. The integration is done at fixed T. The FHH equation is not valid at the lower limit, but they write

$$\sigma_{iw} + \sigma_{ig} = -\int_0^{\infty} n \frac{d(\mu - \mu_i)}{dn} dn$$

where the σ's are the surface tensions of the bulk at its interfaces with the wall and with its own vapor. Then

$$\Omega_S = A \left(\int_{n_i}^{\infty} n \frac{d(\mu - \mu_i)}{dn} dn + \sigma_{ig} + \sigma_{iw} \right)$$

and the Landau potential for the system, $\Omega_i = -pV + \Omega_S$ is

$$\Omega_i = -pV + \left[\frac{3}{2} \frac{\Delta C_3^{iw}}{v_i} \left(\frac{\mu_i(T) - \mu}{\Delta C_3^{iw}} \right)^{2/3} + \sigma_{ig} + \sigma_{iw} \right] A$$

where p is the pressure, V is the gas volume.

Melting occurs in the film when the Landau potentials for the solid and liquid states are equal, yielding

$$\frac{(\mu_s(T) - \mu)^{2/3}}{v_s} - \frac{(\mu_l(T) - \mu)^{2/3}}{v_l} = \frac{2}{3} \frac{\delta}{(\Delta C_3^{iw})^{1/3}}$$

where $\delta = \sigma_{sg} + \sigma_{sw} - \sigma_{lg} - \sigma_{lw}$. Taking $\mu_s(T) - \mu_l(T) = \alpha(T - T_t)$ where $\alpha = L/T_t$, L=latent heat of fusion, it follows that for $\delta > 0$, the melting curve will intersect the solid coexistence curve at a temperature

$$T_{DW} = T_t - \frac{1}{\alpha} \left(\frac{2}{3} \frac{\delta v_l}{(\Delta C_3^{iw})^{1/3}} \right)^{3/2}$$

with a slope $d(\mu_s - \mu)/dT \propto (T_{DW} - T)^{1/2}$. $\delta > 0$ implies $T_{DW} < T_t$ and the slope of the melting curve is positive - so that melting in the film occurs at a lower temperature than melting in the bulk.

Two phase transitions must occur along the bulk coexistence curve as $T \to T_t^-$. For $T < T_{DW}$, a solid film wets the substrate. For $T > T_t$, a liquid film grows continuously into bulk liquid. For $T_{DW} < T < T_t$, the film is liquid and the bulk phase is solid. A finite liquid film coexists with bulk solid at $\mu = \mu_s$. Hence, the endpoint of the multilayer melting curve is a dewetting transition, with a fully wetted solid phase at $T < T_{DW}$, and an incompletely wetted liquid film at $T_{DW} < T < T_t$. A wetting transition occurs at T_t.

The data of Lysek et al.[19] agree with these predictions. In particular, since $(T_{DW} - T) \propto (\mu_s - \mu)^{2/3}$ and $(\mu_s - \mu) \propto 1/n^3$, then a plot of $(T_{DW} - T_p)$ vs. $1/n^2$, where T_p is the temperature where melting begins, should yield a straight line. The data are consistent with this result.

Figure 6 shows the latent heat of melting, $L = \int C_N dT$, divided by the number of molecules adsorbed, as a function of N. The latent heat per molecule is seen to go to zero at about 4 layers. If this is, as it seems to be, a critical point, it is the only example of one known to us for any bulk melting transition. On general grounds, melting would be expected to continue to thinner films as a higher order transition, ending on another line of phase transitions. These considerations lead to the conjectured dashed line connecting the melting curve to first layer melting and second layer condensation in Fig. 1. Further work on this region is clearly needed.

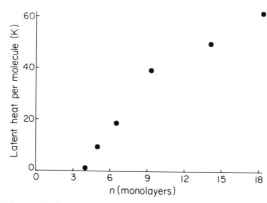

Figure 6. Latent heat vs. coverage from 5-18 layer
specific heat data. (Ref. [3]).

In any case, for films of about 4 layers and more, melting appears to occur in a first order phase transition, the entropy, density, and film thickness all changing discontinuously at constant T and μ. It is a thick-thin transition (thick solid to thin liquid) ending, in the thick film limit, in a (de)wetting transition and in the thin film limit in a (tri)critical point. These are the characteristics of what is known as a "prewetting" transition, although to our knowledge no other example has been observed.

Using this picture, one can understand why film melting cannot be a continuation through the triple point of bulk melting. The essential difference is that film melting is always a three phase equilibrium (solid film, liquid film and gas) whereas bulk melting is a two phase equilibrium. As we follow the melting curve upward in μ toward bulk coexistence, the solid phase approaches infinite thickness, but the liquid phase (at the same μ) approaches finite thickness, in equilibrium with (in principle) an infinite thickness of vapor. In other words, as bulk coexistence is approached from below, the macroscopic phases are solid and gas, so the dominant phase transition is not melting, but rather sublimation. That is why the "melting" curve not only fails to approach T_t, it also merges with the sublimation curve at T_{DW} with zero relative slope.

An alternative picture of how melting proceeds was proposed by ZD to explain data for argon and neon[10] on graphite, as well as other systems. It is properly referred to as surface melting[36] with respect to bulk solids and stratified melting in films, but the two phrases are often confused in the literature. It is expected to occur when a thin layer of liquid melt at the surface of a solid lowers the free energy of the system even at temperatures below the triple point temperature. Melting then occurs via the solid-liquid interface proceeding into the bulk. Adsorbed films would be expected to be stratified below T_t, liquid on top and solid below. As T_t is approached from below at constant N, the liquid-solid interface moves downward consuming the film. Stratified melting is quite different from the transition envisioned for methane/graphite in Fig. 1, where a homogeneous solid film melts completely at constant μ and T.

Using a simple model of stratified melting, and assuming a Van der Waals interaction between adsorbate and substrate, ZD point out that the heat capacity C_N ought to rise as T_t is approached according to $C_N \propto (T_t - T)^{-4/3}$. For finite n, however, this behavior persists

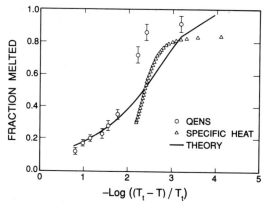

Figure 7. Fraction melted vs. temperature for QENS data (Ref. [12]) and 9.3 layer specific heat data (Ref. [3]). The theory curve comes from the stratified melting model of Ref. [6].

only until a peak temperature, $T_p(n)$, the temperature at which melting is complete and the interface stops moving. It is expected that $T_t - T_p \propto n^{-3}$. Both of these expected behaviors afford means of analyzing heat capacity data for evidence of stratified melting, and ZD report that the data obey these expectations reasonably well for argon and neon adsorbed on graphite. In krypton,[25] Pengra, Zhu and Dash observed specific heat data showing qualitatively different behavior from that of argon or neon and they explain this behavior by adding substrate induced strains to the model. Unfortunately, in these experiments the thermometers were not independently calibrated at the triple point temperatures of the bulk materials, but instead it was assumed that T_t is equal to the peak temperature of a very thick film. As we have seen from the example of methane, that assumption may not be correct.

The methane/graphite data have been carefully analyzed for thermodynamic evidence of stratified melting. C_N has been plotted versus $T_t - T$ and $T_{DW} - T$ to search for evidence of power law behavior, without success. Figure 4 shows the result of testing for $T_t - T_p \propto n^{-3}$. Clearly these data are not consistent with a simple stratified melting model.

Recently, Bienfait et al. have put forth evidence, from QENS measurements of molecular mobility, that stratified melting does occur in methane films adsorbed on MgO[26] (which is assumed to expose the (100) face of methane, rather than the (111) close-packed face expected on graphite), and then on methane/graphite as well.[12] Measurements of the NMR transverse relaxation time T_2, which is also a measure of molecular mobility, did not give any indication of the phenomenon,[6] but the NMR data may lack the temperature resolution to be compared successfully to the QENS data. What the QENS data actually show is an onset of increased mobility in part of the film at temperatures a few degrees below T_t. The authors assume that the film is stratified and the mobile part is on top.

In reality, the heat capacity data for methane/graphite shown in Fig. 5 also give evidence of an onset of melting below T_t, because the peaks are much broader than one would expect for the first order melting transition of the Fig. 1 phase diagram. It is assumed that this broadening is due to surface imperfections, such as pores and cracks where melting might be induced to occur at lower temperature than it would on an ideal flat surface (we return later to a more detailed discussion of melting and capillary condensation). We have attempted in Fig. 7 to compare the heat capacity data and the QENS data regarding what fraction of the melting has occurred at any $T < T_t$. The QENS data are taken from the analysis of Bienfait et al.,[12] normalized by the maximum amount they find melted (or mobile) at higher T, about 6 layers out of 10 adsorbed. To make the heat capacity data self-normalizing, we take the fraction of the latent heat for each curve consumed up to the given temperature, i.e., $\int_0^T C_N dT / \int_0^\infty C_N dT$. We plot the results for the 9.3 layer film. As the figure shows, the two sets of data agree reasonably well, suggesting they may both result from the same phenomenon.

We also plot the results of a thermodynamic analysis of stratified melting due to Pettersen et al.[6] In their analysis all films are regarded as thin slabs of bulk matter in Van der Waals

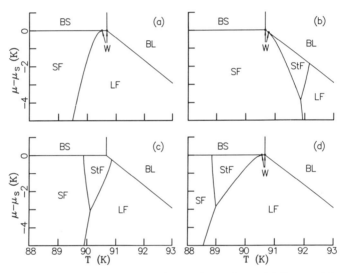

Figure 8. Phase diagrams from the thermodynamic model discussed in the text. SF = solid film, LF = liquid film, BS = bulk solid, BL = bulk liquid, StF = stratified film, and W denotes dewetting and wetting transitions. μ_s is the chemical potential of bulk solid. (a) $\delta = 0.21$ K/Å^2, $\delta' \le 0.07$ K/Å^2; (b) $\delta = -0.5$ K/Å^2, $\delta' = -0.1$ K/Å^2; (c) $\delta = 0.1$ K/Å^2, $\delta' = 0.2$ K/Å^2; (d) $\delta = 0.45$ K/Å^2, $\delta' = 0.35$ K/Å^2. (Ref. [6].)

fields due to the other phases present. Using techniques due to Landau[27] and de Gennes,[28] they calculate the grand canonical potential, Ω, as a function of μ and T for a system consisting of a liquid film on top of a solid film which is adsorbed on a semi-infinite substrate. One contribution to Ω comes from the presence of condensed phases in the region of the phase diagram where the vapor is the lowest energy state. The extra energy cost of forming the wrong phase is $\rho_i[\mu_i(T) - \mu]Z_i$, where Z_i is the thickness of the stratum of the film in phase i. Moreover, each particle in the system feels a potential due to the perturbation of the local van der Waals field because far away, the medium is replaced by some other phase or substrate. The interaction between two semi-infinite media is subsumed in the surface tension. The resulting surface excess Landau potential is

$$\frac{\Omega^{sm}}{A} = (\rho_s - \rho_l)\frac{\Delta C_3^{sw}}{2Z_s^2} + \rho_l\frac{\Delta C_3^{ls}}{2Z_l^2}$$
$$+ \rho_l\frac{\Delta C_3^{sw}}{2(Z_s + Z_l)^2} - \rho_l[\mu - \mu_l(T)]Z_l - \rho_s[\mu - \mu_s(T)]Z_s + \sigma_{sw} + \sigma_{sl} + \sigma_{lg}.$$

Ω is minimized with respect to Z_s and Z_l, the thicknesses of the solid and liquid strata. The resulting expressions which relate (Z_l, Z_s) to (μ, T) are

$$\rho_l[\mu - \mu_l(T)] = -\rho_l\left(\frac{\Delta C_3^{ls}}{Z_l^3} + \frac{\Delta C_3^{sw}}{(Z_s + Z_l)^3}\right),$$

$$\rho_s[\mu - \mu_s(T)] = -\left((\rho_s - \rho_l)\frac{\Delta C_3^{sw}}{Z_s^3} + \rho_l\frac{\Delta C_3^{sw}}{(Z_s + Z_l)^3}\right).$$

These can be solved for the thickness of the liquid layer as a function of temperature and we plot in Figure 8 the prediction of this model for a 10 layer film, assuming the values calculated by Cheng and Cole[32] for the Van der Waals interaction constants. The model fits the QENS data very well at lower temperatures. However, close to T_t the heat capacity and QENS data both disagree with the prediction for stratified melting.

In an attempt to sort out whether stratified melting or first order melting from homogeneous solid to liquid ought to be expected, Pettersen et al.[6] compare the free energies of a stratified film, a homogeneous solid, and a homogeneous liquid.[29] At each μ and T the state of lowest Ω is stable. The result depends on differences among the surface tensions of the various interfaces in the problem, σ_{ij}, where $i, j = s$ (solid), l (liquid), g (gas), or w (wall or substrate). The two combinations that matter are

$$\delta = \sigma_{sg} + \sigma_{sw} - \sigma_{lg} - \sigma_{lw}$$

and

$$\delta' = \sigma_{sg} - \sigma_{lg} - \sigma_{sl}.$$

Positive δ means solid slabs have larger combined upper and lower surface tensions (and hence, more free energy per unit area) than liquid slabs, thus tending to favor the liquid state, pushing the melting curve to lower temperature. Positive δ' means a hypothetical bare solid-gas interface would have higher surface tension than liquid-gas and solid-liquid interfaces combined, thus favoring surface melting of the bulk species.

Figure 8 shows examples of phase diagrams that may be generated using bulk methane parameters and various different combinations of δ and δ'. Figure 8(a) has the form postulated for methane/graphite, including the dewetting transition. Figure 8(c) has the form postulated by ZD for argon and neon. Figures 8(b) and 8(d) show that in principle, stratified melting and triple-point dewetting can occur in the same system.

Figure 8(a) is an excellent quantitative fit to the methane data, including the slope of the melting curve and the intercept temperature, T_{DW}, if $\delta = 0.21K/\text{Å}^2$ (δ is the only adjustable parameter). Of the four surface tensions comprising δ, only σ_{lg} has been measured, $\sigma_{lg} = 13.5 \ \mathrm{K/\mathring{A}^2}$,[34] nearly 100 times bigger than the fitted value of δ. Thus δ is a small difference of large terms. We note also that Fig. 8(a) fits the methane/graphite data for any value of δ' less than $+0.07 \mathrm{K/\mathring{A}^2}$. Since positive δ' is the condition for surface melting in the bulk, this means that methane adsorbed on graphite may not have a region of stratified melting even if bulk methane does undergo surface melting. Presumably this would occur if T_{DW} is lower than the lowest temperature of bulk surface melting. Conversely, Fig. 8(b) is generated assuming $\delta' = -0.1 K/\mathring{A}^2$, i.e., the bulk does not surface melt, but the film does exhibit stratified melting. However, what occurs in this case is a phenomenon that has been called substrate freezing: the surface tensions stabilize a thin layer of solid at the lower interface for $T > T_t$, and the liquid grows to infinite thickness as μ is raised at constant T.

To compare Fig. 8(c) to the data of ZD, Pettersen *et al.* used their model to compute the heat capacity numerically. The results do not obey the equation postulated by ZD, $C_N \propto (T_t - T)^{-4/3}$. However, they do reasonably well obey $C_N \propto (T_p - T)^{-4/3}$ below T_p, where T_p is the temperature of the peak. Since ZD did not measure T_t independently, this may have been the dependence they were actually testing. The calculated heat capacities also obey $T_t - T_p \propto n^{-3}$, but the measured values of $\zeta = (T_t - T_p)n^3$ disagree with those calculated from the model for both argon and neon on graphite.

In summary, the model of Pettersen *et al.* shows that small changes in δ and δ', which are themselves small differences of large numbers, can produce topologically different phase diagrams, including or excluding both stratified melting and triple-point dewetting.

CAPILLARY CONDENSATION

Capillary condensation[38] occurs when the system is able to lower its free energy by filling pores, cracks or other spaces between surfaces with condensed material. Capillary condensation at saturated vapor pressure, such as the rise of a liquid into a capillary, is a familiar phenomenon in nature. We wish here, however, to consider the circumstances in which bulk condensed phase might be induced to form in capillaries below saturated vapor pressure, when the only stable bulk phase is gas. Any real adsorption substrate can presumably be represented reasonably well by flat surfaces plus some distribution of pores, gaps, and corners of various sizes. At temperature T, if the chemical potential of the adsorbate is $\mu < \mu_o(T)$ (μ corresponds to the bulk phase, liquid or solid), then a uniform film of some thickness Z is to be found everywhere on the reservoir of flat surface, and on the walls of the pore and gap (the condition $\mu < \mu_o$ necessarily means that film growth will continue if μ increases; non-wetting, the growth of bulk crystals or droplets, can occur only at $\mu = \mu_o$). Let us suppose that the net Van der Waals potential due to the substrate is $f_i(Z)$ (for example, in the Frenkel-Halsey-Hill approximation, $f_i(Z) = \Delta C_3^{iw}/Z^3$ where ΔC_3^{iw} is constant and $i = s$ or l for solid or liquid adsorbate). Then

$$\mu = \mu_g = \mu_o - f_i(Z) \tag{1}$$

where μ_g is the chemical potential of the gas, and $\mu_o - f_i(Z)$ is that of the adsorbed film in the reservoir. Condensate can form in the pore or gap only if its surface has negative curvature (i.e., is concave), reducing the chemical potential inside by reducing the pressure P according to

$$\delta\mu = v\delta P \tag{2}$$

$$= -v_i \frac{2\sigma_{ig}}{R} \quad \text{(pore)} \tag{2a}$$

$$= -v_i \frac{\sigma_{ig}}{R} \quad \text{(gap)} \tag{2b}$$

where v_i is the molar volume of $i = l$ (liquid) or s (solid) and R is the radius of curvature of the interface. Then, comparing (1) and (2),

$$f_i(Z) = \frac{2\sigma_{ig}}{R} v_i \quad \text{(pore)} \tag{3a}$$

or,

$$f_i(Z) = \frac{\sigma_{ig}}{R} v_i \quad \text{(gap)} . \tag{3b}$$

Once an interface of curvature obeying Eqs.(3) is nucleated, the pore or slab should be able to fill with condensate at constant free energy, the work required to remove molecules from the reservoir being compensated by the reduction in surface free energy in the pore or gap. The work required to remove a molecule from the surface of the film to infinity is just $f_i(Z)$. Filling the pore requires filling a volume $\pi r^2 h$, where h is the distance the curved interface moves. The total work is

$$W = \frac{\pi r^2 h}{v_i} f_i(Z) .$$

The result is to reduce the exposed surface area by $2\pi r h$. The surface eliminated was that of a film, thickness Z, presumably in the same physical state (liquid or solid) as the bulk condensate. Thus its surface tension was approximately σ_{ig}. The reduction in free energy is thus $2\pi r h \sigma_{ig}$. Equating this to W, we find

$$f_i(Z) = \frac{2\sigma_{ig}}{r} v_i \quad \text{(pore)} \tag{4a}$$

or, repeating the analogous calculation for the gap,

$$f_i(Z) = \frac{\sigma_{ig}}{D} v_i \quad \text{(gap)} . \tag{4b}$$

Comparing respectively to Eqs.(3a) and (3b), we find $R = r$ for the pore, or $R = D$ for the gap. The equilibrium interface is thus that of half a sphere in the pore or half a cylinder in the gap, both meeting the wall at zero contact angle, consistent with the fact that the film wets ($\mu < \mu_o$). Quite generally, one expects capillary condensation to occur when

$$f_i(Z) = \frac{\sigma_{ig} v_i}{D} = \frac{\Delta C_3^{iw}}{Z^3}$$

where D is a length characteristic of the size of the region where the condensation takes place. For a cylindrical pore, $D = r/2$; for a gap, D is half the distance between the walls. Cheng and Cole[39] have derived the corresponding result for an oblique corner af arbitrary angle. In that case, D is the distance from the corner to the point where the curved interface is tangent to the flat film. At any given μ, all regions of characteristic size less than D should be filled with capillary condensate. The arguments work equally well for solid and liquid condensate, provided that the solid-gas interface is roughened, so that it has the same equilibrium shape as the liquid gas interface.

For Grafoil, where the average distance between opposing surfaces is approximately $600 - 900$Å, capillary condensation should occur when $Z \approx 4 - 5$ layers. For graphite foam, the corresponding numbers are about $2500 - 5000$Å and $Z \approx 6 - 10$ layers. At this point however, capillary condensation does not merely fill up a few accidental pores – instead it should fill all of the vacant space inside the graphite medium. In other words, if these estimates are correct, all experimental data for nominal film thicknesses greater than about 5 layers in Grafoil or 6-10 layers in foam should not reflect film properties, but rather be completely dominated by capillary condensation.

The heat capacity study of melting by Lysek *et al.* would appear to constitute evidence against capillary condensation, because if the heat capacity peaks they observed are due to the melting of a capillary condensate, they might be expected to extrapolate to the bulk triple point, contrary the indication shown in Fig. 4. However, let us present here the contrary argument and analyze the data for capillary condensation.

We suppose that the behavior of the film at thicknesses less than 5 layers is entirely two dimensional, showing no effect of or connection with bulk melting. Above 5 layers, some of the smaller gaps start to fill with 3D condensate according to Eq.(4b). At 5 layers, using (4b), $f_i(Z) = \mu - \mu_0$ and the value for $\mu - \mu_0$ at 5 layers, obtained from vapor pressure isotherms by HD, we obtain $2D_{5\ layers} \approx 150\text{Å}$. Using Eq.(2b) melting should occur in a gap when

$$\mu_{ol}(T) - v_l \frac{\sigma_{lg}}{D} = \mu_{os}(T) - v_s \frac{\sigma_{sg}}{D} \tag{5}$$

where μ_{oi} is the real or extrapolated value of the bulk phase chemical potential at temperature T. Near T_t, $\mu_{os} - \mu_{ol} = \alpha(T - T_t)$ where $\alpha = L/T_t$ and L is the bulk latent heat of fusion. Thus, melting in the gap takes place at temperature T_m given by

$$\frac{L(T_m - T_t)}{T_t} = \frac{v_s \sigma_{sg} - v_l \sigma_{lg}}{D} \tag{6}$$

Because σ_{sg} is not known, we do not know with certainty even the sign of $v_s \sigma_{sg} - v_l \sigma_{lg}$. In other words, we do not know whether capillary condensation would produce melting at temperatures above or below T_t. However, assuming the Lysek *et al.* data are produced by capillary condensation, we observe that when melting occurs at $2D \approx 150\text{Å}$ estimated above, $v_s \sigma_{sg} - v_l \sigma_{lg} \approx -69 K\text{Å}$ yielding $\sigma_{sg} \approx 13 K/\text{Å}^2$, which is a plausible value. Using these values, we can estimate the largest separation of the gaps in the material from the extrapolated maximum melting temperature of 90.48K (called T_{DW} in the dewetting picture above). This gives $2D_{max} \approx 620\text{Å}$. This maximum gap width is inconsistent with the average gap spacing obtained from the measurement of the specific area, of graphite foam ($2D_{avg} \approx 2500 - 5000\text{Å}$), as well as gap spacing obtained from electron micrographs[30] of graphite foam.

On the other hand, it can also be argued that the peaks in the 18-layer specific heat data, which occur at about 90.48K, are due to the melting of capillary condensed bulk in 600Å pores, whose existence is plausible. Then data taken at higher coverages would then show peaks at higher temperatures corresponding to bigger pores being filled. This would invalidate the dewetting analysis since the heat capacity signal would then be due to a combination of bulk and film, and a proper accounting of the capillary condensed part would have to be carried out before the film signal could be analyzed. This analysis would require a knowledge of the pore size distribution in the foam.

However, the shapes of the heat capacity peaks in Fig. 5 are wrong for capillary condensation. Since the $v_s \sigma_{sg} - v_l \sigma_{lg}$ is negative, then the rising (low temperature) sides of the peaks may be due to melting in pores and gaps smaller than those producing the top of the peak, but larger gaps should not yet be filled, so the heat capacity should fall abruptly from the peak to its background value. The broadening above the peak temperature, and especially the fact that part of each peak extends above T_t, is inconsistent with the capillary condensation picture.

The NMR measurements of PG also give evidence that these data are not due to capillary condensation. For one thing, the T_2 data indicate that melting at $T \approx 90K$ continues below 4 layers, down to 1.8 layers or less, contradicting our assumption that melting begins when capillary condensation does. In addition PG's T_2 measurements should be sensitive to capillary condensation for precisely the same reason they are sensitive to non-wetting. If bulk (capillary condensed) and thin film regions coexist, these parts of the system should have widely different T_2's. The authors explicitly discuss examining their data for the expected evidence, which they failed to find. Instead the data are consistent with uniform film growth to a thickness of 51 layers, as described above. This result would correspond to filling about 400Å of the average 600 − 900Å spacing between surfaces.

To summarize this section, we find that a picture based on capillary condensation leads to plausible numerical values for some parameters of the system (e.g., σ_{sg}), but cannot account for why the melting curve does not extrapolate to a temperature much closer to T_t. A good deal of contrary evidence also exists. Presumably, if capillary condensation did not occur in the reported experiments, its nucleation had to be impeded by a free energy barrier, making uniform film growth a long-lived metastable state. Hysteresis in adsorption isotherms is a signature of capillary condensation.[38] This kind of behavior has been observed in other systems[31][33] and might plausibly be expected in Grafoil and graphite foam, where the geometry might impede the smooth growth of condensate with uniform negative surface curvature. If this is the case, it should be possible in future experiments to induce capillary condensation by using different filling procedures.

On the other hand, if capillary condensation has occurred, and the contrary evidence can be accounted for satisfactorily, then, as we have seen, these experiments might afford a means of measuring σ_{sg} for methane and other substances.

CONCLUSION

As of this date, we propose to leave the phase diagram, Fig. 1, temporarily intact. The evidence makes it clear that changes are likely to be needed, but it is not yet clear what those changes are. The value of T_R may require modification. The question of wetting by solid methane at low temperature is far from clear. It may well be that methane creates a mutually commensurate trilayer rather than the bilayer proposed in the diagram. Finally it is possible that the entire melting curve is an artifact of capillary condensation. If it is not, there is a tricritical point in the melting transition that must be examined for evidence of the nature of the crossover from 2D to 3D behavior. Clearly, there is much work still to be done.

Acknowledgement

This work was supported by DOE contract no. DE-FG03-85ER45192. M. La Madrid acknowledges support from the Continental Oil Fund.

REFERENCES

[1] D. L. Goodstein, J. J. Hamilton, M. J. Lysek, and G. Vidali, Surf. Sci. **148**, 187 (1984).

[2] M. Wortis, in *Fundamental Problems in Statistical Mechanics VI*, ed. E.G.D. Cohen (North-Holland, Amsterdam, 1985), p.87.

[3] M. S. Pettersen, M. J. Lysek, D. L. Goodstein, Surf. Sci. **175**, 141 (1986).

[4] A summary of papers giving first layer data can be found in Refs. [21-24] of Hamilton and Goodstein, Ref. [9].

[5] R. Pandit, M Schick, and M. Wortis, Phys. Rev. B **26**, 5112 (1982); M. J. de Oliveira and R. B. Griffiths, Surf. Sci. **71**, 687 (1978).

[6] M. S. Pettersen, M. J. Lysek, D. L. Goodstein, Phys. Rev. B **40**, 4938 (1989).

[7] R. Pandit and M. E. Fisher, Phys. Rev. Lett. **51**, 1772 (1983).

[8] C. Ebner, Phys. Rev. B **28**, 2890 (1983); See also, J. Kahng and C. Ebner, preprint; M. W. Conner and C. Ebner, Phys. Rev. B **36**, 3683 (1987); C. Ebner, in *Chemistry and Physics of Solid Surfaces VI*, edited by R. Vanselow and R. Howe (Springer-Verlag, New York, 1986), p.581.

[9] J. J. Hamilton and D. L. Goodstein, Phys. Rev. B **28**, 3838 (1983).

10] D.-M. Zhu and J. G. Dash, Phys. Rev. B **38**, 11673 (1988); D.-M. Zhu and J. G. Dash, Phys. Rev. Lett. **60**, 432 (1988); D.-M. Zhu and J. G. Dash, Phys. Rev. Lett. **57**, 2959 (1986).

[11] Y. Lahrer and F. Angerand, Europhys. Lett. **7**, 447 (1988).

[12] M. Bienfait, P. Zeppenfeld, J. M. Gay, J. P. Palmari, Surf. Sci. **226**, 327 (1990).

[13] J. Krim, J. M. Gay, J. Suzanne, E. Lerner, J. Physique **47**, 1757 (1986).

[14] H. K. Kim, Q. M. Zhang, M.H.W. Chan, J. Chem. Soc., Faraday Trans. 2, **82**, 1647 (1986).

[15] H. S. Nham and G. B. Hess, Langmuir **5**, 575 (1989).

[16] A. Inaba, and J. A. Morrison, Chem. Phys. Lett. **124**, 361 (1986).

[17] A. Thomy and X. Duval, J. Chim. Phys. **66**, 1966 (1969); **67**, 286 (1970); **67**, 1101 (1970).

[18] J. Krim, J. G. Dash, and J. Suzanne, Phys. Rev. Lett. **52**, 640 (1984).

[19] M. J. Lysek, M. S. Pettersen, and D. L. Goodstein, Phys. Lett. A **115**, 340 (1986).

[20] M. S. Pettersen and D. L. Goodstein, Surf. Sci. **209**, 455 (1989).

[21] J. H. Quateman and M. Bretz, Phys. Rev. B 29, 1159 (1984).

[22] D. A. Huse, Phys. Rev. B **29**, 6985 (1984); F. T. Gittes and M. Schick, Phys. Rev. B **30**, 209 (1984).

[23] J. Z. Larese, M. Harada, L. Passell, J. Krim, S. Satija, Phys. Rev. B **37**, 4735 (1988).

[24] J. M. Phillips and C. D. Hruska, Phys. Rev. B **39**, 5425 (1989).

[25] D. B. Pengra, D. M. Zhu and J. G. Dash, preprint (1990).

[26] M. Bienfait, Europhys. Lett. **4**, 79 (1987).

[27] L. D. Landau and E. N. Lifshitz, in *Statistical Physics*, 3rd ed., edited by E. M. Lifshitz, L. P. Pitaevskii, J. B. Sykes, and M. J. Kearsley (Pergamon, Oxford, 1982), Pt. I.

[28] P. G. deGennes, J. Phys. (Paris) Lett. **42**, 1377 (1981).

[29] G. An and M. Schick, Phys. Rev. B **37**, 7534 (1988) and Ebner Ref. [8] have obtained related phase diagrams studying lattice gas models.

[30] M. Dowell and R. Howard, Carbon **24**, 311 (1986).

[31] D. S. W. Kwoh, Ph.D. Thesis, California Institute of Technology, 1979.

[32] E. Cheng and M. W. Cole, Phys. Rev. B **38**, 987 (1988).

[33] W. F. Saam and M. W. Cole, Phys. Rev. B **11**, 1086 (1975).

[34] *A Compendium of the Properties of Materials at Low Temperature (Phase I), Pt. I. Properties of Fluids*; V. J. Johnson, ed.; WADD Technical Report 60-56; Wright Air Development Division, 1960.

[35] G. Zimmerli and M.H.W. Chan, preprint (1990).

[36] Recent reviews of surface melting appear in J. G. Dash, Contemporary Physics, **30**, 89 (1989); and J. F. van der Veen, B. Pluis, A. W. Denier van der Gon in *Chemistry and Physics of Solid Surfaces VII*, eds. R. Vanselow and R. F. Howe (Springer-Verlag, Berlin, Heidelberg (1988), p.455.

[37] Discussions of the location of the specific heat bump relative to T_R are given in the chapters by den Nijs, Villain, and Lapujoulade. It is suggested that there are no special reasons for the bump to be at, before, or after T_R.

[38] This is also discussed in the book by R. Defay, I. Prigogine, A. Bellemans, and D.H. Everett, *Surface Tension and Adsorption*, (John Wiley and Sons, Inc., New York, New York, 1966).

[39] E. Cheng and Milton W. Cole, Phys. Rev. B, **41**, 9650 (1990).

KINETICS OF OVERLAYER GROWTH

G.–C. Wang, J.–K. Zuo and T.–M. Lu

Department of Physics
Rensselaer Polytechnic Institute
Troy, NY 12180–3590

1.   INTRODUCTION

Atoms or molecules adsorbed on crystalline surfaces often form interesting equilibrium structures depending upon the balance between adsorbate–adsorbate and adsorbate–substrate interactions.   These various structures are functions of temperature and coverage and can lead to phase diagrams with phase boundaries separating ordered, disordered, and phase coexistence regions. These are commonly studied using low energy electron diffraction (LEED) and grazing incidence X–ray diffraction.   While equilibrium properties of overlayers have been extensively studied in the past few decades, much less is known about the kinetics of growth processes and overlayer ordering, especially the relaxation from a state far from equilibrium.[1] This irreversible and highly nonlinear phenomenon occurs frequently in solids, for examples, during crystal and alloy growth.[2]   Only recently, considerable effort has been focused on understanding the dynamics of overlayer domain growth during the relaxation from an extreme non–equilibrium state to an equilibrium   state.[3,4]   This   process   involves   fundamental   questions   in non–equilibrium statistics. The interesting and important issues of the dynamics include the law that describes the domain growth, the value of the growth exponents, the form of the dynamical scaling function (self–similar growth) of the non–equilibrium structure factor, and the relevant parameters for determining the possible   dynamical   universality   classes   such   as   ground   state   degeneracy, conservation or non–conservation of order parameter and density, temperature, interaction potential and randomness (impurities and vacancies).

Most of the theoretical studies of non–equilibrium phase transitions and ordering kinetics are carried out by computer simulation except a few cases which can be solved analytically. The study covers a wide range of phenomena such as phase separation, spinodal decomposition, and order–disorder transitions. In the computer simulations of ordering kinetics, a system is quenched from a disordered state to an unstable state below a critical temperature. Right after the quench the system is far from its equilibrium state characterized by the temperature to which the quench is held. Then the system will undergo a spontaneous disorder to order phase transition or an ordering process. Initially, small ordered domains nucleate and grow. Later, the domains interconnect and grow in order to reduce the excess free energy associated with the domain boundaries. Eventually, the system will reach its final ordered equilibrium state. In general, the system can order in several thermodynamically equivalent ordered structures (p degeneracy) separated by domain boundaries. The approach to equilibrium is governed by the competition between the different ordered domains which are nucleated simultaneously. The dynamics of the boundary network and their instabilities and fluctuations, and the thermodynamic forces between different ordered domains, govern the way the equilibrium is reached.

There is a general agreement for Ising model with non—conserved order parameter and ground state degeneracy p = 2. Lifshitz[5] and Allen and Cahn[6] (LAC) have derived the growth law using both a phenomenological theory and a field theoretical treatment. They showed that the average domain size R(t) grows as t^n where the growth exponent n = 1/2 for d > 1.

For higher ground state degeneracy (p>2), the subject has been controversial. Lifshitz[5] first argued that the growth kinetics might slow down for a system with p $\geq$ d + 1, where d is the space dimension. Using the same line of argument, Safran[7] showed that domains may become pinned if p is $\geq$ d + 1. Early computer simulations[8-11] also showed the slow—down of the growth kinetics for large p systems. For example, for the Q—state Potts model, simulations[8-10] showed that the growth exponent n decreased from 0.5 to 0.41 as Q increased from 2 to 30. Lattice gas model simulations[9] showed that for p = 4, the growth exponent became 1/3 instead of 1/2. However, recent more accurate calculations[12-15] have shown that the degeneracy of the ground state does not affect the growth kinetics, and the growth exponent should be 1/2 independent of p. The results has been confirmed in both the lattice gas model simulations[14,15] and Q—state Potts model simulations.[12,13] It has been found also that this curvature driven (n = 1/2) law is independent of the density conservation and interaction potential.

For conserved order parameter, Lifshitz and Slyozov[16] assumed the growth is governed by diffusion through the domains and have shown that $R(t) \sim t^{1/3}$. This 1/3 growth exponent is known to be independent of the details of the dynamical model but less is known about its dependence on ground state degeneracy.

Another important concept in the dynamics is the dynamical scaling property. The scaling or self—similarity means that during a growth, the domain morphology of the system at some later time looks very much like that at an earlier time under the transformation of a length scale. In mathematical terms, this implies scaling of pair—correlation function or the structure factor S(k,t) which contains a time invariant scaling function F(x), i.e., $S(k,t) = \bar{R}^d(t)F(x)$, where **k** is the momentum transfer vector, d is the spatial dimensionality, $x = k/\bar{k}$ and $1/\bar{k}$ is proportional to the average domain size R(t). Most systems studied by simulations[12,17-20] and theories[21-24] exhibit the dynamical scaling.

It is well known that impurities can affect both the equilibrium[25] and nonequilibrium physical behavior of a statistical system, such as the lowering of the critical dimension d_ℓ (no long—range order at any temperature for dimension d < d_ℓ), the growth exponent, and the scaling behavior in the kinetics of domain growth. How do the growth law and scaling property change if random impurities are introduced into a pure 2—D system? Recent theory and Monte Carlo simulations on quenched randomness in Random Field Ising Model[26-34] and quenched site dilution in Ising and Potts models[35] indicate a slowing down of the growth due to the randomness. That is, it reduces the growth exponent from n = 1/2 to n < 1/2. It also would break self—similar scaling at high fields and at long times.[27,31] This is because random fields can give rise to many local free energy minima and dilute the domain structure. Thus, two adjacent domains meet at only a few points instead of along a line as in the zero—field case. This diluted or roughened boundary greatly reduces or eliminates the curvature driving force for the domain growth. At the same time, the random impurities would pin the domain walls and allow the domains to grow no larger than a maximum size $\bar{R}_m$. Thus, when the domain size

R(t) gets close to $\bar{R}_m$, the size distribution will be forced to change to the distribution of the relative impurity separation. Therefore, a breakdown of scaling should occur.

As mentioned before, 2–D overlayers on solid surfaces form a variety of phases with different symmetries and ground state degeneracies, and can be used as a testing ground for the growth laws and scaling behavior predicted by theories and simulations. In this paper we shall focus on a 2–D system which contains two–fold ground state degeneracy and a non–conserved order parameter. We shall present detailed experimental results on the growth kinetics[36] and scaling behavior[37] in a chemisorbed oxygen overlayer on a W(112) surface using the time resolved, high resolution LEED technique.[38] The effects of random impurities (nitrogen) on the growth and scaling of the oxygen overlayer will also be presented.[39]

2.   PURE OXYGEN OVERLAYER ON W(112) SURFACE

Oxygen chemisorbed on a W(112) surface forms a (2x1) superstructure over a wide range of submonolayer coverage and temperature.[40] A representation of the adsorbed layer is shown in Fig. 1.

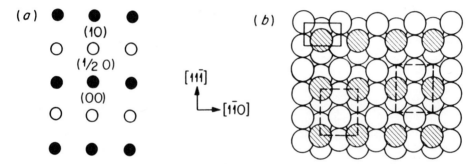

Fig. 1.   (a) Schematic LEED pattern of W(112) (2x1)–O at half–monolayer coverage. Solid circles, integral–order beams; open circles, half–order beams. (b) Schematic drawing of the (2x1)–O layer on a W(112) surface. Open circles, W atoms; hatched circles, oxygen atoms; solid–line rectangle, (1x1) substrate unit mesh; dashed–line rectangles, (2x1) superlattice unit mesh. There are two possible positions of (2x1) unit mesh which have antiphase relationship with respect to each other.

The structures of both the clean W(112) surface and (2x1)–O covered W(112) surface at 0.5ML coverage have been determined by a comparison between LEED I–V profiles and a dynamical LEED model calculation.[41] The results show a nonreconstruction surface and a simple overlayer, respectively. There are two possible positions of (2x1) unit mesh which have antiphase relationship with respect to each other. Therefore, the O/W(112) system can be considered as a prototypical system which possesses two–fold ground state degeneracy.[42-44] The equilibrium properties of this system have been studied in detail. Figure 2 shows the experimentally determined temperature versus coverage phase diagram (T–θ) of the W(112) (2x1)–O system for $0 < \theta < 1$. Each data point, representing a transition temperature at a fixed coverage, was obtained from the inflection point of the intensity versus temperature curve of the half–order beams measured by LEED. The data points form a second–order phase boundary which separates the low–temperature (2x1) phase and the high–temperature lattice–gas phase. The oxygen overlayer is not mobile for T < 225 K, the region under the dotted line.

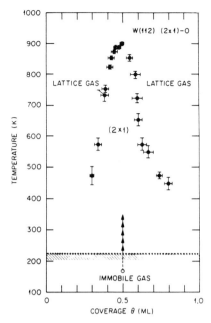

Fig. 2.   Phase diagram of the W(112) (2x1)–O overlayer.

For half–a–monolayer coverage, this system undergoes a second–order order–disorder phase transition at T_c = 899 K. According to the classification scheme suggested by Domany et al.[45], this transition belongs to the Ising universality class. The order parameter has only one component and is nonconserved. The values of the critical exponents are consistent with the 2–D Ising model.[43],[44]

2.1 Kinetic of Domain Growth in W(112) (2x1)–O Overlayer

The dynamics of the (2x1) domain growth was studied with use of an "up–quenched" technique.[46],[47] In this technique, oxygen was adsorbed on the surface at low temperature in a form of random lattice gas (immobile) and the system was then quenched up to an elevated temperature to form ordered (2x1) domains.

The experiment was performed in an UHV chamber equipped with a HRLEED and a Cylindrical Mirror Auger Analyzer. The base pressure was $6.5 - 8 \times 10^{-11}$ torr. The preparation and cleaning of the W(112) surface, the sample mounting as well as the temperature measurement have been described elsewhere.[43] The terrace width of the clean W(112) surface was estimated from the half–order beam profiles (using a Gaussian distribution) to be about 410 ± 10 Å and 115 ± 5 Å along the [11$\bar{1}$] and [1$\bar{1}$0] directions, respectively. The quantitative evaluation will be discussed later in Section 3.3. A typical 0.5 ML oxygen coverage requires 0.4 Langmuir (1L = 10^{-6} torr sec.) exposure with an oxygen partial pressure of ~ 1×10^{-9} torr. The oxygen atoms were adsorbed at ~ 350K after a flash–cleaning of the W(112) substrate at ~ 2430 K. At this adsorption temperature a nearly random 2–D lattice gas was formed. The sample was subsequently up–quenched and held at a temperature T where oxygen atoms were mobile. The time required for the up–quenching to a predetermined temperature was typically less than a few seconds. While the (2x1) domains were growing, the angular distribution (profile) of the (1/2 0) superlattice beam intensity, which is a measure of the structure factor S($\mathbf{k}$,t), was measured as a function of time. Depending on the up–quenching temperature, the

458

entire LAC growth period could be as long as 10^2–10^3 seconds from the time of quench.

Figure 3 shows the development of the peak intensity of the (1/2 0) beam vs time from the 0.5 ML (2x1)–O overlayer at different up–quenching temperatures.[36] A very weak peak intensity I_0 at t = 0 just after the quench has been substracted from the time dependent peak intensity $I_p(t)$ in Fig. 3. From Fig. 3 we see that on a log–log scale the peak intensity grows linearly with time in the early stage and then levels off. The peak intensity I_P is a measure of the mean–squared domain size,[17-20,48] i.e., $I_P \propto \bar{R}^2$. Assuming a power growth law $\bar{R}^2 \propto t^{2n}$ in the early stage, the exponent 2n extracted from the slope of the log–log plot in Fig. 3 was found to be 1.01 ±0.02 which is independent of up–quenching temperature. This result remarkably demonstrates the LAC curvature driven growth law that has been reported previously.[46]

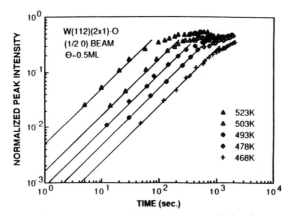

Fig. 3.   The time–dependent peak intensity of the (1/2 0) beam diffracted from W(112) (2x1)–O overlayer at 0.5 ML coverage is plotted for different up–quenching temperatures. Each curve was normalized to the maximum intensity from a well–annealed overlayer. The incident electron energy is 55 eV. The solid lines represent the power–law fit in which the slope was found to be 1.01 ± 0.02.

2.2   Self–Similar Domain Growth (Scaling) in the W(112) (2x1)–O Overlayer

In the study of scaling we have measured the angular profiles in both the [1$\bar{1}$0] and [11$\bar{1}$] directions at different evolving times. A low uniform background intensity ($\lesssim$4% of the peak intensity) near t = 0 was observed in the measured angular profile. As the peak intensity was developing, the ratio of the background intensity to the peak intensity became smaller and eventually approached zero. This low background intensity has been substracted from the measured angular profiles. All the profiles presented here have been deconvoluted with an effective instrument response function which has a narrow width ($\sim$0.012±0.002 Å^{-1}).[49] The deconvoluted profiles reflects the true structure factor $S(k,t)$. Considering $\bar{R}^2 \propto I_P$ and taking $\bar{k}$ as the full–width at half–maximum (FWHM) of the angular profile, we obtained the measured scaling function $F(x) = S(k,t)/I_P$. This equation implies that if the normalized angular profiles at different evolving times are plotted in the $x = k/\bar{k}$ coordinate (not in the k coordinate), i.e., scaled by its own FWHM then all the profiles should superpose on each other. It is known that the shape of the

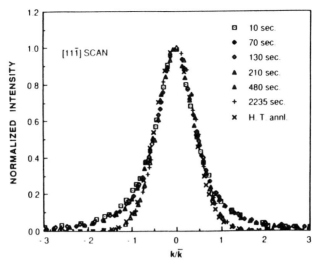

Fig. 4. The scaled angular profiles of the (1/2 0) beam for various times at T = 493 K. Scans are along the [11$\bar{1}$] direction. The scaling holds only for t < 240 sec. where the power—law growth applies. The H.T. annl. means a profile was obtained after the overlayer has annealed to ~ 670 K. The vertical error bar is less than twice the size of the symbol due to the statistical fluctuations of intensity in the initial times.

angular distribution of intensity reflects domain size distribution. Scaling law implies that, although the average domain size increases with time, the functional form of the size distribution does not change. Figure 4 is the plot of the scaling function $F_2(x)$ at various times scanned along the [11$\bar{1}$] direction. The scaling function $F_1(x)$ along the [1$\bar{1}$0] direction is not shown here. The up—quenching temperature was 493K. As seen from the figure, within the LAC growth regime (<240 sec., see Fig.3), the profiles scale very well. Even more remarkable is that $F_1(x)$ and $F_2(x)$ almost coincide with each other as shown in Fig. 5, despite the fact that the average domain sizes in these two directions at any given time were different (before the scaling, i.e., in the k coordinate, the FWHM of a profile in the [1$\bar{1}$0] direction is at least two times wider than that in the [11$\bar{1}$] direction at the same time). Similar agreement was obtained for other up—quenching temperatures. Furthermore, the scaling profiles obtained at different up—quenching temperatures are identical, too. Therefore, $F_1(x) = F_2(x) = F(x)$ is a universal function independent of direction and up—quenching temperature despite the fact that the detailed microscopic interactions among the oxygen atoms in the [1$\bar{1}$0] and [11$\bar{1}$] directions are not isotropic.[42,43,50-52]

The scaling behavior in the kinetics of domain growth has been a subject of a variety of theoretical discussions.[21-24] For the system with two—fold ground state degeneracy, Kawasaki et al.[21] first described scaling of the correlation function by considering the time—dependent Ginzburg—Landau model in a weak—coupling, long time limit. Ohta et al.[22] using the LAC interface curvature—driven model, have derived an approximate, but explicit expression for the scaling function F(x). The best—fitted profile using Ohta's expression for the scaling function is plotted as the solid curve in Fig. 5. The fitting shows good agreement with experiment in the small $|x|$ (<1.0) regime but deviates somewhat in the tails.

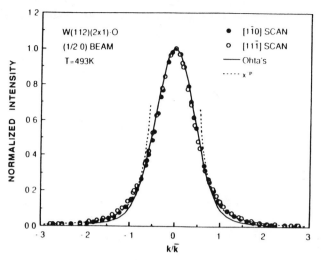

Fig. 5.   Typical scaled angular profiles from Fig. 4 are presented. Filled circles: a scan along the [1$\bar{1}$0] direction at t = 215 sec. Open circles: a scan along the [11$\bar{1}$] direction at t = 210 sec. The solid curve represents a fit of the theory predicted by Ohta et al. The dashed curve represents the fit of x^{-p} with p = 2.9±0.1 in the limit of $|x| \geq 1.0$.

The theory predicted by Kawasaki et al. has essentially the same scaling behavior as that of Ohta et al., although they were based on different approaches. Subsequently, Mazenko and Valls[23] numerically calculated the scaling function based on the renormalization–group method combined with Monte Carlo simulations.   Their prediction for the scaling function of this system with a non–conserved order parameter is rather similar to that of Ohta et al. except for the intermediate values of the scaling variable, i.e., the tails of the scaling function calculated from the renormalization group theory is higher than that of Ohta et al. Therefore, the measured scaling function in Fig. 5 may be in better agreement with the results of Mazenko et al. All these theories exhibit a Gaussian behavior in the small x regime and a Porod's law[53] behavior in the large x regime. That is, for large x, $F(x) \approx x^{-p}$, where p=d+1=3 (d is dimensionality). We have also fitted the tail part ($|x| \geq 1.0$) of our scaling profiles using Porod's law and found p = 2.9±0.1. This is done by ignoring the data for $|x| < 1.0$. As seen in Fig. 5, the fitting is very good for $|x| \geq 1.0$, which indicates that the dimensional dependence of the exponent p in Porod's law is consistent with the theories.

The angular profile of the (1/2 0) beam was also measured beyond the LAC growth regime (>240 sec., see Fig. 3). For the profiles scanned along the [1$\bar{1}$0] direction, the line shape is almost unchanged for t > 240 sec., but there is small deviation when the sample was annealed to a higher temperature (670K) to accelerate the growth process.   The data are not shown here.   An apparent breakdown of scaling was observed in the [11$\bar{1}$] direction for t > 240 sec., as shown in Fig. 4. Furthermore, the function F(x) in the [11$\bar{1}$] direction behaves quite similarly for t $\gtrsim$ 480 sec. including that of the well annealed profile.   After the annealing the domains approached their ultimate sizes which were limited by the surface heterogeneity of the substrate. The breakdown of scaling is an indication of the termination of the curvature–driven process.   One possible origin of the observed termination of the curvature–driven process might be due to the finite size effect. Previous Monte–Carlo simulations[54,55] have estimated that finite size effects become important when $\bar{R} \simeq 0.4R_f$, where R_f is the finite area used in the simulations. We estimated that the average domain sizes were $\sim 0.51R_1$ and $0.60R_1$ in the [1$\bar{1}$0] direction, and $0.70R_2$ and $0.91R_2$ in the [11$\bar{1}$] direction for t $\sim$ 240 sec. and 2230 sec., respectively, where R_1 and R_2 are the average terrace width in the [1$\bar{1}$0] and [11$\bar{1}$] directions, respectively. Therefore, the breakdown of scaling around t = 240 sec. should result from the finite size effect. This is because when domains grow large enough and reach the interaction range from steps, the distribution of domain sizes will be forced to change into the distribution of terrace width. As a result, the function F(x) changed shape. We must point out that even for an ideal surface where the growth is not limited by finite size and a later growth regime may exist. There is no reason to believe that the function F(x) obtained in the later stage should scale the same way as that in the LAC growth regime. It would therefore be desirable to have a surface with a larger average terrace width so that the behavior of the function F(x) in the unconstrained later stage of growth can be studied unambiguously.

In summary, the self–similar scaling in the (2x1) domain growth of oxygen overlayer on W(112) surface at 0.5 ML coverage was observed by the analysis of time resolved HRLEED angular profiles of the (1/2 0) beam. The scaling function extracted from the angular profiles was found to be a universal function independent of the growth direction and the up–quenching temperature, and was compared with recent analytical theories and Monte Carlo simulations. An obvious breakdown of scaling in the [11$\bar{1}$] direction was observed after the peak intensity levels off. This was attributed to the finite size effect dictated by the surface heterogeneity.

3.   NITROGEN DOPED OXYGEN OVERLAYER ON W(112) SURFACE

In the study of effects of impurity on the growth kinetics, we predosed the surface with 0.5% to 5% ML of nitrogen before the oxygen was adsorbed using the same method as described in Section 2. The angular profiles of the (1/2 0) beam for the well–annealed sample were observed to broaden considerably (accompanied by a reduction of the peak intensity) in the case of N–contaminated O layers.[36] The broadening of the angular profiles indicated that the (2x1) domain size in the mixed overlayer did not reach the same size as in the pure O–overlayer case. This result can be interpreted as an evidence of the destruction of the overlayer long–range order by random N impurities. The nitrogen impurities are believed to adsorb on the surface randomly. This conclusion is based on the fact that the FWHM of (1/2 0) beam for the well–annealed mixed overlayer is always broader than that of the well–annealed pure oxygen overlayer at all oxygen coverage.[56] Also, the degree of broadening increases with increasing amount of predosed nitrogen impurities. If small amount of preadsorbed nitrogen clustered on the surface, then the FWHM of the (1/2 0) beam obtained from the mixed O+N overlayer, especially for the low oxygen coverage case, should not broaden as compared with that obtained from the pure O overlayer. The broadening of the (1/2 0) beam and the large reduction of the peak intensity are the results of the diffraction from smaller finite–size domains which are stabilized by the random nitrogen atoms. These impurities may create static random fields[31,57,58] and disturb the periodic surface potential to give local minima of the free energy which obscure the equilibrium state of the overlayer. The system is then a 2–D physical realization of the well known random–field Ising model which has a lower critical dimension of 2, and has been a subject of intense discussions recently.[24,59-62]

3.1   Kinetic of Domain Growth in W(112) (2x1)–(O+N) Overlayer

A stronger support for the above interpretation of the random–field effects comes from the observation of the dramatic alteration of the domain growth kinetics from the simple LAC growth law for the O+N mixed overlayer.

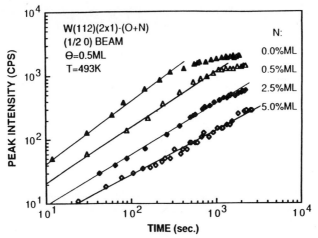

Fig. 6. The time dependence of the (1/2 0) beam peak intensity diffracted from the W(112) (2x1)–(N+O) mixed overlayers at half–monolayer coverage is plotted for different N content at the same up–quenching temperature of 493 K. The solid lines represent the effective power–law fit. The slope changes as a function of N content and an apparent deviation from the Lifshitz–Allen–Cahn growth law is observed.

A dramatic change of the growth kinetics was observed when the system was predosed with nitrogen impurities as shown in Fig. 6. These curves were obtained for different nitrogen doses at a fixed up–quenching temperature of 493 K.[31,57,63] Following the suggestion of previous computer simulations, we were able to define an effective exponent 2n for the power growth law, $\bar{R}^2 \sim t^{2n}$. From the data we deduced that 2n is $\simeq$ 0.88, 0.76 and 0.70 for nitrogen doses of 0.005, 0.025 and 0.05 ML, respectively. These exponents again do not depend on the up–quenching temperature. One is then able to estimate the value of the random–field strength h as a function of the nitrogen concentration by comparing our value of n to the results of computer simulation.[31,57] The h was found to be 0.18 to 0.48 which corresponds to the nitrogen doses of 0.5% to 5% ML. The random field we used so far was thus small.

3.2 Self–Similar Domain Growth (Scaling) in the W(112) (2x1)–(O+N) Overlayer

To study the scaling behavior, we have measured the angular profiles of the (1/2 0) superlattice beam in both the [1$\bar{1}$0] and [1$\bar{1}$1] directions at different times. A uniform low background intensity ($\sim$ 5% of the peak intensity) around the (1/2 0) beam position at the initial times was observed. The ratio of the background intensity to the peak intensity became small as the peak intensity was developing. This very low background has been subtracted from the profiles. All profiles presented here have been deconvoluted with an effective resolution function ($\sim$0.012±0.002Å^{-1}) containing the instrumental broadening and other corrections for mosaic spread and strain etc., measured from the clean W(112) surface.[49]

In the case of (O+N) overlayer, the scaling behavior in the power–law growth regime are quite similar to that in the pure O overlayer case. However, there is a major difference in the scaling behavior between the two cases in the level–off regime. As we shall see later, the breakdown of scaling after the cross–over point in the (O+N) overlayer case is mainly due to the N impurity pinning effect. In the power—law growth regime the scaling was observed in both $[1\bar{1}0]$ and the $[11\bar{1}]$ directions for the N impurity dose up to 5% ML (or equivalent to the random field strength of 0.48) which is the heaviest dose used in the present experiments. Experimentally, we did not dose beyond 5% ML because for such a heavy dose the intensity of the superlattice beam is too low for a meaningful measurement. Figure 7 is a plot of the scaling profile $F_2(x)$ as a function of time in the $[11\bar{1}]$ direction.

Scaling profile $F_1(x)$ along the $[1\bar{1}0]$ direction is not shown here. The up—quenching temperature was 493K and the N impurity dose was chosen at ~ 2.5% ML where the growth includes a cross—over regime in the observation period of the experiment. As seen from these figures, within the power—law growth regime (t< 2000 sec.), profiles scale very well in both directions. Also, for a fixed N impurity dose, the functions $F_1(x)$ and $F_2(x)$ almost coincide with each other as shown in Fig. 8 even though the average domain size in these two directions are quite different at any given time. The $F(x)$'s are also identical for different up—quenching temperatures within the experimental uncertainty. A similar agreement was achieved for other fixed N impurity doses.

Further, we found that there is no obvious deviation among the scaling profiles obtained for different N impurity doses (not shown here). The above observations are consistent with Monte Carlo simulations[31] and the theoretical prediction.[27] Monte Carlo simulations showed that in the earlier time regime, scaling still holds if the field strength was not too large and all profiles obtained with small fields can be scaled to the zero—field profile. The theory predicted that the real—space correlation function $S(\mathbf{r},t)$ had a very weak dependence on the random field h explicitly, so that small fields are not strong enough to break the scaling law at short times. However, we still can not rule out the possibility that the scaling profile could depend on the random field h implicitly through, for example, the inverse correlation length. Thus, the $F_1(x) = F_2(x) = F(x)$ in our experiments can be viewed as an isotropic universal function which does not depend strongly on the random field strength. The scaling in both directions, and $F_1(x) = F_2(x)$, implies that in the power—law growth regime self—similar growth occurs isotropically with an invariant domain size distribution which has the same functional form in the two

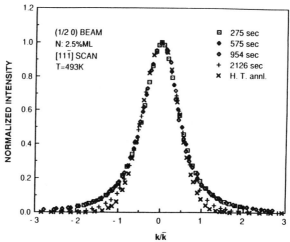

Fig. 7.   The time dependence of the scaled angular profiles of the (1/2 0) beam, diffracted from the W(112) (2x1)—(O+N) overlayer at 0.5 ML coverage, are plotted for an N dose of ~ 2.5% ML for T = 493 K. Scans are along the $[11\bar{1}]$ direction. The scaling holds only in the power—law growth regime. The H.T. annl. means a profile was obtained after the overlayer has been annealed to ~670 K.

directions even in the presence of impurities. Another way to state this result is that as long as the growth can be described by a power–law, scaling will hold. Theory and Monte Carlo simulations also support this conclusion.[27,31]

The angular profile of the (1/2 0) beam was also measured around the cross–over point (see Fig. 6). In the [111] direction, an apparent deviation of the above scaling behavior was observed around ~ 2000 sec. during growth, as shown inFig. 7. A further deviation occurred when the sample was annealed to ~670K, shown as crosses in Fig. 7. After annealing, the domains approached their ultimate sizes around $\bar{R}_m$, limited by the distribution of N impurities. The breakdown of scaling is an indication of a change of domain size distribution which results from the "finite size effects" induced by the impurities. These random impurities dilute or roughen the interfaces so that eventually, the curvature–driven mechanism is eliminated. The details of domain size distribution will be discussed in the next section.

Recent theory has predicted that the equilibrium structure factor in the presence of random fields should have the form:[64]

$$S(\mathbf{k}) = \frac{A}{(\zeta^2 + k^2)^2} + \frac{B}{(\zeta^2 + k^2)}, \tag{1}$$

i.e., the sum of a Lorentzian squared (L^2) and a Lorentzian (L), where ζ is the inverse correlation length. The first term in Eq. (1) describes the scattering from the microdomains broken by impurities. The second term in Eq. (1) is the scattering from the fluctuation of local short–range order. This L^2+L form actually is a result of the Ornstein–Zernike approximation where the Bragg term describing the long–range order is now replaced by the L^2 term due to the loss of the long–range order.

Also, the L^2+L form with different percentages between the L^2 and L terms characterizes different domain morphology at different stages of growth. The larger

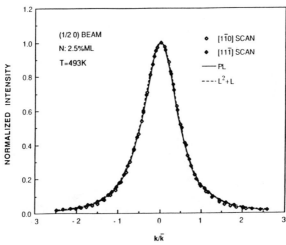

Fig. 8. The isotropically scaled angular profiles of the (1/2 0) beam are presented. Open diamonds: a scan along the [110] direction. Solid diamonds: a scan along the [111] direction. The solid curve represents a fit of the PL form with m = 1.5 ± 0.1. The dashed curve represents a fit of L^2 + L from Eq. (1).

percentage of the L² term implies a more compact domain morphology, where the ordered domains are randomly distributed and are interconnected. The larger percentage of the L term describes a more disordered domain morphology with rougher boundaries.

We can learn the following facts from the relative percentage between L² and L terms obtained by fitting Eq. (1), to the measured (1/2 0) beam profiles: 1). In the scaling region the L² term makes a constant fractional contribution of about 83% to the peak intensity at x = 0 for different N impurity doses. In this region the domains are not fully grown. There are many antiphase boundaries between small domains. 2). In the equilibrium state the contribution of the L² terms to the peak intensity has raised to more than ~ 97%. Many of the antiphase boundaries in the overlayer have been either reduced or eliminated and the domains have coalesced. Thus, the previous structure with small domains has changed into a structure containing microdomains broken only by the impurities. Therefore, the long–range order breaks down. Interestingly, we found that a power Lorentzian (PL) form $\propto A'/(\zeta^2 + k^2)^m$ also describes our data well with the best–fitted parameters m = 1.5 ± 0.1 for the profiles obtained in the power–law growth region, as shown in Fig. 8. The scaling function F(x) with m = 1.5 exhibits the asymptotic behavior of Porod's law[53], i.e., $F(x) \sim x^{-(d+1)}$ (d+1 = 3). This is the well known consequence of scattering from a compact domain morphology. This result further confirms the conclusion drawn from the L² + L fittings.

3.3 Breakdown of Self–Similar Domain Growth (Scaling) Due to the Change of Domain Size Distributions in Different Growth Regimes

As the domain growth proceeds, the ordered domains are evolving to reduce their boundary energies (curvatures). Eventually, a stage is reached at which the rate of growth slows down. After this cross–over period, the size distribution of ordered domains will change. There are two possible physical reasons that may lead to this cross–over. (1) If the growth of the overlayer occurs on a stepped surface with finite terrace widths, and the overlayer on each terrace is thermodynamically independent, then the ordering will be limited within each terrace. When the domains have grown large enough to be affected by the existence of the step, the rate of growth may slow down and the domain size distribution will be forced to approach the terrace width distribution. This mechanism is called the finite size effect due to substrate heterogeneity (finite size terraces). Previous Monte Carlo simulations have estimated[54,65,66] that the finite size effects may become important when $R \approx 0.4\ R_f$, where R_f is the finite terrace width of the substrate. (2) For a system with random impurities, random impurities roughen the domain boundaries and reduce the curvature driving forces which systematically slows down the growth for the entire time regime including the initial and the late stages. This behavior is greatly different from the above mentioned mechanism. Eventually, the rate of growth slows down when the domains are interconnected at the roughened interfaces caused by impurities. At the same time, the sizes are turned into a distribution around R_m which is determined by the impurity distribution. This may be considered as a kind of "finite size effect" due to impurities. No matter which mechanism dominates, the macroscopic manifestations are the leveling–off of the peak intensity and the breakdown of scaling.

The size distribution can be determined from the angular profiles of the superlattice beams. Since the domains on different terraces are independent[56], the diffraction from the domains on different terraces are incoherent.[67-69] But on the same terrace there are two kinds of equivalent (2x1)–O domains with a translational antiphase relationship in the [11̄1] direction for the W(112) (2x1)–O overlayer. Hence, the two kinds of domains will in principle contribute interferences at the respective superlattice beams. To a first approximation, we have neglected the interference among the antiphase domains within a terrace, and simply expressed the superlattice beam intensity as the sum of intensities scattered from individual domains. This approximation is expected to work better when the domains are small but becomes worse as the domains grow bigger. We have tried three different size distributions in the fittings of the angular profiles, i.e., Gaussian, cut–off Gaussian, and Geometric distribution.

In a pure O overlayer case, the result of the fitting showed that the size distribution in the power–law growth regime (t < 240 sec.) is a cut–off Gaussian in both the [1$\bar{1}$0] and the [11$\bar{1}$] directions. After the cross–over point (t> 240 sec.), the cut–off Gaussion distribution changes into a Gaussian distribution in the [11$\bar{1}$] direction. The size distributions of the (2x1)–O domains in the well–annealed pure O overlayer at 0.5 ML coverage reflect the terrace width distributions of the W(112) substrate.

Figure 9 shows the time evolution of the best–fit size distribution at an nitrogen impurity dose of 2.5% ML in the [11$\bar{1}$] direction. Comparing these results with the peak intensity–vs–time plot in Fig. 6, we see that the size distribution is a cut–off Gaussian in the power–law growth regime and this distribution turned into a Gaussian in the [11$\bar{1}$] direction in the leveling–off regime. This cross–over of size distribution is mainly due to the impurity effects mentioned above, which prevent the domains from growing larger than $\bar{R}_m$. Therefore, larger domains are most probably pinned, while smaller domains, having a smaller probability of meeting impurities, continue growing until their sizes approach a distribution around $\bar{R}_m$. This explains why the size distribution changes from the original cut–off Gaussian to a Gaussian peaked at $\bar{R}_m$. Also shown in Fig. 9 as x are the domain size distributions in the (O+N) overlayers after these overlayers have been annealed at a high temperature (~ 670K). Note that along the [11$\bar{1}$] direction, the domain size distribution is still a Gaussian but it has a larger average domain size, $\bar{R} \sim 0.80 R_2$, (peak around 330Å) as compared with the average domain size $\bar{R} \sim 0.54R_2$, ($\bar{R} \sim$ 220Å) obtained in the region of the breakdown of scaling (t ~ 2126 sec.). This indicates that high temperature annealing speeds up domain growth. However, the average domain size does not reach the average domain size in the pure overlayer

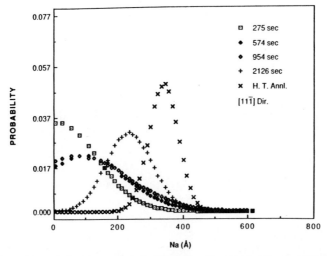

Fig. 9.   The time evolution of the best–fit domain–size distribution for the (O+N) mixed overlayer preadsorbed with ~ 2.5% ML N impurity for T = 493 K. The domain–size distributions in the [11$\bar{1}$] direction.

case ($\sim 410\mathring{A}$) due to the impurity pinning effect. Along the [$1\bar{1}0$] direction, the domain size distribution has not drastically changed, but the average domain size increases after a high temperature annealing. Again, the average domain size in the higher temperature annealed (O+N) overlayer, $R \sim 0.66R_1$, ($R \sim 80\mathring{A}$) is smaller than the average domain size in the pure overlayer case ($\sim 110\mathring{A}$) due to the effect of impurity pinning.

4.   CONCLUSIONS

In this paper, we have presented in detail the time resolved, high resolution LEED studies of the growth and scaling of oxygen chemisorbed on a W(112) surface with and without the random nitrogen impurities. For this system of two–fold ground state degeneracy with a non–conserved order parameter, we found 1) the growth law obeys $R(t) \sim t^n$. The value of n was found to decrease from 1/2 as the density of impurity is increased from zero. 2) An isotropic scaling of the oxygen domains was observed with and without the random nitrogen impurities in the power law growth regime. This isotropic universal function is independent of up–quenching temperature and is also practically independent of the impurity doses or random fields (in the small random fields regime). 3) Also the breakdown of scaling was observed after the termination of the power law growth regime. The breakdown of the scaling in a pure oxygen system was caused by a finite size effect due to the surface heterogeneity. The breakdown of the scaling in the domain growth containing impurities was caused by the change of domain size distribution due to the impurity (random fields) effects which pinned the domain growth.

There have been several experimental studies which focused on the domain growth kinetics for the case of a non–conserved order parameter with $p \geq d + 1$. The experiments which demonstrated the n = 1/2 growth kinetics were for Ag deposited on Ge(111)[70] with p > 3 and for the SbCℓ_5–intercalated graphite system[71] with p = 7. The other two experiments of O/W(110)[47] (p = 4 or 8) and S/Mo(110)[72] (p = 3) all give an n value close to 1/3. Therefore it appears that there is a discrepancy between some of the experimental results on growth kinetics for $3 \leq p \leq 8$ and the theoretical predictions which favors 1/2 for all p. For the O/W(110) system, the scaling was also observed.[73]

In the case of a conserved order parameter, where the system separates into high and low density phases, a recent LEED experiment of O/W(110)p(2x1) + p(2x2) has shown[74] that n = 1/3 which is consistent with the prediction by Lifshitz and Slyozov theory.[16] The limited experimental results on 2–D growth kinetics seem to support the two universality classes with n = 1/2 for non–conserved order parameter and n = 1/3 for conserved order parameter. Other parameters such as ground state degeneracy, temperature and detail interaction potential seem irrelevant in the classification of the universality class.

We feel that substantial amount of experimental work is still needed to better understand the growth kinetics, especially for high degeneracy systems, for examples, p = 6 for N_2 (or CO)/Graphite[75] and p = 8 for N_2 (or CO) + Ar/Graphite.[76] Also, the effects of various randomness (impurities and vacancies) and imperfection on the growth and scaling for p > 3 systems needs to be investigated systematically using high resolution time resolved diffraction or direct imaging (Low Energy Electron Microscopy[77] and Scanning Tunneling Microscopy[78]) techniques.

ACKNOWLEDGEMENT

This research was supported by the NSF under Grant No. DMR–8607369.

REFERENCES

1. For review, see J.D. Gunton, M. San Miguel, and P.S. Sahni, in "Phase Transitions and Critical Phenomena", C. Domb and J.L. Lebowitz, ed., Vol. 8, Academic Press, New York (1983).
2. For example, see "Dynamics of Ordering Process in Condensed Matter", S. Komura and H. Furukawa, ed., Plenum Press, New York (1988).
3. O.G. Mouritsen in "Kinetics of Ordering and Growth at Surfaces", M.G. Lagally, ed., Plenum Press, New York (1990) and references therein.
4. K. Heinz, in "Kinetics of Interface Reactions", M. Grunze and M.J. Kreuzer, ed., Springer–Verlag (1987), p. 202.
5. I.M. Lifshitz, Sov. Phys. JETP 15:939 (1962) [J. Eksptl. Theoret. Phys. (USSR) 42: 1354 (1962)].
6. S.M. Allen and J.W. Cahn, Acta Metall. 27:1085 (1979).
7. S.A. Safran, Phys. Rev. Lett. 46:1581 (1981).
8. P.S. Sahni, D.J. Srolovitz, G.S. Grest, M.P. Anderson, and S.A. Safran, Phys. Rev. B 28:2705 (1983).
9. M.P. Anderson, D.J. Srolovitz, G.S. Grest, and P.S. Sahni, Acta Metall. 32:783 (1984).
10. D.J. Srolovitz, M.P. Anderson, P.S. Sahni, and G.S. Grest, Acta Metall. 32:793 (1984).
11. A. Sadiq and K. Binder, Phys. Rev. Lett. 51:674 (1983).
12. S. Kumar, J.D. Gunton, and K.K. Kaski, Phys. Rev. B 35:8517 (1987).
13. G.S. Grest, M.P. Anderson, and D.J. Srolovitz, Phys. Rev. B 38:4752 (1988).
14. H.C. Fogedby and O.G. Mouritsen, Phys. Rev. B 37:5962 (1988).
15. T. Ala–Nissila and J.D. Gunton, in "Kinetics of Interface Reactions", M. Grunze and H.J. Kreuzer ed., Springer Verlag, Berlin (1987), p. 253.
16. I.M. Lifshitz and V.V. Slyozov, J. Phys. Chem. Solid 19:35 (1961).
17. P.S. Sahni, G. Dee, J.D. Gunton, M. Phani, J.L. Lebowitz, and M.H. Kalos, Phys. Rev. B 24:410 (1981).
18. K. Kaski, M.C. Yalabik, J.D. Gunton, and P.S. Sahni, Phys. Rev. B 28:5263 (1983).
19. J.D. Gunton, in "Kinetics of Interface Reactions", M. Grunze and H.J. Kreuzer, ed., Springer–Verlag, Berlin (1987), p. 239.
20. K. Binder, Ber. Bunsenges. Phys. Chem. 90:257 (1986); V.P. Zhadanov, Surf. Sci. 194:L100 (1988).
21. K. Kawasaki, M.C. Yalabik, and J.D. Gunton, Phys. Rev. A 17:455 (1978).
22. T. Ohta, D. Jasnow, and K. Kawasaki, Phys. Rev. Lett. 49:1223 (1982).
23. G.F. Mazenko and O.T. Valls, Phys. Rev. B 30:6732 (1984).
24. H. Tomita, Prog. Theor. Phys. 75:482 (1986).
25. Y. Imry and S.–K. Ma, Phys. Rev. Lett. 35:1399 (1975).
26. J. Villain, Phys. Rev. Lett. 52:1543 (1984); G. Grinstein and J.F. Fernandez, Phys. Rev. B 29:6389 (1984).
27. Martin Grant and J.D. Gunton, Phys. Rev. B 29:6266 (1984).
28. R. Bruinsma and G. Aeppli, Phys. Rev. Lett. 52:1547 (1984).
29. M. Grant and J.D. Gunton, Phys. Rev. B 35:4922 (1987).
30. E.T. Gawlinski, K. Kaski, M. Grant, and J.D. Gunton, Phys. Rev. Lett. 53:2266 (1984).
31. E.T. Gawlinski, S. Kumar, M. Grant, J.D. Gunton, and K. Kaski, Phys. Rev. B 32:1575 (1985).
32. Gary S. Grest and David J. Srolovitz, Phys. Rev. B 32:3014 (1985).
33. Scott R. Anderson, Phys. Rev. B 36:8435 (1987).
34. D.J. Srolovitz and G.N. Hassold, Phys. Rev. B 35:6902 (1987).
35. O.G. Mouritsen and P.J. Shah, in "Kinetics of Ordering and Growth at Surfaces", M.G. Lagally, ed., Plenum Press, New York (1990).
36. J.–K. Zuo, G.–C. Wang, and T.–M. Lu, Phys. Rev. Lett. 60:1053 (1988).
37. J.–K. Zuo, G.–C. Wang, and T.–M. Lu, Phys. Rev. B 39:9432 (1989); J.–K. Zuo and G.–C. Wang, J. Vac. Sci. Technol. A 7:2155 (1989).
38. U. Scheithauer, G. Meyer, and M. Henzler, Surf. Sci. 178:441 (1986).
39. J.–K. Zuo, G.–C. Wang, and T.–M. Lu, Phys. Rev. B 40:524 (1989).

40. C.C. Chang and L.H. Germer, Surf. Sci. 8:115 (1967).
41. H.L. Davis and G.–C. Wang, Bull. Am. Phys. Soc. 29:221 (1984).
42. G.–C. Wang and T.–M. Lu, Phys. Rev. B 28:6795 (1983).
43. G.–C. Wang and T.–M. Lu, Phys. Rev. B 31:5918 (1985).
44. J.–K. Zuo and G.–C. Wang, Phys. Rev. B 41:7078 (1990).
45. E. Domany, M. Shick, J.S. Walker, and R.B. Griffiths, Phys. Rev. B 18:2209 (1978); ibid 20:3828 (1979).
46. G.–C. Wang and T.–M. Lu, Phys. Rev. Lett. 50:2014 (1983).
47. P.K. Wu, J.H. Perepezko, J.T. McKinney, and M.G. Lagally, Phys. Rev. Lett. 51:1577 (1983); P.K. Wu, M.C. Tringides, and M.G. Lagally, Phys. Rev. B 39:7595 (1989).
48. M.C. Tringides, P.K. Wu, W. Moritz, and M.G. Lagally, Ber. Bunsenges. Phys. Chem. 90:277 (1986).
49. J.–K. Zuo and G.–C. Wang, Surf. Sci. 194:L77 (1988).
50. J.M. Pimbley, T.–M. Lu, and G.–C. Wang, J. Vac. Sci. Technol. 4:1357 (1986).
51. G. Ertl and M. Plancher, Surf. Sci. 48:364 (1975).
52. M.C. Tringides and R. Gomer, J. Chem. Phys. 84:4049 (1986).
53. G. Porod, in "Small Angle X–Ray Scattering", Glatter and O. Kratky, ed., Academic Press, New York (1982).
54. A. Sadiq and K. Binder, J. Stat. Phys. 35:517 (1984).
55. E.T. Gawlinski, M. Grant, J.D. Gunton, and K. Kaski, Phys. Rev. B 31:281 (1985).
56. J.–K. Zuo and G.–C. Wang, J. Vac. Sci. Technol. A6:649 (1988).
57. E.T. Gawlinski, K. Kaski, M. Grant, and J.D. Gunton, Phys. Rev. Lett. 53:2266 (1984).
58. J. Villian, J. Phys. (Paris) 43:L551 (1982).
59. Y. Imry and S.–K. Ma, Phys. Rev. Lett. 35:1399 (1975).
60. For review, J. Villian, in "Scaling Phenomena in Disordered Systems", R. Pynn and A. Skjetorp, ed., Plenum Press, New York, (1985).
61. R.J. Birgeneau, R.A. Cowley, G. Shirane, and H. Yoshizawa, Phys. Rev. Lett. 54:2174 (1985) and references therein.
62. D.P. Belanger, A.R. King, and V. Jaccarino, Phys. Rev. B 31:4538 (1985) and references therein.
63. D. Chowdhury and D. Stauffer, Z. Phys. B 60:249 (1985).
64. E. Pytte, Y. Imry, and D. Mukamel, Phys. Rev. Lett. 46:1173 (1981); K. Binder, Y. Imry, and E. Pytte, Phys. Rev. B 24:6736 (1981); D. Mukamel and E. Pytte, ibid. 25:4779 (1982).
65. E.T. Gawlinski, M. Grant, J.D. Gunton, and K. Kaski, Phys. Rev. B 24:281 (1985).
66. A. Sadiq and K. Binder, Phys. Rev. Lett. 51:674 (1983).
67. T.–M. Lu, L.–H. Zhao, M.G. Lagally, G.–C. Wang, and J.E. Houston, Surf. Sci. 122:519 (1982); G.–C. Wang and T.–M. Lu, ibid. 122:L635 (1982).
68. L.–H. Zhao, T.–M. Lu, and M.G. Lagally, Appl. Surf. Sci. 11/12:634 (1982).
69. C.C. Chang, PhD Thesis, Cornell University (1967).
70. H. Busch and M. Henzler, Phys. Rev. B 41:4891 (1990).
71. H. Homma and R. Clarke, Phys. Rev. Lett. 52:629 (1984).
72. W. Witt and E. Bauer, Ber. Bunsenges. Phys. Chem. 90:248 (1986).
73. M.C. Tringides, P.K. Wu, and M.G. Lagally, Phys. Rev. Lett. 59:315 (1987).
74. M.C. Tringides, Growth Kinetics of O/W(110) at High Coverage, preprint.
75. H. You, S.C. Fain, S. Satija, and L. Passell, Phys. Rev. Lett. 56:244 (1986); H. You and S.C. Fain, Phys. Rev. B 34:7840 (1988); R.D. Diehl, in "The Time Domain in Surface and Structural Dynamics", G.J. Long and F. Grandjean, ed., Kluwer Academic Publishers (1988), p. 439.
76. E.J. Nicol, C. Kallin, and A.J. Berlinskez, Phys. Rev. B 38:556 (1988).
77. E. Bauer and W. Telieps, Scanning Microscopy Supplement 1:99 (1987).
78. G. Binnig, H. Rohrer, C. Gerber, and E. Weibel, Phys. Rev. Lett. 58:120 (1983).

CONFERENCE SUMMARY:

WHERE FROM?   WHERE TO?

Michael Wortis
Physics Department
Simon Fraser University
Burnaby, BC   V5A 1S6, Canada

I.   INTRODUCTION

The last NATO Advanced Study Institute on "Phase Transitions in Surface Films"[1] was held at Erice in 1979. Much has happened during last eleven years. The talks at this conference have aimed to touch on most of the major developments. Written versions of some but not all of the talks appear in this volume. Other material was covered in poster sessions, not reported here. It is not my purpose to try to review the material covered. Rather, I would like to present a kind of overview and commentary. Although this commentary is certainly based on the many fine presentations I have heard over the course of the last two weeks, I certainly take responsibility for any misunderstandings and misinterpretations that may have crept in. My view is a personal one and a little distant now, as I have been working over the last couple of years in other areas.

What follows divides into three parts: The first part presents some comments on the developments of the last eleven years. The second part consists of a series of remarks (sermons? worries?) motivated by points raised in discussion here at Erice which seemed to me in need of some clarification or resolution. The third part is an attempt at least to pose the question, "Where are we headed next?"

II.   WHAT HAVE WE ACCOMPLISHED IN THE LAST 11 YEARS?

> "Ibergekumene tsores iz gut tsu dertseylin."
> (Troubles overcome are good to tell.)
> --Yiddish Proverb
> from Primo Levi, "The Periodic Table"

Table 1 shows my attempt to pigeonhole the talks which we have heard under a set of topics. These topics are not claimed to be unique or even particularly well chosen. Furthermore, many speakers touched on several topics, so I have had to exercise judgement somewhat arbitrarily in forcing talks into a single topic or spreading them over several. Nevertheless, I think that a certain pattern emerges. The following comments are an attempt to identify the main lines or themes of that pattern.

Phase Transitions in Surface Films 2
Edited by H. Taub *et al.*, Plenum Press, New York, 1991

Table 1. Talks and Topics at Erice 1990. Numbers in parentheses
indicate which one of two unrelated talks by the same author.

Methods	Monolayers	Multilayers	Melting/Wetting
Torzo	Comsa(1)	Hess	Tosatti
Krim	Chan(1)	Larher(1)	Levi
Lauter	Coulomb	Coulomb	Dietrich
Bienfait	Bruch	La Madrid	Bienfait
Comsa(2)	Diehl	Chan(2)	Dash
van der Veen	Gay	Lauter	Fisher
	Lauter	Landau	van der Veen(2)
	Landau		Larher(2)
Reconstruction	Roughening	Dynamics	Other
Villain	Lapuloulade	Bienfait	Reiter
van der Veen(2)	van Beijeren	Lauter	Dutta
den Nijs	den Nijs	Bruch	Leibler
	Kern	Comsa(2)	
		Wang	

A. New/Improved Experimental Probes

Had we been asked in 1979 to identify a single crucial
development of the the previous ten years of work on adsorbed
films, I think many of us would have responded with a single word:
grafoil. It was the development of this clean, coherent, high-
area substrate which permitted for the first time the well-
characterized physical measurements on which progress in our field
has been based. In a similar way, I would claim that the single
dominant theme of the last eleven years, as it has emerged from
this conference, is the development of new and improved
experimental probes. The thermodynamic probes of vapor pressure
and heat capacity, originally developed by Halsey,[2] Thomy and
Duval,[3] Dash,[4] and others, have continued to be refined in both
precision and speed, as has been reported to us by Larher and
Chan. In addition, however, some new and important probes have
come into their own, each with its own special strengths. In this
connection, I mention the ellipsometric technique reported by
Hess, which allows single-crystal substrates to be used instead of
powders, avoiding some of the problems of capillary condensation.
Neutrons--especially quasi-elastic neutron scattering, as reported
on by Bienfait and Lauter--now allow us to see regions where
capillary condensation has occurred and, thus, to study and to
correct for this ubiquitous phenomenon. X-rays--especially
synchrotron radiation sources--presently provide a tool of truly
marvelous precision for studying delicate ordering phenomena such
as two-dimensional melting. LEED and RHEED have been refined so
as to provide usable probes sensitive to complex surface ordering
and layer growth (Fain, Diehl). The techniques of ion
backscattering (van der Veen) and ^{4}He-ion scattering (Comsa)
provide strongly enhanced first-layer sensitivity, which has
proved crucial in elucidating details of surface melting and
surface mobility. These developments in technique are still
continuing apace. We have, in particular, heard exciting
presentations by Krim on the quartz crystal microbalance and by
Torzo on the graphite fiber microbalance.

B. Monolayer Phases

Monolayer phases, phase diagrams, and phase transitions
dominated discussion at the last Erice conference. These systems
continue to be of interest, but I think it is fair to say that we

have at a generic level a pretty good idea of the kinds of things which go on in simple monolayer systems such as rare gases on graphite (Chan), including vapor, liquid, epitaxial registered ($\sqrt{3}\times\sqrt{3}$), and a variety of incommensurate (IC) phases (solid, striped, domain-wall, and Novaco-McTague rotated), as described by Comsa, Villain, Bruch, Gay, and den Nijs. In some simple situations, we can even predict the occurrence and properties of such phases from atomic interactions (Bruch, Landau).

C. <u>New Substrates (New Symmetries) and New Adsorbates</u>

Much of the early work in physisorption was done on the rare gas/graphite systems or with simple diatomics like N_2, O_2, or H_2. The last decade has seen a great flowering of new substrates often with different symmetry, such as the (square) (100) face of MgO (Coulomb), CaI_2 (Larher), the metals Pt, Rh, Ir, Cu, Ru, Ni, Cd, Pb, Au, Ag, Al, etc. (Comsa, Diehl, etc.), and even semiconductors like Ge and Si (van der Veen). In addition, new adsorbates with more complex structure or different bonding properties have begun to be extensively explored. Examples are polar molecules like NH_3, C_2H_3F, HCl (Larher), alkalais like Cs, Na, and K (Diehl), and the hydrocarbons CH_4, C_2H_4, C_2H_6, etc.

D. <u>Convergence</u>

Partly as a consequence of the shift of attention to these new systems, a phenomenon of "convergence" has occurred. In the old days workers focussed on physisorption or chemisorption but not usually both, and the field was populated mainly by a small group of surface physicists and surface chemists. Now, we have seen the establishment of connections with metal physics, semiconductor physics, chemical physics and chemical engineering. At the same time, the close parallels between the physics of adsorbed films and that of interfaces, including clean surfaces (solid/vapor interfaces), binary-fluid mixtures (Dietrich), graphite intercalates (Reiter), and even grain boundaries have been recognized and exploited. Finally, the impressive development of model calculations and techniques has proved to be a unifying influence. These calculations use statistical mechanics, density-functional theory, etc., and have now reached a stage where quantitative as well as qualitative (e.g., "universal") comparison with experiments is often possible (Landau, Dietrich). Far from diluting our field, I think this convergence has brought to us a wealth of new ideas, new systems and new techniques. It bodes well for the future.

E. <u>Multilayers</u>

As our understanding of monolayers has increased, work has increasingly begun to shift to multilayers, including thick films, wetting, surface melting, and the like. Multilayer phase diagrams have begun to be mapped out, at least for "simple" systems (Hess, Larher, Lauter). Indeed, it is a remarkable sign of progress that Hess can now refer to the behavior of C_2H_4/graphite as "boring"!

Our old friend, the ^{4}He/graphite system, has finally but only recently been followed carefully from the monolayer regime up into the region of surface superfluidity (Chan), thus completing the work started by Frederikse[5] and Bretz[6] and making contact with Reppy's[7] beautiful experiments which demonstrated the Kosterlitz-Thouless character of the d=2 superfluid transition. On the other hand, I do not want to leave the impression that this fascinating area is anywhere near to being completely understood. On the contrary, there is a lot of unfinished business. A few years ago, the work of Goodstein[8] and his group made it look as if CH_4/graphite was understood in the multilayer region right up to coexistence. More recent work by the same group suggests evidence of capillary condensation, and many of the features of the earlier phase diagram are again in doubt, as reported to us by La Madrid. It appears that even Ar/graphite (Hess) holds some new surprises (den Nijs, see Sec. III.D).

F. The Effective Interface Potential

The concept that has emerged as central in thinking about thick multilayers and wetting (and corresponding interface phenomena such as surface melting) is that of the so-called effective interface potential,

$$V_{eff}(\ell) = (\Delta\mu)\ell + \frac{H}{\ell^2} + V_{short}(\ell),$$

which describes the effective interaction between the wall and a film (or interfacial layer) of average thickness ℓ. As described to us by Tosatti, Levi, and Dietrich, the $\Delta\mu$ term represents the (positive) energy cost of building thickness ℓ of a film of the "wrong" phase, $\Delta\mu = \mu_c - \mu$ below the coexistence curve. The term H/ℓ^2 arises from van der Waals interactions (for forces of strictly finite range this would be replaced by an exponential), and $V_{short}(\ell)$ reflects the possibility of short-range atomic-core interactions, electronic bonding, etc. The minimum of $V_{eff}(\ell)$ determines the equilibrium thickness of the film. As coexistence is approached, $\Delta\mu \to 0^+$, and, for (effective) Hamaker constant $H>0$, the minimum moves off to infinity (wetting), unless the short-range attraction $V_{eff}(\ell)$ pins the interface near the substrate. For $H<0$, the film does not (completely) wet. Dietrich has shown how this simple idea can be developed in a fully quantitative manner.

G. Film-Interface Equivalence

There is a complete formal equivalence between surface films and interfaces. Both are "sandwiches," consisting of two bulk phases separated by an inhomogeneous region of "filling." In the case of adsorption, the filling is the film and the "bread" is the substrate on one side and the vapor on the other. In the case of the interface, the "bread" consists of two coexisting bulk phases A and B (e.g., the solid and its vapor) and the filling is the interfacial region between them, where profiles of density and other physical quantities depart from and interpolate between their bulk values. The only difference between the two cases is that in adsorption the wall is normally inert and the "system" consists of at least two chemically separate constituents, the substrate and the adsorbate. For the interface, the coexisting phases A and B may be chemically identical or different in composition. (Indeed, when a second species is added to the system with the property that it prefers to segregate to the interface, interesting things can happen, e.g., surfactant behavior, microemulsions, etc. (Dutta, Leibler).) The equivalence between film and interface phases and phase behavior is illustrated in Table 2 below. A general recognition of this parallelism has developed during the last decade. As emphasized by Villain and van der Veen, there are now numerous examples of commensurate reconstructions such as (2x1) or (3x1) and, of course, more complicated ones like the Si (7x7). Incommensurate

474

Table 2. Equivalence between phases/processes of adsorbed films and solid/gas interfaces

Adsorbed Film	Solid/vapor interface
ordinary "film" (vapor or liquid)	"normal" interface profile
registered, epitaxial phase (e.g., $\sqrt{3}\times\sqrt{3}$)	surface reconstructed phase (commensurate)
incommensurate solid phase	surface reconstructed phase (incommensurate)
layering	layer-by-layer development of interface profile
wetting	surface melting

reconstructions are less common but not unknown. One example is the hexagonal so-called "5x20" reconstruction of Au(001).[9] I think it is reasonable at this point to hypothesize that any behavior which you can find for an adsorbed film will have its precise analogue for an appropriately chosen interfacial system, and vice versa.

H. Roughening and Faceting

A good deal of attention has focussed recently on the roughening of close-packed crystal surfaces and the corresponding faceting, where a low-temperature facet of the equilibrium crystal shape disappears (van Beijeren). Such transitions appear to be detectable both by scattering methods (Lapujoulade, Kern) and via direct observation of the equilibrium crystal shapes (e.g., ^{4}He crystals in contact with the superfluid at less than 2°K and small metal crystals (van Beijeren)).

I. Critical Phenomena

In earlier years, there was a great explosion of interest in adsorbed films as vehicles for studying two-dimensional critical phenomena, especially the "universal" critical behavior of simple models like the Ising, XY, 3-state Potts models, etc. I detect in Table 1 some shift of interest away from studies of universal critical behavior and into global and systematic studies of particular systems (Sec. II.C). From one point of view, this is simply a consequence of success: There is at this point a remarkable correspondence between experimentally determined exponents and the properties of corresponding statistical-mechanics models. I know at this point of no major disagreement between experiment and theory. (Even the rather subtle issue of two-dimensional melting and the predicted hexatic phase is in reasonable shape, although it would be nice to be able to predict reliably the order of the transition and to understand in a quantitative way the influence of substrate corrugation in modifying the predicted smooth-substrate behavior.) So, are we finished with adsorption as a probe of two-dimensional phase transitions? I will address this issue in Sec. III.D; however, for the moment let me make two remarks. First, as Fisher has emphasized in his lecture, scaling functions carry useful information, are universal up to various scale factors, and have for the most part not been explored experimentally. Second, I believe that there are new phases and new phase transitions out there waiting to be found. Both den Nijs and Villain have given examples. In particular, den Nijs's "disordered flat" phase may well have found its realization in Hess's Ar/graphite data and, if not there, then probably elsewhere. In short, I think that the disappearance of interest in critical phenomena in this field, while perhaps inevitable in the long run, is presently premature.

J. Dynamics

It is clear that the study of film and interface dynamics is a growing field. This includes both time-dependent fluctuations about equilibrium (Bienfait, Lauter, Bruch) and dynamics of driven systems far from equilibrium, as reported here by Wang and Comsa.

K. Computer Capability

No overview of the last ten years would be complete without a comment on the vast increase of computer capability, eloquently reported by Landau at this meeting. In an obvious way, this permits theorists to solve (or at least to develop numerically precise information about) simple models, thus allowing more realistic model fits to specific experiments. This leaves us free to think about physical effects rather than calculational techniques, which, I suppose, is good. However, the larger message is more complex: Why should we be interested in the properties of lattice models possessing six adjustable interaction parameters available to tailor-fit experiments? First, this capability will help at the experimental level to disentangle intrinsic effects from extrinsic effects, involving, e.g., finite sizes, boundaries, etc. We have already seen models used to compute complex LEED patterns to compare with experiment. Having on-line on-site capability of this type is beginning to allow real-time interaction with experiments. Secondly, as larger and more complex simulations become possible, the computation, itself, becomes the experiment. The output is data, just as much in need of interpretation as any other experimental data only with the considerable simplification that we and not nature control the input physics. Will this put experimentalists out of business? Of course not! But, it should force us to think more carefully about why we are doing each experiment (and each computation). The object of experiments and computations is, after all, not just to verify what we know but to discover what we do not know.

III. REMARKS, SERMONS, WORRIES

A. Surface and Interface Thermodynamics (Sermon #1)

It is important to define carefully what we are talking about.[10] Imagine for the sake of definiteness that our "sample" is a collection of identical atoms (or molecules) enclosed in a region Ω (the "box") by inert walls (see Fig. 1). Each atom interacts with the walls via a potential v_{aw} (i.e., $v_{atom-wall}$) and with other atoms via a potential v_{aa} (i.e., $v_{atom-atom}$), where both potentials depend on the atomic positions, orientations, etc. The atoms are assumed to be in contact with external thermal and material reservoirs, so the system is characterized by a temperature T and a chemical potential μ. The free energy of the sample is some function $F_{sample}(\mu, T; v_{aa}, v_{aw}, \Omega)$. This is what you measure in the lab, if you are an experimentalist. If you are a theorist, then you can imagine calculating it by taking the logarithm of an appropriate (grand) partition function. Notice that F_{sample} depends on the potentials v_{aa} and v_{aw} and on the size and shape of the box Ω. The function $F_{sample}(\mu, T; v_{aa}, v_{aw}, \Omega)$ contains all kinds of interesting thermodynamic information about

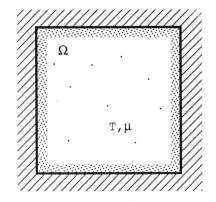

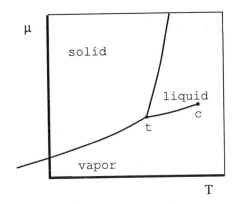

Fig. 1. Box with inert walls. T
and μ have been adjusted
here so that a dense film
forms at the walls.

Fig. 2. Generic phase diagram
showing lines of first-
order coexistence.
Triple (t) and critical
(c) points are indicated.

the material in the box. But, it does not distinguish between the
bulk material and the film and it does not contain any *sharp*
information about phases and phase transitions. Indeed, it may
surprise you to learn that, under quite general conditions, F_{sample}
is a completely analytic function of its arguments μ and T!

In order to talk about phase transitions *sharply*, it is
necessary to take the so-called thermodynamic limit. This is
accomplished for the bulk material by dividing out the box volume
$V(\Omega)$ and taking the limit of a sequence of larger and larger boxes
of similar shape,

$$f_b(\mu,T;v_{aa}) = \lim_{\Omega \to \infty}[F_{sample} / V(\Omega)] \;.$$

Under suitable conditions, this limit exists and the resultant
function, f_b, the bulk free energy per unit volume, depends only on
the arguments shown. Dependences on the wall potential v_{aw} and on
the box Ω have disappeared. f_b contains all possible
thermodynamic information about the bulk contents of the box but
no information whatsoever about the adsorbed film. To talk in a
similarly precise way about the properties of the film, it is
necessary to subtract from F_{sample} the free energy $V(\Omega)f_b$ of a
corresponding volume of the pure bulk phase.* Under suitable
conditions the difference scales as the total wall area $A(\Omega)$, so
the quotient,

$$f_s(\mu,T;v_{aa},v_{aw}) = \lim_{\Omega \to \infty}[(F_{sample} - V(\Omega)f_b) / A(\Omega)] \;,$$

defines the surface (excess) free energy per unit area of the
film. This depends (of course) on the wall potential but not on
the size and shape of the region Ω.

It is possible to define in a similarly precise manner an
interfacial free energy per unit interface area, $f_i(T;v_{aa})$, for two
coexisting bulk phases (Fig. 2). The procedure is a little more
involved than that for the surface,[10] since it is necessary to

---

*The astute reader will notice that there is some ambiguity in
deciding what volume to assign to the bulk phase. This ambiguity
becomes more evident in dealing with interfaces (below). The
principal conventions are discussed in Ref. 10. Physical results
do not depend on the choice of convention.

adjust conditions inside the box so that a planar interface does, indeed, exist and then to subtract out appropriate bulk and wall (surface) contributions from F_{sample} before dividing by the interface area. Notice that f_i depends on T or μ but not both, since the interface is only defined along the appropriate two-phase-coexistence boundary $\mu(T)$. Interface free energies must be defined separately for each (first-order) bulk phase boundary.

Thus, for example, the solid-vapor interface free energy $f_i^{s/v}(T)$ is only defined for the interval $T \in [0, T_t]$ up to the triple

temperature, while the liquid-vapor interface free energy $f_i^{l/v}(T)$ (which is just the mechanical surface tension) is defined only on the interval $T \in [T_t, T_c]$ between the triple point and the critical point. When one or both of the coexisting phases are anisotropic, then additional variables are needed to specify the relative orientations of the two bulk phases and the interface. For example, at solid-vapor coexistence the interfacial free energy depends on the orientation $\hat{m}$ of the interface relative to the bulk

crystal axes, so $f_i^{s/v}(T, \hat{m})$, which, as Lapuloulade and van Beijeren have reminded us, determines the equilibrium crystal shape.

Unlike F_{sample}, the limiting quantities, $f_b(\mu, T)$, $f_s(\mu, T)$, and $f_i(T)$, can and often do have mathematical singularities. Bulk and surface phase diagrams are nothing but the loci of nonanalyticity of the functions $f_b(\mu, T)$ and $f_s(\mu, T)$, respectively, over the (μ, T) phase plane. The phase space of the solid-vapor interface consists of a sphere of radius T_t with singularities at, e.g., the roughening temperatures $T_R(\hat{m})$ of the principal facets. The liquid-vapor phase space, on the other hand, is just the segment

$T_t \leq T \leq T_c$. $f_i^{l/v}(T)$ exhibits critical behavior at the upper limit, where the surface tension goes to zero.

This exposition should make clear that the surface (film) or interface is a kind of sandwich, in which a thin inhomogeneous region lives between two regions of distinct macroscopic bulk. It follows that the surface (or interface) free energy is singular whenever there is a singularity in either one of the two bounding bulk phases. The converse is not true: There may be singular behavior of the surface with no corresponding singularity in any bulk phase. The reason is simple: The surface is at every point in intimate contact with the bulk, but not conversely. By the same logic, individual layers of a surface film, for example, do not have distinct thermodynamic identities, because each layer is always in intimate contact with all other layers. Thus, a two-layer film cannot be thought of as some kind of coexistence between two distinct one-layer systems. So-called second-layer melting, for example, involves conceptually a change of profile of the whole two-layer system. Of course, quantitatively, the most important changes in the profile of lateral order or of mobility may well be localized in the second layer.

An example will serve to illustrate this point and to provide a warning. Consider a one-layer 2D vapor or 2D liquid film on an inert substrate. Both vapor and liquid phases have the symmetry of the substrate, since the substrate potential modulates the film density, so a vapor-liquid transition (as a function, for example, of chemical potential or of coverage) is a transition between states of the same symmetry (the symmetry of the clean substrate). For clarity let me call such a transition without symmetry change a "condensation." A line of first-order condensation can end in a critical point, because there is no symmetry distinction to prevent continuous passage from one phase into the other. By contrast, a one-layer incommensurate solid (IC) phase has a different symmetry from the vapor (or the liquid), so a line of first-order fluid-IC solid coexistence is not a condensation and can never terminate except in another phase boundary. Now, take a one-layer IC solid film and begin to add a second layer. At low second-layer density, we certainly expect a second-layer 2D vapor, but this "vapor" now has the symmetry of the first IC solid layer and is, thus, symmetry-distinct from the one-layer vapor phase. Furthermore, as the second-layer density increases, a second IC solid layer may form. If formation of this second layer does not involve a change of symmetry (i.e., if the second IC solid layer forms in registry with the first), then the transition from IC-solid-plus-vapor to two-layer IC solid is a condensation and can end at a critical point! In this situation, there is no symmetry distinction between second-layer fluid and IC solid phases.*

B.   Roughening (Sermon #2)

"It's just terminology."   --Sam Fain

There is an issue of terminology which has run through this conference. It does not cause confusion for the experts (van Beijeren, Lapujoulade, Kern), but it does sometimes cause confusion for the rest of us. I suggest that we distinguish with care between the following two phenomena:
"Atomic roughening": A significant number of adatom-vacancy pairs and other short-range structural disorder on an otherwise perfect crystal surface.
"Thermodynamic roughening": Divergence of the interfacial width, $\langle h^2 \rangle \to \infty$, as L (crystal size) goes to infinity, or, equivalently, logarithmic growth of the height-height correlations $\langle [h(\vec{r})-h(\vec{0})]^2 \rangle$ at long distance for a thermodynamically large sample.
Let me be absolutely clear. These phenomena are conceptually unrelated. They do not need to occur together even approximately. When *thermodynamic* roughening occurs, it takes place at a thermodynamically sharp phase transition (Sermon #1). By contrast, *atomic* roughening does not generally occur at a sharp transition: Some adatom-vacancy pairs are present as thermal excitations at any T>0, and it is up to you to decide what is "a significant number." If and when a significant increase of the number of adatom-vacancy pairs does take place at a sharp phase transition, that transition is not necessarily a thermodynamic roughening transition.
The relation between atomic and thermodynamic roughening is in the technical sense non-universal. We all understand, I think, that the true thermodynamic roughening singularity is a very weak nonanalyticity of the free energy ($e^{-\frac{const}{\sqrt{T_R-T}}}$), which does not generally take place at the (smooth) peak of the (interfacial) specific heat. The specific heat peak occurs where the short-range surface disorder is changing rapidly and is, therefore, one

---

*It is possible, of course that the formation of the second IC layer is accompanied by a change of symmetry of the film as a whole. When this happens, then the second layer vapor-solid transition is not a condensation and cannot end at a critical point.

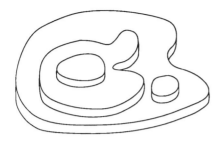

Fig. 3. Islands on a low-index surface. Each island is enclosed by a single atomic step.

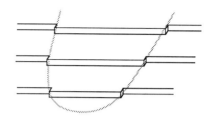

Fig. 4. Island on a vicinal surface. The effective step shown in outline is formed from a set of atomic kinks.

indicator of atomic roughening. Several participants have asked this week whether the roughening anomaly takes place above or below the specific heat peak. That is a question which has a well defined answer for each specific roughening model, but there is no reason to believe that the answer is the same for different models. As far as the thermodynamic roughening transition is concerned, the specific heat peak is a background effect, and background effects are nonuniversal.

The reason for this nonuniversality is simple: Available excitations contribute differently to atomic roughening and to thermodynamic roughening. Thermodynamic roughening of a particular crystal surface occurs when the free energy $f_{step}(T)$ per unit length of a "step" across that surface goes to zero.[*] The connection is made clear physically in Fig. 3. The vanishing of the step free energy makes the cost of large islands (or large depressions), islands-on-islands, etc., go to zero, and it is the corresponding proliferation of these island configurations which leads to the logarithmic growth of the height-height correlations at large distance and to the corresponding divergence of the interfacial width. The step free energy is made up of an energy contribution and an entropy contribution,

$$f_{step}(T) = e_{step}(T) - T s_{step}(T) ,$$

and each kind of excitation contributes to the energy and entropy terms. Excitations include kinks of the step, nearby adatoms/vacancies, nearby voids in the solid (or "atoms" in the nearby gas), overhanging configurations, etc. These same excitations contribute to the specific heat, but they may contribute rather differently. The key point is that the "step" is a very different object on a dense, low-index, high-symmetry surface (Fig. 3) and on a high-index, vicinal surface (Fig. 4). On a high-symmetry surface, the step is an ordinary atomic step.

---

[*]The thoughtful reader may ask at this point whether the orientation of the step on the surface plays a role. Steps with different orientations may and generally do have different free energies per unit length; however, it turns out that, for a given crystal surface $\hat{m}$, the free energy of steps of all orientations vanishes simultaneously at $T_R(\hat{m})$.

On a vicinal surface, on the other hand, the "flat" surface already has a regular array of atomic steps, and the "effective step" which controls roughening is a coherent array of atomic kinks on this preexisting array of atomic steps. Forming a kink on these "effective steps" involves moving atomic kinks, not creating them. The upshot is that a high-symmetry surface roughens by atomic processes which are not qualitatively different from the formation of adatom-vacancy pairs, and there is typically a good deal of atomic roughening well below T_R. On the other hand, a vicinal surface roughens dominantly by formation and displacement of atomic kinks, which is generally less expensive energetically than adatom-vacancy formation and can take place at temperatures well below the roughening temperature of the nearby high-symmetry terraces. From this perspective, it is natural that, for vicinal surfaces, roughening takes place well below the peak of the specific heat; whereas, for high-symmetry surfaces, roughening may occur above or below the specific heat peak but is probably nearby.

C. Surface Melting (Sermon #3)

"Surface disorder and diffusion below T_m" --Title of talk by Michel Bienfait

There are two conceptually distinct phenomena associated with the occurrence of liquid-like behavior at the vapor-solid interface. The first is simply a thermodynamic wetting phenomenon: If at the triple point T_t of a given substance (e.g., Pb(110), see van der Veen's talk, or H_2O, see Dash's talk) the liquid phase wets ("completely wets," if you prefer) a particular solid-vapor interface, then, as T approaches T_t from below along the coexistence curve (see Fig. 2), this solid-vapor interface profile, develops a region of substantial thickness ℓ of material with local bulk-liquid properties. As T approaches T_t, ℓ becomes macroscopic in thickness. Near T_t the usual wetting arguments apply, and it is useful to think in terms of the the effective interface potential $V_{eff}(\ell)$. When forces are van der Waals,

$\ell \sim (T_t - T)^{-\frac{1}{3}}$. When the film is thick, the interface will, of course, exhibit liquid-like mobility, and it will certainly become "rough" as T approaches T_t (if it was not already rough at a lower T_R), because the liquid-vapor interface supports capillary waves. I suggest that we agree to call this interfacial wetting phenomenon "surface melting."

On the other hand, some materials develop significant surface-phonon anharmonicity and atomic roughness as much as 40% below T_t (Kern) and then may show a large, liquid-like surface mobility 10-20% below T_t (Bienfait, van der Veen), in regimes where the interfacial thickness is still on an atomic scale. These onsets may be smooth, without any evident phase transition, or they may occur abruptly via some interfacial phase transition (analogous to layering, disordering, or melting of a thin film). These are thin-interface phenomenon. They contain lots of interesting physics but are entirely distinct from "surface melting," as defined above. Materials exhibiting such behavior may or may not go on to exhibit full "surface melting" close to T_t.

D. Conformal Invariance or "Is there life after the Ising model?"

At this point in history, all surface physicists know quite a lot about the Ising model, the Potts model, the XY model, etc., in two dimensions. We understand what phases these models have, what is the character of the transitions between those phases (including the values of the critical exponents for transitions which are second-order), and what are the connections between these models and the behavior of the films and interfaces we study in the lab. Because we understand these things so well, we are perhaps a little bored. Lest we become too complacent, I think it is important that some words be said at this conference about a subject which, from a theorist's point of view, is certainly the most important development to take place in the understanding of phase transitions in the last ten years.[11]

It has been known for a long time that the critical state is scale-invariant, i.e., that the form of correlations at criticality is not changed when we modify the spatial scale b at which we look at the system, $r \rightarrow r' = b^{-1}r$. It is a direct consequence of this scale invariance that the correlation function between two local operators $\varphi_i(\vec{r})$ and $\varphi_j(\vec{r})$ varies like

$$<\varphi_i(\vec{r}_1)\varphi_j(\vec{r}_2)>_* = \frac{C_{ij}}{r_{12}^{x_{ij}}} ,$$

where the star (*) indicates that the evaluation is at criticality. Here, C_{ij} is called a critical amplitude, $r_{12} = |\vec{r}_1 - \vec{r}_2|$ is much larger than the lattice spacing and the power x_{ij} is a critical exponent. The corresponding relation for the three-point ("triplet") correlations $<\varphi_i(\vec{r}_1)\varphi_j(\vec{r}_2)\varphi_k(\vec{r}_3)>_*$ involves an unknown scaling function.

The hypothesis of conformal invariance is that at criticality correlations are invariant not only under (global) dilations $\vec{r} \rightarrow \vec{r}' = b^{-1}\vec{r}$ but also under the more general group of conformal transformations. This is a much larger group of (local) transformations and appears to be built into the structure of the renormalization group.* Because the group is larger, conformal invariance places more stringent conditions on the critical correlations:

$$<\varphi_i(\vec{r}_1)\varphi_j(\vec{r}_2)>_* = \frac{\delta_{ij}C_i}{r_{12}^{2x_i}}$$

$$<\varphi_i(\vec{r}_1)\varphi_j(\vec{r}_2)\varphi_k(\vec{r}_3)>_* = \frac{C_{ijk}}{r_{12}^{x_i+x_j-x_k} r_{13}^{x_i+x_k-x_j} r_{23}^{x_j+x_k-x_i}} .$$

The exponents x_i, x_j, and x_k are called operator dimensions. The extra information for the pair correlations is not great, but the form of the triplet correlations is now completely determined!

In two space dimensions the conformal group has an infinite number of generators, each analogous to one of the (three) angular momentum operators which generate ordinary, three-dimensional spatial rotations. These generators have well-defined commutation properties with the operators of the field theory which describes

---

*Conformal invariance in effect replaces the global scale factor b by a spatially dependent scale factor $b(\vec{r})$. In the renormalization-group context, this corresponds to the condition that the recursion relations are spatially local, i.e., that the new couplings at the point $\vec{r}'$ depend only on couplings at points $\vec{r}$ in the nearby region. In all applications I am aware of this has turned out to be so.

the long-distance correlations at criticality, and these commutation properties place strong conditions on the operator correlations. In two dimensions, it turns out that these conditions suffice to classify possible critical states. The so-called "central charge" c, which is just the critical amplitude C_i for the local energy-momentum tensor, plays a central role. Indeed, all "unitary" theories with c<1 can be classified by a single integer, which determines c and critical exponents for all operators of the model. Comparison with known exponents shows that ordinary Ising models, Potts models, etc., are of this class, so conformal invariance allows us to compute exactly all the familiar critical exponents in two dimensions, once the appropriate correspondences have been established.

This is very beautiful, but there is more. It turns out that there exist may other conformal field theories with, e.g., c≥1. These presumably correspond to critical states of other two-dimensional statistical mechanics models. The message I want to get across is that there are more kinds of critical points than the familiar ones we all know and love. What physical degrees of freedom will be necessary to produce these new critical points, we do not now know. It takes only a small leap of faith to believe that these systems will have physical realizations, and what better place to look for them than in the rich fabric of surface phases and phase transitions? Indeed, den Nijs has proposed at this conference a model which has a critical point with c=3/2, which may be relevant to surfaces which can both reconstruct and roughen. We have even heard it hypothesized that Ar/graphite may contain this behavior.

E. Theory, Experiment, Simulation (Sermon #4)
   "Combined efforts in experiment, theory, and simulation
                      are needed to progress significantly" --Lapujoulade
   "Women: You can't live with them and you can't live without them" --Anon.

It is a sign of a healthy field, I think, that there is enough ferment--enough excitement--to motivate theorists, experimentalists, and simulators to talk together. That certainly was happening ten years ago in our field. I worried a bit in the early days of this conference that it was happening less now--that we were all a bit wrapped up in the technicalities of our own particular speciality. I feel better now at the end of the conference than I did at the beginning. Dash has challenged the theorists to understand frost heave. Den Nijs has challenged the experimentalists to reinterpret Ar/graphite in the multilayer regime. Lapujoulade has warned us all that the interpretation of roughening experiments will require both theory and simulation. People out there are listening, I hope.

In the end, it is really the students to whom I address myself: It is not always easy to cross those lines. Indeed, it is often hard, frustrating, and time-consuming. People speak different languages and have different aims and priorities. But, I believe you will find it worthwhile, both personally and from the point of view of learning physics. If you are an experimentalist, you will find it worth your while to talk with den Nijs, with Dietrich, and with Fisher. If you are a theorist, you will learn from talking to Dash, to Lauter, and to Chan. We will all benefit from talking with Landau and learning what the simulations can do and what they cannot do. I like to think that I am not a sexist and would happily substitute "men" for "women" in the anonymous quotation above. In a similar spirit, I suggest that you try substituting "theorist," "experimentalist," or "simulator," according to your own community (and the other guy's).

IV. WHERE ARE WE HEADED NEXT?

As scientists, most of us spend our lives deciding what problem/field/effect/material to study next. I suppose that, if we had to write down our criteria for making this decision, the list would include some or most of the following. A new direction should:
* involve interesting qualitative effects
* possess depth and conceptual interest
* lend itself to clean, well-characterized measurements
* have openings toward the future
* have some possible practical utility

I have no special depth, insight or originality in trying to look into the crystal ball. This is a personal list, and I certainly do not claim objectivity. Please feel free to make your own list and to disagree with mine! In any case, listening to your wise and stimulating presentations and cogitating for a few moments in the quiet privacy of my sumptuous Erice accommodation leads me to the following suggestions:

A. Unfinished business

Multilayers: CH_4/graphite, Ar/graphite, others?: These systems now appear to harbor unexpected complications, whose resolution may (or may not) involve conceptual issues.
Capillary condensation: This is an important technical complication for heterogeneous substrates. We now have the ability to study thick-film adsorption both with (via quasielastic neutrons) and without (via ellipsometry) capillary condensation. Of course, characterizing the inhomogeneity (pore-size distribution) will be difficult.
Microscopic aspects of roughening and surface melting: We seem to have the universal features under control; however, we are some distance from being able to make nonuniversal predictions on a microscopic basis.
Incommensurate phases, commensurate-incommensurate transitions: Notwithstanding the beautiful work described to us by Comsa, Bruch, Villain, Gay, and others, it is my feeling that there is still much left to be done here, e.g., orientational phases and phase transitions in thicker films.

B. Ongoing developments, presently visible openings (hazy)

New adsorbates/new substrates: Especially "soft" and structured systems such as liquid-crystalline or polymeric adsorbates. Water might also be included here, if an appropriate substrate can be found which allows the layer structure due to hydration forces to be probed.
New phases and phase transitions of films and interfaces: See Sec. III.D.
Phases of films on vicinal surfaces: Adsorption on stepped surfaces presents manifold possibilities for interplay between terrace geometry and adatom size, shape, and orientation.
Coadsorbed films: Here, again, there are exciting possibilities for interplay between solubility/segregation and in-layer ordering.

484

Interfaces of binary and impure crystals: When an impurity species adsorbs on/segregates to the solid-vapor interface, large changes in interfacial (free) energy are expected relative to the pure material. These energy changes provide powerful tools for modifying crystal shapes and hold the potential, e.g., for allowing specific catalyst-shape design.[12]

This well-developed field has remained until now rather chemical and practical in its orientation. However, the connection to adsorption and other interface problems is close, and I anticipate more cross-fertilization in the future.

Disordered substrates: Adsorption and film structure on heterogeneous and disordered substrates is a technically important issue. I have in mind here both large planar substrates with random defects and porous or even fractal substrates, in which behavior crosses over between thin and thick films. Krim and Chan have provided some examples.

Dynamics: Equilibrium film dynamics has been studied for a long time; however, refined experimental techniques are providing new and provocative data. See, e.g., the presentations of Lauter and of Bienfait. A rapidly developing area is nonequilibrium dynamics. Recent work suggests that, as the driving forces increase, a sequence of growth modes is followed in crystal growth, starting with laminar growth and continuing through into unstable regimes such as dendritic growth and various branching regimes.[13] There is evidence that anisotropy plays an important role in determining such morphological phase diagrams. Some work in this area has been reported to us by Comsa and by Wang. I think there will be much more in the future. Other related matters not touched on at this conference include the statistical properties of nonequilibrium steady states (e.g., so-called driven diffusive systems or DDS[14]).

Engineering applications: Better understanding of films and interfaces will continue to contribute to applications. Talks at this conference have touched on material fabrication (van der Veen) and frost heave (Dash). There are and will be other examples.

C. _The future_ (even hazier!)

New types of interfaces: Our community has so far concentrated its attention dominantly on solid-vapor interface properties. The reason is largely practical, in that the low-density vapor allows relatively easy probing of the interface region. Many of the same ideas carry over, of course, to solid-liquid and solid-solid (e.g., grain boundaries) interfaces. There are active communities working in these areas. I anticipate convergence and cross-fertilization with these fields, especially as new probes (e.g., modern electron microscopy) come on line which are able to probe these "internal" interfaces with increased range and resolution.

More dynamics: I am confident that this will prove to be another growing opening towards the future, especially as new fast instruments are developed for time-resolved measurements.

"Soft" surfaces: Historically, most workers in our community have studied "hard", clean surfaces. Probably, this is a consequence of selecting systems which can be well characterized and precisely measured. Two speakers at this conference have described recent progress in soft systems: Dutta on Langmuir systems and Leibler on membranes. It is clear that these systems will be extraordinarily rich in the future, combining, as they do,

rather complicated molecules and excitations with energies in the
room-temperature range. Inspiration from biological systems and
materials is already evident, and I think there will be increasing
interest in developing a detailed understanding of where physical
mechanisms are biologically relevant and where, on the other hand,
behavior of interest is controlled largely by specific
chemical/biochemical pathways.

D. Closing thoughts

> Some say the world will end in fire,
> Some say in ice.
> From what I've tasted of desire,
> I hold with those who favor fire.
> But if it had to perish twice,
> I think I know enough of hate
> To say that for destruction ice
> Is also great
> And would suffice
>
> --Robert Frost

> "Ice is very difficult" --J.G. Dash
> (and so is water, says S. Leibler)

> "I have to admit that I am pretty confused by a lot of things." --D. Pengra
> (M. Wortis subscribes to this, too!)

As long as there are plenty of good things to be confused
about, our field is not yet near its end!

REFERENCES

1. "Phase Transitions in Surface Films," J.G. Dash and J.
      Ruvalds, eds., Plenum Press, New York (1980).
2. J.H. Singleton and G.D. Halsey, Jr., The adsorption of Argon
      on Xenon layers, J. Phys. Chem. 58:330 (1954).
3. A. Thomy, X. Duval, and J. Regnier, Two dimensional phase
      transitions as displayed by adsorption isotherms on
      graphite and other lamellar solids, Surf. Sci. Rep. 1:1
      (1981).
4. J.G. Dash, "Films on Solid Surfaces," Academic Press, New York
      (1975).
5. H.P.R. Frederikse, PhD Thesis, (Heat capacity of [4]He films on
      rouge), Leyden (1947); On the specific heat of adsorbed
      Helium, Physica 15:860 (1949).
6. M. Bretz, Heat capacity of multilayer He[4] on graphite, Phys.
      Rev. Lett. 31:1447 (1973).
7. J.D. Reppy, Superfluidity in thin helium four films, in:
      "Phase Transitions in Surface Films," J.G. Dash and J.
      Ruvalds, eds., Plenum Press, New York (1980). See also,
      D.J. Bishop and J. D. Reppy, Study of the superfluid
      transition in two-dimensional [4]He films, Phys. Rev. Lett.
      40:1727 (1978).
8. D. Goodstein, J.S. Hamilton, M.J. Lysek, and G. Vidali, Phase
      diagrams of multilayer adsorbed methane, Surf. Sci. 148:187
      (1984).
9. S.G.J. Mochrie, D.M. Zehner, Doon Gibbs and B.M. Ocko,
      Structure and phases of the Au(001) surface: X-ray
      scattering measurements, Phys. Rev. Lett. 64:2925 (1990).
10. R.B. Griffiths, An introduction to the thermodynamics of
      surfaces, in: "Phase Transitions in Surface Films," J.G.

Dash and J. Ruvalds, eds., Plenum Press, New York (1980).

11. J.L. Cardy, conformal invariance, _in_: "Phase Transitions and Critical Phenomena," Vol. 11, C. Domb and J.L. Lebowitz, eds., Academic Press, London (1987).

12. A-C. Shi et al., Perspective on the use of gas adsorption for particle-shape control in supported metal catalysis, _in_: "Microstructure and Properties of Catalysts," M. Treacy, ed., Materials Research Society, Pittsburgh (1988).

13. J.S. Langer, Instabilities and pattern formation in crystal growth, _Rev. Mod. Phys._ 52:1 (1980); D.A. Kessler, J. Koplik, and H. Levine, Pattern selection in fingered growth phenomena, _Adv. Phys._ 37:255 (1988); N.D. Goldenfeld, Dynamics of unstable interfaces, _in_: "Physicochemical Hydrodynamics, M. Verlarde, ed., Plenum Press, New York (1988).

14. S. Katz, J.L. Lebowitz, and H. Spohn, Phase transitions in stationary nonequilibrium states of model lattice systems, _Phys. Rev._ B28:1655 (1983); Nonequilibrium steady states of stochastic lattice-gas models of fast ionic conductors, _J. Stat. Phys._ 34:497 (1984).

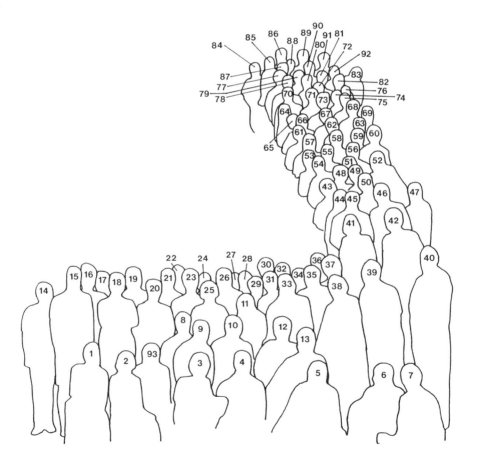

1. D. Pengra
2. M. La Madrid
3. H. Lauter
4. J. Sprosser
5. F. Toigo
6. A. Levi
7. R. Ferrando
8. M. den Nijs
9. J. Dash
10. H. Taub
11. M. Chan
12. S. Fain
13. G. Torzo
14. M. Lysek
15. H. Doebereiner
16. J. van der Veen
17. Hermansfeldt
18. M. Haarmans
19. K. Kern
20. P. Bancel
21. S. Ehrlich
22. G. Mistura
23. G. Gonnella
24. C. Mitsas
25. G. Wang
26. J. Larese
27. T. Angot
28. J. Coulomb
29. J. Dennison
30. S. Dietrich
31. J. Lapujoulade

32. S. Lipson
33. M. Fisher
34. A. Migone
35. G. Comsa
36. J. Phillips
37. A. Khater
38. S. Moss
39. D. Landau
40. M. Wortis
41. H. Xu
42. A. Hoss
43. B. Dev
44. G. Langie
45. D. Fisher
46. U. Romahn
47. P. Dutta
48. E. Conrad
49. A. Mitus
50. Y. Larher
51. J. Esteves
52. ?
53. N. Kalkan
54. B. Joós
55. G. Mazzeo
56. P. Eng
57. H. van Pinxteren
58. R. Diehl
59. R. Phelps
60. H. Ernst
61. J. Villain
62. G. Hess

63. Y. Girard
64. Z. Li
65. J. Krim
66. M. Finney
67. T. Moeller
68. F. Grey
69. U. Albrecht
70. M. Flatté
71. C. Glover
72. R. Chiarello
73. C. Walden
74. L. Bruch
75. S. Chandavarkar
76. R. Suter
77. A. Zerrouk
78. A. Bensaoula
79. M. Hamichi
80. N. Shrimpton
81. H. van Beijeren
82. J. Palmeri
83. V. Etgens
84. V. Eden
85. J. Suzanne
86. A. Colli
87. B. Salanon
88. N. Papanicolaou
89. G. Reiter
90. U. Volkmann
91. E. Polturak
92. M. Bienfait
93. G. London

PARTICIPANTS

U. ALBRECHT, Fak.f.Physik, Universität Konstanz, Universitätsstr., D-7750 Konstanz, Germany

T. ANGOT, Université d'Aix-Marseille II, Faculté des Sciences de Luminy, Département de Physique, Case 901, 13288 Marseille, France

P. BANCEL, I.B.M., T. J. Watson Research Center, P. O. Box 218, Yorktown Heights, NY 10598, USA

A. BENSAOULA, Space Vacuum Epitaxy Center, Room 724, Physics Department, University of Houston, Houston, TX 77204-5507, USA

M. BIENFAIT, Centre de Recherche sur les Mécanismes de la Croissance Cristalline (CRMC$_2$), CNRS, Campus de Luminy, Case 913, 13288 Marseille Cedex 9, France

L. W. BRUCH, Department of Physics, University of Wisconsin, Madison, 1150 University Avenue, Madison, WI 53706, USA

M. H. W. CHAN, Department of Physics, Pennsylvania State University, 104 Davey Lab, University Park, PA 16802, USA

S. CHANDAVARKAR, Department of Physics, University of Liverpool, Liverpool L69 3BX, United Kingdom

R. P. CHIARELLO, Department of Physics, Northeastern University, Boston, MA 02115, USA

A. COLLI, Physik Department E10, Fakultät für Physik, Technische Universität München James-Franck Straße 1, D-8046 Garching, Germany

G. COMSA, Insitut für Grenzflächenforschung und Vakuumphysik, Kernforschungsanlage Jülich, 5170 Jülich, Germany

E. CONRAD, Department of Physics & Astronomy, 223 Physics Building, University of Missouri-Columbia, Columbia, MO 65211, USA

J. P. COULOMB, Département de Physique, Case 901, Université d'Aix-Marseille II, 70, route Léon-Lachamp, 13288 Marseille Cedex, France

P. DAI, Department of Physics & Astronomy, 223 Physics Building, University of Missouri-Columbia, Columbia, MO 65211, USA

J. G. DASH, Department of Physics, FM-15, University of Washington, Seattle, WA 98195, USA

M. P. DEN NIJS, Department of Physics, FM-15, University of Washington, Seattle, WA 98195, USA

J. R. DENNISON, Department of Physics, Utah State University, Logan, UT 84322-4415, USA

B. N. DEV, Institute of Physics, Sachivalaya Marg, Bhubaneswar - 751 005, India

R. D. DIEHL, Department of Physics, The Pennsylvania State University, 104 Davey Laboratory, University Park, PA 16802, USA

S. DIETRICH, Fachbereich Physik, Bergische Universität Wuppertal, Postfach 100127, D-5600 Wuppertal 1, Germany

H.-G. DOEBEREINER, Department of Physics, Simon Fraser University, Burnaby, B. C. V5A 1S6, Canada

N. DUPONT, CNRS, Laboratoire Maurice Letort, rue de Vandœuvre - B. P. 104, 54600 Villers-les-Nancy, France

P. DUTTA, Materials Research Center, Technical Institute, Room 2033, Northwestern University, Evanston IL 60208-9990, USA

V. EDEN, Department of Physics, FM-15, University of Washington, Seattle, WA 98195, USA

S. N. EHRLICH, NSLS, Bldg. 510 E, Brookhaven National Laboratory, Upton, NY 11973, USA

P. J. ENG, Department of Physics, SUNY at Stony Brook, Stony Brook, NY 11974, USA

H.-J. ERNST, CEN Saclay / SPAS, 91191 Gif-sur-Yvette, France

J. M. C. ESTEVES, Centro de Fisica da Materia Condensada, Av. Prof. Gama Pinto, 2, P-1699 Lisboa Codex, Portugal

V. ETGENS, Bat 209-D LURE, 91405 Orsay Cedex, France

S. C. FAIN, JR., Department of Physics, FM-15, University of Washington, Seattle, WA 98195, USA

N. FAMELI, Dipartimento di Fisica, Università di Padova, via Marzolo 8, I 35131 Padova, Italy

R. FERRANDO, University of Genova, Dipartimento di Fisica, Via Dodecaneso 33, Genova, Italy

M. S. FINNEY, Department of Physics, University of Leicester, University Road, Leicester LE1 7RH, United Kingdom

D. FISHER, Department of Physics, University of Liverpool, Liverpool L69 3BX, United Kingdom

MICHAEL E. FISHER, Institute for Physical Science and Technology, University of Maryland at College Park, College Park, MD 20742-2431, USA

M. FLATTE, Physics Department, University of California, Santa Barbara, CA 93106, USA

Y. GIRARD, Groupe de Physique des Solides, Université Paris VII, Tour 23, 2 Place Jussieu, 75251 Paris Cedex 05, France

G. GIUGLIARELLI, Theoretical Physics, Dipartimento di Fisica, Università di Padova, Via F. Marzolo 8, 35131 Padova, Italy

C. GLOVER, Theory of Condensed Matter Group, Cavendish Laboratory, Madingley Road, Cambridge CB3 OHE, United Kingdom

G. GONNELLA, Università degli Studi di Bari, Facoltà di Scienze, Dipartimento di Fisica, via G. Amendola, 173, 70126 Bari, Italy

F. GREY, Risø National Laboratory, DK-4000 Roskilde, Denmark

M. HAARMANS, Department of Chemistry, Leiden University, P. O. Box 9502, 2300 RA Leiden, The Netherlands

M. HAMICHI, Laboratoire Maurice Letort, Route de Vandœuvre, 54600 Villers les Nancy, France

G. B. HESS, Department of Physics, University of Virginia, J. W. Beams Laboratory of Physics, McCormick Road, Charlottesville, VA 22901, USA

A. HOSS, Kernforschungszentrum Karlsruhe GmbH, Institut f. Nukleare Festkörperphysik - INFP, P. O. Box 3640, D-7500 Karlsruhe, Germany

B. JOOS, Ottawa-Carleton Institute for Physics, University of Ottawa Campus, Ottawa, Ontario K1N 6N5, Canada

N. KALKAN, University of Istanbul, Faculty of Science, Department of Physics, 34459, Vezneciler, Istanbul, Turkey

K. KERN, Insitut für Grenzflächenforschung und Vakuumphysik, Kernforschungsanlage, Jülich, 5170 Jülich, Germany

A. KHATER, Director, Laboratoire de Physique des Matèriaux, Université du Maine, 72017 Le Mans Cedex, France

J. KRIM, Department of Physics, Northeastern University, Boston, MA 02115, USA

M. LA MADRID, 114-36 Sloan Annex, California Institute of Technology, Pasadena, CA 91125, USA

D. P. LANDAU, Department of Physics and Astronomy, University of Georgia, Athens, Georgia 30602, USA

G. LANGIE, Katholieke Universiteit Leuven, Lab. Vaste Stof-Fysika en Magnetisme, Celestijnenlaan 200D, 3030 Leuven, Belgium

J. LAPUJOULADE, CEN SACLAY, IRF-DPHG-PAS, F-91191 Gif-sur-Yvette Cedex, France

J. Z. LARESE, Department of Chemistry, Building 555, Brookhaven National Laboratory, Upton, NY 11973, USA

Y. LARHER, Département de Physico-chemie, Centre d'Etudes Nucléaires de Saclay, Boite Postal No. 2, 91190 Gif-sur-Yvette Cedex, France

H. J. LAUTER, Institut Laue-Langevin, 156X, 38042 Grenoble Cedex, France

S. LEIBLER, CEN Saclay, IRF-DPHG-SPT, 91191 Gif-sur-Yvette Cedex, France

A. C. LEVI, International School for Advanced Studies, Strada Costiera 11, I-34014 Trieste, Italy

Z.-Y. LI, Department of Physics, University of Liverpool, Liverpool L69 3BX, United Kingdom

S. G. LIPSON, Physics Department, Technion - Israel Institute of Technology, Haifa 32000, Israel

M. J. LYSEK, California Institute of Technology, Condensed Matter Physics 114-36, Pasadena, CA 91125, USA

G. MAZZEO, S.I.S.S.A. - I.S.A.S., International School of Advanced Studies, Strada Costiera 11, I-34014 Trieste, Italy

A.D. MIGONE, Department of Physics, Southern Illinois University at Carbondale, Carbondale, IL 62901-4401, USA

G. MISTURA, Department of Physics, 104 Davey Laboratory, Penn State University, University Park, PA 16802, USA

C. L. MITSAS, Aristotle University of Thessaloniki, Department of Physics, Solid State Section 313 1, 540 06 Thessaloniki, Greece

A. C. MITUS, Department of Theoretical Physics, Universität Saarbrücken, 6600 Saarbrücken, Germany

T. MOELLER, Institut Laue Langevin, B.P. 156X, 38042 Grenoble Cedex, France

S. C. MOSS, Physics Department, University of Houston, 4800 Calhoun, Houston, TX 77204, USA

J. P. PALMERI, CEN - Saclay, F-91191 Gif-sur-Yvette Cedex, France

N. PAPANICOLAOU, University of Ioannina, Department of Physics, P. O. Box 1186, GR - 451 10 Ioannina, Greece

D. B. PENGRA, Department of Physics, FM-15, University of Washington, Seattle, WA 98195, USA

R. PHELPS, Department of Physics, 366 LeConte Hall, University of California, Berkeley, Berkeley, CA 94720, USA

J. M. PHILLIPS, Department of Physics, University of Missouri-Kansas City, Kansas City, Missouri 64110, USA

E. POLTURAK, Department of Physics, Technion - I. I. T., Haifa 32000, Israel

G. REITER, Department of Physics, University of Houston, Houston, TX 77204-5504, USA

U. ROMAHN, Institut für Nukleare Festkörperphysik, Kernforschungszentrum, Karlsruhe, GmbH, INFP, Postfach 3640, D-7500 Karlsruhe 1, Germany

B. SALANON, DPHG/PAS CEN Saclay, 91191 Gif-sur-Yvette Cedex, France

J. M. P. R. SEBROSA, Centro de Fisica da Materia Condensada, Av. Prof. Gama Pinto, 2, P-1699 Lisboa Codex, Portugal

N. D. SHRIMPTON, Department of Chemistry, The Pennsylvania State University, 152 Davey Laboratory, University Park, PA 16802, USA

J. SPROSSER, Service de Physique des Atomes et des Surfaces, C.E.N. de Saclay, 91191 Gif-sur-Yvette Cedex, France

R. M. SUTER, Department of Physics, Carnegie-Mellon University, Pittsburgh, PA 15213, USA

J. SUZANNE, Département de Physique, Case 901, Université d'Aix-Marseille II, 70, route Léon-Lachamp, 13288 Marseille Cedex, France

H. TAUB, Department of Physics & Astronomy, 223 Physics Building, University of Missouri-Columbia, Columbia, MO 65211, USA

F. TOIGO, Dipartimento di Fisica, Università di Padova, via Marzolo 8, I 35131 Padova, Italy

G. TORZO, Dipartimento di Fisica, Università di Padova, via Marzolo 8, I 35131 Padova, Italy

E. TOSATTI, International Center for Theoretical Physics, P. O. Box 586, I-34100 Trieste, Italy

H. VAN BEIJEREN, Instituut voor Theoretische Fysica, Rijksuniversiteit Utrecht, Princeton Plein, P. O. Box 80.006, NL 3508 TA UTRECHT, The Netherlands

J. F. VAN DER VEEN, Foundation for Fundamental Research on Matter (FOM), Institute for Atomic and Molecular Physics, Kruislaan 407, 1098 SJ Amsterdam, The Netherlands

H. VAN PINXTEREN, Foundation for Fundamental Research on Matter (FOM), Institute for Atomic and Molecular Physics, Kruislaan 407, 1098 SJ Amsterdam, The Netherlands

J. VILLAIN, CENG, DRF-SPH-MDN, F-38041 Grenoble, France

U. G. VOLKMANN, Johannes Gutenberg - Universität Mainz, Fachbereich Physik, Institut für Physik, Staudinger Weg 7, D-6500 Mainz, Germany

C. J. WALDEN, H. H. Wills Physics Laboratory, University of Bristol, Royal Fort, Tyndall Avenue, Bristol BS8 1TL, United Kingdom

G.-C. WANG, Physics Department, Rensselaer Polytechnic Institute, VTE, Troy, NY 12180, USA

M. WORTIS, Department of Physics, Simon Fraser University, Burnaby, BC V5A 1S6, Canada

H. XU, University Chemical Laboratory, University of Cambridge, Lensfield Road, Cambridge CB2 1EW, United Kingdom

A. ZERROUK, School of M.A.P.S. Postgraduate P/H, University of Sussex, Brighton BN1 9QH, United Kingdom

INDEX

Na on Ru(0001), 93
Ne on graphite, 1, 72, 93, 369
  heat capacity, 369
  isotherms, volumetric, 369
  layer critical points, 369
  phase diagram, 1
  RHEED, 369
  roughening transition, 370
Neutron scattering, 135-136, 153 (*see also* MgO)
  for melting studies, quasielastic, 167
  for multilayer film studies
    elastic, 359, 371, 379-382, 443
    inelastic, 359
    quasielastic, 264, 359, 440, 447, 472
      (*see also* surface-induced melting, neutron scattering, quasielastic)
  profile analysis of diffraction patterns, 162
Ni(110) surface roughening, 249
Nitrogen
  on graphite, 75
    heat capacity, 6, 380
    isotherm, volumetric, 8
    multilayer studies, 380
    phase diagram, 6, 8
    phonons, 139
  liquid monolayer on quartz crystal oscillator, 176
  on gold, slip time, 176-177
Novaco-McTague rotation, 87
Nuclear magnetic resonance (NMR)
  for multilayer film studies, 359
  measurements of T_2, mobility, 440, 447
  measurements of T_1, 442
Nucleation theory, classical, 290

Octadecanoic acid on water, 187
Organic monolayers, 183-200 (*see also* Langmuir films and Lipid monolayers on water, and molecules, *n*-alkane)
Oxygen
  on graphite, 380-383
  on W(112), chemisorbed, 457
    high-resolution LEED, 457
    random impurities, effect of, 457

Paraffin monolayers on water, 192
Pb
  clusters on Ge(111) and Si(111), 90
  (110) surface melting, 294, 313
Pd(110) surface roughening, 249
Pentadecanoic acid on water, 185, 187
Phase diagrams, 1-10, 99-100, 105, 135
  (*see also* alkali metal adsorption)
Phase transitions, 11, 13, 36
  conformal invariance, *see* Conformal field theories
  commensurate-commensurate
    continuous, 106
    first-order, 106, 109

Phase transitions (continued)
  commensurate-commensurate (continued)
    phase coexistence, 106
    uniaxial compression, 106
  commensurate-incommensurate, 6, 33, 47, 136, 484
    relation to chiral 4-state clock model, 258
    satellite diffraction peaks, 140
  conformal field theories of, 260
  deconstruction of surfaces, 202, 260
    of Au(110) surface, 210
  dewetting, 444
  drying, 405
  incommensurate melting, 247
  Kosterlitz-Thouless type, 226, 249
  lattice gas (*see also* lattice gas models)
    He^4 on graphite, 16
    Kr on graphite, 18
    H on Pd(100), 21
  layering, 408 (*see also* Layering, Wetting)
  liquid-vapor, 1
  liquid-solid on surface of quartz oscillator, 179
  on MgO, 115
  multilayer adsorption, 28
  order-disorder, 5, 109
  orientational ordering of molecules, 28, 30-31, 153, 380
  Pokrovsky-Talapov type, 264
  preroughening, 247, 281
  rotational, 83
  roughening, 210, 213-215, 217 (*see also* Ising model and roughening)
  sliding friction as a probe of, 179
  thin-thick, 409, 440 (*see also* Wetting)
  wetting, 391
Phonon
  anharmonicity, D_2 on graphite, 140
  density of states, HD and D on graphite, 138
  dispersion relation (D_2 on graphite), 138
  gap, 136
  surface, 60
  Einstein oscillator, 61
  hybridization splitting, 62
  line-width broadening, 62
  phonon-roton dispersion curve, 148
  ripplon, 148
  zone boundary, 136
Phospholipid monolayers on water, 191
Physical adsorption, 67, 113, 154
Polymorphism, 153
Potts model
  three-state, 6, 8
  four-state, 254
  six-state, 444
Power law
  coexistence boundary, 2
Prewetting, *see* Wetting, prewetting